Spin Geometry

PRINCETON MATHEMATICAL SERIES

1. The Classical Groups, By HERMANN WEYL
3. An Introduction to Differential Geometry, By LUTHER PFAHLER EISENHART
4. Dimension Theory, By W. HUREWICZ and H. WALLMAN
6. The Laplace Transform, By D. V. WIDDER
7. Integration, By EDWARD J. MCSHANE
8. Theory of Lie Groups: I, By C. CHEVALLEY
9. Mathematical Methods of Statistics, By HARALD CRAMÉR
10. Several Complex Variables, By S. BOCHNER and W. T. MARTIN
11. Introduction to Topology, By S. LEFSCHETZ
12. Algebraic Geometry and Topology, Edited by R. H. FOX, D. C. SPENCER, and A. W. TUCKER
14. The Topology of Fibre Bundles, By NORMAN STEENROD
15. Foundations of Algebraic Topology, By SAMUEL EILENBERG and NORMAN STEENROD
16. Functionals of Finite Riemann Surfaces, By MENAHEM SCHIFFER and DONALD C. SPENCER.
17. Introduction to Mathematical Logic, Vol. I, By ALONZO CHURCH
19. Homological Algebra, By H. CARTAN and S. EILENBERG
20. The Convolution Transform, By I. I. HIRSCHMAN and D. V. WIDDER
21. Geometric Integration Theory, By H. WHITNEY
22. Qualitative Theory of Differential Equations, By V. V. NEMYTSKII and V. V. STEPANOV
23. Topological Analysis, By GORDON T. WHYBURN (revised 1964)
24. Analytic Functions, By AHLFORS, BEHNKE and GRAUERT, BERS, et al.
25. Continuous Geometry, By JOHN VON NEUMANN
26. Riemann Surfaces, By L. AHLFORS and L. SARIO
27. Differential and Combinatorial Topology, Edited by S. S. CAIRNS
28. Convex Analysis, By R. T. ROCKAFELLAR
29. Global Analysis, Edited By D. C. SPENCER and S. IYANAGA
30. Singular Integrals and Differentiability Properties of Functions, By E. M. STEIN
31. Problems in Analysis, Edited by R. C. GUNNING
32. Introduction to Fourier Analysis on Euclidean Spaces, By E. M. STEIN and G. WEISS
33. Étale Cohomology, By J. S. MILNE
34. Pseudodifferential Operators, By MICHAEL E. TAYLOR
36. Representation Theory of Semisimple Groups: An Overview Based on Examples, By ANTHONY W. KNAPP
37. Foundations of Algebraic Analysis, By MASAKI KASHIWARA, TAKAHIRO KAWAI, and TATSUO KIMURA. Translated by GORO KATO
38. Spin Geometry, By H. BLAINE LAWSON and MARIE-LOUISE MICHELSOHN

Spin Geometry

H. BLAINE LAWSON, JR.
and
MARIE-LOUISE MICHELSOHN

PRINCETON UNIVERSITY PRESS
PRINCETON, NEW JERSEY
1989

Published by Princeton University Press, 41 William Street,
Princeton, New Jersey 08540
In the United Kingdom: Princeton University Press,
Chichester, West Sussex

Library of Congress Cataloging-in-Publication Data

Lawson, H. Blaine.
Spin geometry/H. B. Lawson and M.-L. Michelsohn.
p. cm.—(Princeton mathematical series: 38)
Includes index.
ISBN 0-691-08542-0 (alk. paper):
1. Nuclear spin—Mathematics. 2. Geometry. 3. Topology.
4. Clifford algebras. 5. Mathematical physics. I. Michelsohn,
M.-L. (Marie-Louise), 1941– II. Title. III. Series.
QC793.3.S6L39 1989
539.7'25—dc20 89-32544

Princeton University Press books are printed on acid-free paper
and meet the guidelines for permanence and durability of the
Committee on Production Guidelines for Book Longevity
of the Council on Library Resources

Printed in the United States of America

Second Printing, with
errata sheet, 1994

5 7 9 10 8 6

http://pup.princeton.edu

ISBN-13: 978-0-691-08542-5
ISBN-10: 0-691-08542-0

For Christie, Didi,
Michelle, and Heather

Contents

Preface

In the late 1920's the relentless march of ideas and discoveries had carried physics to a generally accepted relativistic theory of the electron. The physicist P.A.M. Dirac, however, was dissatisfied with the prevailing ideas and, somewhat in isolation, sought for a better formulation. By 1928 he succeeded in finding a theory which accorded with his own ideas and also fit most of the established principles of the time. Ultimately this theory proved to be one of the great intellectual achievements of the period. It was particularly remarkable for the internal beauty of its mathematical structure which not only clarified much previously mysterious phenomena but also predicted in a compelling way the existence of an electron-like particle of negative energy. Indeed such particles were subsequently found to exist and our understanding of nature was transformed.

Because of its compelling beauty and physical significance it is perhaps not surprising that the ideas at the heart of Dirac's theory have also been discovered to play a role of great importance in modern mathematics, particularly in the interrelations between topology, geometry and analysis. A great part of this new understanding comes from the work of M. Atiyah and I. Singer. It is their work and its implications which form the focus of this book.

It seems appropriate to sketch some of the fundamental ideas here. In searching for his theory, Dirac was faced, roughly speaking, with the problem of finding a Lorentz-invariant wave equation $D\psi = \lambda\psi$ compatible with the Klein-Gordon equation $\Box\psi = \lambda\psi$ where $\Box = (\partial/\partial x_0)^2 - (\partial/\partial x_1)^2 - (\partial/\partial x_2)^2 - (\partial/\partial x_3)^2$. Causality required that D be first order in the "time" coordinate x_0. Of course by Lorentz invariance there could be no preferred time coordinate, and so D was required to be first-order in all variables. Thus, in essence Dirac was looking for a first-order differential operator whose square was the laplacian. His solution was to replace the complex-valued wave function ψ with an n-tuple $\Psi = (\psi_1, \ldots, \psi_n)$ of such functions. The operator D then became a first-order system of the form

$$D = \sum_{\mu=0}^{3} \gamma_\mu \frac{\partial}{\partial x_\mu}$$

where $\gamma_0, \ldots, \gamma_3$ were $n \times n$-matrices. The requirement that

$$D^2 = \begin{pmatrix} \square & & \\ & \ddots & \\ & & \square \end{pmatrix}$$

led to the equations

$$\gamma_\nu \gamma_\mu + \gamma_\mu \gamma_\nu = \pm 2\delta_{\nu\mu}.$$

These were easily and explicitly solved for small values of n, and the analysis was underway.

This construction of Dirac has a curious and fundamental property. Lorentz transformations of the space-time variables $(x_0, \ldots, x_3)$ induce linear transformations of the n-tuples Ψ which are determined only up to a sign. Making a consistent choice of sign amounts to passing to a non-trivial 2-fold covering $\tilde{L}$ of the Lorentz group L. That is, in transforming the Ψ's one falls upon a representation of $\tilde{L}$ which does not descend to L.

The theory of Dirac had another interesting feature. In the presence of an electromagnetic field the Dirac Hamiltonian contained an additional term added on to what one might expect from the classical case. There were strong formal analogies with the additional term one obtains by introducing internal spin into the mechanical equations of an orbiting particle. This "spin" or internal magnetic moment had observable quantum effects. The n-tuples Ψ were thereby called *spinors* and this family of transformations was called the *spin representation.*

This physical theory touches upon an important and general fact concerning the orthogonal groups. (We shall restrict ourselves for the moment to the positive definite case.) In the theory of Cartan and Weyl the representations of the Lie algebra of SO_n are essentially generated by two basic ones. The first is the standard n-dimensional representation (and its exterior powers). The second is constructed from the representations of the algebra generated by the γ_μ's as above (the *Clifford algebra* associated to the quadratic form defining the orthogonal group). This second representation is called the spin representation. It does not come from a representation of the orthogonal group, but only of its universal covering group, called Spin_n. It plays a key role in an astounding variety of questions in geometry and topology: questions involving vector fields on spheres, immersions of manifolds, the integrality of certain characteristic numbers, triality in dimension eight, the existence of complex structures, the existence of metrics of positive scalar curvature, and perhaps most basically, the index of elliptic operators.

In the early 1960s general developments had led mathematicians to consider the problem of finding a topological formula for the index of any elliptic operator defined on a compact manifold. This formula was to generalize the important Hirzebruch-Riemann-Roch Theorem already established in the complex algebraic case. In considering the problem, Atiyah and Singer noted that among all manifolds, those whose SO_n-structure could be simplified to a $Spin_n$-structure had particularly suggestive properties. Realizing that over such spaces one could carry out the Dirac construction, they produced a globally defined elliptic operator canonically associated to the underlying riemannian metric. The index of this operator was a basic topological invariant called the $\hat{A}$-genus, which was known always to be an integer in this special class of spin manifolds. (It is not an integer in general.) Twisting the Dirac-type operator with arbitrary coefficient bundles led, with some sophistication, to a general formula for the index of any elliptic operator.

Atiyah and Singer went on to understand the index in the more proper setting of K-theory. This led in particular to the formulation of certain KO-invariants which have profound applications in geometry and topology. These invariants touch questions unapproachable by other means. Their study and elucidation was a principal motivation for the writing of this tract.

It is interesting to note in more recent years there has been another profound and beautiful physical theory whose ideas have come to the core of topology, geometry and analysis. This is the non-abelian gauge field theory of C. N. Yang and R. L. Mills which through the work of S. Donaldson and M. Freedman has led to astonishing results in dimension four. Yang-Mills theory can be plausibly considered a highly non-trivial generalization of Dirac's theory which encompasses three fundamental forces: the weak, strong, and electromagnetic interactions. This theory involves modern differential geometry in an essential way. The theory of connections, Dirac-type operators, and index theory all play an important role. We hope this book can serve as a modest introduction to some of these concepts.

H. B. LAWSON AND M.-L. MICHELSOHN
Stony Brook

Acknowledgments

This book owes much to the fundamental work of Michael Atiyah and Iz Singer. Part of our initial motivation in writing the book was to give a leisurely and rounded presentation of their results.

The authors would like to express particular gratitude to Peter Landweber, Jean-Pierre Bourguignon, Misha Katz, Haiwan Chen and Peter Woit, each of whom read large parts of the original manuscript and made a number of important suggestions. We are also grateful to the National Science Foundation and the Brazilian C. N. Pq. for their support during the writing of this book.

H. B. LAWSON AND M.-L. MICHELSOHN

This author would like to take this opportunity to express her deep appreciation to Mark Mahowald who held out a hand when one was so dearly needed.

M.-L. MICHELSOHN

Spin Geometry

Introduction

Over the past two decades the geometry of spin manifolds and Dirac operators, and the various associated index theorems have come to play an increasingly important role both in mathematics and in mathematical physics. In the area of differential geometry and topology they have become fundamental. Topics like spin cobordism, previously considered exotic even by topologists, are now known to play an essential role in such classical questions as the existence or non-existence of metrics of positive curvature. Indeed, the profound methods introduced into geometry by Atiyah, Bott, Singer and others are now indispensible to mathematicians working in the field. It is the intent of this book to set out the fundamental concepts and to present these methods and results in a unified way.

A principal theme of the exposition here is the consistent use of Clifford algebras and their representations. This reflects the observed fact that these algebras emerge repeatedly at the very core of an astonishing variety of problems in geometry and topology.

Even in discussing riemannian geometry, the formalism of Clifford multiplication will be used in place of the more conventional exterior tensor calculus. There is a philosophical justification for this bias. Recall that to any vector space V there is naturally associated the exterior algebra $\Lambda^* V$, and this association carries over directly to vector bundles. Applied to the tangent bundle of a smooth manifold, it gives the de Rham bundle of exterior differential forms. In a similar way, to any vector space V equipped with a quadratic form q, there is associated the Clifford algebra $C\ell(V, q)$, and this association carries over directly to vector bundles equipped with fibre metrics. In particular, applied to the tangent bundle of a smooth riemannian manifold, it gives a canonically associated bundle of algebras, called the Clifford bundle. As a vector bundle it is isomorphic to the bundle of exterior forms. However, the Clifford multiplication is strictly richer than exterior multiplication; it reflects the inner symmetries and basic identities of the riemannian structure. In fact fundamental curvature identities will be derived here in the formalism of Clifford multiplication and applied to some basic problems.

Another justification for our approach is that the Clifford formalism gives a transparent unification of all the fundamental elliptic complexes in differential geometry. It also renders many of the technical arguments involved in applying the Index Theorem quite natural and simple.

This point of view concerning Clifford bundles and Clifford multiplication is an implicit, but rarely an explicit theme in the writing of Atiyah and Singer. The authors feel that for anyone working in topology or geometry it is worthwhile to develop a friendly, if not intimate relationship with spin groups and Clifford modules. For this reason we have used them explicitly and systematically in our exposition.

The book is organized into four chapters whose successive themes are algebra, geometry, analysis, and applications. The first chapter offers a detailed introduction to Clifford algebras, spin groups and their representations. The concepts are illuminated by giving some direct applications to the elementary geometry of spheres, projective spaces, and low-dimensional Lie groups. K-theory and KR-theory are then introduced, and the fundamental relationship between Clifford algebras and Bott periodicity is established.

In the second chapter of this book, the algebraic concepts are carried over to define structures on differentiable manifolds. Here one enters properly into the subject of spin geometry. Spin manifolds, spin cobordism, and spinor bundles with their canonical connections are all discussed in detail, and a general formalism of Dirac bundles and Dirac operators is developed. Hodge-de Rham Theory is reviewed in this formalism, and each of the fundamental elliptic operators of riemannian geometry is derived and examined in detail.

Special emphasis is given here to introducing the notion of a $C\ell_k$-linear elliptic operator and discussing its index. This index lives in a certain quotient of the Grothendieck group of Clifford modules. For the fundamental operators (which are discussed in detail here) it is one of the deepest and most subtle invariants of global riemannian geometry. The systematic discussion of $C\ell_k$-linear differential operators is one of the important features of this book.

In the last section of Chapter II a universal identity of Bochner type is established for any Dirac bundle, and the classical vanishing theorems of Bochner and Lichnerowicz are derived from it.

This seems an appropriate time to make some general observations about spin geometry. To begin it should be emphasized that spin geometry is really a special topic in riemannian geometry. The central concept of a spin manifold is often considered to be a topological one. It is just a manifold with a simply-connected structure group. This is understood systematically as follows. On a general differentiable n-manifold ($n \geq 3$), the tangent bundle has structure group GL_n. The manifold is said to be oriented if the structure group is reduced to GL_n^+ (the connected component of the identity). The manifold is said to be spin if the structure group GL_n^+ can be "lifted" to the universal covering group $\widetilde{GL}_n \to GL_n^+$. This approach is perfectly correct, but there is a hidden obstruction to the viability of the concept: namely, *the group* $\widetilde{GL}_n$ *(for* $n \geqq 3$*) has no finite dimensional representations that do not come from* GL_n^+. This means that in terms of standard tensor calculus, nothing has been gained by this refinement of the structure.

However, if one passes from GL_n to the maximal compact subgroup O_n, that is, if one introduces a riemannian metric on the manifold, the story is quite different. An orientation corresponds to reducing the structure group to SO_n, and a spin structure corresponds to then lifting the structure group to the universal covering group $Spin_n \to SO_n$. Maximal compact subgroups are homotopy equivalent to the Lie groups which contain them, and there is essentially no topological difference in viewing spin structures this way. However, *there do exist finite dimensional representations of* $Spin_n$ *which are not lifts of representations of* SO_n. Over a spin manifold one can thereby construct certain new vector bundles, called bundles of spinors, which do not exist over general manifolds. Their existence allows the introduction of certain important analytic tools which are not generally available, and these tools play a central role in the study of the global geometry of the space. It is, by the way, an important fact that this construction is metric-dependent; *the bundle of spinors itself depends in an essential way on the choice of riemannian structure on the manifold.*

These observations lead one to suspect that there must exist a local spinor calculus, like the tensor calculus, which should be an important component of local riemannian geometry. A satisfactory formalism of this type has not yet been developed. However, the spinors bundles have yielded profound relations between local riemannian geometry and global topology.

The main tools by which we access the global structure of spin manifolds are the various index theorems of Atiyah and Singer. These are presented and proved in Chapter III of the book. They include not just the standard G-Index Theorem but also the Index Theorem for Families and the $C\ell_k$-Index Theorem (for $C\ell_k$-linear elliptic operators). There are in existence today many elegant proofs of index theorems which use the methods of the heat equation. These do not apply to the $C\ell_k$-Index Theorem however, because of the non-local nature of this index. For this reason our exposition follows the "softer," or more topological, arguments given in the original proofs.

Chapter IV of the book is concerned with applications of the theory. There is no attempt to be exhaustive; such an attempt would be pointless and nearly impossible. We have tried however to demonstrate the broad range of problems in which the considerations of spin geometry can be effectively implemented.

It is of some historical interest to note that while Dirac did essentially use Clifford modules in the construction of his wave operator, he was not really responsible for what is commonly called the "Dirac operator" in riemannian geometry. The construction of this operator is due to Atiyah and Singer and is, in our estimation, one of their great achievements. It required for its discovery an understanding of the subtle geometry of spin manifolds and a recognition of the central role it would play in the general theory of elliptic operators. Even the formidable Élie Cartan, who sensed the importance of the question and, of course, authored the general theory of spinors and who was not unaware of the fundamentals of global analysis, never reached the point of defining this operator in the proper context of spin manifolds. In keeping with historical developments we shall call the general construction of operators from modules over the Clifford bundle, the Dirac construction, and we shall call the specific operator so defined on the spinor bundle, the Atiyah-Singer operator.

It is this operator which in a very specific sense generates all elliptic operators over a spin manifold. It introduces a direct relationship between curvature and topology which exists only under the spin hypothesis. The $C\ell_k$-linear version of this operator carries an index in KO-theory. In fact its index gives a basic ring homomorphism $\Omega_*^{\mathrm{Spin}} \to KO^{-*}(\mathrm{pt})$ which generalizes to KO-theory the classical $\hat{A}$-genus. The applications of this to geometry include the fact that half the exotic spheres in dimensions one and two (mod 8) do not carry metrics of positive scalar curvature.

The presentation in this book is aimed at readers with a knowledge of elementary geometry and topology. Important things, such as the concept of spin manifolds and the theory of connections, are developed from basic definitions. The Atiyah-Singer index theorems are formulated and proved assuming little more than a knowledge of the Fourier inversion formula. There are several appendices in which principal bundles, classifying spaces, Thom isomorphisms, and spin manifolds are discussed in detail.

The references to theorems and equations within each chapter are made without reference to the chapter itself (e.g., 2.7 or (5.9)). References to other chapters are prefaced by the chapter number (e.g., III.2.7 or (IV.5.9)).

CHAPTER I

Clifford Algebras, Spin Groups and Their Representations

The object of this chapter is to present the algebraic ideas which lie at the heart of spin geometry. The central concept is that of a Clifford algebra. This is an algebra naturally associated to a vector space which is equipped with a quadratic form. Within the group of units of the algebra there is a distinguished subgroup, called the spin group, which, in the case of the positive definite form on $\mathbb{R}^n$ ($n > 2$), is the universal covering group of SO_n.

It is a striking (and not commonplace) fact that Clifford algebras and their representations play an important role in many fundamental aspects of differential geometry. These include such diverse topics as Hodge-de Rham Theory, Bott periodicity, immersions of manifolds into spheres, families of vector fields on spheres, curvature identities in riemannian geometry, and Thom isomorphisms in K-theory. The effort invested in becoming comfortable with this algebraic formalism is well worthwhile.

Our discussion begins in a very general algebraic context but soon moves to the real case in order to keep matters simple and in the domain of most interest. In §§7 and 8 we present some applications of the purely algebraic theory to topology and to the appearence of exceptional phenomena in the theory of Lie groups.

The last part of the chapter is devoted to K-theory. Basic definitions are given and fundamental results are reviewed. The discussion culminates with the Atiyah-Bott-Shapiro isomorphisms which directly relate the periodicity phenomena in Clifford algebras to the classical Bott Periodicity Theorems. In particular, explicit isomorphisms are given between $K^{-*}(\mathrm{pt}) = \bigoplus_n K(S^n)$ (and $KO^{-*}(\mathrm{pt}) = \bigoplus_n KO(S^n)$) and a certain quotient of the ring of Clifford modules. Section 10 is concerned with KR-theory which later plays a role in the index theorem for families of real elliptic operators. This is a bigraded theory and the corresponding Atiyah-Bott-Shapiro isomorphism entails representations of Clifford algebras $C\ell_{r,s}$ for quadratic forms of indefinite signature.

§1. Clifford Algebras

Let V be a vector space over the commutative field k and suppose q is a quadratic form on V. The **Clifford algebra** $C\ell(V,q)$ associated to V and

q is an associative algebra with unit defined as follows. Let

$$\mathscr{T}(V) = \sum_{r=0}^{\infty} \bigotimes^r V$$

denote the tensor algebra of V, and define $\mathscr{I}_q(V)$ to be the ideal in $\mathscr{T}(V)$ generated by all elements of the form $v \otimes v + q(V)1$ for $v \in V$. Then the Clifford algebra is defined to be the quotient

$$\mathrm{C}\ell(V,q) \equiv \mathscr{T}(V)/\mathscr{I}_q(V).$$

There is a natural embedding

$$V \hookrightarrow \mathrm{C}\ell(V,q) \tag{1.1}$$

which is the image of $V = \bigotimes^1 V$ under the canonical projection

$$\pi_q : \mathscr{T}(V) \longrightarrow \mathrm{C}\ell(V,q). \tag{1.2}$$

We prove that $\pi_q|_V$ is injective as follows. We say that an element $\varphi \in \mathscr{T}(V)$ is of **pure degree** s if $\varphi \in \bigotimes^s V$. (Every element of $\mathscr{T}(V)$ is a finite sum of elements of pure degree.) We want to show that any element $\varphi \in \mathscr{I}_q(V) \cap V$ is zero. Any such element can be written as a finite sum $\varphi = \sum a_i \otimes (v_i \otimes v_i + q(v_i)) \otimes b_i$ where we may assume that the a_i's and b_i's are of pure degree. Since $\varphi \in V = \bigotimes^1 V$, we conclude that $\sum a_{i'} \otimes (v_{i'} \otimes v_{i'}) \otimes b_{i'} = 0$, where this sum is taken over those indices with $\deg a_i + \deg b_i$ maximal. This equation implies, by contraction with q, that $\sum a_{i'} q(v_{i'}) \cdot b_{i'} = 0$. Proceeding inductively, we prove that $\varphi = 0$.

The algebra $\mathrm{C}\ell(V,q)$ is generated by the vector space $V \subset \mathrm{C}\ell(V,q)$ (and the identity 1) subject to the relations:

$$v \cdot v = -q(v)1 \tag{1.3}$$

for $v \in V$. If the characteristic of k is not 2, then for all $v,w \in V$,

$$v \cdot w + w \cdot v = -2q(v,w) \tag{1.4}$$

where $2q(v,w) \equiv q(v + w) - q(v) - q(w)$ is the polarization of q. The relations (1.3) can be used to give the following universal characterization of the algebra.

Proposition 1.1. *Let $f : V \to \mathscr{A}$ be a linear map into an associative k-algebra with unit, such that*

$$f(v) \cdot f(v) = -q(v)1 \tag{1.5}$$

for all $v \in V$. Then f extends uniquely to a k-algebra homomorphism $\tilde{f} : \mathrm{C}\ell(V,q) \to \mathscr{A}$. Furthermore, $\mathrm{C}\ell(V,q)$ is the unique associative k-algebra with this property.

Proof. Any linear map $f: V \to \mathscr{A}$ extends to a unique algebra homomorphism $\tilde{f}: \mathscr{T}(V) \to \mathscr{A}$. Property (1.5) implies that $\tilde{f} = 0$ on $\mathscr{I}_q(V)$, and so $\tilde{f}$ descends to $C\ell(V,q)$. Suppose now that $\mathscr{C}$ is an associative k-algebra with unit and that $i: V \hookrightarrow \mathscr{C}$ is an embedding with the property that any linear map $f: V \to \mathscr{A}$ ($\mathscr{A}$ as above) with property (1.5) extends uniquely to an algebra homomorphism $\tilde{f}: \mathscr{C} \to \mathscr{A}$. Then the isomorphism from $V \subset C\ell(V,q)$ to $i(V) \subset \mathscr{C}$ clearly induces an algebra isomorphism $C\ell(V,q) \xrightarrow{\approx} \mathscr{C}$. ■

This characterization of Clifford algebras is extremely useful. It shows, for example, that they are functorial in the following sense. Given a morphism $f:(V,q) \to (V',q')$, i.e., a k-linear map $f: V \to V'$ between vector spaces which preserves the quadratic forms ($f^*q' = q$), there is, by Proposition 1.1, an induced homomorphism $\tilde{f}: C\ell(V,q) \to C\ell(V',q')$. Given another morphism $g:(V',q') \to (V'',q'')$, we see from the uniqueness in Proposition 1.1, that $\widetilde{g \circ f} = \tilde{g} \circ \tilde{f}$.

A particular consequence of this is that the orthogonal group $O(V,q) \equiv \{f \in GL(V): f^*q = q\}$ extends canonically to a group of automorphisms of $C\ell(V,q)$. We shall see later that this embedding

$$O(V,q) \subset \mathrm{Aut}(C\ell(V,q)) \tag{1.6}$$

actually lies in the subgroup of inner automorphisms.

An element here of particular importance is the automorphism

$$\alpha: C\ell(V,q) \longrightarrow C\ell(V,q) \tag{1.7}$$

which extends the map $\alpha(v) = -v$ on V. Since $\alpha^2 = Id$, there is a decomposition

$$C\ell(V,q) = C\ell^0(V,q) \oplus C\ell^1(V,q) \tag{1.8}$$

where $C\ell^i(V,q) = \{\varphi \in C\ell(V,q): \alpha(\varphi) = (-1)^i\varphi\}$ are the eigenspaces of α. Clearly, since $\alpha(\varphi_1\varphi_2) = \alpha(\varphi_1) \cdot \alpha(\varphi_2)$, we have that

$$C\ell^i(V,q) \cdot C\ell^j(V,q) \subseteq C\ell^{i+j}(V,q) \tag{1.9}$$

where the indices are taken modulo 2. An algebra with a decomposition (1.8) satisfying (1.9) is called a **$\mathbb{Z}_2$-graded algebra.** Note that $C\ell^0(V,q)$ is a subalgebra of $C\ell(V,q)$. It is called the **even part** of $C\ell(V,q)$. The subspace $C\ell^1(V,q)$ is called the **odd part.** It is an observation of Atiyah, Bott and Shapiro that this $\mathbb{Z}_2$-grading plays an important role in the analysis and application of Clifford algebras.

There exist some elementary and important relationships between the Clifford algebra $C\ell(V,q)$ of a space and its exterior algebra Λ^*V (whose definition is, of course, independent of the quadratic form q). There is a natural filtration $\tilde{\mathscr{F}}^0 \subset \tilde{\mathscr{F}}^1 \subset \tilde{\mathscr{F}}^2 \subset \ldots \subset \mathscr{T}(V)$ of the tensor algebra, which is defined by

$$\tilde{\mathscr{F}}^r \equiv \sum_{s \leqq r} \bigotimes^s V,$$

and has the property that $\tilde{\mathscr{F}}^r \otimes \tilde{\mathscr{F}}^{r'} \subseteq \tilde{\mathscr{F}}^{r+r'}$. If we set $\mathscr{F}^i = \pi_q(\tilde{\mathscr{F}}^i)$ we obtain a filtration $\mathscr{F}^0 \subset \mathscr{F}^1 \subset \mathscr{F}^2 \subset \ldots \subset C\ell(V,q)$ of the Clifford algebra, which also has the property that

$$\mathscr{F}^r \cdot \mathscr{F}^{r'} \subseteq \mathscr{F}^{r+r'} \tag{1.10}$$

for all r,r'. This makes $C\ell(V,q)$ into a **filtered algebra.** It follows from (1.10) that the multiplication map descends to a map $(\mathscr{F}^r/\mathscr{F}^{r-1}) \times (\mathscr{F}^s/\mathscr{F}^{s-1}) \to (\mathscr{F}^{r+s}/\mathscr{F}^{r+s-1})$ for all r,s. Setting $\mathscr{G}^* \equiv \bigoplus_{r \geqq 0} \mathscr{G}^r$ where $\mathscr{G}^r \equiv \mathscr{F}^r/\mathscr{F}^{r-1}$, we obtain the **associated graded algebra.**

Proposition 1.2. *For any quadratic form q, the associated graded algebra of $C\ell(V,q)$ is naturally isomorphic to the exterior algebra $\Lambda^* V$.*

Proof. The map $\bigotimes^r V \xrightarrow{\pi_r} \mathscr{F}^r \to \mathscr{F}^r/\mathscr{F}^{r-1}$, which is given by $v_{i_1} \otimes \cdots \otimes v_{i_r} \mapsto [v_{i_1} \cdots v_{i_r}]$ clearly descends to a map $\Lambda^r V \to \mathscr{F}^r/\mathscr{F}^{r-1}$ by property (1.4). (When the characteristic of k is 2, we use the fact that $v \cdot w + w \cdot v = 0$.) This map is evidently surjective and is easily seen to give a homomorphism of graded algebras $\Lambda^* V \to \mathscr{G}^*$.

To see that this map is injective we proceed as follows. The kernel of $\bigotimes^r V \to \mathscr{G}^r$ consists of the r-homogeneous pieces of elements $\varphi \in \mathscr{I}_q(V)$ of degree $\leqq r$. Any such φ can be written as a finite sum $\varphi = \sum a_i \otimes (v_i \otimes v_i + q(v_i)) \otimes b_i$ where $v_i \in V$ and where we may assume that the a_i and b_i are of pure degree with $\deg a_i + \deg b_i \leqq r - 2$. The r-homogeneous part of φ is then of the form $\varphi_r = \sum a_i \otimes v_i \otimes v_i \otimes b_i$ (where $\deg a_i + \deg b_i = r - 2$ for each i). Since $v_i \wedge v_i = 0$ for all i, we see that the image of φ in the exterior algebra is zero. Hence the map $\Lambda^r V \to \mathscr{G}^r$ is injective. ∎

Proposition 1.2 says that Clifford multiplication is an enhancement of exterior multiplication which is determined by the form q. Note that $C\ell(V,0) \cong \Lambda^* V$.

Proposition 1.3. *There is a canonical vector space isomorphism*

$$\Lambda^* V \xrightarrow{\approx} C\ell(V,q) \tag{1.11}$$

compatible with the filtrations.

Remark 1.4. The map (1.11) is, of course, *not* an isomorphism of algebras unless $q = 0$. The point here is that the map is canonical. Thus we may speak of the embeddings

$$\Lambda^r V \subset C\ell(V,q) \qquad \text{for all } r \geqq 0. \tag{1.12}$$

Proof. We define a map of the r-fold direct product $f: V \times \cdots \times V \to C\ell(V,q)$ by setting

$$f(v_1, \ldots, v_r) = \frac{1}{r!} \sum_{\sigma} \text{sign}(\sigma) v_{\sigma(1)} \cdots v_{\sigma(r)} \tag{1.13}$$

where the sum is taken over the symmetric group on r elements. (If the characteristic of k is not zero, one must drop the factor $1/r!$.) Clearly f determines a linear map $\tilde{f}: \Lambda^r V \to C\ell(V,q)$ whose image lies in $\mathscr{F}^r$. The composition of $\tilde{f}$ with the projection $\mathscr{F}^r \to \mathscr{F}^r/\mathscr{F}^{r-1}$ is easily seen to be the map discussed in the proof of Proposition 1.2. Hence $\tilde{f}$ is injective, and the direct sum of these maps (1.11) is an isomorphism.

We now take up the question of tensor products. Recall that if $\mathscr{A}$ and $\mathscr{B}$ are algebras with unit over k, then the tensor product of the algebras $\mathscr{A} \otimes \mathscr{B}$ is the algebra whose underlying vector space is the tensor product of $\mathscr{A}$ and $\mathscr{B}$ and whose multiplication is given (on simple elements) by the rule $(a \otimes b) \cdot (a' \otimes b') = (aa') \otimes (bb')$. If, however,

$$\mathscr{A} = \mathscr{A}^0 \oplus \mathscr{A}^1 \qquad \text{and} \qquad \mathscr{B} = \mathscr{B}^0 \oplus \mathscr{B}^1$$

are $\mathbb{Z}_2$-graded algebras, then we can introduce a second "$\mathbb{Z}_2$-graded" multiplication, determined by the rule

$$(a \otimes b) \cdot (a' \otimes b') = (-1)^{\deg(b)\deg(a')}(aa') \otimes (bb') \tag{1.14}$$

whenever b and a' are of pure degree (even or odd). The resulting algebra is called the **$\mathbb{Z}_2$-graded tensor product** and is denoted $\mathscr{A} \hat{\otimes} \mathscr{B}$.

The $\mathbb{Z}_2$-graded tensor product is again $\mathbb{Z}_2$-graded with

$$(\mathscr{A} \hat{\otimes} \mathscr{B})^0 = \mathscr{A}^0 \otimes \mathscr{B}^0 + \mathscr{A}^1 \otimes \mathscr{B}^1$$

$$(\mathscr{A} \hat{\otimes} \mathscr{B})^1 = \mathscr{A}^1 \otimes \mathscr{B}^0 + \mathscr{A}^0 \otimes \mathscr{B}^1.$$

It also carries a filtration $\mathscr{F}^0 \subset \mathscr{F}^1 \subset \mathscr{F}^2 \subset \ldots \subset \mathscr{A} \hat{\otimes} \mathscr{B}$, where

$$\mathscr{F}^r \equiv \sum_{i+j=r} \mathscr{F}^i(\mathscr{A}) \hat{\otimes} \mathscr{F}^j(\mathscr{B}).$$

The importance of the $\mathbb{Z}_2$-graded tensor product for Clifford algebras is evident from the following proposition.

Proposition 1.5. *Let $V = V_1 \oplus V_2$ be a q-orthogonal decomposition of the vector space V (i.e., $q(v_1 + v_2) = q(v_1) + q(v_2)$ for all $v_1 \in V_1$ and $v_2 \in V_2$). Then there is a natural isomorphism of Clifford algebras*

$$C\ell(V,q) \longrightarrow C\ell(V_1,q_1) \hat{\otimes} C\ell(V_2,q_2)$$

where q_i denotes the restriction of q to V_i and where $\hat{\otimes}$ denotes the $\mathbb{Z}_2$-graded tensor product.

Proof. Consider the map $f: V \to C\ell(V_1,q_1) \hat{\otimes} C\ell(V_2,q_2)$ given by $f(v) = v_1 \otimes 1 + 1 \otimes v_2$ where $v = v_1 + v_2$ is the decomposition of v with respect to the splitting $V = V_1 \oplus V_2$. From (1.14) and the q-orthogonality of this splitting we see that $f(v) \cdot f(v) = (v_1 \otimes 1 + 1 \otimes v_2)^2 = v_1^2 \otimes 1 + 1 \otimes v_2^2 = -(q_1(v_1) + q_2(v_2))1 \otimes 1 = -q(v)1 \otimes 1$. Hence, by Proposition 1.1, f extends to an algebra homomorphism $\tilde{f}: C\ell(V,q) \to C\ell(V_1,q_1) \hat{\otimes} C\ell(V_2,q_2)$. The image of $\tilde{f}$ is a subalgebra which contains $C\ell(V_1,q_1) \otimes 1$ and $1 \otimes C\ell(V_2,q_2)$. Therefore, $\tilde{f}$ is surjective. Injectivity follows easily by considering a basis for $C\ell(V,q)$ generated by a basis of V which is compatible with the splitting. ∎

We finish this section by introducing a second fundamental involution on the algebra. The tensor algebra $\mathcal{T}(V)$ has an involution, given on simple elements by the reversal of order, i.e., $v_1 \otimes \cdots \otimes v_r \mapsto v_r \otimes \cdots \otimes v_1$. This map clearly preserves the ideal $\mathcal{I}(V,q)$ and so descends to a map

$$(\quad)^t : C\ell(V,q) \longrightarrow C\ell(V,q) \tag{1.15}$$

called the **transpose.** Note that $(\quad)^t$ is an antiautomorphism, i.e., $(\varphi\psi)^t = \psi^t\varphi^t$.

§2. The Groups Pin and Spin

We now consider the **multiplicative group of units** in the Clifford algebra, which is defined to be the subset

$$C\ell^\times(V,q) \equiv \{\varphi \in C\ell(V,q) : \exists \varphi^{-1} \text{ with } \varphi^{-1}\varphi = \varphi\varphi^{-1} = 1\} \tag{2.1}$$

This group contains all elements $v \in V$ with $q(v) \neq 0$. When $\dim V = n < \infty$, and k is either $\mathbb{R}$ or $\mathbb{C}$, this is a Lie group of dimension 2^n. In general, there is an associated Lie algebra $\mathfrak{cl}^\times(V,q) = C\ell(V,q)$ with Lie bracket given by

$$[x,y] = xy - yx. \tag{2.2}$$

The group of units always acts naturally as automorphisms of the algebra. That is, there is a homomorphism

$$\mathrm{Ad}: C\ell^\times(V,q) \longrightarrow \mathrm{Aut}(C\ell(V,q)) \tag{2.3}$$

called the **adjoint representation**, which is given by

$$\mathrm{Ad}_\varphi(x) \equiv \varphi x \varphi^{-1}. \tag{2.4}$$

Taking the "derivative" of this gives a homomorphism

$$\mathrm{ad}: \mathfrak{cl}^\times(V,q) \longrightarrow \mathrm{Der}(C\ell(V,q)) \tag{2.5}$$

into the derivations of the algebra, defined by setting

$$\mathrm{ad}_y(x) \equiv [y,x].$$

REMARK 2.1. Suppose V is finite dimensional, and defined over $\mathbb{R}$ or $\mathbb{C}$. Then there is a natural exponential mapping $\exp: \mathcal{cl}^\times(V,q) \to C\ell^\times(V,q)$, defined by setting

$$\exp(y) = \sum_{m=0}^{\infty} \frac{1}{m!} y^m. \tag{2.6}$$

Note that this series converges since for any choice of positive definite inner product on $C\ell(V,q)$, we have $\|xy\| \leq c\|x\|\,\|y\|$ for some $c > 0$. It is easy to see that

$$\frac{d}{dt} \mathrm{Ad}_{\exp(ty)}(x)\Big|_{t=0} = \mathrm{ad}_y(x). \tag{2.7}$$

From this point on we shall assume that *the characteristic of the field k is different from* 2. Under this assumption, we have the following important facts concerning the adjoint representation:

Proposition 2.2. *Let $v \in V \subset C\ell(V,q)$ be an element with $q(v) \neq 0$. Then $\mathrm{Ad}_v(V) = V$. In fact, for all $w \in V$, the following equation holds:*

$$-\mathrm{Ad}_v(w) = w - 2\frac{q(v,w)}{q(v)}v. \tag{2.8}$$

Proof. Since $v^{-1} = -v/q(v)$, we have from (1.4) that

$$\begin{aligned} -q(v)\mathrm{Ad}_v(w) &= -q(v)vwv^{-1} = vwv \\ &= -v^2w - 2q(v,w)v = q(v)w - 2q(v,w)v. \quad \blacksquare \end{aligned}$$

This leads us naturally to consider the subgroup of elements $\varphi \in C\ell^\times(V,q)$ such that $\mathrm{Ad}_\varphi(V) = V$. By Proposition 2.2, this group contains all elements $v \in V$ with $q(v) \neq 0$. Furthermore, we see from equation (2.8) that whenever $q(v) \neq 0$, the transformation Ad_v preserves the quadratic form q. That is, $(\mathrm{Ad}_v^* q)(w) \equiv q(\mathrm{Ad}_v(w)) = q(w)$ for all $w \in V$. Therefore, we define $P(V,q)$ to be the subgroup of $C\ell^\times(V,q)$ generated by the elements $v \in V$ with $q(v) \neq 0$, and observe that there is a representation

$$P(V,q) \xrightarrow{\mathrm{Ad}} \mathrm{O}(V,q) \tag{2.9}$$

where

$$\mathrm{O}(V,q) = \{\lambda \in \mathrm{GL}(V) : \lambda^* q = q\} \tag{2.10}$$

is the **orthogonal group** of the form q. The group $P(V,q)$ has certain important subgroups.

DEFINITION 2.3 The **Pin group** of (V,q) is the subgroup $\mathrm{Pin}(V,q)$ of $\mathrm{P}(V,q)$ generated by the elements $v \in V$ with $q(v) = \pm 1$. The associated **Spin group** of (V,q) is defined by

$$\mathrm{Spin}(V,q) = \mathrm{Pin}(V,q) \cap C\ell^0(V,q).$$

We observe now that the right-hand side of equation (2.8) is just the map $\rho_v : V \to V$ given by reflection across the hyperplane $v^\perp = \{w \in V : q(w,v) = 0\}$. That is, the map ρ_v fixes this hyperplane and maps v to $-v$. Unfortunately, there is a minus sign on the left in equation (2.8). This means, for example, that if $\dim V$ is odd, then Ad_v is always orientation preserving. This defect can be removed by considering the **twisted adjoint representation**

$$\widetilde{\mathrm{Ad}} : C\ell^\times(V,q) \longrightarrow \mathrm{GL}(C\ell(V,q))$$

defined by setting

$$\widetilde{\mathrm{Ad}}_\varphi(y) = \alpha(\varphi) y \varphi^{-1}. \tag{2.11}$$

Clearly, $\widetilde{\mathrm{Ad}}_{\varphi_1\varphi_2} = \widetilde{\mathrm{Ad}}_{\varphi_1} \circ \widetilde{\mathrm{Ad}}_{\varphi_2}$ and $\widetilde{\mathrm{Ad}}_\varphi = \mathrm{Ad}_\varphi$ for even elements φ (i.e., for $\varphi \in C\ell^0(V,q)$). Furthermore, from (2.8) we have

$$\widetilde{\mathrm{Ad}}_v(w) = w - 2\frac{q(v,w)}{q(v)} v. \tag{2.12}$$

We then define the subgroup

$$\widetilde{\mathrm{P}}(V,q) \equiv \{\varphi \in C\ell^\times(V,q) : \widetilde{\mathrm{Ad}}_\varphi(V) = V\}. \tag{2.13}$$

It is clear that $\mathrm{P}(V,q) \subset \widetilde{\mathrm{P}}(V,q)$. Furthermore, we have the following.

Proposition 2.4. *Suppose that V is finite dimensional and that q is nondegenerate. Then the kernel of the homomorphism*

$$\widetilde{\mathrm{P}}(V,q) \xrightarrow{\widetilde{\mathrm{Ad}}} \mathrm{GL}(V)$$

is exactly the group $k^\times$ of non-zero multiples of 1.

Proof. Choose a basis $\{v_1, \ldots, v_n\}$ for V such that $q(v_i) \neq 0$ for all i and $q(v_i,v_j) = 0$ for all $i \neq j$. Suppose $\varphi \in C\ell^\times(V,q)$ is in the kernel of $\widetilde{\mathrm{Ad}}$, that is, suppose φ has the property that $\alpha(\varphi)v = v\varphi$ for all $v \in V$. Write $\varphi = \varphi_0 + \varphi_1$, where φ_0 is even and φ_1 is odd, and observe that

$$\begin{aligned} v\varphi_0 &= \varphi_0 v \\ -v\varphi_1 &= \varphi_1 v \end{aligned} \tag{2.14}$$

for all $v \in V$. The terms φ_0 and φ_1 can be written as polynomial expressions in $v_1, \ldots, v_n$. Successive use of the fact (1.4) that $v_i v_j = -v_j v_i - 2q(v_i,v_j)$ shows that φ_0 can be expressed as $\varphi_0 = a_0 + v_1 a_1$ where a_0 and a_1 are polynomial expressions in $v_2, \ldots, v_n$. Applying α shows that a_0 is even and a_1 is odd. Setting $v = v_1$ in (2.14), we see that

$$\begin{aligned} v_1 a_0 + v_1^2 a_1 &= a_0 v_1 + v_1 a_1 v_1 \\ &= v_1 a_0 - v_1^2 a_1. \end{aligned}$$

Hence, $v_1^2 a_1 = -q(v_1)a_1 = 0$, and so $a_1 = 0$. This implies that φ_0 does not involve v_1. Proceeding inductively, we see that φ_0 does not involve any of the terms $v_1, \ldots, v_n$ and so $\varphi_0 = t \cdot 1$ for $t \in k$.

The analogous argument can now be applied to φ_1. Write $\varphi_1 = a_1 + v_1 a_0$, where a_0 and a_1 do not involve v_1. Note that a_1 is odd and a_0 is even; and therefore, from (2.14), $-v_1 a_1 - v_1^2 a_0 = a_1 v_1 + v_1 a_0 v_1 = -v_1 a_1 + v_1^2 a_0$. Hence, $a_0 = 0$ and so φ_1 is independent of v_1. By induction, φ_1 is independent of $v_1, \ldots, v_n$ and so $\varphi_1 = 0$.

Now we have $\varphi = \varphi_0 + \varphi_1 = t \cdot 1 \in k$. But $\varphi \neq 0$, so $\varphi \in k^\times$. ■

Note that this proposition requires the **twisted** adjoint representation and not the adjoint representation. The minus sign in (2.14) is crucial to the proof.

Proposition 2.4 is false if we do not assume that q is non-degenerate. To see this, consider the extreme case $C\ell(V,0) = \Lambda^* V$. For all $v_1, v_2 \in V$, we have $1 + v_1 v_2 \in C\ell^\times(V,0)$. In fact, $(1 + v_1 v_2)^{-1} = 1 - v_1 v_2$. However, for any $v \in V$, we see that $\alpha(1 + v_1 v_2) v (1 + v_1 v_2)^{-1} = (1 + v_1 v_2) \cdot v(1 - v_1 v_2) = v$. Hence, the kernel of the homomorphism includes many non-scalar terms.

We now introduce the **norm** mapping $N: C\ell(V,q) \to C\ell(V,q)$ defined by setting

$$N(\varphi) \equiv \varphi \cdot \alpha(\varphi^t). \tag{2.15}$$

Here φ^t denotes the transpose of φ introduced in (1.15). It is easy to see that $\alpha(\varphi^t) = (\alpha(\varphi))^t$. Note that

$$N(v) = q(v) \qquad \text{for } v \in V. \tag{2.16}$$

The importance of the norm is evident from the following proposition.

Proposition 2.5. *Suppose that V is finite dimensional and that q is non-degenerate. Then the restriction of N to the group $\tilde{P}(V,q)$ gives a homomorphism*

$$N: \tilde{P}(V,q) \longrightarrow k^\times \tag{2.17}$$

into the multiplicative group of non-zero multiples of the identity in $C\ell(V,q)$.

Proof. To begin we observe that $N(\tilde{P}(V,q)) \subset k^\times$. Choose $\varphi \in \tilde{P}(V,q)$ and recall that by definition, $\alpha(\varphi) v \varphi^{-1} \in V$ for all $v \in V$. Applying the transpose antiautomorphism, which is the identity on V, we see that

$$(\varphi^t)^{-1} v \alpha(\varphi^t) = \alpha(\varphi) v \varphi^{-1}.$$

Hence,

$$\begin{aligned}\varphi^t\alpha(\varphi)v\varphi^{-1}(\alpha(\varphi^t))^{-1} &= \alpha[\alpha(\varphi^t)\varphi]v[\alpha(\varphi^t)\varphi]^{-1}\\ &= \widetilde{\mathrm{Ad}}_{\alpha(\varphi^t)\varphi}(v) = v\end{aligned}$$

for all $v \in V$. Hence, $\alpha(\varphi^t)\varphi$ is in the kernel of $\widetilde{\mathrm{Ad}}$. It is easy to check that $\alpha(\varphi^t)$ belongs to $\tilde{\mathrm{P}}(V,q)$, and therefore so does $\alpha(\varphi^t)\varphi$. Hence, by Proposition 2.4 we have $\alpha(\varphi^t)\varphi \in k^\times$. Applying α shows that $\varphi^t\alpha(\varphi) = N(\varphi^t) \in k^\times$. Since the transpose antiautomorphism preserves $\tilde{\mathrm{P}}(V,q)$, we conclude that $N(\varphi) \in k^\times$ for all $\varphi \in \tilde{\mathrm{P}}(V,q)$.

We now observe that if $\varphi,\psi \in \tilde{\mathrm{P}}(V,q)$, then $N(\varphi\psi) = \varphi\psi\alpha(\psi^t)\alpha(\varphi^t) = \varphi N(\psi)\alpha(\varphi^t) = \varphi\alpha(\varphi^t)N(\psi) = N(\varphi)N(\psi)$. Thus, N is a homomorphism on $\tilde{\mathrm{P}}(V,q)$. ∎

Continuing to assume that $\dim V < \infty$ and q is non-degenerate, we have the following.

Corollary 2.6. *The transformations* $\widetilde{\mathrm{Ad}}_\varphi : V \to V$ *for* $\varphi \in \tilde{\mathrm{P}}(V,q)$ *preserve the quadratic form* q. *Hence, there is a homomorphism*

$$\widetilde{\mathrm{Ad}} : \tilde{\mathrm{P}}(V,q) \longrightarrow \mathrm{O}(V,q) \tag{2.18}$$

Proof. To begin we note that $N(\alpha\varphi) = N(\varphi)$ for $\varphi \in \tilde{\mathrm{P}}(V,q)$ since $N(\alpha\varphi) = \alpha(\varphi)\varphi^t = \alpha N(\varphi) = N(\varphi)$. Consequently, if we set

$$V^\times = \{v \in V : q(v) \neq 0\}, \tag{2.19}$$

then for each $v \in V^\times(\subset \tilde{\mathrm{P}}(V,q))$, we have $N(\widetilde{\mathrm{Ad}}_\varphi(v)) = N(\alpha(\varphi)v\varphi^{-1}) = N(\alpha\varphi)N(v)N(\varphi)^{-1} = N(\varphi)N(\varphi)^{-1}N(v) = N(v)$. Since $N(\varphi) = q(v)$ for $v \in V$ (cf. (2.16)), we see that $\widetilde{\mathrm{Ad}}_\varphi$ preserves all non-zero q-lengths. Applying $\widetilde{\mathrm{Ad}}_{\varphi^{-1}}$ now shows that $\mathrm{Ad}_\varphi(V^\times) = V^\times$ and so $\widetilde{\mathrm{Ad}}_\varphi$ leaves invariant the set of vectors of zero q-length. Thus, $\widetilde{\mathrm{Ad}}_\varphi$ is q-orthogonal. ∎

We now return to the group $\mathrm{P}(V,q) \subseteq \tilde{\mathrm{P}}(V,q)$ and observe that by definition

$$\mathrm{P}(V,q) = \{v_1 \cdots v_r \in C\ell(V,q) : v_1 \ldots ,v_r \text{ is a finite sequence from } V^\times\}. \tag{2.20}$$

Recall that the twisted adjoint representation gives a homomorphism $\widetilde{\mathrm{Ad}} : \mathrm{P}(V,q) \to \mathrm{O}(V,q)$ such that

$$\widetilde{\mathrm{Ad}}_{v_1 \cdots v_r} = \rho_{v_1} \circ \cdots \circ \rho_{v_r} \tag{2.21}$$

where

$$\rho_v(w) = w - 2\frac{q(w,v)}{q(v)}v \tag{2.22}$$

is **reflection across** $v^{\perp}$. Thus the image of $\mathrm{P}(V,q)$ under $\widetilde{\mathrm{Ad}}$ is exactly the group generated by reflections. It is an important and classical result that this is always the entire orthogonal group.

Theorem 2.7 (Cartan-Dieudonné). *Let q be a non-degenerate quadratic form on a finite dimensional vector space V. Then every element $g \in \mathrm{O}(V,q)$ can be written as a product of r reflections*

$$g = \rho_{v_1} \circ \cdots \circ \rho_{v_r}$$

where $r \leqq \dim(V)$.

We refer the reader to Artin's book [1] for the general proof. In the special case where $V = \mathbb{R}^n$ and $q(x) = \|x\|^2$ is the standard norm, this theorem is easily proved by putting the orthogonal matrix g in "diagonal" form:

$$\begin{pmatrix} \boxed{R_{\theta_1}} & & & & & & \\ & \boxed{R_{\theta_2}} & & & & & \\ & & \ddots & & & & \\ & & & \boxed{R_{\theta_k}} & & & \\ & & & & \pm 1 & & \\ & & & & & \ddots & \\ & & & & & & \pm 1 \end{pmatrix}$$

where each R_{θ_i} is a 2×2 rotation matrix (which can be expressed as a product of two reflections).

Theorem 2.7 says that the homomorphism $\widetilde{\mathrm{Ad}} : \mathrm{P}(V,q) \to \mathrm{O}(V,q)$ is surjective. Furthermore, we could consider the group $\mathrm{SP}(V,q) = \mathrm{P}(V,q) \cap C\ell^0(V,q)$ and, since $\dim V$ is finite, the **special orthogonal group**

$$\mathrm{SO}(V,q) = \{\lambda \in \mathrm{O}(V,q) : \det(\lambda) = 1\}.$$

Theorem 2.7 also says that the homomorphism $\widetilde{\mathrm{Ad}} : \mathrm{SP}(V,q) \to \mathrm{SO}(V,q)$ is surjective. To see this, we first show that $\det(\rho_v) = -1$ for any $v \in V$. To prove this, choose a basis $v_1, \ldots, v_n$ such that $v_1 = v$ and $q(v,v_j) = 0$ for $j \geqq 2$. It follows from the definition that $\rho_v(v_1) = -v_1$ and $\rho_v(v_j) = v_j$ for $j \geqq 2$, and so $\det(\rho_v) = -1$ as claimed. Thus from Theorem 2.7 we conclude that

$$\mathrm{SO}(V,q) = \{\rho_{v_1} \circ \cdots \circ \rho_{v_r} : q(v_j) \neq 0 \text{ and } r \text{ is even}\}. \tag{2.23}$$

From the definition (cf. (2.20)) we see that $\mathrm{SP}(V,q) = \{v_1 \cdots v_r \in \mathrm{P}(V,q) : r$ is even$\}$. The surjectivity of $\widetilde{\mathrm{Ad}} : \mathrm{SP}(V,q) \to \mathrm{SO}(V,q)$ follows immediately (see (2.21)).

We now return to the groups Pin and Spin. Recall that these are the groups generated by the generalized unit sphere $S = \{v \in V : q(v) = \pm 1\}$ in V. That is,

$$\mathrm{Pin}(V,q) = \{v_1 \cdots v_r \in \mathrm{P}(V,q) : q(v_j) = \pm 1 \text{ for all } j\} \tag{2.24}$$

and

$$\mathrm{Spin}(V,q) = \{v_1 \cdots v_r \in \mathrm{Pin}(V,q) : r \text{ is even}\}. \tag{2.25}$$

In light of the above it is natural to ask whether the homomorphism $\widetilde{\mathrm{Ad}}$ restricted to $\mathrm{Pin}(V,q)$ and $\mathrm{Spin}(V,q)$ maps *onto* $\mathrm{O}(V,q)$ and $\mathrm{SO}(V,q)$ respectively. This seems quite likely since at a glance one can see that

$$\rho_{tv} = \rho_v \tag{2.26}$$

for any non-zero scalar $t \in k$, and so one should be able to renormalize any $v \in V^\times$ to have q-length ± 1. Of course since q is quadratic, $q(tv) = t^2 q(v)$, and the equation $t^2 q(v) = \pm 1$, i.e., the equation $t^2 = \pm a$ for a given a, may or may not be solvable in a general field k. If $k = \mathbb{R}$ or $\mathbb{C}$, of course, it is always solvable. If $k = \mathbb{Q}$ (the rational numbers) it is very often not solvable. (The group $\mathbb{Q}^\times/(\mathbb{Q}^\times)^2$ is infinitely generated.) If k is a finite field of characteristic $\neq 2$, then $k^\times/(k^\times)^2 \cong \mathbb{Z}_2$ and -1 may or may not lie in $(k^\times)^2$. In the cases where $\widetilde{\mathrm{Ad}}$ is not surjective, we still have the following general fact, which is interesting because the group $\mathrm{SO}(V,q)$ is often almost a simple group (see Artin [1]). (The reader interested only in the real and complex cases can skip this proposition.)

Proposition 2.8. *Each of the images* $\widetilde{\mathrm{Ad}}(\mathrm{Pin}(V,q))$ *and* $\widetilde{\mathrm{Ad}}(\mathrm{Spin}(V,q))$ *is a normal subgroup of* $\mathrm{O}(V,q)$.

Proof. Recall (cf. (1.6)) that from the universal property of $C\ell(V,q)$, the action of $\mathrm{O}(V,q)$ on V extends to automorphisms of $C\ell(V,q)$. It is easy to see that these automorphisms commute with α. Suppose then that we have $v,w \in V$ with $q(v) \neq 0$ and choose $g \in \mathrm{O}(V,q)$. Then $\widetilde{\mathrm{Ad}}_{g(v)}(w) = \alpha(gv)w(gv)^{-1} = g(\alpha v)wg(v^{-1}) = g(\alpha(v)g^{-1}(w)v^{-1}) = g\widetilde{\mathrm{Ad}}_v(g^{-1}w)$. Consequently, we have that

$$\widetilde{\mathrm{Ad}}_{gv} = g \circ \widetilde{\mathrm{Ad}} \circ g^{-1} \tag{2.27}$$

for all $v \in V$ with $q(v) \neq 0$ and for all $g \in \mathrm{O}(V,q)$. The proposition now follows immediately from (2.24), (2.25) and (2.27). ■

We now come to the main result of this section. We are primarily interested in the real and complex cases, so we shall focus on fields k that have the property discussed above. We shall say that a field k of characteristic $\neq 2$ is **spin** if at least one of the two equations $t^2 = a$ and $t^2 = -a$ can be solved in k for each non-zero element $a \in k^\times$. That is, k is spin if $k^\times = (k^\times)^2 \cup (-(k^\times)^2)$. The fields $\mathbb{R}$, $\mathbb{C}$ and $\mathbb{Z}_p$ for p a prime with $p \equiv 3 (\mathrm{mod}\ 4)$, are spin.

Theorem 2.9. *Let V be a finite-dimensional vector space over a spin field k, and suppose q is a non-degenerate quadratic form on V. Then there are short exact sequences*

$$0 \longrightarrow F \longrightarrow \mathrm{Spin}(V,q) \xrightarrow{\widetilde{\mathrm{Ad}}} \mathrm{SO}(V,q) \longrightarrow 1 \tag{2.28}$$

$$0 \longrightarrow F \longrightarrow \mathrm{Pin}(V,q) \xrightarrow{\widetilde{\mathrm{Ad}}} \mathrm{O}(V,q) \longrightarrow 1 \tag{2.29}$$

where

$$F = \begin{cases} \mathbb{Z}_2 = \{1,-1\} & \text{if } \sqrt{-1} \notin k \\ \mathbb{Z}_4 = \{\pm 1, \pm\sqrt{-1}\} & \text{otherwise.} \end{cases}$$

These sequences hold for general fields provided that $\mathrm{SO}(V,q)$ *and* $\mathrm{O}(V,q)$ *are replaced by appropriate normal subgroups of* $\mathrm{O}(V,q)$.

Proof. Suppose $\varphi = v_1 \cdots v_r \in \mathrm{Pin}(V,q)$ is in the kernel of $\widetilde{\mathrm{Ad}}$. Then $\varphi \in k^\times$ by Proposition 2.4, and so $\varphi^2 = N(\varphi) = N(v_1) \cdots N(v_r) = \pm 1$. This establishes the kernel of $\widetilde{\mathrm{Ad}}$ in both cases. The surjectivity of the homomorphisms follows from Theorem 2.7, the fact that $\rho_v = \rho_{tv}$, and the fact that since k is spin, any $v \in V^\times$ can be renormalized to have q-length 1. ■

It is interesting to observe that if k is a spin field, then either $\tilde{\mathrm{P}}(V,q) = \mathrm{P}(V,q)$ or $\tilde{\mathrm{P}}(V,q)/\mathrm{P}(V,q) \cong \mathbb{Z}_2$. The proof (which the reader may skip) is as follows. Since $\mathrm{P}(V,q)$ is generated by $V^\times$, we know that $t^2 q(v) \in \mathrm{P}(V,q)$ for all $t \in k^\times$ and $v \in V^\times$. Since k is spin, this implies that $\mathrm{P}(V,q)$ contains $(k^\times)^2$ or $-(k^\times)^2$ (and possibly more). In fact, if we set $k_0^\times = \{t \in k^\times : t \cdot 1 \in \mathrm{P}(V,q)\}$, then from the above and from the definition of a spin field, we see that $k^\times = k_0^\times \cup (-k_0^\times)$. Thus, $k^\times / k_0^\times = 0$ or $\mathbb{Z}_2$. Now we have the sequence

$$k_0^\times \subseteq \mathrm{P}(V,q) \subseteq \tilde{\mathrm{P}}(V,q) \xrightarrow{\widetilde{\mathrm{Ad}}} \mathrm{O}(V,q)$$

where $k^\times = \ker(\widetilde{\mathrm{Ad}})$ and where $\widetilde{\mathrm{Ad}}(P(V,q)) = \mathrm{O}(V,q)$. It follows that $\mathrm{O}(V,q) \cong \tilde{\mathrm{P}}(V,q)/k^\times \cong \mathrm{P}(V,q)/k_0^\times$. It then follows without difficulty that $\tilde{\mathrm{P}}(V,q)/\mathrm{P}(V,q) \cong k^\times / k_0^\times \cong 0$ or $\mathbb{Z}_2$ as claimed.

We now examine the real case in some detail. Let V be an n-dimensional vector space over $\mathbb{R}$, and suppose q is a non-degenerate quadratic form on V. Then we may choose a basis for $V \cong \mathbb{R}^n$ so that

$$q(x) = x_1^2 + \ldots + x_r^2 - x_{r+1}^2 - \ldots - x_{r+s}^2 \tag{2.30}$$

where $r + s = n$ and $0 \le r \le n$. It is standard notation to write: $q_{r,s} \equiv q$, $\mathrm{O}_{r,s} \equiv \mathrm{O}(V,q)$ and $\mathrm{SO}_{r,s} \equiv \mathrm{SO}(V,q)$. In accordance we write

$$\mathrm{Pin}_{r,s} \equiv \mathrm{Pin}(V,q) \quad \text{and} \quad \mathrm{Spin}_{r,s} \equiv \mathrm{Spin}(V,q). \tag{2.31}$$

Similarly, it is conventional to write $\mathrm{O}_n \equiv \mathrm{O}_{n,0} \cong \mathrm{O}_{0,n}$ and $\mathrm{SO}_n \equiv \mathrm{SO}_{n,0} \cong \mathrm{SO}_{0,n}$. Thus, we set

$$\mathrm{Pin}_n \equiv \mathrm{Pin}_{n,0} \quad \text{and} \quad \mathrm{Spin}_n \equiv \mathrm{Spin}_{n,0}. \tag{2.32}$$

We also write $\mathrm{P}_{r,s} \equiv \mathrm{P}(V,q)$ and $\tilde{\mathrm{P}}_{r,s} \equiv \tilde{\mathrm{P}}(V,q)$, and note from the paragraph above that

$$\mathrm{P}_{r,s} = \tilde{\mathrm{P}}_{r,s}. \tag{2.33}$$

It is a classical fact (cf. Helgason [1]) that SO_n is connected and that $\mathrm{SO}_{r,s}$, for $r,s \geqq 1$, has exactly two connected components. It is also a classical fact that $\pi_1(\mathrm{SO}_n) \cong \mathbb{Z}_2$ for $n \geqq 3$ and $\pi_1(\mathrm{SO}^0_{r,s}) \cong \pi_1(\mathrm{SO}_r) \times \pi_1(\mathrm{SO}_s)$ for all r,s. Hence, $\pi_1(\mathrm{SO}^0_{1,r}) = \pi_1(\mathrm{SO}^0_{r,1}) = \mathbb{Z}_2$ and $\pi_1(\mathrm{SO}^0_{r,s}) = \mathbb{Z}_2 \times \mathbb{Z}_2$ for all $r,s \geqq 3$. (Here $\mathrm{SO}^0_{r,s}$ denotes the connected component of the identity.)

The main result of this section is the following.

Theorem 2.10. *There are short exact sequences*

$$0 \longrightarrow \mathbb{Z}_2 \longrightarrow \mathrm{Spin}_{r,s} \longrightarrow \mathrm{SO}_{r,s} \longrightarrow 1$$
$$0 \longrightarrow \mathbb{Z}_2 \longrightarrow \mathrm{Pin}_{r,s} \longrightarrow \mathrm{O}_{r,s} \longrightarrow 1$$

for all (r,s). Furthermore, if $(r,s) \neq (1,1)$, these two-sheeted coverings are non-trivial over each component of $\mathrm{O}_{r,s}$. In particular, in the special case

$$0 \longrightarrow \mathbb{Z}_2 \longrightarrow \mathrm{Spin}_n \xrightarrow{\xi_0} \mathrm{SO}_n \longrightarrow 1 \tag{2.34}$$

the map $\xi_0 \equiv \widetilde{\mathrm{Ad}}$ represents the universal covering homomorphism of SO_n for all $n \geqq 3$.

Proof. The exact sequences are a direct consequence of Theorem 2.9. The kernel in each case is explicitly given by $\mathbb{Z}_2 = \{1,-1\}$. To prove that the coverings are non-trivial, it suffices to join -1 to 1 by a path in $Spin_{r,s}$. Choose orthogonal vectors $e_1,e_2 \in \mathbb{R}^n$ with $q(e_1) = q(e_2) = \pm 1$. (This is possible since $(r,s) \neq (1,1)$.) Then $\gamma(t) = \pm\cos(2t) + e_1e_2\sin(2t) = (e_1\cos t + e_2\sin t)(e_2\sin t - e_1\cos t)$ does the job. ∎

The above argument also shows that restricting $\widetilde{\mathrm{Ad}}$ to the identity component $\mathrm{Spin}^0_{r,1}$ of $\mathrm{Spin}_{r,1}$ gives the universal covering homomorphism

$$0 \longrightarrow \mathbb{Z}_2 \longrightarrow \mathrm{Spin}^0_{r,1} \xrightarrow{\xi} \mathrm{SO}^0_{r,1} \longrightarrow 1 \tag{2.35}$$

for all $r \geqq 3$.

§3. The Algebras $C\ell_n$ and $C\ell_{r,s}$

We shall now study the Clifford algebras $C\ell_{r,s} \equiv C\ell(V,q)$ where $V = \mathbb{R}^{r+s}$ and

$$q(x) = x_1^2 + \ldots + x_r^2 - x_{r+1}^2 - \ldots - x_{r+s}^2. \tag{3.1}$$

Of particular interest are the cases

$$C\ell_n \equiv C\ell_{n,0} \qquad \text{and} \qquad C\ell_n^* \equiv C\ell_{0,n}. \tag{3.2}$$

One reason for studying these algebras is the following. As seen in §2, the algebra $C\ell_{r,s}$ contains the groups $\mathrm{Spin}_{r,s}$ and $\mathrm{Pin}_{r,s}$, and so any representation of the algebra $C\ell_{r,s}$ restricts to a representation of these groups which is non-trivial on the element -1. (Such representations are therefore *not* induced from representations of $O_{r,s}$ or $SO_{r,s}$.)

These algebras have a simple classical presentation:

Proposition 3.1. *Let $e_1, \ldots, e_{r+s}$ be any q-orthonormal basis of $\mathbb{R}^{r+s} \subset C\ell_{r,s}$. Then $C\ell_{r,s}$ is generated (as an algebra) by $e_1, \ldots, e_{r+s}$ subject to the relations*

$$e_ie_j + e_je_i = \begin{cases} -2\delta_{ij} & \text{if } i \leq r \\ +2\delta_{ij} & \text{if } i > r. \end{cases} \tag{3.3}$$

Proof. This follows easily from the discussion in §1. ∎

We also have a pretty decomposition in terms of the $\mathbb{Z}_2$-graded tensor product.

Proposition 3.2. *There is an isomorphism*

$$C\ell_{r,s} \cong C\ell_1 \hat{\otimes} \cdots \hat{\otimes} C\ell_1 \hat{\otimes} C\ell_1^* \hat{\otimes} \cdots \hat{\otimes} C\ell_1^* \tag{3.4}$$

where $C\ell_1$ appears r times and $C\ell_1^*$ appears s times on the right in (3.4).

Proof. Decompose $\mathbb{R}^{r+s}$ into one-dimensional q-orthogonal subspaces and apply Proposition 1.5 inductively. ∎

It is not difficult to see that as algebras over $\mathbb{R}$,

$$C\ell_1 \cong \mathbb{C} \quad \text{and} \quad C\ell_1^* \cong \mathbb{R} \oplus \mathbb{R}. \tag{3.5}$$

It follows immediately that $\dim_{\mathbb{R}}(C\ell_{r,s}) = 2^{r+s}$. Proposition 3.2 is, however, not so useful if we wish to represent $C\ell_{r,s}$ as a matrix algebra. For this it is more useful to find decompositions in terms of ungraded tensor products. We shall do this in the next section.

For the remainder of this section we shall examine some of the general properties of the algebras $C\ell_{r,s}$. We begin with a discussion of the volume element. Choose an orientation for $\mathbb{R}^{r+s}$ and let $e_1, \ldots, e_{r+s}$ be any positively-oriented, q-orthonormal basis. Then the associated **(oriented) volume element** is defined to be

$$\omega = e_1 \cdots e_{r+s} \tag{3.6}$$

If $e'_1, \ldots, e'_{r+s}$ is any other such basis, then $e'_i = \sum_j g_{ij}e_j$ for $g = ((g_{ij})) \in SO_{r,s}$. From (3.3) we easily see that $e'_1 \cdots e'_{r+s} = \det(g)e_1 \cdots e_{r+s} = e_1 \cdots e_{r+s}$. Hence the definition (3.6) is independent of the choice of the basis.

Proposition 3.3. *The volume element* (3.6) *in* $C\ell_{r,s}$ *has the following basic properties. Let* $n = r + s$. *Then*

$$\omega^2 = (-1)^{\frac{n(n+1)}{2}+s}, \tag{3.7}$$

$$v\omega = (-1)^{n-1}\omega v \qquad \text{for all } v \in \mathbb{R}^n, \tag{3.8}$$

In particular, if n *is odd, then the element* ω *is central in* $C\ell_{r,s}$. *If* n *is even, then*

$$\varphi\omega = \omega\alpha(\varphi) \tag{3.9}$$

for all $\varphi \in C\ell_{r,s}$.

Proof. Choose a q-orthonormal basis and apply the relations (3.3). ■

We note that property (3.7) can be rewritten as

$$\omega^2 = \begin{cases} (-1)^s & \text{if } n \equiv 3 \text{ or } 4 \pmod 4 \\ (-1)^{s+1} & \text{if } n \equiv 1 \text{ or } 2 \pmod 4 \end{cases} \tag{3.7$'$}$$

We now make the following elementary but important observation.

Lemma 3.4. *Suppose the volume element* ω *in* $C\ell_{r,s}$ *satisfies* $\omega^2 = 1$, *and set*

$$\pi^+ = \frac{1}{2}(1 + \omega) \qquad \text{and} \qquad \pi^- = \frac{1}{2}(1 - \omega). \tag{3.10}$$

Then π^+ *and* π^- *satisfy the relations*

$$\pi^+ + \pi^- = 1 \tag{3.11}$$

$$(\pi^+)^2 = \pi^+ \qquad \text{and} \qquad (\pi^-)^2 = \pi^- \tag{3.12}$$

$$\pi^+\pi^- = \pi^-\pi^+ = 0. \tag{3.13}$$

Proof. This is a trivial consequence of the fact that $\omega^2 = 1$. ■

This leads to two basic but important facts:

Proposition 3.5. *Suppose that the volume element* ω *in* $C\ell_{r,s}$ *satisfies* $\omega^2 = 1$, *and that* $r + s$ *is odd. Then* $C\ell_{r,s}$ *can be decomposed as a direct sum*

$$C\ell_{r,s} = C\ell_{r,s}^+ \oplus C\ell_{r,s}^- \tag{3.14}$$

of isomorphic subalgebras, where $C\ell_{r,s}^\pm = \pi^\pm \cdot C\ell_{r,s} = C\ell_{r,s} \cdot \pi^\pm$ *and where* $\alpha(C\ell_{r,s}^\pm) = C\ell_{r,s}^\mp$.

Proof. Since $r + s$ is odd, we know from Proposition 3.3 that ω is central. Hence π^+ and π^- are central and the decomposition (3.14) into ideals follows directly from (3.11), (3.12) and (3.13). Since ω is an odd element, $\alpha(\pi^{\pm}) = \pi^{\mp}$ and so $\alpha(C\ell_{r,s}^{\pm}) = C\ell_{r,s}^{\mp}$. Since α is an automorphism, we conclude that these two ideals are isomorphic. ∎

Proposition 3.6. *Suppose that the volume element ω in $C\ell_{r,s}$ satisfies $\omega^2 = 1$ and that $r + s$ is even. Let V be any $C\ell_{r,s}$-module (i.e., V is a real vector space with an algebra homomorphism $C\ell_{r,s} \to \text{Hom}(V, V)$). Then there is a decomposition*

$$V = V^+ \oplus V^- \tag{3.15}$$

into the $+1$ and -1 eigenspaces for multiplication by ω. In fact,

$$V^+ = \pi^+ \cdot V \quad \text{and} \quad V^- = \pi^- \cdot V,$$

and for any $e \in \mathbb{R}^{r+s}$ with $q(e) \neq 0$, module multiplication by e gives isomorphisms

$$e: V^+ \longrightarrow V^- \qquad \text{and} \qquad e: V^- \longrightarrow V^+. \tag{3.16}$$

Proof. The decomposition (3.15) is a direct consequence of (3.11), (3.12) and (3.13), together with the observation that

$$\omega \cdot \pi^{\pm} = \pm \pi^{\pm}.$$

The isomorphisms (3.16) follow directly from the facts that by (3.8),

$$\begin{cases} e\pi^+ = \dfrac{1}{2} e(1 + \omega) = \dfrac{1}{2}(1 - \omega)e = \pi^- e \\ e\pi^- = \pi^+ e \end{cases}$$

and $e \cdot e = -q(e) \cdot 1$. ∎

REMARK. The above construction will prove useful when we are dealing with vector bundles in the next chapter.

We now come to an important and basic fact. Recall the even-odd decomposition $C\ell_{r,s} = C\ell_{r,s}^0 \oplus C\ell_{r,s}^1$ given in (1.8), where the subalgebra $C\ell_{r,s}^0$ is the fixed-point set of the automorphism α.

Theorem 3.7. *There is an algebra isomorphism*

$$C\ell_{r,s} \cong C\ell_{r+1,s}^0 \tag{3.17}$$

for all r,s. In particular,

$$C\ell_n \cong C\ell_{n+1}^0 \tag{3.18}$$

for all n.

Proof. Choose a q-orthonormal basis $e_1, \ldots, e_{r+s+1}$ of $\mathbb{R}^{r+s+1}$ so that $q(e_i) = 1$ for $1 \leqq i \leqq r+1$ and $q(e_i) = -1$ for $i > r+1$. Let $\mathbb{R}^{r+s} = \operatorname{span}\{e_i | i \neq r+1\}$ and define a map $f: \mathbb{R}^{r+s} \to C\ell^0_{r+1,s}$ by setting

$$f(e_i) = e_{r+1}e_i$$

for $i \neq r+1$, and extending linearly. For $x = \sum_{i \neq r+1} x_i e_i$, we have that:

$$\begin{aligned} f(x)^2 &= \sum_{i,j} x_i x_j e_{r+1} e_i e_{r+1} e_j \\ &= \sum_{i,j} x_i x_j e_i e_j \\ &= x \cdot x = -q(x) \cdot 1 \end{aligned}$$

since $e_{r+1} \cdot e_{r+1} = -1$ and $e_{r+1}e_i = -e_i e_{r+1}$ for $i \neq r+1$. It follows from the universal property (Proposition 1.1) that f extends to an algebra homomorphism

$$\tilde{f}: C\ell_{r,s} \longrightarrow C\ell^0_{r+1,s}.$$

Checking $\tilde{f}$ on a linear basis shows that $\tilde{f}$ is an isomorphism. ∎

We now specialize to the case of $C\ell_n$.

Proposition 3.8. *Let $L: C\ell_n \to C\ell_n$ be the linear map defined by setting*

$$L(\varphi) = -\sum_j e_j \varphi e_j \tag{3.19}$$

where $e_1, \ldots, e_n$ is any orthonormal basis of $\mathbb{R}^n$. Set $\tilde{L} = \alpha \circ L$. Then the eigenspaces of $\tilde{L}$ are the canonical images of $\Lambda^p \equiv \Lambda^p\mathbb{R}^n$ in $C\ell_n$. In fact

$$\tilde{L}|_{\Lambda^p} = (n-2p)\mathrm{Id} \tag{3.20}$$

for $p = 0, \ldots, n$.

Proof. It suffices to consider $\varphi = e_1 \cdots e_p$. Then

$$\begin{aligned} L(\varphi) &= -\sum_{j=1}^{p} e_j e_1 \cdots e_p e_j - \sum_{j=p+1}^{n} e_j e_1 \cdots e_p e_j \\ &= -\sum_{j=1}^{p} (-1)^{p-1} e_j^2 e_1 \cdots e_p - \sum_{j=p+1}^{n} (-1)^p e_j^2 e_1 \cdots e_p \\ &= (-1)^{p-1} p e_1 \cdots e_p + (-1)^p (n-p) e_1 \cdots e_p \\ &= (-1)^p (n-2p) e_1 \cdots e_p = (n-2p)\alpha(\varphi) \quad ∎ \end{aligned}$$

Under the canonical isomorphism $C\ell_n \cong \Lambda^*\mathbb{R}^n$, Clifford multiplication has a nice interpretation. Using the inner product on $\mathbb{R}^n$ we can identify $\mathbb{R}^n$ with its dual. We can thereby talk about the **interior product** or **contraction** in $\Lambda^* \mathbb{R}^n$. For $v \in \mathbb{R}^n$, this is a linear map $(v \llcorner): \Lambda^p\mathbb{R}^n \to \Lambda^{p-1}\mathbb{R}^n$

given on simple vectors by

$$v \llcorner (v_1 \wedge \ldots \wedge v_p) \equiv \sum_{i=1}^{p} (-1)^{i+1} \langle v_i, v \rangle v_1 \wedge \ldots \wedge \hat{v}_i \wedge \ldots \wedge v_p \tag{3.21}$$

where $(\hat{\cdot})$ indicates deletion. This gives a skew-derivation of the algebra, i.e., $v \llcorner (\varphi \wedge \psi) = (v \llcorner \varphi) \wedge \psi + (-1)^p \varphi \wedge (v \llcorner \psi)$ for any $\varphi \in \Lambda^p \mathbb{R}^n$. It is not difficult to see that $v \llcorner (v \llcorner) = 0$ for any $v \in \mathbb{R}^n$. Hence, by universality the interior product extends to all elements of $\Lambda^* \mathbb{R}^n$, i.e., to a bilinear map $\Lambda^* \mathbb{R}^n \times \Lambda^* \mathbb{R}^n \to \Lambda^* \mathbb{R}^n$.

Proposition 3.9. *With respect to the canonical isomorphism $C\ell_n \cong \Lambda^* \mathbb{R}^n$, Clifford multiplication between $v \in \mathbb{R}^n$ and any $\varphi \in C\ell_n$ can be written as*

$$v \cdot \varphi \cong v \wedge \varphi - v \llcorner \varphi \tag{3.22}$$

Proof. Choose an orthonormal basis $e_1, \ldots, e_n$ for $\mathbb{R}^n$ with $v = te_1$ for some $t \in \mathbb{R}$. Let $\varphi = e_{i_1} \cdots e_{i_p}$ for $i_1 < \cdots < i_p$. Then

$$v \cdot \varphi = \begin{cases} -te_{i_2} \cdots e_{i_p} \cong (v \wedge - v \llcorner)\varphi & \text{if } i_1 = 1 \\ te_1 e_{i_1} \cdots e_{i_p} \cong (v \wedge - v \llcorner)\varphi & \text{if } i_1 > 1. \end{cases}$$

Since (3.22) holds on an additive basis of $C\ell_n$, it holds in general. ■

§4. The Classification

In this section we shall give an explicit description of the algebras $C\ell_{r,s}$ as matrix algebras over $\mathbb{R}$, $\mathbb{C}$, or $\mathbb{H}$ (= quaternions). With little difficulty the reader can check the first few cases:

$$\begin{array}{ll} C\ell_{1,0} = \mathbb{C} & C\ell_{0,1} = \mathbb{R} \oplus \mathbb{R} \\ C\ell_{2,0} = \mathbb{H} & C\ell_{0,2} = \mathbb{R}(2) \\ \multicolumn{2}{c}{C\ell_{1,1} = \mathbb{R}(2)} \end{array} \tag{4.0}$$

where $\mathbb{R}(2)$ denotes the algebra of 2×2 real matrices.

The key facts to the classification are the following:

Theorem 4.1. *There are isomorphisms*

$$C\ell_{n,0} \otimes C\ell_{0,2} \cong C\ell_{0,n+2} \tag{4.1}$$

$$C\ell_{0,n} \otimes C\ell_{2,0} \cong C\ell_{n+2,0} \tag{4.2}$$

$$C\ell_{r,s} \otimes C\ell_{1,1} \cong C\ell_{r+1,s+1} \tag{4.3}$$

for all $n,r,s \geqq 0$.

Note that here we are using the *ungraded* tensor product.

Proof. Let $e_1, \dots, e_{n+2}$ be an orthonormal basis for $\mathbb{R}^{n+2}$ in the standard inner product, and let $q(x) = -\|x\|^2$. Let $e'_1, \dots, e'_n$ denote standard generators for $C\ell_{n,0}$ and let e''_1, e''_2 denote standard generators for $C\ell_{0,2}$ (in the sense of Proposition 3.1). Define a map $f: \mathbb{R}^{n+2} \to C\ell_{n,0} \otimes C\ell_{0,2}$ by setting

$$f(e_i) = \begin{cases} e'_i \otimes e''_1 e''_2 & \text{for } 1 \leqq i \leqq n \\ 1 \otimes e''_{i-n} & \text{for } i = n+1, n+2 \end{cases}$$

and extending linearly. Note that for $1 \leq i,j \leq n$, we have $f(e_i)f(e_j) + f(e_j)f(e_i) = (e'_i e'_j + e'_j e'_i) \otimes (-1) = 2\delta_{ij} 1 \otimes 1$; and for $n+1 \leqq \alpha, \beta \leqq n+2$ we have $f(e_\alpha)f(e_\beta) + f(e_\beta)f(e_\alpha) = 1 \otimes (e''_{\alpha-n} e''_{\beta-n} + e''_{\beta-n} e''_{\alpha-n}) = 2\delta_{\alpha\beta} 1 \otimes 1$. Also we see that $f(e_i)f(e_\alpha) + f(e_\alpha)f(e_i) = 0$. It follows that $f(x)f(x) = \|x\|^2 1 \otimes 1$ for all $x \in \mathbb{R}^{n+2}$. Hence, by the universal property (Proposition 1.1), f extends to an algebra homomorphism $\tilde{f}: C\ell_{0,n+2} \to C\ell_{n,0} \otimes C\ell_{0,2}$. Since $\tilde{f}$ maps onto a set of generators for $C\ell_{n,0} \otimes C\ell_{0,2}$, it must be surjective. Then, since $\dim C\ell_{0,n+2} = \dim C\ell_{n,0} \otimes C\ell_{0,2}$, we conclude that $\tilde{f}$ must be an isomorphism. This proves (4.1). The proof of (4.2) is entirely analogous.

For (4.3) we proceed in a similar manner. We choose a q-orthogonal basis $e_1, \dots, e_{r+1}, \varepsilon_1, \dots, \varepsilon_{s+1}$ for $\mathbb{R}^{r+s+2}$ such that $q(e_i) = 1$ and $q(\varepsilon_j) = -1$ for all i,j. We then let $e'_1, \dots, e'_r, \varepsilon'_1, \dots, \varepsilon'_s$ and e''_1, ε''_1 be corresponding bases for $\mathbb{R}^{r+s}$ and $\mathbb{R}^2$, and we define a map $f: \mathbb{R}^{r+s+2} \to C\ell_{r,s} \otimes C\ell_{1,1}$ by setting

$$f(e_i) = \begin{cases} e'_i \otimes e''_1 \varepsilon''_1 & \text{for } 1 \leq i \leq r \\ 1 \otimes e''_1 & \text{for } i = r+1, \end{cases}$$

and

$$f(\varepsilon_j) = \begin{cases} \varepsilon'_j \otimes e''_1 \varepsilon''_1 & \text{for } 1 \leq j \leq s \\ 1 \otimes \varepsilon''_1 & \text{for } j = s+1, \end{cases}$$

and then extending linearly. We now apply Proposition 1.1 and complete the argument as in the previous cases. ■

To apply this basic proposition we shall need the following elementary facts concerning the tensor products of algebras over $\mathbb{R}$. For $K = \mathbb{R}$, $\mathbb{C}$ or $\mathbb{H}$, we denote by $K(n)$ the algebra of $n \times n$-matrices with entries in K.

Proposition 4.2.

$$\mathbb{R}(n) \otimes \mathbb{R}(m) \cong \mathbb{R}(nm) \qquad \textit{for all } n,m. \tag{4.4}$$

$$\mathbb{R}(n) \otimes_{\mathbb{R}} K \cong K(n) \qquad \textit{for } K = \mathbb{C} \text{ or } \mathbb{H} \textit{ and for all } n. \tag{4.5}$$

$$\mathbb{C} \otimes_{\mathbb{R}} \mathbb{C} \cong \mathbb{C} \oplus \mathbb{C} \tag{4.6}$$

$$\mathbb{C} \otimes_{\mathbb{R}} \mathbb{H} \cong \mathbb{C}(2) \tag{4.7}$$

$$\mathbb{H} \otimes_{\mathbb{R}} \mathbb{H} \cong \mathbb{R}(4). \tag{4.8}$$

Proof. The isomorphisms (4.4) and (4.5) are obvious. The isomorphism $\mathbb{C} \oplus \mathbb{C} \to \mathbb{C} \otimes \mathbb{C}$ is determined by sending

$$(1,0) \longrightarrow \tfrac{1}{2}(1 \otimes 1 + i \otimes i),$$

$$(0,1) \longrightarrow \tfrac{1}{2}(1 \otimes 1 - i \otimes i).$$

For the isomorphism (4.7) we consider $\mathbb{H}$ as a $\mathbb{C}$-module under left scalar multiplication, and we define an $\mathbb{R}$-bilinear map $\Phi : \mathbb{C} \times \mathbb{H} \to \mathrm{Hom}_{\mathbb{C}}(\mathbb{H},\mathbb{H})$ by setting $\Phi_{z,q}(x) \equiv zx\bar{q}$. This extends (by the universal property of $\otimes$) to an $\mathbb{R}$-linear map $\Phi : \mathbb{C} \otimes_{\mathbb{R}} \mathbb{H} \to \mathrm{Hom}_{\mathbb{C}}(\mathbb{H},\mathbb{H}) \cong \mathbb{C}(2)$. Since $\Phi_{z,q} \circ \Phi_{z',q'} = \Phi_{zz',qq'}$, we have that $\tilde{\Phi}$ is an algebra homomorphism. Checking $\tilde{\Phi}$ on a natural basis shows that it is injective. Hence, since $\dim_{\mathbb{R}}(\mathbb{C} \otimes_{\mathbb{R}} \mathbb{H}) = \dim_{\mathbb{R}}(\mathbb{C}(2))$, $\tilde{\Phi}$ is an isomorphism.

The isomorphism (4.8) is proved similarly. Consider the $\mathbb{R}$-bilinear map $\Psi : \mathbb{H} \times \mathbb{H} \to \mathrm{Hom}_{\mathbb{R}}(\mathbb{H},\mathbb{H}) \cong \mathbb{R}(4)$ given by setting $\Psi_{q_1,q_2}(x) \equiv q_1 x \bar{q}_2$. The resulting $\mathbb{R}$-linear map $\Psi : \mathbb{H} \otimes_{\mathbb{R}} \mathbb{H} \to \mathrm{Hom}_{\mathbb{R}}(\mathbb{H},\mathbb{H})$ is an algebra homomorphism between algebras of the same dimension. The injectivity of $\tilde{\Phi}$ can be checked on a natural basis for $\mathbb{H} \otimes \mathbb{H}$. ■

We now come to the first main result of the section. Before stating the result, we make the observation that for any (r,s), the complexification of the algebra $C\ell_{r,s}$ is just the Clifford algebra (over $\mathbb{C}$) corresponding to the complexified quadratic form, i.e., $C\ell_{r,s} \otimes_{\mathbb{R}} \mathbb{C} \cong C\ell(\mathbb{C}^{r+s}, q \otimes \mathbb{C})$. (This follows easily from Proposition 1.1.) However, all non-degenerate quadratic forms on $\mathbb{C}^n$ are equivalent over $C\ell_n(\mathbb{C})$. Hence, sctting

$$q_{\mathbb{C}}(z) = \sum_{j=1}^{n} z_j^2$$

and defining

$$\mathbb{C}\ell_n \equiv C\ell(\mathbb{C}^n, q_{\mathbb{C}}), \tag{4.9}$$

we have that

$$\mathbb{C}\ell_n \cong C\ell_{n,0} \otimes_{\mathbb{R}} \mathbb{C} \cong C\ell_{n-1,1} \otimes_{\mathbb{R}} \mathbb{C} \cong \cdots \cong C\ell_{0,n} \otimes_{\mathbb{R}} \mathbb{C} \tag{4.10}$$

Theorem 4.3. *For all $n \geq 0$, there are "periodicity" isomorphisms*

$$C\ell_{n+8,0} \cong C\ell_{n,0} \otimes C\ell_{8,0} \tag{4.11}$$

$$C\ell_{0,n+8} \cong C\ell_{0,n} \otimes C\ell_{0,8} \tag{4.12}$$

$$\mathbb{C}\ell_{n+2} \cong \mathbb{C}\ell_n \otimes_{\mathbb{C}} \mathbb{C}\ell_2 \tag{4.13}$$

where

$$C\ell_{8,0} = C\ell_{0,8} = \mathbb{R}(16) \tag{4.14}$$

$$\mathbb{C}\ell_2 = \mathbb{C}(2). \tag{4.15}$$

Therefore, by using the identities (4.4) *and* (4.5), *all the algebras* $C\ell_{n,0}$, $C\ell_{0,n}$ *and* $\mathbb{C}\ell_n$ *can be easily deduced from the following table.*

Table I

	1	2	3	4	5	6	7	8
$C\ell_{n,0}$	$\mathbb{C}$	$\mathbb{H}$	$\mathbb{H} \oplus \mathbb{H}$	$\mathbb{H}(2)$	$\mathbb{C}(4)$	$\mathbb{R}(8)$	$\mathbb{R}(8) \oplus \mathbb{R}(8)$	$\mathbb{R}(16)$
$C\ell_{0,n}$	$\mathbb{R} \oplus \mathbb{R}$	$\mathbb{R}(2)$	$\mathbb{C}(2)$	$\mathbb{H}(2)$	$\mathbb{H}(2) \oplus \mathbb{H}(2)$	$\mathbb{H}(4)$	$\mathbb{C}(8)$	$\mathbb{R}(16)$
$\mathbb{C}\ell_n$	$\mathbb{C} \oplus \mathbb{C}$	$\mathbb{C}(2)$	$\mathbb{C}(2) \oplus \mathbb{C}(2)$	$\mathbb{C}(4)$	$\mathbb{C}(4) \oplus \mathbb{C}(4)$	$\mathbb{C}(8)$	$\mathbb{C}(8) \oplus \mathbb{C}(8)$	$\mathbb{C}(16)$

Proof. From (4.1) and (4.2) we see that for any n, we have $C\ell_{n+8,0} \cong C\ell_{n,0} \otimes C\ell_{0,2} \otimes C\ell_{2,0} \otimes C\ell_{0,2} \otimes C\ell_{2,0}$. Using (4.0) and Proposition 4.2 we see that $C\ell_{n+8,0} \cong C\ell_{n,0} \otimes \mathbb{H} \otimes \mathbb{H} \otimes \mathbb{R}(2) \otimes \mathbb{R}(2) \cong C\ell_{n,0} \otimes \mathbb{R}(4) \otimes \mathbb{R}(4) \cong C\ell_{n,0} \otimes \mathbb{R}(16)$. This establishes (4.11). The periodicity (4.12) is proved similarly. To prove (4.13), note from (4.10) that $\mathbb{C}\ell_{n+2} \cong C\ell_{n+2,0} \otimes \mathbb{C} \cong C\ell_{n,0} \otimes C\ell_{0,2} \otimes \mathbb{C} \cong \mathbb{C}\ell_n \otimes_{\mathbb{C}} \mathbb{C}\ell_2$.

Using the isomorphisms (4.1) and (4.2), and the facts (4.4) to (4.8), one can work out the first two rows of the table in "criss-cross" fashion (starting with the initial data (4.0)). The third row of the table now follows by taking the tensor product of corresponding terms in either of the first two rows with $\mathbb{C}$. ■

Combining Table I with the fundamental periodicity isomorphism (4.3) and the fact that $C\ell_{1,1} \cong \mathbb{R}(2)$, we achieve the complete classification in Table II.

By now the reader has probably noticed some of the intrinsic beauty of this constellation of algebras and its interrelationships. There are some observations one can make from the table that are interesting exercises to prove. For example,

$$C\ell_{r,s} \cong C\ell_{r-4,s+4} \tag{4.16}$$

$$C\ell_{r,s+1} \cong C\ell_{s,r+1} \qquad \text{(symmetry about the axis } y = x + 1). \tag{4.17}$$

REMARK. The above classification reduces the Clifford algebras to familiar matrix algebras over $K = \mathbb{R}$, $\mathbb{C}$ or $\mathbb{H}$. Of course it is also useful to think of this result as introducing hidden and unexpected structure in the algebras $K(2^m)$. This information can be quite interesting as we shall see when we discuss vector fields on spheres in §8.

Table II. $C\ell_{r,s}$ in the box (r,s)

8	$\mathbb{R}(16)$	$\mathbb{C}(16)$	$\mathbb{H}(16)$	$\mathbb{H}(16) \oplus \mathbb{H}(16)$	$\mathbb{H}(32)$	$\mathbb{C}(64)$	$\mathbb{R}(128)$	$\mathbb{R}(128) \oplus \mathbb{R}(128)$	$\mathbb{R}(256)$
7	$\mathbb{C}(8)$	$\mathbb{H}(8)$	$\mathbb{H}(8) \oplus \mathbb{H}(8)$	$\mathbb{H}(16)$	$\mathbb{C}(32)$	$\mathbb{R}(64)$	$\mathbb{R}(64) \oplus \mathbb{R}(64)$	$\mathbb{R}(128)$	$\mathbb{C}(128)$
6	$\mathbb{H}(4)$	$\mathbb{H}(4) \oplus \mathbb{H}(4)$	$\mathbb{H}(8)$	$\mathbb{C}(16)$	$\mathbb{R}(32)$	$\mathbb{R}(32) \oplus \mathbb{R}(32)$	$\mathbb{R}(64)$	$\mathbb{C}(64)$	$\mathbb{H}(64)$
5	$\mathbb{H}(2) \oplus \mathbb{H}(2)$	$\mathbb{H}(4)$	$\mathbb{C}(8)$	$\mathbb{R}(16)$	$\mathbb{R}(16) \oplus \mathbb{R}(16)$	$\mathbb{R}(32)$	$\mathbb{C}(32)$	$\mathbb{H}(32)$	$\mathbb{H}(32) \oplus \mathbb{H}(32)$
4	$\mathbb{H}(2)$	$\mathbb{C}(4)$	$\mathbb{R}(8)$	$\mathbb{R}(8) \oplus \mathbb{R}(8)$	$\mathbb{R}(16)$	$\mathbb{C}(16)$	$\mathbb{H}(16)$	$\mathbb{H}(16) \oplus \mathbb{H}(16)$	$\mathbb{H}(32)$
3	$\mathbb{C}(2)$	$\mathbb{R}(4)$	$\mathbb{R}(4) \oplus \mathbb{R}(4)$	$\mathbb{R}(8)$	$\mathbb{C}(8)$	$\mathbb{H}(8)$	$\mathbb{H}(8) \oplus \mathbb{H}(8)$	$\mathbb{H}(16)$	$\mathbb{C}(32)$
2	$\mathbb{R}(2)$	$\mathbb{R}(2) \oplus \mathbb{R}(2)$	$\mathbb{R}(4)$	$\mathbb{C}(4)$	$\mathbb{H}(4)$	$\mathbb{H}(4) \oplus \mathbb{H}(4)$	$\mathbb{H}(8)$	$\mathbb{C}(16)$	$\mathbb{R}(32)$
1	$\mathbb{R} \oplus \mathbb{R}$	$\mathbb{R}(2)$	$\mathbb{C}(2)$	$\mathbb{H}(2)$	$\mathbb{H}(2) \oplus \mathbb{H}(2)$	$\mathbb{H}(4)$	$\mathbb{C}(8)$	$\mathbb{R}(16)$	$\mathbb{R}(16) \oplus \mathbb{R}(16)$
0	$\mathbb{R}$	$\mathbb{C}$	$\mathbb{H}$	$\mathbb{H} \oplus \mathbb{H}$	$\mathbb{H}(2)$	$\mathbb{C}(4)$	$\mathbb{R}(8)$	$\mathbb{R}(8) \oplus \mathbb{R}(8)$	$\mathbb{R}(16)$
	0	**1**	**2**	**3**	**4**	**5**	**6**	**7**	**8**

§5. Representations

Most of the important applications of Clifford algebras come through a detailed understanding of their representations. This understanding follows rather easily from the classification given in §4.

We begin with a general definition. Let V be a vector space over a field k and let q be a quadratic form on V.

DEFINITION 5.1. Let $K \supseteq k$ be a field containing k. Then a ***K*-representation** of the Clifford algebra $C\ell(V,q)$ is a k-algebra homomorphism

$$\rho : C\ell(V,q) \longrightarrow \mathrm{Hom}_K(W,W)$$

into the algebra of linear transformations of a finite dimensional vector space W over K. The space W is called a **$C\ell(V,q)$-module** over K. We shall often simplify notation by writing

$$\rho(\varphi)(w) \equiv \varphi \cdot w \tag{5.1}$$

for $\varphi \in C\ell(V,q)$ and $w \in W$, when no confusion is likely to occur. The product $\varphi \cdot w$ in (5.1) is often referred to as **Clifford multiplication.**

Note. By a ***k*-algebra homomorphism** we mean a k-linear map ρ which satisfies the property $\rho(\varphi\psi) = \rho(\varphi) \circ \rho(\psi)$ for all $\varphi,\psi \in C\ell(V,q)$.

We shall be interested in K-representations of $C\ell_{r,s}$ where $K = \mathbb{R}$, $\mathbb{C}$ or $\mathbb{H}$. Note that a complex vector space is just a real vector space W together with a real linear map $J : W \to W$ such that $J^2 = -\mathrm{Id}$. A complex representation of $C\ell_{r,s}$ is just a real representation $\rho : C\ell_{r,s} \to \mathrm{Hom}_{\mathbb{R}}(W,W)$ such that

$$\rho(\varphi) \circ J = J \circ \rho(\varphi) \tag{5.2}$$

for all $\varphi \in C\ell_{r,s}$. Thus the image of ρ commutes with the subalgebra $\mathrm{span}\{\mathrm{Id},J\} \cong \mathbb{C}$. (This algebra is called a "commuting subalgebra" for ρ.)

Strictly analogous remarks apply to quaternionic representations of $C\ell_{r,s}$. Here the real vector space W carries three real linear transformations I, J and K such that

$$I^2 = J^2 = K^2 = -\mathrm{Id}$$

$$IJ = -JI = K, \qquad JK = -KJ = I, \qquad KI = -IK = J.$$

This makes W into an $\mathbb{H}$-module. A representation $\rho : C\ell_{r,s} \to \mathrm{Hom}_{\mathbb{R}}(W,W)$ is quaternionic if

$$\rho(\varphi) \circ I = I \circ \rho(\varphi), \qquad \rho(\varphi) \circ J = J \circ \rho(\varphi), \qquad \rho(\varphi) \circ K = K \circ \rho(\varphi) \tag{5.3}$$

for all $\varphi \in C\ell_{r,s}$. That is, ρ has a commuting subalgebra $\mathrm{span}_{\mathbb{R}}\{\mathrm{Id},I,J,K\}$ isomorphic to $\mathbb{H}$.

REMARK 5.2. Any complex representation of $C\ell_{r,s}$ automatically extends to a representation of $C\ell_{r,s} \otimes_{\mathbb{R}} \mathbb{C} \cong \mathbb{C}\ell_{r+s}$. Any quaternionic representation of $C\ell_{r,s}$ is automatically complex (by restricting to $\mathbb{C} \subset \mathbb{H}$). Of course the complex dimension of any $\mathbb{H}$-module is even.

The above remarks will prove useful when we carry these constructions over to vector bundles in Chapter II.

We now come to the notion of irreducibility.

DEFINITION 5.3. Let $V, q, k \subseteq K$, be as in definition 5.1. A K-representation $\rho: C\ell(V,q) \to \mathrm{Hom}_K(W,W)$ will be said to be **reducible** if the vector space W can be written as a non-trivial direct sum (over K).

$$W = W_1 \oplus W_2$$

such that $\rho(\varphi)(W_j) \subseteq W_j$ for $j = 1,2$ and for all $\varphi \in C\ell(V,q)$. Note that in this case we can write

$$\rho = \rho_1 \oplus \rho_2$$

where $\rho_j(\varphi) \equiv \rho(\varphi)|_{W_j}$ for $j = 1,2$. A representation is called **irreducible** if it is not reducible.

It is more conventional to call a representation "irreducible" if it has the property that there are no proper invariant subspaces. However, since $C\ell_n$ is the algebra of a finite group (see the discussion following Proposition 5.15), the two concepts are easily seen to agree in this case.

Proposition 5.4. *Every K-representation ρ of a Clifford algebra $C\ell(V,q)$ can be decomposed into a direct sum $\rho = \rho_1 \oplus \cdots \oplus \rho_m$ of irreducible representations.*

Proof. If ρ is reducible, it can be decomposed as a direct sum $\rho = \rho_1 \oplus \rho_2$. If either ρ_1 or ρ_2 is reducible, ρ can be further decomposed. This process must stop because of the finite dimensionality of the module. ∎

We shall be interested here, of course, only in equivalence classes of representations.

DEFINITION 5.5. Two representations $\rho_j: C\ell(V,q) \to \mathrm{Hom}_K(W_j, W_j)$ for $j = 1,2$ are said to be **equivalent** if there exists a K-linear isomorphism $F: W_1 \to W_2$ such that $F \circ \rho_1(\varphi) \circ F^{-1} = \rho_2(\varphi)$ for all $\varphi \in C\ell(V,q)$.

From §4 we know that every algebra $C\ell_{r,s}$ is of the form $K(2^m)$ or $K(2^m) \oplus K(2^m)$ for $K = \mathbb{R}$, $\mathbb{C}$ or $\mathbb{H}$. The representation theory of such algebras is particularly simple.

Theorem 5.6. Let $K = \mathbb{R}$, $\mathbb{C}$ or $\mathbb{H}$, *and consider the ring $K(n)$ of $n \times n$ K-matrices as an algebra over $\mathbb{R}$. Then the natural representation ρ of $K(n)$ on the vector space K^n is, up to equivalence, the only irreducible real representation of $K(n)$.*

The algebra $K(n) \oplus K(n)$ has exactly two equivalence classes of irreducible real representations. They are given by

$$\rho_1(\varphi_1,\varphi_2) \equiv \rho(\varphi_1) \qquad \textit{and} \qquad \rho_2(\varphi_1,\varphi_2) = \rho(\varphi_2)$$

acting on K^n.

Proof. This follows from the classical fact that the algebras $K(n)$ are simple and that simple algebras have only one irreducible representation up to equivalence. See Lang [1]. ∎

From the classification of §4 (see Table II) we immediately conclude the following:

Theorem 5.7. *Let $\nu_{r,s}$ denote the number of inequivalent irreducible real representations of $C\ell_{r,s}$, and let $\nu_n^{\mathbb{C}}$ denote the number of inequivalent irreducible complex representatons of $\mathbb{C}\ell_n$. Then*

$$\nu_{r,s} = \begin{cases} 2 & \text{if } r + 1 - s \equiv 0 \pmod 4 \\ 1 & \text{otherwise} \end{cases}$$

and

$$\nu_n^{\mathbb{C}} = \begin{cases} 2 & \text{if } n \text{ is odd} \\ 1 & \text{if } n \text{ is even.} \end{cases}$$

This is a good time to recall (cf. Theorem 3.7) that there are isomorphisms

$$C\ell_{r,s} \cong C\ell^0_{r+1,s} \tag{5.4}$$

for all r,s, and consequently

$$\mathbb{C}\ell_n \cong \mathbb{C}\ell^0_{n+1} \tag{5.5}$$

for all n. Since

$$\mathrm{Spin}_{r,s} \subset C\ell^0_{r,s} \subset \mathbb{C}\ell^0_{r+s} \tag{5.6}$$

we see that it is the irreducible representations of $C\ell_{r-1,s}$ and $\mathbb{C}\ell_{r+s-1}$ that are relevant to constructing irreducible real and complex representations of $\mathrm{Spin}_{r,s}$.

From this point on we shall restrict our attention to the algebras $C\ell_n = C\ell_{n,0}$ (and $\mathbb{C}\ell_n = C\ell_n \otimes_{\mathbb{R}} \mathbb{C}$) in order to simplify the exposition. Corresponding facts for the general case $C\ell_{r,s}$ are easy to deduce if the reader is interested. We shall begin with a summary of information easily deduced from the classification theorem 4.3.

We begin with some definitions. For each n, let $d_n = \dim_{\mathbb{R}}(W)$ where W is an irreducible $\mathbb{R}$-module for $C\ell_n$. Similarly, let $d_n^{\mathbb{C}} = \dim_{\mathbb{C}}(W')$ where W' is an irreducible complex module for $C\ell_n$ (and therefore for $\mathbb{C}\ell_n =$

$C\ell_n \otimes_{\mathbb{R}} \mathbb{C}$). Let $K_n = \mathbb{R}$, $\mathbb{C}$ or $\mathbb{H}$ denote the maximal commuting subalgebra for an irreducible real representation of $C\ell_n$. Thus if $K_n = \mathbb{C}$, this representation is automatically complex. If $K_n = \mathbb{H}$, it is automatically quaternionic. (Note that in the cases where $C\ell_n$ has two distinct irreducible representations, d_n, $d_n^{\mathbb{C}}$ and K_n are the same for both.)

An object which will be of interest later on is the following. Let $\mathfrak{M}_n$ (or $\mathfrak{M}_n^{\mathbb{C}}$) denote the Grothendieck group of equivalence classes of irreducible real (respectively, complex) representations of $C\ell_n$. This is merely the free abelian group generated by the distinct irreducible representations over $\mathbb{R}$ (or $\mathbb{C}$). Since any representation can be decomposed into irreducibles, it naturally corresponds to an element in this group (with positive coefficients).

Theorem 5.8. *For* $1 \leq n \leq 8$, *the elements* $v_n \equiv v_{n,0}$, $v_n^{\mathbb{C}}$, d_n, $d_n^{\mathbb{C}}$, K_n, $\mathfrak{M}_n$ and $\mathfrak{M}_n^{\mathbb{C}}$ *(defined above) are as given in* Table III.

Table III

n	$C\ell_n$	v_n	d_n	K_n	$\mathfrak{M}_n$	$\mathbb{C}\ell_n$	$v_n^{\mathbb{C}}$	$d_n^{\mathbb{C}}$	$\mathfrak{M}_n^{\mathbb{C}}$
1	$\mathbb{C}$	1	2	$\mathbb{C}$	$\mathbb{Z}$	$\mathbb{C} \oplus \mathbb{C}$	2	1	$\mathbb{Z} \oplus \mathbb{Z}$
2	$\mathbb{H}$	1	4	$\mathbb{H}$	$\mathbb{Z}$	$\mathbb{C}(2)$	1	2	$\mathbb{Z}$
3	$\mathbb{H} \oplus \mathbb{H}$	2	4	$\mathbb{H}$	$\mathbb{Z} \oplus \mathbb{Z}$	$\mathbb{C}(2) \oplus \mathbb{C}(2)$	2	2	$\mathbb{Z} \oplus \mathbb{Z}$
4	$\mathbb{H}(2)$	1	8	$\mathbb{H}$	$\mathbb{Z}$	$\mathbb{C}(4)$	1	4	$\mathbb{Z}$
5	$\mathbb{C}(4)$	1	8	$\mathbb{C}$	$\mathbb{Z}$	$\mathbb{C}(4) \oplus \mathbb{C}(4)$	2	4	$\mathbb{Z} \oplus \mathbb{Z}$
6	$\mathbb{R}(8)$	1	8	$\mathbb{R}$	$\mathbb{Z}$	$\mathbb{C}(8)$	1	8	$\mathbb{Z}$
7	$\mathbb{R}(8) \oplus \mathbb{R}(8)$	2	8	$\mathbb{R}$	$\mathbb{Z} \oplus \mathbb{Z}$	$\mathbb{C}(8) \oplus \mathbb{C}(8)$	2	8	$\mathbb{Z} \oplus \mathbb{Z}$
8	$\mathbb{R}(16)$	1	16	$\mathbb{R}$	$\mathbb{Z}$	$\mathbb{C}(16)$	1	16	$\mathbb{Z}$

For n > 8 *these elements can be computed from the following facts, which hold for all m,k* $\geqq 1$.

$$v_{m+8k} = v_m \qquad v_{m+2k}^{\mathbb{C}} = v_m^{\mathbb{C}} \tag{5.7}$$

$$d_{m+8k} = 2^{4k} d_m \qquad d_{m+2k}^{\mathbb{C}} = 2^k d_m \tag{5.8}$$

$$\mathfrak{M}_{m+8k} \cong \mathfrak{M}_m \qquad \mathfrak{M}_{m+2k}^{\mathbb{C}} \cong \mathfrak{M}_m^{\mathbb{C}} \tag{5.9}$$

$$K_{m+8k} \cong K_m. \tag{5.10}$$

Proof. This is a direct consequence of Theorem 4.3. ∎

We shall now consider the key role played by the volume element in determining irreducible representations. Recall from §3 that the volume element in $C\ell_n$ is defined as

$$\omega = e_1 \cdots e_n \tag{5.11}$$

where $e_1, \dots, e_n$ is an orthonormal basis of $\mathbb{R}^n$. It is well defined up to sign and is fixed after a choice of orientation on $\mathbb{R}^n$. In the complex case we have a corresponding element $\omega_{\mathbb{C}} \in \mathbb{C}\ell_n$ given by

$$\omega_{\mathbb{C}} = i^{\left[\frac{n+1}{2}\right]}\omega, \tag{5.12}$$

called the **complex volume element.** Note that when $n = 2m$, we have

$$\omega_{\mathbb{C}} = i^m e_1 \cdots e_{2m}$$

(Note also that $\omega_{\mathbb{C}} = \omega$ only in dimensions seven and eight modulo 8. Other conventions for a complex volume element are possible. This one is particularly useful in studying elliptic operators.)

Recall from Proposition 3.3 that if n is odd, then ω and $\omega_{\mathbb{C}}$ are **central.** Furthermore, by (3.7)′ we have that

$$\omega^2 = 1 \qquad \text{if } n \equiv 3 \text{ or } 4 \pmod 4, \tag{5.13}$$

$$(\omega_{\mathbb{C}})^2 = 1 \qquad \text{for all } n. \tag{5.14}$$

Thus there are algebra decompositions

$$C\ell_n = C\ell_n^+ \oplus C\ell_n^- \qquad \text{for } n \equiv 3 \pmod 4 \tag{5.15}$$

$$\mathbb{C}\ell_n = \mathbb{C}\ell_n^+ \oplus \mathbb{C}\ell_n^- \qquad \text{for } n \text{ odd} \tag{5.16}$$

where

$$C\ell_n^\pm = (1 \pm \omega)C\ell_n \quad \text{and} \quad \mathbb{C}\ell_n^\pm = (1 \pm \omega_{\mathbb{C}})\mathbb{C}\ell_n. \tag{5.17}$$

(see Proposition 3.5). These decompositions correspond to the ones given in Table III.

Proposition 5.9. *Let $\rho : C\ell_n \to \mathrm{Hom}_{\mathbb{R}}(W, W)$ be any irreducible real representation where $n = 4m + 3$. Then either*

$$\rho(\omega) = \mathrm{Id} \quad \text{or} \quad \rho(\omega) = -\mathrm{Id}.$$

Both possibilities can occur, and the corresponding representations are inequivalent. (They represent the two generators of $\mathfrak{M}_n$.)

The analogous statements are true in the complex case for $\mathbb{C}\ell_n$, n odd.

Proof. Since $\rho(\omega)^2 = \rho(\omega^2) = \mathrm{Id}$, we can decompose W into $W = W^+ \oplus W^-$ where W^+ and W^- are the $+1$ and -1 eigenspaces for $\rho(\omega)$ respectively. Since ω is central, the spaces W^+ and W^- are $C\ell_n$-invariant. By irreducibility either $W^+ = W$ or $W^- = W$. This proves the first statement.

The inequivalence of representations ρ_+ and ρ_- with $\rho_\pm(\omega) = \pm\mathrm{Id}$ is evident, since if $F: W \to W'$ is an isomorphism and if $\rho(\omega): W \to W$ is a scalar multiple of Id, then $F \circ \rho(\omega) \circ F^{-1}$ is the same scalar multiple of Id.

To see that both possibilities exist we take irreducible factors of $C\ell_n$ acting on $C\ell_n^+$ and on $C\ell_n^-$ by multiplication from the left.

The complex case is proved in the analogous manner by using $\omega_{\mathbf{C}}$. ∎

Proposition 5.10. *Let $\rho: C\ell_n \to \mathrm{Hom}_{\mathbf{R}}(W,W)$ be an irreducible real representation where $n = 4m$, and consider the splitting*

$$W = W^+ \oplus W^-$$

where $W^\pm = (1 \pm \rho(\omega)) \cdot W$ (as in Proposition 3.6). Then each of the subspaces W^+ and W^- is invariant under the even subalgebra $C\ell_n^0$. Under the isomorphism (3.18) $C\ell_n^0 \cong C\ell_{n-1}$, these spaces correspond to the two distinct irreducible real representations of $C\ell_{n-1}$.

The analogous statements are true in the complex case for $\mathbb{C}\ell_n$, n even.

Proof. The invariance of W^+ and W^- under $C\ell_n^0$ is evident from the fact that ω commutes with everything in $C\ell_n^0$ (see (3.9)). Under the isomorphism $C\ell_{n-1} \xrightarrow{\approx} C\ell_n^0$ given in (3.18), we see that the volume element $\omega' = e_1 \cdots e_{n-1}$ of $C\ell_{n-1}$ goes to the volume element $\omega \in C\ell_n^0$. (To see this note that $(e_1e_n) \cdots (e_{n-1}e_n) = (-1)^{\frac{1}{2}(n-1)(n-2)} e_1 \cdots e_{n-1}(e_n)^{n-1} = e_1 \cdots e_n$ since $n = 4m$.) It follows that $\omega' \cong \mathrm{Id}$ on W^+ and $\omega' \cong -\mathrm{Id}$ on W^-. Hence, by Proposition 5.9 these representations of $C\ell_{n-1}$ are inequivalent.

The complex case is proved in the same manner using the volume form $\omega_{\mathbf{C}}$ for $\mathbb{C}\ell_n$. ∎

The representations of the algebras $C\ell_n$ give rise to important representations of certain groups.

Consider the spin group

$$\mathrm{Spin}_n \subset C\ell_n^0 \subset C\ell_n. \tag{5.18}$$

DEFINITION 5.11. The **real spinor representation** of Spin_n is the homomorphism

$$\Delta_n : \mathrm{Spin}_n \longrightarrow \mathrm{GL}(S)$$

given by restricting an irreducible real representation $C\ell_n \to \mathrm{Hom}_{\mathbf{R}}(S,S)$ to $\mathrm{Spin}_n \subset C\ell_n^0 \subset C\ell_n$.

Proposition 5.12. *When $n \equiv 3 \pmod 4$ this definition of Δ_n is independent of which irreducible representation of $C\ell_n$ is used. For $n \not\equiv 0 \pmod 4$ the representation Δ_n is either irreducible or a direct sum of two equivalent irreducible representations. (The second possibility occurs exactly when $n \equiv 1$ or $2 \pmod 8$.) In the other cases there is a decomposition*

$$\Delta_{4m} = \Delta_{4m}^+ \oplus \Delta_{4m}^- \tag{5.19}$$

where Δ_{4m}^+ and Δ_{4m}^- are inequivalent irreducible representations of Spin_{4m}.

Proof. Recall that if $n \equiv 3 \pmod 4$, then the automorphism $\alpha: C\ell_n \to C\ell_n$ interchanges the factors $C\ell_n^+$ and $C\ell_n^-$ (since $\alpha(\omega) = -\omega$). Consequently, $C\ell_n^0$ sits diagonally in the decomposition $C\ell_n = C\ell_n^+ \oplus C\ell_n^-$, i.e.,

$$C\ell_n^0 = \{(\varphi, \alpha(\varphi)) \in C\ell_n^+ \oplus C\ell_n^- : \varphi \in C\ell_n^+\} \tag{5.20}$$

The two irreducible representations of $C\ell_n$ differ by the automorphism α, and are clearly equivalent when restricted to $C\ell_n^0$. This proves the first statement of the proposition.

It is evident from Table III that the restriction of an irreducible real representation of $C\ell_n$ to $C\ell_n^0 \cong C\ell_{n-1}$ is still irreducible if $n \equiv 3, 5, 6$, or $7 \pmod 8$, and must be two copies of an irreducible representation when $n \equiv 1$ or $2 \pmod 8$. When $n \equiv 0 \pmod 4$, we know from Proposition 5.10 that the restriction to $C\ell_n^0$ splits into two inequivalent irreducible representations. To complete the proof we observe that any irreducible representation of $C\ell_n^0$ restricts to an irreducible representation of Spin_n because Spin_n contains an additive basis for $C\ell_n^0$. ■

REMARK 5.13. Note that the spin representations are complex for $n \equiv 2$ or $6 \pmod 8$, and are quaternionic for $n \equiv 3, 4$ or $5 \pmod 8$. (The maximal commuting algebra is determined by $C\ell_{n-1} \cong C\ell_n^0$.)

The analysis above carries over to the complex case.

DEFINITION 5.14. The **complex spin representation** of Spin_n is the homomorphism

$$\Delta_n^{\mathbb{C}}: \mathrm{Spin}_n \longrightarrow \mathrm{GL}_{\mathbb{C}}(S)$$

given by restricting an irreducible complex representation $\mathbb{C}\ell_n \to \mathrm{Hom}_{\mathbb{C}}(S,S)$ to $\mathrm{Spin}_n \subset C\ell_n^0 \subset \mathbb{C}\ell_n$.

Proposition 5.15. *When n is odd, this definition of $\Delta_n^{\mathbb{C}}$ is independent of which irreducible representation of $\mathbb{C}\ell_n$ is used. Furthermore, when n is odd, the representation $\Delta_n^{\mathbb{C}}$ is irreducible. When n is even, there is a decomposition*

$$\Delta_{2m}^{\mathbb{C}} = \Delta_{2m}^{\mathbb{C}+} \oplus \Delta_{2m}^{\mathbb{C}-} \tag{5.21}$$

into a direct sum of two inequivalent irreducible complex representations of Spin_n.

Proof. The proof is entirely analogous to that of Proposition 5.12. ■

It should be pointed out that the spin representations defined above do not descend to the group $\mathrm{SO}_n = \mathrm{Spin}_n/\mathbb{Z}_2$ since $\Delta_n(-1) = -\mathrm{Id}$.

It is worthwhile noting that representations of $C\ell_n$ also give rise to representations of the **Clifford group.** This is the finite group $F_n \subset C\ell_n^\times$

generated by an orthonormal basis $e_1, \ldots, e_n$ of $\mathbb{R}^n$. It can be presented by the abstract elements $e_1, \ldots, e_n, -1$ subject to the relation that -1 is central and that $(-1)^2 = 1$, $(e_i)^2 = -1$ and $e_i e_j = (-1)e_j e_i$ for all $i \neq j$. The Clifford algebra is nearly the group algebra $\mathbb{R}F_n$ of F_n. More explicitly

$$C\ell_n \cong \mathbb{R}F_n/\mathbb{R} \cdot \{(-1) + 1\}$$

It is clear that representations of $C\ell_n$ correspond exactly to linear representations of F_n such that (-1) acts by $-\mathrm{Id}$.

This group enables us to draw an important conclusion:

Proposition 5.16. *Let $C\ell_n \to \mathrm{Hom}_{\mathbb{R}}(W,W)$ be a real representation of $C\ell_n$. Then there exists an inner product $\langle \cdot,\cdot \rangle$ on W such that Clifford multiplication by unit vectors $e \in \mathbb{R}^n$ is orthogonal, i.e., such that*

$$\langle e \cdot w, e \cdot w' \rangle = \langle w, w' \rangle \tag{5.22}$$

for all $w,w' \in W$ and for all $e \in \mathbb{R}^n$ with $\|e\| = 1$. If K_n ($= \mathbb{R}$, $\mathbb{C}$ or $\mathbb{H}$) is a commuting subalgebra for the representation, then the inner product can be chosen to be K_n-invariant, that is, so that J is orthogonal when $K_n = \mathbb{C}$ and so that I, J and K are each orthogonal if $K_n = \mathbb{H}$.

In particular, the *spin representations Δ_n are unitary if $n \equiv 2$ or* 6 (mod 8) *and symplectic* if $n \equiv 3$, 4 or 5 (mod 8).

Proof. Choose a K_n-invariant inner product and average it over the finite group F_n. Note that if $e = \sum a_j e_j$ where $\sum a_j^2 = 1$, then

$$\langle ew, ew \rangle = \sum a_j^2 \langle e_j w, e_j w \rangle + \sum_{i \neq j} a_i a_j \langle e_i w, e_j w \rangle = \langle w, w \rangle$$

since $\langle e_i w, e_i w \rangle = \langle w, w \rangle$ and for $i \neq j$, $\langle e_i w, e_j w \rangle = \langle e_j e_i w, -w \rangle = \langle e_i e_j w, w \rangle = -\langle e_j w, e_i w \rangle = 0$. For the last statement, recall that Δ_n comes from a representation of $C\ell_n^0 \cong C\ell_{n-1}$. ■

Corollary 5.17. *Let $\langle \cdot,\cdot \rangle$ be the metric discussed in Proposition* 5.16. *Then for any $v \in \mathbb{R}^n$,*

$$\langle v \cdot w, w' \rangle = -\langle w, v \cdot w' \rangle \tag{5.23}$$

for all $w,w' \in W$. That is, Clifford multiplication by any vector $v \in \mathbb{R}^n$ is a skew-symmetric transformation of W.

Proof. Assume $v \neq 0$. Then $\langle v \cdot w, w' \rangle = \langle (v/\|v\|) \cdot v \cdot w, (v/\|v\|) \cdot w' \rangle = (1/\|v\|^2)\langle v^2 \cdot w, v \cdot w' \rangle = -\langle w, v \cdot w' \rangle$. ■

It is worth noting that the irreducible representions of $C\ell_{2n}$ have a particularly nice description. Introduce on $\mathbb{C}^n$ the standard hermitian metric

$$(z,\zeta) \equiv \sum_{j=1}^{n} z_j \bar{\zeta}_j. \tag{5.24}$$

and use this inner product to define a complex linear contraction map

$$(v \llcorner) : \Lambda^p_{\mathbb{C}}\mathbb{C}^n \longrightarrow \Lambda^{p-1}_{\mathbb{C}}\mathbb{C}^n$$

for $v \in \mathbb{C}^n$ by formula (3.21). We then define $f_v : \Lambda^*_{\mathbb{C}}\mathbb{C}^n \to \Lambda^*_{\mathbb{C}}\mathbb{C}^n$ by setting

$$f_v(\varphi) = v \wedge \varphi - v \llcorner \varphi. \tag{5.25}$$

Then since $v \wedge v = 0$, $(v \llcorner)(v \llcorner) = 0$, and $v \llcorner (v \wedge \varphi) = \|v\|^2\varphi - v \wedge (v \llcorner \varphi)$, we see that

$$f_v \circ f_v(\varphi) = -\|v\|^2\varphi. \tag{5.26}$$

Note that since the inner product is $\mathbb{C}$-*anti*linear in the second variable, the map $v \to f_v$ is only $\mathbb{R}$-linear. Nevertheless, writing $\mathbb{R}^{2n} \cong \mathbb{C}^n$, we see from universality (Proposition 1.1) that property (5.26) determines a unique extension of f to a representation

$$\tilde{f} : \mathrm{C}\ell_{2n} \longrightarrow \mathrm{Hom}_{\mathbb{C}}(\Lambda^*\mathbb{C}^n, \Lambda^*\mathbb{C}^n). \tag{5.27}$$

Since the complex dimension of this representation is 2^n, we see that it must be the irreducible one.

We now make some remarks concerning tensor products. Suppose W is a K-module for $\mathrm{C}\ell_n$ (where $K = \mathbb{R}$ or $\mathbb{C}$) and let V be any vector space over K. Then $W \otimes_K V$ is also a K-module for $\mathrm{C}\ell_n$ where by definition

$$\varphi \cdot (w \otimes v) = (\varphi w) \otimes v. \tag{5.28}$$

Therefore, if W_1 and W_2 are K-modules for $\mathrm{C}\ell_n$, then $W_1 \otimes_K W_2$ is a K-module for $\mathrm{C}\ell_n$ in two distinct ways. We set

$$\lambda_\varphi(w_1 \otimes w_2) = (\varphi w_1) \otimes w_2,$$
$$\rho_\varphi(w_1 \otimes w_2) = w_1 \otimes (\varphi w_2).$$

Then λ and ρ are commuting representations of $\mathrm{C}\ell_n$. Furthermore, the product

$$\Phi_{\varphi_1,\varphi_2}(w_1 \otimes w_2) = (\varphi_1 w_1) \otimes (\varphi_2 w_2)$$

is a representation of the (ungraded) tensor product $\mathrm{C}\ell_n \otimes \mathrm{C}\ell_n$.

Proposition 5.18. *Let* $\mathbb{C}\ell_{2n} \to \mathrm{Hom}_{\mathbb{C}}(S,S)$ *be an irreducible complex representation of* $\mathbb{C}\ell_{2n}$. *Then the tensor product representation of* $\mathbb{C}\ell_{2n} \otimes_{\mathbb{C}} \mathbb{C}\ell_{2n}$ *on* $S \otimes_{\mathbb{C}} S$ *is equivalent to the representation* Φ *on* $\mathbb{C}\ell_{2n}$ *itself given by setting*

$$\Phi_{\varphi_1,\varphi_2}(\varphi) = \varphi_1 \cdot \varphi \cdot \varphi_2^t$$

Proof. Since $\mathbb{C}\ell_{2n} \otimes_{\mathbb{C}} \mathbb{C}\ell_{2n} \cong \mathbb{C}\ell_{4n}$ (see Theorem 4.3), we see that $S \otimes_{\mathbb{C}} S$ must be an irreducible module for reasons of dimension. Since $\dim_{\mathbb{C}}(S \otimes S) = 2^{2n} = \dim_{\mathbb{C}}(\mathbb{C}\ell_{2n})$, the representations must be equivalent. ∎

Corollary 5.19. *Let $\rho_n : \mathrm{Spin}_n \to \mathrm{SO}(\mathbb{R}^n)$ denote the standard n-dimensional representation of* Spin_n. *Then in the complex representation ring of* Spin_{2m} (cf. Adams [1]) *we have the equation*

$$\begin{aligned}(\Delta_{2m}^{\mathbb{C}+} + \Delta_{2m}^{\mathbb{C}-}) &\cdot (\Delta_{2m}^{\mathbb{C}+} + \Delta_{2m}^{\mathbb{C}-}) \\ &= 2(1 + \rho_{2m}^{\mathbb{C}} + \Lambda^2\rho_{2m}^{\mathbb{C}} + \ldots + \Lambda^{m-1}\rho_{2m}^{\mathbb{C}}) + \Lambda^m\rho_{2m}^{\mathbb{C}}\end{aligned}$$

where $\rho_{2m}^{\mathbb{C}}$ denotes the complexification of ρ_{2m}.

Proof. The tensor product $\Delta_{2m}^{\mathbb{C}} \otimes \Delta_{2m}^{\mathbb{C}}$ is obtained by embedding Spin_{2m} into $\mathbb{C}\ell_{2m} \otimes \mathbb{C}\ell_{2m}$ diagonally ($g \mapsto g \otimes g$) and restricting the tensor product representation. By Proposition 5.18 this is equivalent to the adjoint representation $\Phi_g(\varphi) = g\varphi g^t = g\varphi g^{-1} = \mathrm{Ad}_g(\varphi)$. Under the correspondence $\mathbb{C}\ell_{2n} \cong \Lambda_{\mathbb{C}}^*\mathbb{C}^{2m}$, this representation is equivalent to $(1 + \rho_{2m} + \Lambda^2\rho_{2m} + \ldots + \Lambda^{2m}\rho_{2m}) \otimes \mathbb{C}$. However, by Hodge duality $\Lambda^p\rho_{2m} \cong \Lambda^{2m-p}\rho_{2m}$. ∎

Results analogous to Proposition 5.18 and Corollary 5.19 hold for the algebras $C\ell_{8m}$ and for the real spin representation $\Delta_{8m} = \Delta_{8m}^+ + \Delta_{8m}^-$.

Note that the tensor product of irreducible real representations of $C\ell_n$ and $C\ell_8$ gives an irreducible real representation of $C\ell_{n+8} \cong C\ell_n \otimes C\ell_8$. Similarly the complex tensor product of irreducible complex representations of $\mathbb{C}\ell_n$ and $\mathbb{C}\ell_2$ gives an irreducible complex representation of $\mathbb{C}\ell_{n+2} \cong \mathbb{C}\ell_n \otimes \mathbb{C}\ell_2$. In general, however, $C\ell_n \otimes C\ell_m$ is not a Clifford algebra. Thus, to find a multiplicative structure in the representations of Clifford algebras it is natural to consider the category of $\mathbb{Z}_2$-graded modules. A **$\mathbb{Z}_2$-graded module** for $C\ell_n$ is a module W with a decomposition $W = W^0 \oplus W^1$ such that

$$C\ell_n^i \cdot W^j \subseteq W^{(i+j)(\mathrm{mod}\ 2)}$$

for $0 \leqq i, j \leqq 1$.

Proposition 5.20. *There is an equivalence between the category of $\mathbb{Z}_2$-graded modules over $C\ell_n$ and the category of ungraded modules over $C\ell_{n-1}$. It is defined by passing from the graded module $W^0 \oplus W^1$ over $C\ell_n$ to the module W^0* over $C\ell_n^0 \cong C\ell_{n-1}$.

Proof. The inverse procedure is given by assigning to a $C\ell_n^0$-module W^0, the $\mathbb{Z}_2$-graded module

$$W \equiv C\ell_n \otimes_{C\ell_n^0} W^0$$

(Left multiplication by $C\ell_n$ on $C\ell_n$ makes W into a $\mathbb{Z}_2$-graded module.) The remainder of the proof is straightforward. ∎

There is a natural definition of the $\mathbb{Z}_2$-graded tensor product of $\mathbb{Z}_2$-graded modules $W = W^0 \oplus W^1$ and $V = V^0 \oplus V^1$ over $C\ell_n$ and $C\ell_m$

respectively. We set

$$(W \hat{\otimes} V)^0 = W^0 \otimes V^0 + W^1 \otimes V^1$$
$$(W \hat{\otimes} V)^1 = W^0 \otimes V^1 + W^1 \otimes V^0.$$

The action of $C\ell_n \hat{\otimes} C\ell_m$ on $W \hat{\otimes} V$ is given by

$$(\varphi \otimes \psi) \cdot (w \otimes v) \equiv (-1)^{pq}(\varphi w) \otimes (\psi v)$$

where $\deg(\psi) = p$ and $\deg(w) = q$. Under the isomorphism $C\ell_{n+m} \cong C\ell_n \hat{\otimes} C\ell_m$, induced from mapping $\mathbb{R}^n \oplus \mathbb{R}^m \to C\ell_{n+m}$

$$\begin{cases} e \longmapsto e \otimes 1 & \text{if } e \in \mathbb{R}^n \subset C\ell_n \\ e' \longmapsto 1 \otimes e' & \text{if } e' \in \mathbb{R}^m \subset C\ell_m, \end{cases}$$

$V \hat{\otimes} W$ becomes a $\mathbb{Z}_2$-graded module over $C\ell_{n+m}$. This construction holds for either real or complex modules.

In analogy with the above we define $\hat{\mathfrak{M}}_n$ ($\hat{\mathfrak{M}}_n^{\mathbb{C}}$) to be the Grothendieck group of real (complex) $\mathbb{Z}_2$-graded modules over $C\ell_n$. Note that by Proposition 5.20 there are natural isomorphisms

$$\hat{\mathfrak{M}}_n \cong \mathfrak{M}_{n-1} \qquad \text{and} \qquad \hat{\mathfrak{M}}_n^{\mathbb{C}} \cong \mathfrak{M}_{n-1}^{\mathbb{C}}. \tag{5.29}$$

The arguments just given have established the following:

Proposition 5.21. *There are natural pairings*

$$\hat{\mathfrak{M}}_n \otimes_{\mathbb{Z}} \hat{\mathfrak{M}}_m \longrightarrow \hat{\mathfrak{M}}_{n+m} \tag{5.30}$$
$$\hat{\mathfrak{M}}_n^{\mathbb{C}} \otimes_{\mathbb{Z}} \hat{\mathfrak{M}}_m^{\mathbb{C}} \longrightarrow \hat{\mathfrak{M}}_{n+m}^{\mathbb{C}} \tag{5.31}$$

induced by the $\mathbb{Z}_2$-graded tensor product. These pairings are associative and give $\hat{\mathfrak{M}}_* \equiv \bigoplus_{n \geq 0} \hat{\mathfrak{M}}_n$ and $\hat{\mathfrak{M}}_*^{\mathbb{C}} \equiv \bigoplus_{n \geq 0} \hat{\mathfrak{M}}_n^{\mathbb{C}}$ the structure of graded rings.

These pairings are important in the relation of Clifford algebras to real and complex K-theory (see §9).

§6. Lie Algebra Structures

This section shall be concerned with the Lie algebra of Spin_n. Recall that the group of units $C\ell_n^\times$ is a Lie group with Lie algebra $\mathfrak{cl}_n^\times \equiv (C\ell_n, [\cdot,\cdot])$ where $[\varphi, \psi] \equiv \varphi \cdot \psi - \psi \cdot \varphi$. There is an exponential mapping $\exp : \mathfrak{cl}_n^\times \to C\ell_n^\times$ given by the standard series (see Remark 2.1). The group Spin_n is an explicitly defined, compact subgroup of $C\ell_n^\times$. We shall now investigate its associated Lie subalgebra $\mathfrak{spin}_n$ in $C\ell_n$.

Recall that there are canonical embeddings $\Lambda^p\mathbb{R}^n \subset C\ell_n$ for all p.

Proposition 6.1. *The Lie subalgebra of* $(C\ell_n, [\cdot,\cdot])$ *corresponding to the subgroup* $\mathrm{Spin}_n \subset C\ell_n^\times$ *is*

$$\mathfrak{spin}_n = \Lambda^2\mathbb{R}^n. \tag{6.1}$$

In particular, $\Lambda^2\mathbb{R}^n$ *is closed under the bracket operation.*

Proof. The Lie subalgebra $\mathfrak{spin}_n$ is the vector subspace of $C\ell_n$ spanned by the tangent vectors to the submanifold Spin_n at 1. Fix an orthonormal basis $e_1, \dots, e_n$ of $\mathbb{R}^n$ and consider for each pair $i < j$, the curve $\gamma(t) \equiv (e_i \cos t + e_j \sin t) \cdot (-e_i \cos t + e_j \sin t) = (\cos^2 t - \sin^2 t) + 2e_ie_j \sin t \cos t = \cos(2t) + \sin(2t)e_ie_j$. This curve lies in Spin_n by definition of Spin_n, and its tangent vector at $\gamma(0) = 1$ is $2e_ie_j$. Hence, $\mathfrak{spin}_n$ contains the vector subspace $\mathrm{span}_{\mathbb{R}}\{e_ie_j\}_{i<j} = \Lambda^2\mathbb{R}^n$. Since $\dim_{\mathbb{R}}(\mathfrak{spin}_n) = n(n-1)/2$, we conclude they are equal. ∎

We now recall that the Lie algebra of the orthogonal group SO_n is exactly the space

$$\mathfrak{so}_n = \{\lambda : \mathbb{R}^n \longrightarrow \mathbb{R}^n : \lambda \text{ is linear and skew-symmetric}\} \tag{6.2}$$

There is a natural isomorphism $\Lambda^2\mathbb{R}^n \xrightarrow{\approx} \mathfrak{so}_n$ induced by associating to a pair of vectors $v,w \in \mathbb{R}^n$, the skew-symmetric endomorphism "$v \wedge w$" defined by

$$(v \wedge w)(x) \equiv \langle v,x\rangle w - \langle w,x\rangle v, \tag{6.3}$$

and then extending by universality. Note that $e_i \wedge e_j$, for $i < j$, corresponds to the elementary skew-symmetric (i,j) matrix:

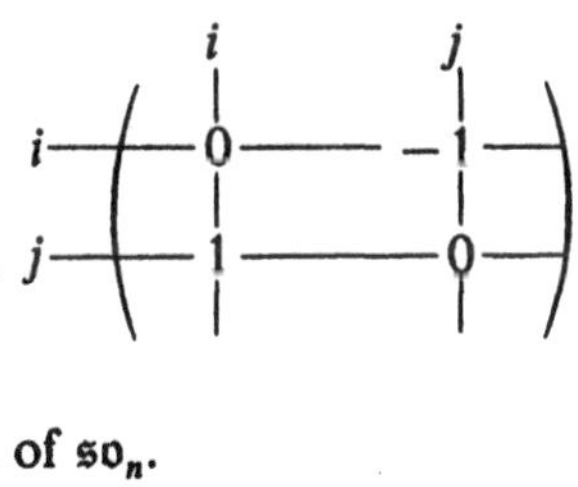

This is a standard basis of $\mathfrak{so}_n$.

Recall now that the adjoint representation gives a surjective homomorphism

$$\mathrm{Spin}_n \xrightarrow{\xi_0} \mathrm{SO}_n.$$

(Since $\mathrm{Spin}_n \subset C\ell_n^0$, we have $\widetilde{\mathrm{Ad}}|_{\mathrm{Spin}_n} = \mathrm{Ad}|_{\mathrm{Spin}_n}$.) This induces an associated Lie algebra isomorphism

$$\mathfrak{spin}_n \xrightarrow{\Xi_0} \mathfrak{so}_n \tag{6.4}$$

Proposition 6.2. *The Lie algebra isomorphism (6.4) induced by the adjoint representation is given explicitly on basis elements $\{e_i e_j\}_{i<j}$ by*

$$\Xi_0(e_i e_j) = 2e_i \wedge e_j. \tag{6.5}$$

Consequently for $v,w \in \mathbb{R}^n$,

$$\Xi_0^{-1}(v \wedge w) = \frac{1}{4}[v,w] \tag{6.6}$$

Note. This factor of $\frac{1}{4}$ plays a delicate role in geometric applications.

Proof. Consider the curve $\gamma(t) = \cos(t) + \sin(t)e_i e_j$ in Spin_n with $\gamma(0) = 1$ and $\gamma'(0) = e_i e_j$. Then

$$\Xi_0(e_i e_j) = \frac{d}{dt}\xi_0(\gamma(t))\bigg|_{t=0}$$

and to compute this we apply it to a vector $x \in \mathbb{R}^n$. Since

$$\xi_0(\gamma(t))(x) = \gamma(t)x\gamma(t)^{-1},$$

and since $(\gamma^{-1})'(0) = -\gamma'(0) = -e_i e_j$, we have that

$$\begin{aligned}\Xi_0(e_i e_j)(x) &= e_i e_j x - x e_i e_j \\ &= e_i e_j x + (e_i x + 2\langle e_i, x\rangle)e_j \\ &= e_i e_j x - e_i e_j x - 2\langle e_j, x\rangle e_i + 2\langle e_i, x\rangle e_j \\ &= 2(e_i \wedge e_j)(x)\end{aligned}$$

To prove (6.6), note that on basis elements, $\frac{1}{4}[e_i, e_j] = \frac{1}{4}(e_i e_j - e_j e_i) = \frac{1}{2}e_i e_j$. ∎

This Proposition has the following immediate corollary:

Corollary 6.3. *Let $\Delta : \text{Spin}_n \to \text{SO}(W)$ be a representation obtained by restriction of a representation $C\ell_n \to \text{Hom}(W,W)$ of the Clifford algebra $C\ell_n \supset \text{Spin}_n$. Let $\Delta_* : \mathfrak{so}_n \to \mathfrak{so}(W)$ be the associated representation of the Lie algebra (obtained by first pulling back $\mathfrak{so}_n$ to the double covering via Ξ_0^{-1}). Then on the elementary transformations $v \wedge w \in \mathfrak{so}_n$,*

$$\Delta_*(v \wedge w) = \tfrac{1}{4}[v,w] \cdot \tag{6.7}$$

where the dot indicates Clifford module multiplication on W.

In terms of the standard basis $\{e_i \wedge e_j\}_{i<j}$

$$\Delta_*(e_i \wedge e_j) = \tfrac{1}{2}e_i e_j \tag{6.8}$$

Suppose now that $C\ell_n \to \text{Hom}(W,W)$ is a complex representation of $C\ell_n$, and fix an element $w \in W$. The subgroup

$$G_w \equiv \{g \in \text{Pin}_n : gw = w\} \tag{6.9}$$

is called the **isotropy group** of w. Its Lie algebra is the subalgebra

$$\mathfrak{g}_w \equiv \{\varphi \in \mathfrak{spin}_n : \varphi \cdot w = 0\} \tag{6.10}$$

Two elements $w,w' \in W$ are considered to be different as spinors (or more precisely, to have **distinct orbit types**) if their isotropy groups G_w and $G_{w'}$ are not conjugate in Pin_n. One crude measure of this difference is the following:

DEFINITION 6.4. The **rank** of the (generalized) spinor w is the rank of the Lie group G_w. This is the dimension of a maximal torus in G_w or, equivalently, of a maximal abelian subalgebra of $\mathfrak{g}_w$.

In a compact Lie group, every abelian subgroup is contained in a maximal torus, and all maximal tori are conjugate (cf. Adams [1]). Hence any maximal torus T_w of G_w is contained in a maximal torus T of Pin_n which we can assume to be the following standard one associated to a fixed orthonormal basis $\{e_1, \ldots, e_n\}$ of $\mathbb{R}^n$:

$$T \equiv \left\{\prod_{k=1}^{[n/2]} (\cos\theta_k + \sin\theta_k e_{2k-1}e_{2k}) : 0 \leqq \theta_k < 2\pi \text{ for each } k\right\}$$

The Lie algebra of T is given by

$$\mathfrak{t} \equiv \left\{\sum_{k=1}^{[n/2]} \lambda_k e_{2k-1}e_{2k} : \lambda_k \in \mathbb{R} \text{ for each } k\right\}$$

We now use our distinguished orthonormal basis to decompose the module W. For each k, $1 \leqq 2k \leqq n$, define

$$\omega_k = -ie_{2k-1}e_{2k}$$

and note that

$$\omega_1, \ldots, \omega_{[n/2]} \qquad \text{pairwise commute,} \tag{6.11}$$

$$\omega_k^2 = 1 \qquad \text{for each } k, \tag{6.12}$$

$$\omega_k e_k = -e_k\omega_k \qquad \text{for each } k. \tag{6.13}$$

Suppose now that $V \subset W$ is a linear subspace which is e_{2k-1}-invariant and e_{2k}-invariant for some fixed k. Then by (6.12) we know that $V = V_+ \oplus V_-$ where $V_\pm = (1 \pm \omega_k)V$ are the ± 1 eigenspaces of ω_k on V. Furthermore, by (6.13) we see that $e_k V_+ = V_-$ and $e_k V_- = V_+$. In particular, $\dim V_\pm = \frac{1}{2}\dim V$.

We shall now use this process to decompose the module W. We begin with the decomposition $W = W_+ \oplus W_-$ by ω_1. Since ω_1 commutes with $e_3, e_4, \ldots, e_n$, we see that each of the subspaces W_+ and W_- is $e_3, \ldots, e_n$-invariant. Hence we can similarly decompose each subspace W_+ and W_- by ω_2 to get $W_+ = W_{++} \oplus W_{+-}$ and $W_- = W_{-+} \oplus W_{--}$. Each subspace $W_{\pm\pm}$ is $e_5, \ldots, e_n$-invariant. Continuing inductively, we produce a decomposition

$$W = \bigoplus W_{\pm \cdots \pm} = \bigoplus_{\alpha} W_{\alpha} \tag{6.14}$$

where $\dim W_\alpha = (\dim W)/2^{[n/2]}$ for each α and where $\alpha = (\alpha_1, \ldots, \alpha_{[n/2]})$ ranges over the $2^{[n/2]}$-possibilities having $\alpha_k = +$ or $-$ for each k. (Note that if W is an irreducible module, then $\dim_{\mathbb{C}} W_\alpha = 1$ for all α.)

The maximal torus T preserves each subspace W_α. In fact, given $g = \prod (\cos\theta_k + i\sin\theta_k\omega_k) \in T$, we see that

$$g\big|_{W_\alpha} = e^{i\Sigma\alpha_k\theta_k} \equiv e^{i\langle\alpha,\theta\rangle}$$

Similarly for $\varphi = \sum \lambda_k e_{2k-1}e_{2k} = i\sum \lambda_k\omega_k \in \mathfrak{t}$ we have

$$\varphi\big|_{W_\alpha} = i\sum \alpha_k\lambda_k \equiv i\langle\alpha,\lambda\rangle \tag{6.15}$$

The set of vectors $\frac{1}{2}\alpha \in \mathfrak{t}^*$ are called the **weights** of the representation. The $\frac{1}{2}$ occurs because in general theory the weights are normalized by relating them to the weights of the adjoint representation.

We now return to our given spinor $w \in W$. With respect to the decomposition (6.14) we write

$$w = \sum_\alpha w_\alpha$$

From (6.10) and (6.15) we conclude the following:

Proposition 6.5. *The maximal abelian subalgebra of G_w is*

$$\mathfrak{t}_w = \left\{ i\sum \lambda_k\omega_k : \langle\alpha,\lambda\rangle = 0 \text{ for all } \alpha \text{ such that } w_\alpha \neq 0 \right\}$$

Corollary 6.6. $\operatorname{rank} w = [n/2] - \dim \operatorname{span}_{\mathbb{R}}\{\alpha : w_\alpha \neq 0\}$

If $w = w_\alpha$ for some α, i.e., if all but one component vanish, then w is clearly of maximal rank. Those elements that take the simple form $w = w_\alpha$ for some choice of orthonormal basis in $\mathbb{R}^n$, are called **pure.** Pure spinors are related to complex structures, twistors spaces and calibrations. They will be discussed in detail in Chapter IV.

§7. Some Direct Applications to Geometry

In this section we shall use the classification of Clifford modules given above to construct families of pointwise linearly independent vector fields on spheres, projective spaces and other elliptic space forms. We shall also apply the methods to study the "hyperplane" bundle over complex and quaternionic projective space. This allows us to estimate the geometric dimension of $T\mathbb{P}^n(\mathbb{C})$. In almost all cases the families constructed in this manner are maximal.

We begin with the following observation:

Proposition 7.1. *Suppose* $\mathbb{R}^{N+1}$ *is a module for the algebra* $C\ell_n$. *Then there exist* n *pointwise linearly independent tangent vector fields on the sphere* S^N *and also on the projective space* $\mathbb{P}^N(\mathbb{R}) = S^N/\mathbb{Z}_2$.

Proof. Choose an inner product in $\mathbb{R}^{N+1}$ so that Clifford multiplication by unit vectors in $\mathbb{R}^n$ is orthogonal (see Proposition 5.16). Let $S^N = \{x \in \mathbb{R}^{N+1} : \|x\|^2 = 1\}$. Choose a basis $v_1, \ldots, v_n$ for $\mathbb{R}^n$, and to each v_j associate the vector field V_j on $\mathbb{R}^{N+1}$ defined by

$$V_j(x) \equiv v_j \cdot x \qquad j = 1, \ldots, n$$

(where the dot denotes Clifford multiplication). Since the linear transformation $x \mapsto v \cdot x$ is skew-symmetric (see Corollary 5.17), we have that $\langle V_j(x), x\rangle \equiv \langle v_j x, x\rangle \equiv 0$. Hence, the vector fields V_j are tangent to S^N. It remains to show that $V_1, \ldots, V_n$ are pointwise linearly independent. Fix $x \in S^N$ and consider the linear map $i_x : \mathbb{R}^n \to T_x S^N \subset \mathbb{R}^{N+1}$ given by

$$i_x(v) = v \cdot x$$

The image of i_x is the linear span of $V_1(x), \ldots, V_n(x)$, so it suffices to prove that i_x is injective. However if $i_x v = v \cdot x = 0$, then $v \cdot v \cdot x = -\|v\|^2 x = 0$ and so $v = 0$.

Since $V_j(-x) = -V_j(x)$, these vector fields descend to (pointwise linearly independent) vector fields on $\mathbb{P}^N(\mathbb{R})$. ∎

The question now is: given an integer N, what is the largest number of independent vector fields on S^N that can be constructed in this manner? That is, what is the largest integer n such that $\mathbb{R}^{N+1}$ is a $C\ell_n$-module? We recall that the dimension of an irreducible $C\ell_n$-module is always a power of 2. Hence, we want to find the largest power of 2 which divides $N + 1$. That is, we write $N + 1 = p2^m$ where p is odd, and then we consult Table III to find the largest n such that $d_n = 2^m$. The result is the following classical result of Radon and Hurwitz.

Theorem 7.2. *On the sphere* S^N *(and on the projective space* $\mathbb{P}^N(\mathbb{R})$*) there exist* n *pointwise linearly independent vector fields where* n *is computed as follows. Write* $N + 1 = 2^{4a+b}(2t + 1)$, $0 \le b \le 3$. *Then*

$$n = 8a + 2^b - 1. \tag{7.1}$$

Proof. One need only check this when $a = 0$, and then note that for each increase of n by 8 the dimension of the vector space for an irreducible representation of $C\ell_n$ increases by 2^4. Note that when N is even, the number of such vector fields is zero as it must be since the Euler characteristic is non-zero in this case. Note also that this construction gives three vector fields on S^3, seven on S^7 and eight on S^{15}. ∎

One of the deep results of algebraic topology is the following:

Theorem 7.3 (J. F. Adams [2]). *The number of vector fields constructed on S^N above is the largest possible number of pointwise linearly independent vector fields that can exist on S^N.*

It is worth noting that this construction also gives rise to vector fields on many elliptic space forms. If the representation of $C\ell_n$ on $\mathbb{R}^{N+1}$ is complex or quaternionic, then Clifford multiplication by $v \in \mathbb{R}^n$ commutes with complex scalar multiplication. Therefore, if $\beta = e^{2\pi i/p}$ is a pth root of unity, then the vector field $V(x) \equiv v \cdot x$ on S^N has the equivariance property

$$V(\beta x) = \beta \cdot V(x) \tag{7.2}$$

for all $x \in S^N$. This means precisely that the vector field $V(x)$ is invariant under the diffeomorphism $S^N \to S^N$ given by scalar multiplication by β. The $\mathbb{Z}_p$-action on S^N generated by β is free. From (7.2) we conclude that V descends to a vector field on the quotient $S^N/\mathbb{Z}_p$ which is a lens space of simple type $L^N(p) = L^N(p;1,\dots,1)$.

Analogous remarks hold when the representation of $C\ell_n$ is quaternionic. Here we may replace $\mathbb{Z}_p$ by any finite multiplicative subgroup of $\mathbb{H}$. Such subgroups are constructed as follows. The unit sphere $S^3 \subset \mathbb{H}$ is a Lie group isomorphic to Spin_3 (see the paragraph below). Let $\xi_0 : S^3 \to \mathrm{SO}_3$ be the 2-fold covering homomorphism. Then for any finite subgroup $\Gamma_0 \subset \mathrm{SO}_3$, the inverse image $\Gamma = \xi_0^{-1}(\Gamma_0)$ is a finite subgroup of $S^3 \subset \mathbb{H}$. Of course, the symmetry groups of the regular polygons, the so-called dihedral groups, and the symmetry groups of the Platonic solids give many examples of finite subgroups of SO_3. The ξ_0^{-1}-images of dihedral groups are called binary dihedral groups. There are also the **binary tetrahedral group,** the **binary octahedral group,** and the **binary icosahedral group,** corresponding to the ξ_0^{-1}-images of the symmetry groups of the tetrahedron, octahedron and icosahedron, respectively.

From our remarks above we conclude the following two theorems.

Theorem 7.4. *On each simple lens space $L^N(p) = S^N/\mathbb{Z}_p$, for $p \geq 1$, there exist k pointwise linearly independent vector fields where, if $N + 1 = 2^m(2t + 1)$, then*

$$k = 2m - 1$$

Moreover, this is the maximal number of pointwise linearly independent vector fields possible on $L^N(p)$ if $m \equiv 1$ or 2 modulo 4.

Theorem 7.5. *There exist q pointwise linearly independent vector fields on S^N/Γ where Γ is any finite subgroup of $S^3 \subset \mathbb{H}$ and where if $N + 1 =$*

$2^m(2t+1)$ *and* $m = 4a + b$, $2 \le b \le 5$, *then*

$$q = \begin{cases} 8a + b + 1 & \text{if } b \neq 5 \\ 8a + b + 2 & \text{if } b = 5 \end{cases}.$$

Moreover, this is the maximal number of pointwise linearly independent vector fields possible on the elliptic space form S^N/Γ *if* m *is congruent to* 1 *or* 2 *modulo* 4.

Similar constructions can be made for complex and quaternionic projective space. Of course in these cases the Euler characteristic of the manifold is not zero, and so every tangent vector field must vanish somewhere. However, we can pass to the stabilized bundle $TX \oplus \mathbb{R}^m$ where $\mathbb{R}^m$ denotes the trivial bundle of dimension m. Clearly there exists some bundle E over X so that $TX \oplus \mathbb{R}^m \cong E \oplus \mathbb{R}^{m+n-k}$ where $n = \dim X$, $k = \dim_{\mathbb{R}} E$ and k is as small as possible (for any choice of m). The dimension of E is called the **geometric dimension** of TX.

Let $\mathbb{P}^n(K)$ denote the n-dimensional projective space over the field K. For $K = \mathbb{R}$, $\mathbb{C}$ or $\mathbb{H}$ there is a **tautological** K**-line bundle** $\eta \to \mathbb{P}^n(K)$ whose fibre above a point $[\ell] \in \mathbb{P}^n(K)$ is the one-dimensional linear subspace $\ell \subset K^{n+1}$ corresponding to $[\ell]$. For $K = \mathbb{R}$ or $\mathbb{C}$ we have the following fact:

$$T(\mathbb{P}^n(K)) \oplus K \cong \underbrace{\eta^* \oplus \cdots \oplus \eta^*}_{(n+1)\text{-times}} \tag{7.3}$$

where η^* is the dual bundle

$$\eta^* \equiv \mathrm{Hom}_K(\eta, K).$$

To see this, we first show that

$$T_{[\ell]}\mathbb{P}^n(K) \cong \mathrm{Hom}_K(\ell, \ell^\perp). \tag{7.4}$$

Each one-dimensional K-linear subspace $\ell \subset K^{n+1}$ can be canonically identified with the (K-linear) orthogonal projection map $\pi_\ell : K^{n+1} \to \ell$. Note that $\pi_\ell^2 = \pi_\ell$, $\pi_\ell \pi_{\ell^\perp} = \pi_{\ell^\perp}\pi_\ell = 0$ and $\pi_\ell + \pi_{\ell^\perp} = \mathrm{Id}$. Let ℓ_t, $|t| < \varepsilon$, be a smooth family of K-lines with $\ell_0 = \ell$, and set $\dot\pi = (d/dt)\pi_{\ell_t}|_{t=0}$. Then by deriving the identities above we have that $\dot\pi_\ell \pi_\ell + \pi_\ell \dot\pi_\ell = \dot\pi_\ell$, $\dot\pi_\ell \pi_{\ell^\perp} + \pi_\ell \dot\pi_{\ell^\perp} = \dot\pi_{\ell^\perp}\pi_\ell + \pi_{\ell^\perp}\dot\pi_\ell = 0$, and $\dot\pi_\ell + \dot\pi_{\ell^\perp} = 0$. It follows that $\dot\pi_\ell \pi_{\ell^\perp} = 0$ and $\pi_\ell \dot\pi_\ell = 0$. Hence $\dot\pi_\ell \in \mathrm{Hom}_K(\ell, \ell^\perp)$. On the other hand it is easy to construct an elementary basis of $\mathrm{Hom}_K(\ell, \ell^\perp)$ as tangent vectors $\dot\pi_\ell$ for curves $\ell(t)$ with $\ell(0) = \ell$.

From the exact sequence

$$0 \longrightarrow \ell \longrightarrow K^{n+1} \longrightarrow \ell^\perp \longrightarrow 0$$

we obtain the exact sequence

$$0 \longrightarrow \mathrm{Hom}_K(\ell, \ell) \longrightarrow \mathrm{Hom}_K(\ell, K^{n+1}) \longrightarrow \mathrm{Hom}_K(\ell, \ell^{\perp}) \longrightarrow 0. \quad (7.5)$$

From (7.4) this gives an exact sequence of bundles over $\mathbb{P}^n(K)$:

$$0 \longrightarrow \mathrm{Hom}_K(\eta, \eta) \longrightarrow \underbrace{\eta^* \oplus \cdots \oplus \eta^*}_{(n+1) \text{ times}} \longrightarrow T\mathbb{P}^n(K) \longrightarrow 0 \quad (7.5)$$

which holds for $K = \mathbb{R}$, $\mathbb{C}$ or $\mathbb{H}$. When $K = \mathbb{R}$ or $\mathbb{C}$, we have $\mathrm{Hom}_K(\eta, \eta) \cong \eta^* \otimes \eta \cong K$ ($\cong$ the trivial K-line bundle), and this establishes (7.3).

Suppose now that $L: K^{n+1} \to K^m$ is a K-linear map (for any $m \geqq 1$). Then L defines a section of $\eta^* \oplus \cdots \oplus \eta^*$ (m times) as follows. Fix a K-line $\ell \subset K^{n+1}$. Then $L|_{\ell}: \ell \to K^m$ is exactly m K-linear functions on ℓ. Thus, we have an embedding

$$\mathrm{Hom}_K(K^{n+1}, K^m) \subset \underbrace{\Gamma(\eta^* \oplus \cdots \oplus \eta^*)}_{m \text{ times}} \quad (7.7)$$

into the space of sections of the bundle $\eta^* \oplus \cdots \oplus \eta^*$. A section corresponding to a K-linear map L is nowhere zero if and only if $L(x) \neq 0$ for any $x \neq 0$. Similarly, $L_1, \ldots, L_p$ give pointwise linearly independent sections of $\eta^* \oplus \cdots \oplus \eta^*$ if and only if $L_1(x), \ldots, L_p(x)$ are linearly independent at each non-zero $x \in K^{n+1}$. The K-representations of $C\ell_p$ give us precisely p K-linear maps $L_1, \ldots, L_p: K^N \to K^N$ which satisfy this condition of pointwise independence. Thus, by consulting Table III, we have the following result analogous to that of Theorem 7.4 (cf. Lawson-Michelsohn [1]).

Theorem 7.6. *Let $(n+1)\eta^* = \eta^* \oplus \cdots \oplus \eta^*$ denote $(n+1)$-copies of the "hyperplane" bundle over complex projective n-space $\mathbb{P}^n(\mathbb{C})$. If $n + 1 = 2^m(2t+1)$, then there exist k sections of $(n+1)\eta^*$ where*

$$k = 2m - 1.$$

Therefore

$$\gamma\delta_n \leq 2n - 2m + 3$$

where $\gamma\delta_n$ denotes the geometric dimension of the tangent bundle $T\mathbb{P}^n(\mathbb{C})$.

We also have the following result which is analogous to that of Theorem 7.5.

Theorem 7.7. *Let $(n+1)\xi^* = \xi^* \oplus \cdots \oplus \xi^*$ denote $(n+1)$-copies of the "hyperplane" bundle over quaternionic projective n-space $\mathbb{P}^n(\mathbb{H})$. If $n + 1 =$*

$2^m(2t+1)$ *then there exist k sections of* $(n+1)\xi^*$ *where*

$$k = \begin{cases} 2m-3 & \textit{if } m \equiv 0 \pmod 4 \\ 2m-2 & \textit{if } m \equiv 3 \pmod 4 \\ 2m-1 & \textit{otherwise.} \end{cases}$$

Therefore

$$gd_n \leq \begin{cases} 4n-2m+7 & \textit{if } m \equiv 0 \pmod 4 \\ 4n-2m+6 & \textit{if } m \equiv 3 \pmod 4 \\ 4n-2m+5 & \textit{otherwise} \end{cases}$$

where gd_n *denotes the geometric dimension of the bundle* $T\mathbb{P}^n(\mathbb{H}) \oplus \mathrm{Hom}_{\mathbb{H}}(\xi,\xi)$.

Note that we are unable to make a conclusion about the geometric dimension of the tangent bundle itself because $\mathrm{Hom}_{\mathbb{H}}(\xi,\xi)$ is not trivial.

§8. Some Further Applications to the Theory of Lie Groups

It is interesting to note that the classification of Clifford modules gives an immediate proof of certain well-known isomorphisms between low-dimensional Lie groups. It also leads to the fact that $S^7 = \mathrm{Spin}_7/G_2$ and to the principal of triality for Spin_8. We would like to thank Reese Harvey for pointing this out to us.

We begin with some classical definitions. Let $\mathbb{C}^n$ and $\mathbb{H}^n$ carry the standard "hermitian" inner products

$$(x,y) \equiv \sum_{j=1}^{n} x_j \bar{y}_j \tag{8.1}$$

where for a quaternion $x = x_0 + \underline{i}x_1 + \underline{j}x_2 + \underline{k}x_3$, the conjugate is defined by $\bar{x} = x_0 - \underline{i}x_1 - \underline{j}x_2 - \underline{k}x_3$. Then we have the following definitions of the classical groups:

$$\mathrm{U}_n = \{g \in \mathrm{Hom}_{\mathbb{C}}(\mathbb{C}^n,\mathbb{C}^n) : (gx,gy) = (x,y) \text{ for all } x,y \in \mathbb{C}^n\}.$$

$$\mathrm{SU}_n = \{g \in \mathrm{U}_n : \det_{\mathbb{C}}(g) = 1\}$$

$$\mathrm{Sp}_n = \{g \in \mathrm{Hom}_{\mathbb{H}}(\mathbb{H}^n,\mathbb{H}^n) : (gx,gy) = (x,y) \text{ for all } x,y \in \mathbb{H}^n\}$$

Under the natural isometries $\mathbb{R}^{4n} \cong \mathbb{C}^{2n} \cong \mathbb{H}^n$ (and $\mathbb{R}^{2n} \cong \mathbb{C}^n$) one finds that

$$\mathrm{Sp}_n \subset \mathrm{SU}_{2n} \quad \text{and} \quad \mathrm{U}_n \subset \mathrm{SO}_{2n}. \tag{8.2}$$

Elementary linear algebra proves that

$$\begin{cases} \dim(\mathrm{SO}_n) = \frac{1}{2}n(n-1) \\ \dim(\mathrm{SU}_n) = n^2 - 1 \\ \dim(\mathrm{Sp}_n) = n(2n+1), \end{cases} \tag{8.3}$$

and it is not difficult to see that each of these groups is connected.

When the integers p,q,r are sufficiently large (say $\geqq 7$), the groups Spin_p, SU_q, Sp_r are all distinct. However, in low dimensions there are certain exceptional isomorphisms between these groups. These isomorphisms are easily deduced from the Dynkin diagrams. However, the approach requires the non-trivial classification of compact, simply-connected Lie groups. The same isomorphisms can also be deduced from Table II, which was comparatively easy to establish.

Theorem 8.1. *There exist the following isomorphisms between low-dimensional Lie groups:*

$$\mathrm{Spin}_3 \cong \mathrm{SU}_2 \cong \mathrm{Sp}_1$$

$$\mathrm{Spin}_4 \cong \mathrm{Spin}_3 \times \mathrm{Spin}_3$$

$$\mathrm{Spin}_5 \cong \mathrm{Sp}_2$$

$$\mathrm{Spin}_6 \cong \mathrm{SU}_4$$

Furthermore, Spin_7 *has a faithful 8-dimensional real representation, and* Spin_8 *has three inequivalent 8-dimensional real representations.*

Proof. Recall that $\mathrm{Spin}_n \subset C\ell_n^0 \cong C\ell_{n-1}$. Furthermore any representation of $C\ell_{n-1}$ on $\mathbb{C}^N$ or $\mathbb{H}^N$ can be assumed to have the property that, when restricted to the group Spin_n, it preserves the hermitian inner product (8.1). Consequently, since $C\ell_2 \cong \mathbb{H}$ has a faithful one-dimensional quaternionic representation, we get an injection $\mathrm{Spin}_3 \hookrightarrow \mathrm{Sp}_1$. The first isomorphism follows easily. Since $C\ell_3 \cong \mathbb{H} \oplus \mathbb{H}$, we get an injective map $\mathrm{Spin}_4 \hookrightarrow \mathrm{Sp}_1 \times \mathrm{Sp}_1$. Since $\dim(\mathrm{Spin}_4) = \dim(\mathrm{Sp}_1 \times \mathrm{Sp}_1)$ and Spin_4 is connected, we get the second isomorphism. Since $C\ell_4 \cong \mathbb{H}(2)$, we get an injection $\mathrm{Spin}_5 \hookrightarrow \mathrm{Sp}_2$ and the third isomorphism follows. Since $C\ell_5 \cong \mathbb{C}(4)$, we get an injection $\mathrm{Spin}_6 \hookrightarrow \mathrm{U}_4$. By the simplicity of (the Lie algebra of) Spin_6, or by Lemma 8.5 below, we see that Spin_6 must lie in the kernel of the homomorphism $\det_{\mathbb{C}}: \mathrm{U}_4 \to \mathrm{U}_1$. Thus, we have $\mathrm{Spin}_6 \hookrightarrow \mathrm{SU}_4$ and the fourth isomorphism follows.

The existence of a representation of Spin_7 on $\mathbb{R}^8$ is obvious since $C\ell_6 \cong \mathbb{R}(8)$. Furthermore, since $C\ell_7 \cong \mathbb{R}(8) \oplus \mathbb{R}(8)$, we see that the two spin representations Δ_8^+ and Δ_8^- of Spin_8 are on $\mathbb{R}^8$. There is also the adjoint representation Ad of Spin_8 on $\mathbb{R}^8$. To see that these representations are all distinct, it suffices to consider the central elements. Set

$$\mathscr{Z} = \{1, -1, \omega, -\omega\} \cong \mathbb{Z}_2 \oplus \mathbb{Z}_2$$

where ω denotes the oriented volume element of $C\ell_7 \cong C\ell_8^0$. This group lies in the center of Spin_8. (In fact it is the center.) From Propositions 5.9 and 5.10 we know that $\Delta_8^+(\omega) = \mathrm{Id}$ and $\Delta_8^-(\omega) = -\mathrm{Id}$. Since $\Delta_8^\pm$ come from representations of $C\ell_7$ we have

$$\Delta_8^+(\omega) = \mathrm{Id}, \qquad \Delta_8^+(-\omega) = -\mathrm{Id}, \qquad \Delta_8^+(-1) = -\mathrm{Id} \tag{8.4}$$

$$\Delta_8^-(\omega) = -\mathrm{Id}, \qquad \Delta_8^-(-\omega) = \mathrm{Id}, \qquad \Delta_8^-(-1) = -\mathrm{Id}. \tag{8.5}$$

From the definition it is clear that if Ad denotes the adjoint representation of $C\ell_8$ on $\mathbb{R}^8$, restricted to $\mathrm{Spin}_8 \subset C\ell_8^0 \cong C\ell_7$, then

$$\mathrm{Ad}(-1) = \mathrm{Id}.$$

Recall now that equivalent irreducible representations must agree on central elements. (Note, for example, that if there exists an isomorphism $F: \mathbb{R}^8 \to \mathbb{R}^8$ with the property that $F \circ \Delta_8^+(g) \circ F^{-1} = \Delta_8^-(g)$ for all $g \in \mathrm{Spin}_8$, then in particular, $\Delta_8^+(g) = \Delta_8^-(g)$ for all $g \in \mathscr{Z}$.) Consequently the three representations Δ_8^+, Δ_8^- and Ad are inequivalent. ∎

It is an interesting and often useful fact that two 8-dimensional representations of $C\ell_7$ can be explicitly generated using the Cayley numbers. Recall that the **Cayley numbers** $\mathbb{O}$ can be defined as pairs of quaternions with multiplication given by

$$(a,b) \cdot (c,d) = (ac - \bar{d}b,\ da + b\bar{c}). \tag{8.6}$$

The multiplication so defined is neither commutative nor associative. However, every non-zero element has a multiplicative inverse. Furthermore, given a Cayley number $x = (a,b)$, we write $\bar{x} = (\bar{a}, -b)$ and define real and imaginary parts of x by setting

$$\mathrm{Re}(x) = \tfrac{1}{2}(x + \bar{x}); \qquad \mathrm{Im}(x) = \tfrac{1}{2}(x - \bar{x})$$

An inner product on $\mathbb{O}$ is defined by $\langle x,y \rangle = \mathrm{Re}(x\bar{y})$. It has the property that $|xy| = |x|\,|y|$ for all $x,y \in \mathbb{O}$ (where $|x|^2 = \langle x,x \rangle$ as usual). An important fact concerning the Cayley numbers is that *any subalgebra of* $\mathbb{O}$ *generated by two elements is associative.*

We now consider $\mathbb{R}^7 = \mathrm{Im}(\mathbb{O})$ and $\mathbb{R}^8 = \mathbb{O}$ with the above inner product. For any $v \in \mathrm{Im}(\mathbb{O})$ we define a linear endomorphism λ_v of $\mathbb{R}^8$ by setting

$$\lambda_v(x) = v \cdot x \tag{8.7}$$

for $x \in \mathbb{O} = \mathbb{R}^8$. From the associativity of the algebra generated by x and v, or by a direct computation from (8.6), we see that

$$\lambda_v^2 = -\|v\|^2 \mathrm{Id} \tag{8.8}$$

for all $v \in \mathrm{Im}(\mathbb{O}) = \mathbb{R}^7$. Consequently, from the universal property (Proposition 1.1), we know that λ extends to a representation

$$\lambda : C\ell_7 \longrightarrow \mathrm{Hom}_{\mathbb{R}}(\mathbb{R}^8, \mathbb{R}^8) \tag{8.9}$$

of $C\ell_7$. For dimensional reasons this representation must be irreducible. The other irreducible 8-dimensional representation

$$\rho = \lambda \circ \alpha$$

is generated by the mapping $\rho_v(x) = -v \cdot x$. Observe that under the conjugation map $c(x) = \bar{x}$, ρ_v becomes equivalent to right multiplication by v, i.e., $\tilde{\rho}_v = c \circ \rho_v \circ c$ is given by

$$\tilde{\rho}_v(x) = \overline{(-v \cdot \bar{x})} = -\bar{\bar{x}} \cdot \bar{v} = x \cdot v$$

(since for $v \in \mathrm{Im}\, \mathbb{O}$ we have $\bar{v} = -v$).

These two representations are equivalent when restricted to Spin_7, but they are inequivalent on $\mathrm{Spin}_8 \subset C\ell_8^0 \cong C\ell_7$.

We now consider the action of Pin_7 on $\mathbb{R}^8$ given by the representation λ above. From (8.7) it is clear that the orbit of 1 contains all elements $e \in \mathrm{Im}(\mathbb{O})$ with $|e| = 1$. That is, this orbit contains the "equator" $S^6 = S^7 \cap \mathrm{Im}(\mathbb{O})$. It also contains $e \cdot S^6$, which is a great sphere passing through the "north pole" 1, for each $e \in S^6$.

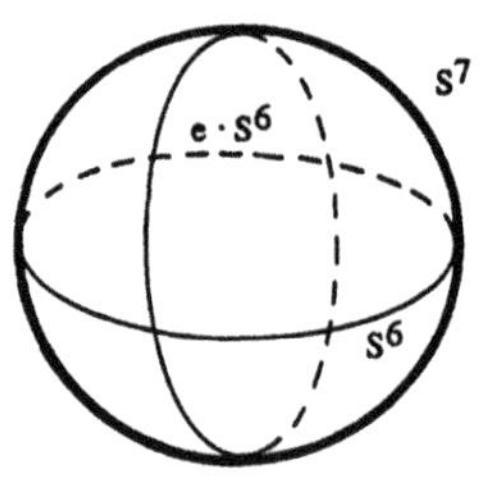

Since the orbit $\mathrm{Pin}_7 \cdot 1$ is a compact embedded submanifold of S^7, we conclude that it is S^7. It now follows that the orbit of Spin_7 must be 7-dimensional and, hence, also equal to S^7. We have proved the result of A. Borel.

Theorem 8.2. *The 8-dimensional spin representation of* Spin_7 *is transitive on the unit sphere* S^7.

With a little more work it is possible to prove that the isotropy subgroup $\{g \in \mathrm{Spin}_7 : \lambda_g(1) = 1\}$ of $1 \in S^7$ is exactly the group $G_2 \equiv \mathrm{Aut}(\mathbb{O})$ (see Harvey-Lawson [3] for example). Hence, we have the diffeomorphism

$$S^7 \cong \mathrm{Spin}_7/G_2.$$

We pass on now to the group Spin_8. Recall from Theorem 8.1 that this group has three distinct homomorphisms $\Delta_8^+, \Delta_8^-, \mathrm{Ad} : \mathrm{Spin}_8 \to \mathrm{SO}_8$ with kernels isomorphic to $\mathbb{Z}_2$. Writing the center $\mathscr{Z} = \{1, -1, \omega, -\omega\}$ as before, we have the following table (cf. (8.4) and (8.5)).

g	ω	$-\omega$	-1
$\Delta_8^+(g)$	Id	$-$Id	$-$Id
$\Delta_8^-(g)$	$-$Id	Id	$-$Id
$\mathrm{Ad}(g)$	$-$Id	$-$Id	Id

(8.10)

We now consider the pair of homomorphisms Δ_8^+ and Δ_8^-. From general covering space theory there must be a lifting of Δ_8^- over Δ_8^+ which takes the identity to itself, i.e., a map $\sigma : \mathrm{Spin}_8 \to \mathrm{Spin}_8$ such that the diagram

$$\begin{array}{ccc} & & \mathrm{Spin}_8 \\ & \overset{\sigma}{\nearrow} & \downarrow{\scriptstyle \Delta_8^+} \\ \mathrm{Spin}_8 & \xrightarrow[\Delta_8^-]{} & \mathrm{SO}_8 \end{array} \tag{8.11}$$

commutes and $\sigma(1) = 1$. The map σ satisfies the condition

$$\sigma(g_1 g_2) = \sigma(g_1)\sigma(g_2) \tag{8.12}$$

for all g_1, g_2 in a small neighborhood of 1, since both Δ_8^+ and Δ_8^- are group homomorphisms. In fact, the relation (8.12) holds for all $g_1, g_2 \in \mathrm{Spin}_8$. To see this note that both sides of (8.12) are well defined on $\mathrm{Spin}_8 \times \mathrm{Spin}_8$ and that the set where (8.12) holds is both open and closed. (Alternatively, one could use the fact that (8.12) is an equation between real analytic maps.) Thus, σ is a group homomorphism. Since σ is a covering map between simply-connected spaces, it must be injective, and so σ is a group automorphism.

It is now clear from (8.11) that σ carries the kernel of Δ_8^- onto the kernel of Δ_8^+. Since $\ker(\Delta_8^-) = \{1, -\omega\}$ and $\ker(\Delta_8^+) = \{1, \omega\}$, and since $\sigma(1) = 1$, we conclude that

$$\sigma(-\omega) = \omega$$

Since ω and $-\omega$ are central, σ must be an *outer automorphism.*

Lifting Ad over the homomorphisms Δ_8^+ and Δ_8^- and applying the arguments above we construct outer automorphisms τ^+ and τ^- of Spin_8 with the property that

$$\tau^+(-1) = \omega \quad \text{and} \quad \tau^-(-1) = -\omega. \tag{8.13}$$

By definition we have that $\Delta_8^\pm \circ \tau^\pm = \mathrm{Ad}$ and $\Delta_8^+ \circ \sigma = \Delta_8^-$. Furthermore, the associated Lie algebra maps $(\Delta_8^\pm)_*$ and Ad_* are isomorphisms, and we have that

$$\tau_*^\pm = (\Delta_*^\pm)^{-1} \circ \mathrm{Ad}_* \quad \text{and} \quad \tau_* = (\Delta_8)_*^{-1} \circ (\Delta_8^-)_*. \tag{8.14}$$

At this point the automorphisms σ and $\tau^\pm$ are only defined modulo inner automorphisms. To get concrete representatives we must choose concrete representatives for the maps $\Delta_8^\pm$ and Ad.

We shall give such an explicit construction for σ. Choose an orthonormal basis $e_1, \ldots, e_8$ of $\mathbb{R}^8$ and recall that the Lie algebra $\mathfrak{spin}_8 = \Lambda^2\mathbb{R}^8$ has an orthonormal basis $\{e_i \cdot e_j\}_{i<j}$. The map $C\ell_7 \xrightarrow{\approx} C\ell_8^0$ is induced by the assignment

$$e_j \longrightarrow e_j e_8 \qquad j = 1, \ldots, 7. \tag{8.15}$$

Consequently, the preimage of $\mathfrak{spin}_8$ under this map is just

$$\mathfrak{spin}_8 \cong \mathbb{R}^7 \oplus \Lambda^2\mathbb{R}^7 \subset C\ell_7 \tag{8.16}$$

where $e_i \mapsto e_i e_8$ for $1 \le i \le 7$ and where $e_i e_j \mapsto e_i e_8 e_j e_8 = e_i e_j$ for $1 \leqq i < j \leqq 7$.

We now consider the two representations λ and ρ of $C\ell_7$ on $\mathbb{R}^8 = \mathbb{O}$ that were constructed above. Since $\lambda_v(x) = v \cdot x$ and $\rho_v(x) = -v \cdot x$ for $v \in \mathbb{R}^7$, we see that

$$\begin{cases} \lambda_\varphi = -\rho_\varphi & \text{for } \varphi \in \mathbb{R}^7 \\ \lambda_\varphi = \rho_\varphi & \text{for } \varphi \in \Lambda^2\mathbb{R}^7. \end{cases} \tag{8.17}$$

Now restricted to $\mathfrak{spin}_8 \cong \mathbb{R}^7 \oplus \Lambda^2\mathbb{R}^7$ we have that $\lambda = (\Delta_8^+)_*$ and $\rho = (\Delta_8^-)_*$. Consequently, $\sigma_* = (\Delta_8^+)_*^{-1} \circ (\Delta_8^-)_*$ has the property that

$$\sigma_*(\varphi) = \begin{cases} -\varphi & \text{if } \varphi \in \mathbb{R}^7 \\ \varphi & \text{if } \varphi \in \Lambda^2\mathbb{R}^7. \end{cases} \tag{8.18}$$

In particular, $(\sigma_*)^2 = \mathrm{Id}$ and so by exponentiation we have that

$$\sigma^2 = \mathrm{Id}. \tag{8.19}$$

We now let O_8 and I_8 denote respectively the groups of outer and inner automorphisms of Spin_8. There is a natural homomorphism

$$O_8/I_8 \longrightarrow \mathrm{Aut}(\mathscr{Z}). \tag{8.20}$$

It is easy to see that $\mathrm{Aut}(\mathscr{Z}) \cong \mathrm{Aut}(\mathbb{Z}_2 \oplus \mathbb{Z}_2) \cong S_3$. In fact, we have naturally that $\mathrm{Aut}(\mathscr{Z}) \cong \mathrm{Perm}(-1, \omega, -\omega)$. Since $\sigma(-\omega) = \omega$ and $\sigma(\mathscr{Z}) = \mathscr{Z}$, we have that either $\sigma(-1) = -1$ or $\sigma(-1) = -\omega$. Since $\sigma^2 = \text{Id}$, we conclude that

$$\sigma(-1) = -1 \quad \text{and} \quad \sigma(\omega) = -\omega.$$

This information, together with (8.13), easily proves the following:

Theorem 8.3. *The homomorphism* $O_8/I_8 \to \text{Aut}(\mathscr{X}) \cong S_3$ *is surjective. In particular, since* O_8/I_8 *is a finite group* (cf. *Helgason* [1]) *there exists an element in* O_8/I_8 *of order three.*

Those readers similar with the classification of Lie groups know that this map $O_8/I_8 \to S_3$ is reflected in the representation of O_8/I_8 as the group of symmetries of the Dynkin diagram of Spin_8. This representation is faithful, and so $O_8/I_8 \cong S_3$.

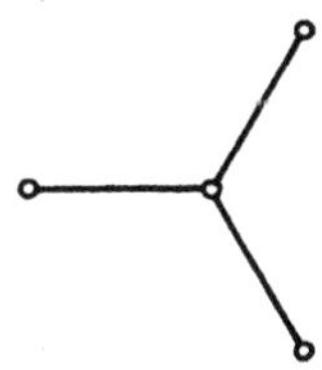

It can be shown that in fact there exists a non-trivial element $A \in O_8$ such that

$$A^3 = \text{Id}.$$

This element is called the **triality automorphism.** Continuing with calculations as above, this automorphism can be constructed explicitly.

It has most likely occurred to the reader that the methods of Theorem 8.1 can be applied more generally to the groups $\text{Spin}_{r,s}$ for any r and s. Indeed, this does produce further exceptional isomorphisms between low-dimensional Lie groups. We recall the following classical groups.

$$\text{SL}_n(\mathbb{R}) = \{g \in \text{Hom}_{\mathbb{R}}(\mathbb{R}^n,\mathbb{R}^n) : \det_{\mathbb{R}}(g) = 1\}$$

$$\text{SL}_n(\mathbb{C}) = \{g \in \text{Hom}_{\mathbb{C}}(\mathbb{C}^n,\mathbb{C}^n) : \det_{\mathbb{C}}(g) = 1\}$$

$$\text{SL}_n(\mathbb{H}) = \{g \in \text{Hom}_{\mathbb{H}}(\mathbb{H}^n,\mathbb{H}^n) : \det_{\mathbb{C}}(g) = 1\}$$

For the last definition we have fixed a presentation $\mathbb{H}^n = (\mathbb{C}^{2n},J)$ where $J:\mathbb{C}^{2n} \to \mathbb{C}^{2n}$ is $\mathbb{C}$-antilinear and $J^2 = -\text{Id}$. We observe that the complex determinant is in fact real-valued on $\text{Hom}_{\mathbb{H}}(\mathbb{H}^n,\mathbb{H}^n) = \{g \in \text{Hom}_{\mathbb{C}}(\mathbb{C}^{2n},\mathbb{C}^{2n}): g \circ J = J \circ g\}$. To see this, let $c:\mathbb{C}^{2n} \to \mathbb{C}^{2n}$ denote complex conjugation and set $c(g) = c \circ g \circ c$. Then since $c^2 = 1$ and since g and J commute, we have $\det_{\mathbb{C}}(c(g) \circ c \circ J) = \det_{\mathbb{C}}(c \circ g \circ J) = \det_{\mathbb{C}}(c \circ J \circ g)$. Hence, $\det_{\mathbb{C}}(c(g)) = \det_{\mathbb{C}}(g)$ and since $\det_{\mathbb{C}}(c(g)) = \overline{\det_{\mathbb{C}}(g)}$, the determinant is real.

Now, each of the groups $SL_n(K)$ is connected, and elementary linear algebra shows that

$$\begin{cases} \dim_{\mathbb{R}}(SL_n(\mathbb{R})) = n^2 - 1 \\ \dim_{\mathbb{R}}(SL_n(\mathbb{C})) = 2(n^2 - 1) \\ \dim_{\mathbb{R}}(SL_n(\mathbb{H})) = 4n^2 - 1. \end{cases} \tag{8.21}$$

Recall that for $r,s > 0$, the group $\mathrm{Spin}_{r,s}$ has two connected components. We let $\mathrm{Spin}^0_{r,s}$ denote the component containing the identity. It is obvious that there is an isomorphism

$$\mathrm{Spin}_{r,s} \cong \mathrm{Spin}_{s,r},$$

and an elementary calculation shows that

$$\dim(\mathrm{Spin}_{r,s}) = \frac{1}{2}(r + s)(r + s - 1) \tag{8.22}$$

for all r,s. A careful look at Table II (p. 29) now proves the following. Let $\widetilde{SL}_n(\mathbb{R}) \to SL_n(\mathbb{R})$ denote the (2-sheeted) universal covering group for $n \geqq 3$.

Theorem 8.4. *There exist the following isomorphisms between low-dimensional Lie groups.*

$$\begin{aligned} \mathrm{Spin}^0_{2,1} &\cong SL_2(\mathbb{R}) \\ \mathrm{Spin}^0_{3,1} &\cong SL_2(\mathbb{C}) \\ \mathrm{Spin}^0_{5,1} &\cong SL_2(\mathbb{H}) \\ \mathrm{Spin}^0_{2,2} &\cong SL_2(\mathbb{R}) \times SL_2(\mathbb{R}) \\ \mathrm{Spin}^0_{3,3} &\cong \widetilde{SL}_4(\mathbb{R}) \end{aligned}$$

Furthermore $\mathrm{Spin}_{4,4}$ *has three inequivalent 8-dimensional real representations.*

Proof. Recall from Theorem 3.7 that for $r \geqq 1$ we have

$$\mathrm{Spin}_{r,s} \subset C\ell^0_{r,s} \cong C\ell_{r-1,s}.$$

Consequently, from Table II, we have the following embeddings: $\mathrm{Spin}_{2,1} \subset GL_2(\mathbb{R})$, $\mathrm{Spin}_{3,1} \subset GL_2(\mathbb{C})$, $\mathrm{Spin}_{5,1} \subset GL_2(\mathbb{H}) \times GL_2(\mathbb{H})$, $\mathrm{Spin}_{2,2} \subset GL_2(\mathbb{R}) \times GL_2(\mathbb{R})$ and $\mathrm{Spin}_{3,3} \subset GL_4(\mathbb{R}) \times GL_4(\mathbb{R})$, where $GL_n(K)$ denotes the set of invertible elements in $\mathrm{Hom}_K(K^n,K^n)$. We now observe that in each of these embeddings the identity component $\mathrm{Spin}^0_{r,s}$ is actually contained in the subgroup $SL_n(K)$ or in $SL_n(K) \times SL_n(K)$ for the latter cases. This is an immediate consequence of the following lemma.

Lemma 8.5. *Let* $\rho: C\ell_{r-1,s} \to \text{Hom}_K(V,V)$ *be any* K*-representation for* $K = \mathbb{R}$ *or* $\mathbb{C}$. *If* $r + s \geqq 3$, *then*

$$\det_K(\rho(g)) = \pm 1$$

for all $g \in \text{Spin}_{r,s} \subset C\ell_{r-1,s}$.

Proof. If $r + s \geqq 3$, then every element $g \in \text{Spin}_{r,s}$ can be written as a product $g = g_1 \cdots g_m$ where $g_j \in \text{Spin}_{r,s}$ satisfies

$$g_j^2 = \pm 1 \tag{8.23}$$

for all j. To see this, write $g = v_1 \cdots v_{2m}$ where $v_j \in \mathbb{R}^{r+s}$ and $q(v_j) = \pm 1$. Write $v_1 v_2 = \mp v_1 v v v_2$ where $v \in \mathbb{R}^{r+s}$ satisfies $q(v) = \pm 1$ and where $q(v,v_1) = q(v,v_2) = 0$. Set $g_1 = v_1 v$ and $g_2 = \mp v v_2$. Then $g_1^2 = v_1 v v_1 v = -v_1^2 v^2 = \pm 1$, and similarly, $g_2^2 = \pm 1$. Of course $v_1 v_2 = g_1 g_2$. Continuing in this manner proves our claim.

Since $\dim_K(V)$ is even, we see that for $g = g_1 \cdots g_m$ as above, we have

$$\begin{aligned}[\det_K(\rho g)]^2 &= \left[\prod_j \det_K(\rho g_j)\right]^2 = \prod_j \det_K(\rho g_j^2) \\ &= \prod_j \det_K(\pm \text{Id}) = 1. \quad \blacksquare\end{aligned}$$

The first, second and fourth isomorphisms of Theorem 8.4 now follow from dimension considerations (cf. (8.21) and (8.22)).

For the fifth isomorphism, consider the image of $\text{Spin}_{3,3}^0$ under the two projections $\text{SL}_4(\mathbb{R}) \times \text{SL}_4(\mathbb{R}) \rightrightarrows \text{SL}_4(\mathbb{R})$. At least one of these must be non-trivial and, therefore, locally injective by the simplicity of the Lie algebra. Since $\dim(\text{Spin}_{3,3}) = \dim(\text{SL}_4(\mathbb{R})) = 15$, this projection is a covering map $\text{Spin}_{3,3}^0 \to \text{SL}_4(\mathbb{R})$. Since $\text{Spin}_{3,3}^0$ is simply connected, it lifts to an isomorphism $\text{Spin}_{3,3}^0 \xrightarrow{\approx} \widetilde{\text{SL}}_4(\mathbb{R})$. The third isomorphism is proved similarly.

Looking more closely we can see that the two projections $\text{Spin}_{3,3}^0 \rightrightarrows \text{SL}_4(\mathbb{R})$ are both non-trivial and inequivalent. To prove this we consider the volume element $\omega = e_1 \cdots e_6$ where $e_1^2 = e_2^2 = e_3^2 = -e_4^2 = -e_5^2 = -e_6^2 = -1$ (see Proposition 3.3). This element is central in $\text{Spin}_{3,3}$ and satisfies $\omega^2 = 1$. It clearly lies in $\text{Spin}_3 \times \text{Spin}_3 \subset \text{Spin}_{3,3}$ and is therefore connected to the identity. The module $V \cong \mathbb{R}^8$ for $C\ell_{3,3}$ decomposes as $V \cong V^+ \oplus V^-$ where $V^\pm = (1 \pm \omega)V$ are invariant subspaces for $\text{Spin}_{3,3}$. Since $\omega = \pm\text{Id}$ on $V^\pm$, we see that the representations are inequivalent. They are each non-trivial, since otherwise we would have the identity $\omega = 1$ (or $\omega = -1$) in $\text{Spin}_{3,3}$, which is clearly false.

To prove the final statement we again consider the volume element $\omega = e_1 \cdots e_8$ in $\text{Spin}_{4,4}$. As above we see that ω is central in $\text{Spin}_{4,4}$ and is connected to the identity. Since $\omega^2 = 1$, the module $W \cong \mathbb{R}^{16}$ for $C\ell_{4,4}$

decomposes as $W = W^+ \oplus W^-$ where $W^\pm = (1 \pm \omega)W$ are each invariant under $\mathrm{Spin}_{4,4}$. Let $\Delta^\pm$ denote these two 8-dimensional representations, and consider the central subgroup

$$\mathscr{Z} = \{1, -1, \omega, -\omega\} \cong \mathbb{Z}_2 \oplus \mathbb{Z}_2$$

in $\mathrm{Spin}_{4,4}$. Then Δ^+, Δ^- and Ad have precisely the values given in the table (8.10) derived above for the case of Spin_8. It follows that these three 8-dimensional representations of $\mathrm{Spin}_{4,4}$ are distinct. ∎

The above arguments actually prove slightly more. Let

$$\mathrm{SL}_n^*(K) = \{g \in \mathrm{Hom}_K(K^n, K^n) : \det_{K'}(g) = \pm 1\}$$

where $K' = K$ if $K = \mathbb{R}$ or $\mathbb{C}$ and $K' = \mathbb{C}$ if $K = \mathbb{H}$. There is a short exact sequence

$$1 \longrightarrow \mathrm{SL}_n(K) \longrightarrow \mathrm{SL}_n^*(K) \xrightarrow{\det_{K'}} \mathbb{Z}_2 \longrightarrow 0.$$

From the proof of Theorem 8.4 we can actually conclude that

$$\mathrm{Spin}_{2,1} \cong \mathrm{SL}_2^*(\mathbb{R})$$
$$\mathrm{Spin}_{3,1} \cong \mathrm{SL}_2^*(\mathbb{C})$$
$$\mathrm{Spin}_{5,1} \cong \mathrm{SL}_2^*(\mathbb{H})$$

Note. For further results of the type given in Theorem 8.4, the reader is referred to the beautiful book of Reese Harvey [1].

§9. *K*-Theory and the Atiyah-Bott-Shapiro Construction

In this section we shall present the essentials of K-theory in a fashion which will be useful later in our discussion of the Atiyah-Singer Index Theorem (Chap. III). Following these basics we shall present the construction of Atiyah, Bott and Shapiro which relates the Grothendieck groups of real Clifford modules to the KO-theory of spheres and that of complex Clifford modules to the K-theory of spheres. Their fundamental isomorphisms explicitly identify Bott periodicity with the periodicity phenomena in the theory of Clifford algebras. For an elaboration of the details presented here, the reader is referred to Atiyah-Bott-Shapiro [1] and Karoubi [2].

Throughout this section all spaces will be assumed to be compact. If X is any such space, we denote by $V(X)$ the set of all isomorphism classes of complex vector bundles over X. The set $V(X)$ is an abelian semigroup if we define addition by direct sum. We let $F(X)$ be the free abelian group generated by the elements of $V(X)$ and let $E(X)$ be the subgroup of $F(X)$

generated by elements of the form $[V] + [W] - ([V] \oplus [W])$ where $+$ denotes addition in $F(X)$ and $\oplus$ denotes addition in $V(X)$.

DEFINITION 9.1. The **K-group** of X is defined to be the quotient

$$K(X) = F(X)/E(X).$$

Note that $K(X)$ is an abelian group. The elements of $K(X)$ are called **virtual bundles.**

If V and W are bundles over X and Y respectively then $V \otimes W$ is a bundle over $X \times Y$. When $X = Y$ the diagonal map $\Delta: X \to X \times X$ can be used to define an **interior tensor product** in $K(X)$ by

$$[u] \cdot [v] \equiv \Delta^*[u \otimes v]. \tag{9.1}$$

This gives us

Proposition 9.2. *The group $K(X)$ has a ring structure with multiplication given by* (9.1).

Let $\alpha: V(X) \to K(X)$ be the composition $V(X) \hookrightarrow F(X) \to F(X)/E(X)$.

Proposition 9.3. *If G is any abelian group and $f: V(X) \to G$ any semi-group homomorphism, then there is a unique homomorphism $\tilde{f}: K(X) \to G$* such that $\tilde{f}\alpha = f$.

Proposition 9.4. *$K(X)$ is the universal group with respect to maps of the type described in Proposition 9.3*

Suppose that $f: X \to Y$ is a continuous map and consider the map $f^*: V(Y) \to V(X)$ given by the induced bundle construction. Since this is a semi-group homomorphism, it descends to a homomorphism f^*: $K(Y) \to K(X)$. One easily checks that K thereby becomes a contravariant functor from the category of compact spaces to the category of abelian groups.

Suppose now that S is any abelian semi-group with unit and let $\Delta: S \to S \times S$ be the diagonal map. This is a semi-group homomorphism. If we let $\mathscr{K}(S)$ be the set of cosets of $\Delta(S)$ in $S \times S$, then $\mathscr{K}(S)$ is a quotient semi-group. Since the interchange of factors in $S \times S$ induces an inverse in $\mathscr{K}(S)$, $\mathscr{K}(S)$ is actually a group.

DEFINITION 9.5. $\mathscr{K}(X)$ is defined to be $\mathscr{K}(V(X))$.

Proposition 9.6. *$\mathscr{K}(X)$ is isomorphic to $K(X)$.*

Proof. For an abelian semi-group S with 0 we define $\beta_S: S \to \mathscr{K}(S)$ to be $s \mapsto (s,0)$ followed by the natural projection $S \times S \to \mathscr{K}(S)$. If $g: S \to T$

is a semi-group homomorphism, then there is induced a map $\mathscr{K}(g):\mathscr{K}(S)\to\mathscr{K}(T)$ so that $\mathscr{K}(g)\circ\beta_S=\beta_T\circ g$. Now let S be $V(X)$ and T be any abelian group. Then β_T is an isomorphism. This shows that $\mathscr{K}(X)$ is universal with respect to semi-group homomorphisms from $V(X)$ to abelian groups. ∎

Corollary 9.7. *Every element of* $K(X)$ *can be represented in the form* $[V]-[W]$ *where* $[V],[W]\in V(X)$.

Lemma 9.8. *Let* $\pi: V\to X$ *be a vector bundle over a compact Hausdorff space* X. *Then for some* N *there is a continuous map* $f: V\to\mathbb{C}^N$ *which is injective and linear on each fibre.*

Proof. Cover X with a finite number of open sets $U_1,\dots,U_r$ over which there exist trivializations $\alpha_j:\pi^{-1}(U_j)\xrightarrow{\approx}U_j\times\mathbb{C}^k$ and set $a_j=pr_j\circ\alpha_j$ where $pr_j: U_j\times\mathbb{C}^k\to\mathbb{C}^k$ is projection. Let $\{\psi_j\}_{j=1}^r$ be a partition of unity subordinate to $\{U_j\}_{j=1}^r$. Then $f\equiv(\psi_1a_1)\oplus\cdots\oplus(\psi_ra_r): V\to\mathbb{C}^k\oplus\cdots\oplus\mathbb{C}^k$ is the desired map. ∎

Corollary 9.9. *To each bundle* V *over* X *there exists a "complementary" bundle* $V^\perp$ *such that* $V\oplus V^\perp$ *is trivial. Hence, each element of* $K(X)$ *can be represented in the form* $[V]-[\tau^N]$, *where* τ^N *is the trivial bundle over* X *of dimension* N.

Proof. Let $f: V\to\mathbb{C}^N$ be the map in Lemma 9.8 and define $V^\perp=\bigcup_{x\in X}f(V_x)^\perp$. Clearly $V\oplus V^\perp\cong X\times\mathbb{C}^N$. Hence, $[V]=[\tau^N]-[V^\perp]$ in $K(X)$ and an arbitrary element $[W]-[V]$ in $K(X)$ can be rewritten as $[W\oplus V^\perp]-[\tau^N]$. ∎

DEFINITION 9.10. We define the **real *K*-ring** for X, $KO(X)$, just as we defined $K(X)$ by replacing $V(X)$ by $V_{\mathbb{R}}(X)$, the set of isomorphism classes of real vector bundles. The same construction and considerations apply. Analogues of the following definitions can also be made for KO-theory. We will assume them without specifically stating them.

We would now like to use the K-groups to define a generalized cohomology theory. For this reason we now consider X to be a space with a distinguished basepoint, $\text{pt}\in X$.

DEFINITION 9.11. The **reduced *K*-ring,** $\tilde{K}(X)$, is defined to be the kernel of the natural projection $K(X)\to K(\text{pt})\cong\mathbb{Z}$, so that $\tilde{K}(X)$ is an ideal of $K(X)$.

In fact the exact sequence

$$0\longrightarrow\tilde{K}(X)\longrightarrow K(X)\longrightarrow K(\text{pt})\longrightarrow 0\tag{9.2}$$

splits in an obvious fashion.

DEFINITION 9.12. Suppose Y is a non-empty closed subset of X. Then we define the **relative K-groups** as follows:

$$K(X,Y) \equiv \tilde{K}(X/Y)$$

where X/Y is taken to have Y as its basepoint.

If Y is empty, X/Y is defined to be the space

$$X^+ \equiv X \cup \{\widetilde{\mathrm{pt}}\} \tag{9.3}$$

where $\widetilde{\mathrm{pt}}$ is a disjoint point which will play the role of basepoint.

OBSERVATION 9.13. On a non-basepointed space X we have the identification

$$K(X) \approx \tilde{K}(X^+) = K(X, \varnothing).$$

DEFINITION 9.14. We define the **wedge** $X \vee Y$ and the **smash product** $X \wedge Y$ of two spaces X,Y with basepoints pt_X and pt_Y by

$$X \vee Y \equiv (X \times \mathrm{pt}_Y) \cup (\mathrm{pt}_X \times Y) \subset X \times Y$$

$$X \wedge Y \equiv X \times Y/X \vee Y.$$

We also define the **(reduced) suspension** $\Sigma(X)$ of X by

$$\Sigma(X) \equiv S^1 \wedge X.$$

Iterating this i times gives us the **i-fold suspension** $\Sigma^i(X)$. For all i there is a homeomorphism $\Sigma^i(X) \approx S^i \wedge X$.

DEFINITION 9.15. When X is a compact basepointed space, or when (X,Y) is a compact pair, we define

$$\tilde{K}^{-i}(X) \equiv \tilde{K}(\Sigma^i(X))$$

$$K^{-i}(X,Y) \equiv \tilde{K}^{-i}(X/Y) \equiv \tilde{K}(\Sigma^i(X/Y)).$$

For spaces X which are not necessarily basepointed we define, in the spirit of 9.13,

$$K^{-i}(X) \equiv K^{-i}(X, \varnothing) \equiv \tilde{K}(\Sigma^i(X^+)).$$

Since the functor K is representable (see Chap. III, Theorem 8.6) there is an exact sequence for basepointed pairs (X,Y)

$$K(X,Y) \longrightarrow \tilde{K}(X) \longrightarrow \tilde{K}(Y)$$

which we may now extend to a Barratt-Puppe sequence (Barratt [1]):

$$\begin{aligned} \cdots \longrightarrow \tilde{K}^{-i-1}(Y) \stackrel{\delta}{\longrightarrow} K^{-i}(X,Y) \longrightarrow \tilde{K}^{-i}(X) \longrightarrow \\ \tilde{K}^{-i}(Y) \longrightarrow \cdots \longrightarrow \tilde{K}^0(X) \longrightarrow \tilde{K}^0(Y). \end{aligned} \tag{9.4}$$

We write $\tilde{K}^{-*}$ for the graded functor $\tilde{K}^{-i}$, $i \geq 0$.

REMARK 9.16. If Y is a retract of X, then for all i

$$0 \longrightarrow K^{-i}(X,Y) \longrightarrow \tilde{K}^{-i}(X) \longrightarrow \tilde{K}^{-i}(Y) \longrightarrow 0$$

is a split short exact sequence and $\tilde{K}^{-i}(X) \cong K^{-i}(X,Y) \oplus \tilde{K}^{-i}(Y)$. This follows from the Barratt-Puppe sequence (9.4).

Now if X and Y are spaces with basepoints then

$$\tilde{K}^{-i}(X \times Y) \cong \tilde{K}^{-i}(X \wedge Y) \oplus \tilde{K}^{-i}(X) \oplus \tilde{K}^{-i}(Y)$$

since X is a retract of $X \times Y$ and Y is a retract of $X \times Y/X$.

We would like now to use the ring structure on $K(X)$ to enable us to define a ring structure on $K^{-*}(X)$.

Proposition 9.17. *Given X and Y and $i,j \geq 0$ there is a pairing $\tilde{K}^{-i}(X) \otimes \tilde{K}^{-j}(Y) \to \tilde{K}^{-i-j}(X \wedge Y)$ which is given by tensor product.*

Proof. For a bundle E on $S^i \wedge X$ and a bundle F on $S^j \wedge Y$ we have the tensor product bundle $E \otimes F$ on $(S^i \wedge X) \times (S^j \wedge Y)$. This induces a pairing

$$\tilde{K}(S^i \wedge X) \otimes \tilde{K}(S^j \wedge Y) \longrightarrow \tilde{K}((S^i \wedge X) \times (S^j \wedge Y)).$$

But $\tilde{K}((S^i \wedge X) \wedge (S^j \wedge Y))$ is the kernel of $\tilde{K}((S^i \wedge X) \times (S^j \wedge Y)) \to \tilde{K}(S^i \wedge X) \oplus \tilde{K}(S^j \wedge Y)$. So we have a pairing $\tilde{K}(S^i \wedge X) \otimes \tilde{K}(S^j \wedge Y) \to \tilde{K}((S^i \wedge X) \wedge (S^j \wedge Y)) = \tilde{K}((S^{i+j}) \wedge X \wedge Y)$ as desired. ∎

Replacing X by X^+ and Y by Y^+ in Proposition 9.17 gives us a pairing $K^{-i}(X) \otimes K^{-j}(Y) \to K^{-i-j}(X \times Y)$. We easily conclude the following.

Corollary 9.18. *The pairing of Proposition 9.17 makes $K^{-*}(\mathrm{pt})$ a graded ring. Furthermore, for any basepointed space (X,pt), this pairing makes $K^{-*}(X)$ a graded module over $K^{-*}(\mathrm{pt})$.*

Thus far everything we have said for K-theory holds equally true for KO-theory. We will now describe periodicity and at this point the descriptions must diverge. We will discuss the basic complex case first and then proceed to the real case. We will simply state the results. A proof based on the theory of Fredholm operators is given in §10 of Chapter III. For alternative proofs the reader is referred to Bott [1], [4].

Bott Periodicity Theorem 9.19 (the complex case). *The ring $K^{-*}(\mathrm{pt})$ is a polynomial algebra generated by an element $\xi \in K^{-2}(\mathrm{pt}) \cong \tilde{K}(S^2)$, i.e., there is a ring isomorphism*

$$K^{-*}(\mathrm{pt}) \cong \mathbb{Z}[\xi]. \tag{9.5}$$

Note. The element ξ is in fact represented by the virtual bundle $\xi \cong [H] - \tau^1 \in \tilde{K}(S^2)$ where H denotes the "tautologically defined" complex line bundle over $S^2 = \mathbb{P}^1(\mathbb{C})$ and τ^1 denotes the trivial line bundle.

The isomorphism (9.5) says in particular that the map $\mu_\xi : K^{-i}(\text{pt}) \to K^{-i-2}(\text{pt})$ induced by multiplication by ξ, is an isomorphism for all i. In this form, the theorem extends to arbitrary compact Hausdorff spaces. Recall from 9.18 that for any pointed space (X,pt) the ring $K^{-*}(X)$ is a module over $K^{-*}(\text{pt})$.

General Bott Periodicity Theorem 9.20 (the complex case). *Let X be any compact Hausdorff space. Then the map*

$$\mu_\xi : K^{-i}(X) \xrightarrow{\approx} K^{-i-2}(X)$$

given by module multiplication by ξ, is an isomorphism for all $i \geqq 0$.

Note. Replacing X by X/Y, we get corresponding isomorphisms

$$\mu_\xi : K^{-i}(X,Y) \longrightarrow K^{-i-2}(X,Y)$$

for any pair (X, Y) of compact Hausdorff spaces.

The situation in KO-theory is slightly more complicated.

Bott Periodicity Theorem 9.21 (the real case). *The ring $KO^{-*}(\text{pt})$ is generated by elements*

$$\eta \in KO^{-1}(\text{pt}), \qquad y \in KO^{-4}(\text{pt}), \qquad x \in KO^{-8}(\text{pt})$$

subject only to the relations

$$2\eta = 0, \qquad \eta^3 = 0, \qquad \eta y = 0, \qquad y^2 = 4x$$

i.e., there is a ring isomorphism

$$KO^{-*}(\text{pt}) \cong \mathbb{Z}[\eta, y, x]/\langle 2\eta, \eta^3, \eta y, y^2 - 4x\rangle \tag{9.6}$$

As before we have that for any space X, $KO^{-*}(X)$ is a $KO^{-*}(\text{pt})$-module.

Theorem 9.22. *Let X be a compact Hausdorff space. Then the map*

$$\mu_x : KO^{-i}(X) \longrightarrow KO^{-i-8}(X)$$

given by module multiplication by $x \in KO^{-8}(\text{pt})$, is an isomorphism for all $i \geqq 0$.

As before there are also isomorphisms $KO^{-i}(X,Y) \xrightarrow{\approx} KO^{-i-8}(X,Y)$ for any pair (X,Y) of compact Hausdorff spaces.

The best explicit representatives for the elements η, x and y in $KO^{-*}(\mathrm{pt})$ are given via the Atiyah-Bott-Shapiro isomorphism. To present this isomorphism, and also to adapt K-theory easily to the theory of elliptic operators, we present now an alternative definition of the groups $K(X,Y)$ and $KO(X,Y)$. The discussion is completely parallel in the real and complex cases. We shall just use the generic term "vector bundle."

We begin with the following definition. We assume throughout that Y is a closed subspace of X.

DEFINITION 9.23. For each integer $n \geqq 1$, consider the set $\mathscr{L}_n(X,Y)$ of elements $\mathbf{V} = (V_0, V_1, \dots, V_n; \sigma_1, \dots, \sigma_n)$ where $V_0, \dots, V_n$ are vector bundles on X and where

$$0 \longrightarrow V_0\big|_Y \xrightarrow{\sigma_1} V_1\big|_Y \xrightarrow{\sigma_2} \cdots \xrightarrow{\sigma_n} V_n\big|_Y \longrightarrow 0$$

is an exact sequence of bundle maps for the restriction of these bundles to Y. Two such elements $\mathbf{V} = (V_0, \dots, V_n; \sigma_1, \dots, \sigma_n)$ and $\mathbf{V}' = (V'_0, \dots, V'_n; \sigma'_1, \dots, \sigma'_n)$ are said to be **isomorphic** if there are bundle isomorphisms $\varphi_i: V_i \to V'_i$ over X so that the diagram

$$\begin{array}{ccc} V_{i-1}\big|_Y & \xrightarrow{\sigma_i} & V_i\big|_Y \\ \Big\downarrow{\scriptstyle \varphi_{i-1}} & & \Big\downarrow{\scriptstyle \varphi_i} \\ V'_{i-1}\big|_Y & \xrightarrow{\sigma'_i} & V'_i\big|_Y \end{array}$$

commutes for each i.

An element $\mathbf{V} = (V_0, \dots, V_n; \sigma_1, \dots, \sigma_n)$ is said to be **elementary** if there is an i such that

(a) $V_i = V_{i-1}$ and $\sigma_i = \mathrm{Id}$
(b) $V_j = \{0\}$ for $j \neq i$ or $i - 1$.

There is an operation of direct sum $\oplus$ defined on the set $\mathscr{L}_n(X,Y)$ in the obvious way. Two elements $\mathbf{V}, \mathbf{V}' \in \mathscr{L}_n(X,Y)$ are defined to be **equivalent** if there exist elementary elements $\mathbf{E}_1, \dots, \mathbf{E}_k, \mathbf{F}_1, \dots, \mathbf{F}_\ell \in \mathscr{L}_n(X,Y)$ and an isomorphism

$$\mathbf{V} \oplus \mathbf{E}_1 \oplus \cdots \oplus \mathbf{E}_k \cong \mathbf{V}' \oplus \mathbf{F}_1 \oplus \cdots \oplus \mathbf{F}_\ell.$$

The equivalence class of an element $(V_0, \dots, V_n; \sigma_1, \dots, \sigma_n)$ will be denoted by $[V_0, \dots, V_n; \sigma_1, \dots, \sigma_n]$. The set of all equivalence classes in $\mathscr{L}_n(X,Y)$ will be denoted by $L_n(X,Y)$.

The set $L_n(X,Y)$ is an abelian group under the operation $\oplus$.

Our first main proposition is the following, whose proof is left to the reader. Consider the natural map $\mathscr{L}_n(X,Y) \to \mathscr{L}_{n+1}(X,Y)$ which associates to each element $(V_0, \dots, V_n; \sigma_1, \dots, \sigma_n)$ the trivially extended element $(V_0, \dots, V_n, 0; \sigma_1, \dots, \sigma_n, 0)$.

Proposition 9.24. *For each $n \geqq 1$, the induced map $L_n(X,Y) \to L_{n+1}(X,Y)$ is an isomorphism.*

We set $L(X,Y) = \varinjlim L_n(X,Y)$. Each inclusion $L_n(X,Y) \to L(X,Y)$ is an isomorphism, and it would have sufficed for many purposes to consider only the case $n = 1$.

Our second main proposition is the following.

Proposition 9.25. *There exists a unique equivalence of functors $\chi : L(X,Y) \to K(X,Y)$ with the property that*

$$\chi([V_0, \ldots, V_n]) = \sum_{k=0}^{n} (-1)^k [V_k] \qquad \text{when } Y = \varnothing. \tag{9.7}$$

Proof. This equivalence will be essentially determined by defining it on $L_1(X,Y)$. Given an element $\mathbf{V} = [V_0, V_1; \sigma] \in L_1(X,Y)$ we associate to it an element $\chi(\mathbf{V}) \in K(X,Y)$ by the following "difference bundle construction". Set $X_k = X \times \{k\}$ for $k = 0,1$ and consider the space $Z = X_0 \cup_Y X_1$ obtained from the disjoint union $X_0 \amalg X_1$ by identifying $y \times \{0\}$ with $y \times \{1\}$ for all $y \in Y$. The natural sequence

$$0 \longrightarrow K(Z,X_1) \xrightarrow{j^*} K(Z) \xrightarrow{i^*} K(X_1) \longrightarrow 0$$

is split exact since there is an obvious retraction

$$\rho : Z \longrightarrow X_1.$$

Furthermore, there is an isomorphism

$$\varphi : K(Z,X_1) \xrightarrow{\approx} K(X,Y)$$

induced by the map of pairs $(X,Y) \to (Z,X_1)$ which identifies X with X_0.

From our element $\mathbf{V} = [V_0, V_1; \sigma]$ we define a vector bundle W over Z (well defined up to isomorphism) by setting $W|_{X_k} \equiv V_k$ and identifying over Y via the isomorphism σ. Setting $W_1 \equiv \rho^*(V_1)$ we have $[W] - [W_1] \in \ker(i^*)$. Hence, there is a unique element $\chi(\mathbf{V}) \in K(X,Y)$ with $j^* \varphi^{-1} \chi(\mathbf{V}) = [W] - [W_1]$. This defines the homomorphism $\chi : L_1(X,Y) \to K(X,Y)$. It clearly has property (9.7).

It is now straightforward to verify that any homomorphism $L_1(X,Y) \to K(X,Y)$ with property (9.7) is an isomorphism, and furthermore, any two such homomorphisms agree. The reader is referred to Atiyah-Bott-Shapiro [1] for details. ■

As a result of this proposition we shall henceforth drop the notation $L(X,Y)$. We will however discuss elements $[V_0, V_1; \sigma] \in K(X,Y)$ whose meaning is now obvious.

For our later discussion of elliptic operators it will be useful to note that the multiplication in $K(X,Y)$ can be realized explicitly in $\mathscr{L}_1(X,Y)$

as follows. Choose $\mathbf{V} = (V_0, V_1; \sigma)$, $\mathbf{W} = (W_0, W_1; \tau) \in \mathscr{L}_1(X,Y)$ and for convenience introduce metrics in each of the bundles. We then define the tensor product $\mathbf{U} = \mathbf{V} \otimes \mathbf{W} \in \mathscr{L}_1(X,Y)$ to be the element $\mathbf{U} = (U_0, U_1; \rho)$ where

$$U_0 = (V_0 \otimes W_0) \oplus (V_1 \otimes W_1) \qquad U_1 = (V_1 \otimes W_0) \oplus (V_0 \otimes W_1)$$

and

$$\rho = \begin{pmatrix} \sigma \otimes 1 & -1 \otimes \tau^* \\ 1 \otimes \tau & \sigma^* \otimes 1 \end{pmatrix}.$$

Under the construction above, this tensor product on $\mathscr{L}_1(X,Y)$ carries over to the standard product on $K(X,Y)$.

This is a convenient time to introduce K-theory for locally compact spaces. K-theory in this setting will be important for certain geometric constructions we shall make in Chapter III.

DEFINITION 9.26. For any locally compact space X we define

$$K_{\text{cpt}}(X) = \tilde{K}(X^+)$$

where $X^+ = X \cup \{\text{pt}\}$ denotes the one point compactification of X. The higher groups are defined by setting

$$K_{\text{cpt}}^{-i}(X) = K_{\text{cpt}}(X \times \mathbb{R}^i)$$

for $i \geqq 0$.

The groups $K_{\text{cpt}}^{-i}(X)$ are functors on the category of locally compact spaces and proper maps. Collectively they comprise the **K-theory of X with compact supports.** They enjoy the multiplicativity properties that we presented above in the compact case. Note incidentally that if X is compact, then $K_{\text{cpt}}^{-i}(X) = K^{-i}(X)$.

Any element in $K_{\text{cpt}}(X)$ can be represented as the formal difference of two bundles E and F on X, each of which is trivialized at infinity (i.e., trivialized outside some compact set in X). In fact, if $\mathcal{O} \subset X$ is an open subset of the locally compact space X, then there is a natural **extension homomorphism** $K_{\text{cpt}}(\mathcal{O}) \to K_{\text{cpt}}(X)$ induced by the map $X^+ \to X^+/(X^+ - \mathcal{O}) = \mathcal{O}^+$. Taking products with $\mathbb{R}^i$ gives extension homomorphisms

$$K_{\text{cpt}}^{-i}(\mathcal{O}) \longrightarrow K_{\text{cpt}}^{-i}(X)$$

for all $i \geqq 0$. Of course for any closed subset $Y \subset X$ we have the functorial restriction homomorphism

$$K_{\text{cpt}}^{-i}(X) \longrightarrow K_{\text{cpt}}^{-i}(Y).$$

The kernel of this map is defined to be the relative group $K_{\text{cpt}}^{-i}(X,Y)$. For any pair (X,Y) where X is a locally compact space and Y is a closed subspace, there is a long exact sequence for the K_{cpt}^{-i}-groups, analogous to (9.4) above. We leave as an exercise for the reader the verification of the following isomorphism:

$$K_{\text{cpt}}^{-i}(X,Y) \cong K_{\text{cpt}}((X - Y) \times \mathbb{R}^i).$$

There exist definitions of "L-type" for the groups $K_{\text{cpt}}(X)$. In analogy with the above we can define $L_n(X)_{\text{cpt}}$ to be the equivalence classes $[V_0,\dots,V_n;\sigma_1,\dots,\sigma_n]$ where $V_0,\dots,V_n$ are vecctor bundles on X, and where $0 \to V_0 \xrightarrow{\sigma_1} V_1 \xrightarrow{\sigma_2} \cdots \xrightarrow{\sigma_n} V_n \to 0$ is an exact sequence of bundle maps defined on the complement of some compact subset in X. As above there are natural isomorphisms $L_1(X)_{\text{cpt}} \xrightarrow{\approx} L_2(X)_{\text{cpt}} \xrightarrow{\approx} \dots K_{\text{cpt}}(X)$. Thus, in particular, *any element of* $K_{\text{cpt}}(X)$ *can be represented by a triple* $[V_0, V_1;\sigma]$ *where* $\sigma: V_0 \xrightarrow{\approx} V_1$ *is a bundle isomorphism defined in a neighborhood of infinity.*

All of the discussion above applies equally well to real bundles and yields groups $KO_{\text{cpt}}^{-i}(X)$ for any locally compact space X.

The Bott Periodicity theorems carry over to locally compact spaces in the following elegant form:

$$\begin{aligned} K_{\text{cpt}}(X) &\cong K_{\text{cpt}}(X \times \mathbb{C}) \\ KO_{\text{cpt}}(X) &\cong KO_{\text{cpt}}(X \times \mathbb{R}^8). \end{aligned} \tag{9.8}$$

These isomorphisms are induced by multiplication by fundamental elements $\xi \in K_{\text{cpt}}(\mathbb{C})$ and $x \in KO_{\text{cpt}}(\mathbb{R}^8)$ respectively. In the last part of this chapter we shall produce explicit representatives for these elements using Clifford modules.

We are now in a position to discuss the isomorphism of Atiyah, Bott and Shapiro. Let $W = W^0 \oplus W^1$ be a $\mathbb{Z}_2$-graded module over the Clifford algebra $C\ell_n \equiv C\ell(\mathbb{R}^n)$. Let $D^n \equiv \{x \in \mathbb{R}^n : \|x\| \le 1\}$ be the unit disk and set $S^{n-1} = \partial D^n$. We now associate to the graded module W, the element

$$\varphi(W) = [E_0,E_1;\mu] \in K(D^n,S^{n-1}) \tag{9.9}$$

where $E_k \equiv D^n \times W^k$ is the trivial product bundle, and where $\mu: E_0 \xrightarrow{\approx} E_1$ is the isomorphism over S^{n-1} given by Clifford multiplication:

$$\mu(x,w) \equiv (x, x \cdot w).$$

One easily checks that the element $\varphi(W)$ depends only on the isomorphism class of the graded module W and, furthermore, that the map $W \mapsto \varphi(W)$ is an additive homomorphism. Hence, (9.9) gives us a homomorphism

$$\varphi: \hat{\mathfrak{M}}_n^{\mathbb{C}} \longrightarrow K(D^n,S^{n-1}) \tag{9.10}$$

where $\hat{\mathfrak{M}}_n^{\mathbb{C}}$ is the Grothendieck group of complex graded $C\ell_n$-modules defined in §5.

Consider for a moment the natural inclusion $i: \mathbb{R}^n \hookrightarrow \mathbb{R}^{n+1}$ given by setting $i(x_1, \ldots, x_n) = (x_1, \ldots, x_n, 0)$. This induces an algebra homomorphism $i_*: C\ell_n \quad C\ell_{n+1}$. Restricting the action from $C\ell_{n+1}$ to $C\ell_n$ thereby induces a homomorphism $i^*: \hat{\mathfrak{M}}^{\mathbb{C}}_{n+1} \to \hat{\mathfrak{M}}^{\mathbb{C}}_n$.

Suppose now that W is a graded $C\ell_n$-module which can be obtained from a $C\ell_{n+1}$-module in the above fashion. This means that the Clifford multiplication of $\mathbb{R}^n$ on W extends to all of $\mathbb{R}^{n+1}$. Hence we may extend the isomorphism μ, defined on $S^{n-1} = \partial D^n$, to all of D^n by setting

$$\mu(x,w) = \left(x, (x + \sqrt{1 - \|x\|^2} e_{n+1}) \cdot w\right)$$

where $e_{n+1} \in \mathbb{R}^{n+1}$ is a unit vector orthogonal to $\mathbb{R}^n$. Since this extended map is an isomorphism over all of D^n, the associated element $\varphi(W)$ must be zero. It now follows that the homomorphism (9.10) descends to a homomorphism

$$\varphi_n: \hat{\mathfrak{M}}^{\mathbb{C}}_n / i^* \hat{\mathfrak{M}}^{\mathbb{C}}_{n+1} \longrightarrow K(D^n, S^{n-1}) \tag{9.11}$$

where $i^*: \hat{\mathfrak{M}}^{\mathbb{C}}_{n+1} \to \hat{\mathfrak{M}}^{\mathbb{C}}_n$ is the restriction map defined above.

Exactly the same construction applied to the real case, gives us a homomorphism

$$\varphi_n: \hat{\mathfrak{M}}_n / i^* \hat{\mathfrak{M}}_{n+1} \longrightarrow KO(D^n, S^{n-1}). \tag{9.12}$$

The isomorphisms (5.29), defined by taking the even part, determine isomorphisms

$$\hat{\mathfrak{M}}^{\mathbb{C}}_n / i^* \hat{\mathfrak{M}}^{\mathbb{C}}_{n+1} \cong \mathfrak{M}^{\mathbb{C}}_{n-1} / i^* \mathfrak{M}^{\mathbb{C}}_n \qquad \text{and} \qquad \hat{\mathfrak{M}}_n / i^* \hat{\mathfrak{M}}_{n+1} \cong \mathfrak{M}_{n-1} / i^* \mathfrak{M}_n$$

Set $Q^{\mathbb{C}}_n = \hat{\mathfrak{M}}^{\mathbb{C}}_n / i^* \hat{\mathfrak{M}}^{\mathbb{C}}_{n+1}$ and $Q_n = \hat{\mathfrak{M}}_n / i^* \hat{\mathfrak{M}}_{n+1}$. The periodicity phenomena (4.11), (4.13) and (5.9) determine periodicity isomorphisms

$$Q^{\mathbb{C}}_{n+2} \cong Q^{\mathbb{C}}_n \qquad \text{and} \qquad Q_{n+8} \cong Q_n.$$

Elementary algebraic arguments show that

$$Q^{\mathbb{C}}_n \cong \begin{cases} \mathbb{Z} & \text{if } n \text{ is even} \\ 0 & \text{if } n \text{ is odd} \end{cases} \qquad Q_n \cong \begin{cases} \mathbb{Z} & \text{if } n \equiv 0 \text{ or } 4 \pmod 8 \\ \mathbb{Z}_2 & \text{if } n \equiv 1 \text{ or } 2 \pmod 8 \\ 0 & \text{otherwise.} \end{cases}$$

As an example consider $Q_2 = \mathfrak{M}_1 / i^* \mathfrak{M}_2$ considered as the quotient of the groups of ungraded modules. Since $C\ell_1 = \mathbb{C}$, the $C\ell_1$-modules are just complex vector spaces, and the isomorphism $\mathfrak{M}_1 \xrightarrow{\approx} \mathbb{Z}$ is generated by taking the complex dimension. Similarly, since $C\ell_2 = \mathbb{H}$, the $C\ell_2$-modules are quaternionic vector spaces and $\mathfrak{M}_2 \xrightarrow{\approx} \mathbb{Z}$ is generated by taking the quaternionic dimension. The map $i^*: \mathfrak{M}_2 \to \mathfrak{M}_1$ is generated by considering a quaternionic vector space to be a complex vector space under restriction of scalars. Clearly this is just the map $\mathbb{Z} \to \mathbb{Z}$ given by multiplication by 2, since the complex dimension is twice the quaternionic dimension. Thus we have $Q_2 \cong \mathbb{Z}_2$.

The reader may find the verification of these isomorphisms in general to be an interesting exercise. Full details are contained in Atiyah-Bott-Shapiro [1].

We recall now from Proposition 5.21 that the graded tensor product of modules gives a multiplication $\hat{\mathfrak{M}}_n^{\mathbb{C}} \otimes \hat{\mathfrak{M}}_m^{\mathbb{C}} \to \hat{\mathfrak{M}}_{n+m}^{\mathbb{C}}$ for all m,n. This makes $(\hat{\mathfrak{M}}_*^{\mathbb{C}}/i^*\hat{\mathfrak{M}}_{*+1}^{\mathbb{C}}) \equiv \bigoplus_{n\geq 0}(\hat{\mathfrak{M}}_n^{\mathbb{C}}/i^*\hat{\mathfrak{M}}_{n+1}^{\mathbb{C}})$ into a graded ring.

On the other hand, by definition we know that

$$K(D^n,S^{n-1}) = \tilde{K}(S^n) = K^{-n}(\mathrm{pt})$$

and therefore the direct sum $\bigoplus_{n\geq 0}K(D^n,S^{n-1}) = K^{-*}(\mathrm{pt})$ is also naturally a graded ring (cf. Corollary 9.18). (The analogous comments apply in the real case.) One of the main results of Atiyah-Bott-Shapiro [1] is the following.

Theorem 9.27 (the Atiyah-Bott-Shapiro Isomorphisms). *The maps* (9.11) *and* (9.12) *defined above induce graded ring isomorphisms:*

$$\varphi_*:(\hat{\mathfrak{M}}_*^{\mathbb{C}}/i^*\hat{\mathfrak{M}}_{*+1}^{\mathbb{C}}) \xrightarrow{\approx} K^{-*}(\mathrm{pt})$$
$$\varphi_*:(\hat{\mathfrak{M}}_*/i^*\hat{\mathfrak{M}}_{*+1}) \xrightarrow{\approx} KO^{-*}(\mathrm{pt}).$$

Since the periodicity of the quotients $\hat{\mathfrak{M}}_*/i^*\hat{\mathfrak{M}}_{*+1}$ is an elementary algebraic fact, this theorem appears to give a new algebraic proof of the Bott Periodicity theorems. However, the argument given in Atiyah-Bott-Shapiro [1] to establish the isomorphisms of 9.27 actually invokes the Bott result. Nevertheless the existence of these isomorphisms is a profound and important fact. It goes a long way towards explaining the fundamental role played by Clifford algebras in the index theory for elliptic operators.

REMARK 9.28. Theorem 9.27 gives us explicit generators for $K^{-*}(\mathrm{pt})$ and $KO^{-*}(\mathrm{pt})$ defined via representations of Clifford algebras. For example, let $S_{\mathbb{C}} = S_{\mathbb{C}}^+ \oplus S_{\mathbb{C}}^-$ be the fundamental $\mathbb{Z}_2$-graded representation space for $\mathbb{C}\ell_{2n}$ where $S_{\mathbb{C}}^{\pm} = (1 \pm \omega_{\mathbb{C}})S_{\mathbb{C}}$. (The sign of the complex volume element $\omega_{\mathbb{C}}$ depends on a choice of orientation in $\mathbb{R}^{2n}$.) There is an isomorphism $\hat{\mathfrak{M}}_{2n}^{\mathbb{C}} \cong \mathbb{Z} \oplus \mathbb{Z}$ with distinguished generators given by $S_{\mathbb{C}}$ and its "flip" $\tilde{S}_{\mathbb{C}}$, the same graded module with the factors interchanged. (This corresponds to a reversal of orientation in $\mathbb{R}^{2n}$.) The generator of $i^*\hat{\mathfrak{M}}_{2n+1}^{\mathbb{C}}$ is $[S_{\mathbb{C}}] + [\tilde{S}_{\mathbb{C}}]$. Hence, the group $K^{-2n}(\mathrm{pt}) \cong K_{\mathrm{cpt}}(\mathbb{R}^{2n}) \cong \mathbb{Z}$ is generated by the element

$$\sigma_{\mathbb{C},n} \equiv [S_{\mathbb{C}}^+,S_{\mathbb{C}}^-;\mu]$$

where $\mu_x: S_{\mathbb{C}}^+ \to S_{\mathbb{C}}^-$ denotes Clifford multiplication by $x \in \mathbb{R}^{2n}$.

The real case is entirely analogous. Let $S = S^+ \oplus S^-$ be the fundamental graded module for $C\ell_{4n}$ where $S^{\pm} = (1 \pm \omega)S$. Then

$$\sigma_{4n} \equiv [S^+, S^-; \mu]$$

is a generator of the group $KO^{-4n}(\text{pt}) \cong KO_{\text{cpt}}(\mathbb{R}^{4n}) \cong \mathbb{Z}$.

Using the structure of Clifford modules we can easily compute that

$$\sigma_{\mathbb{C},n} = (\sigma_{\mathbb{C},1})^n, \qquad \sigma_{8n} = (\sigma_8)^n \qquad \text{and} \qquad 4\sigma_8 = (\sigma_4)^2.$$

§10. *KR*-Theory and the (1,1)-Periodicity Theorem

In this section we present a theory which, in a sense, contains both K-theory and KO-theory. It was invented by Atiyah in the middle 1960s and was motivated in part by the study of indices for families of real operators. (There is a detailed discussion of this at the beginning of §16 in Chapter III.)

The higher groups in this theory carry a natural double-indexing $KR^{r,s}$, $r \geqq 0$, $s \geqq 0$. Interestingly, there is an isomorphism of the Atiyah-Bott-Shapiro type given in §9, which relates $K^{r,s}(\text{pt})$ with real modules over the Clifford algebra $C\ell_{r,s}$. In this sense, KR-theory is the analogue of K-theory and KO-theory suggested by passing from $\mathbb{C}\ell_n$ and $C\ell_n$ to $C\ell_{r,s}$. Basic references for this material are Atiyah [2] and Karoubi [2].

We consider here the category of **Real spaces,** i.e., spaces with involution. This is the category of pairs (X, c_X) where X is compact and $c_X: X \to X$ is a map with $c_X^2 = \text{Id}_X$. The map c_X may itself be the identity. Natural examples of such spaces are provided by complexifications of real algebraic varieties where c is given by complex conjugation. Such an example is $(\mathbb{P}^n(\mathbb{C}), c)$ where in homogeneous coordinates $c([z]) = [\bar{z}]$. The fixed point set here is $\mathbb{P}^n(\mathbb{R}) \subset \mathbb{P}^n(\mathbb{C})$.

DEFINITION 10.1. By a **Real vector bundle** over a Real space (X, c_X) we mean a pair (V, c_V) where $\pi: V \to X$ is a complex vector bundle over X and where $c_V: V \to V$ is an involution such that the diagram

$$\begin{array}{ccc} V & \xrightarrow{c_V} & V \\ {\scriptstyle\pi}\big\downarrow & & \big\downarrow{\scriptstyle\pi} \\ X & \xrightarrow{c_X} & X \end{array}$$

commutes and c_V is $\mathbb{C}$-antilinear on the fibres of V. We denote by $VR(X, c_X)$ the abelian semigroup of isomorphism classes of Real bundles over (X, c_X). Proceeding as in §9 we then define the associated Grothendieck group $KR(X, c_X)$. It is called the **Real-*K*-group** of X. We shall generally drop the explicit mention of the involution c_X and simply write $KR(X)$.

REMARK 10.2. Note that if $c_X = \mathrm{Id}_X$ then we have a natural identification $KO(X) \cong KR(X)$. This identification associates to any real bundle $V_{\mathbb{R}}$ on X, the pair $(V_{\mathbb{R}} \otimes \mathbb{C}, c)$ where c denotes complex conjugation in the fibres. $V_{\mathbb{R}}$ is recovered as the fixed-point set of c.

The groups $\widetilde{KR}(X) \equiv \ker(KR(X) \to KR(\mathrm{pt}))$ and $KR(X,Y) \equiv \widetilde{KR}(X/Y)$ are defined exactly as in §9 and give functors on the obvious categories of spaces. The higher groups: $\widetilde{KR}^{-i}(X) \equiv \widetilde{KR}(\Sigma^i X)$ and $KR^{-i}(X,Y) \equiv \widetilde{KR}(\Sigma^i(X/Y))$, are also defined as in §9. We have the exact sequence for compact pairs (X,Y):

$$\begin{aligned} \dots \longrightarrow KR^{-i}(X,Y) \longrightarrow KR^{-i}(X) \longrightarrow KR^{-i}(Y) \\ \longrightarrow KR^{-i+1}(X,Y) \longrightarrow \dots \end{aligned} \tag{10.1}$$

An interesting facet of KR-theory is that it carries naturally a doubly-indexed family of higher groups. Let $\mathbb{R}^{r,s} \equiv \mathbb{R}^r \oplus \mathbb{R}^s = $ "$\mathbb{R}^r \oplus i\mathbb{R}^s$" be the Real linear space with involution $c(x,y) = (x,-y)$. A basic case is $\mathbb{R}^{1,1} \cong \mathbb{C}$ with complex conjugation. Consider now the compact Real subspaces

$$D^{r,s} \equiv \{(x,y) \in \mathbb{R}^{r,s} : \|x\|^2 + \|y\|^2 \leqq 1\}$$
$$S^{r,s} \equiv \{(x,y) \in \mathbb{R}^{r,s} : \|x\|^2 + \|y\|^2 = 1\},$$

and for any compact Real pair (X,Y) define

$$K^{r,s}(X,Y) \equiv KR(X \times D^{r,s}, X \times S^{r,s} \cup Y \times D^{r,s}). \tag{10.2}$$

(The order of (r,s) here is the opposite of that in Atiyah [2].) Note that it is immediate from the definitions that

$$KR^{i,0} \equiv KR^{-i} \qquad \text{for all } i \geqq 0. \tag{10.3}$$

The sequence (10.1) generalizes to give exact sequences

$$\begin{aligned} \dots \longrightarrow KR^{r,s}(X,Y) \longrightarrow KR^{r,s}(X) \longrightarrow KR^{r,s}(Y) \\ \longrightarrow KR^{r-1,s}(X,Y) \longrightarrow \dots \end{aligned} \tag{10.4}$$

for all $s \geqq 0$. To prove this one establishes the exact sequence for compact triples in KR-theory above, and then applies it to the triple $(X \times D^{0,s}, X \times S^{0,s} \cup Y \times D^{0,s}, X \times S^{0,s})$.

One of the basic facts about $KR^{*,*}$ is that the exterior tensor product induces a bigraded multiplication

$$KR^{r,s}(X,Y) \otimes KR^{r',s'}(X',Y') \longrightarrow KR^{r+r',s+s'}(X'',Y'') \tag{10.5}$$

where $X'' = X \times X'$ and $Y'' = X \times Y' \cup X' \times Y$. In particular, as in §9, the "coefficients" $KR^{*,*}(\mathrm{pt})$ are a bigraded ring, and for any compact space X, $KR^{*,*}(X)$ is a graded $KR^{*,*}(\mathrm{pt})$-module.

Consider for a moment the group $KR^{1,1}(\mathrm{pt}) \equiv KR(D^{1,1}, S^{1,1}) = \widetilde{KR}(\mathbb{P}^1(\mathbb{C}))$. One can show that this group is isomorphic to $\mathbb{Z}$ with generator $\xi \equiv [H] - \tau_1$ where $H \to \mathbb{P}^1(\mathbb{C})$ is the tautological (or "Hopf") complex line bundle over $\mathbb{P}^1(\mathbb{C})$ with its natural Real structure. (The fibre H_ℓ for $\ell \in \mathbb{P}^1(\mathbb{C})$ is the line ℓ itself.)

One of the fundamental results in this theory is the following (see Atiyah [2]):

The (1,1)-Periodicity Theorem 10.3. *Let X be any compact Hausdorff space. Then the map*

$$\mu_\xi : KR^{r,s}(X) \xrightarrow{\approx} KR^{r+1,s+1}(X) \tag{10.6}$$

given by module multiplication by ξ, is an isomorphism for all $r \geqq 0$, $s \geqq 0$. Consequently, for any compact Real pair (X, Y) we have the isomorphism

$$\mu_\xi : KR^{r,s}(X,Y) \xrightarrow{\approx} KR^{r+1,s+1}(X,Y) \tag{10.7}$$

Corollary 10.4. *There exist natural isomorphisms*

$$KR^{r,s} \cong KR^{s-r} \tag{10.8}$$

for all $r \geqq s \geqq 0$ generalizing (10.3).

Using (10.8) we can extend the definition of KR^{-i} to all integers i. The exact sequence (10.1) then extends to infinity in both directions.

REMARK 10.5. There are many further internal symmetries in this theory. Using the multiplication in the fields $\mathbb{R}$, $\mathbb{C}$ and $\mathbb{H}$, Atiyah [2] shows that for any compact space X there are isomorphisms

$$KR(X \times S^{0,p}) \cong KR^{-2p}(X \times S^{0,p}) \tag{10.9}$$

for $p = 1, 2$, and 4. (Recall that $\dim S^{0,p} = p - 1$.) This isomorphism for $p = 1$ gives the complex Bott Periodicity Theorem. From the case $p = 4$ one can deduce the real periodicity theorem.

REMARK 10.6. All of the discussion in §9 concerning the functors L_n and their equivalence with K carries over to the Real situation. Therefore, in particular, elements in $KR(X,Y)$ can be represented by classes of the form $[V_0, V_1; \sigma]$ where V_0 and V_1 are Real bundles on X and where $\sigma : V_0 \to V_1$ is a Real bundle isomorphism defined over Y.

REMARK 10.7. If X is a locally compact space, then the groups $KR_{\mathrm{cpt}}(X)$ are defined exactly as in 9.26. This group is generated by Real bundles on X which are trivialized at ∞. Using the L-construction, it can also be generated by triples $[V_0, V_1; \sigma]$ where V_0 and V_1 are Real bundles on X and where $\sigma : V_0 \to V_1$ is a Real bundle isomorphism defined outside of some compact subset of X.

In this context the higher KR-groups can be written in a particularly nice form:

$$KR^{r,s}_{\mathrm{cpt}}(X) \equiv KR_{\mathrm{cpt}}(X \times \mathbb{R}^{r,s}). \tag{10.10}$$

Of course if X is compact, then $KR^{r,s}_{\mathrm{cpt}}(X) \cong KR^{r,s}(X)$. If we identify $\mathbb{C} = \mathbb{R}^{1,1}$ as before, then the (1,1)-Periodicity Theorem has the particularly nice form:

$$\mathrm{KR}_{\mathrm{cpt}}(X) \cong KR_{\mathrm{cpt}}(X \times \mathbb{C}) \tag{10.11}$$

for any locally compact space X.

In light of the Remark 10.6 above it is natural to examine the Atiyah-Bott-Shapiro construction in this theory. To do this we define the Real Clifford algebra $C\ell(\mathbb{R}^{r,s})$ to be the Clifford algebra generated by $\mathbb{R}^{r,s}$ and the positive definite quadratic form $q(x,y) = \|x\|^2 + \|y\|^2$, taken together with the algebra involution $c: C\ell(\mathbb{R}^{r,s}) \to C\ell(\mathbb{R}^{r,s})$ generated by the involution $(x,y) \mapsto (x,-y)$ of $\mathbb{R}^{r,s}$.

Definition 10.8. By a **Real module** over the algebra $C\ell(\mathbb{R}^{r,s})$ we mean a finite-dimensional complex representation space W for $C\ell_{r,s}$ together with a $\mathbb{C}$-antilinear involution $c: W \to W$ so that

$$c(\varphi \cdot w) = c(\varphi) \cdot c(w) \qquad \text{for all } \varphi \in C\ell(\mathbb{R}^{r,s}) \text{ and all } w \in W. \tag{10.12}$$

If in addition $W = W^0 \oplus W^1$ is a $\mathbb{Z}_2$-graded module with the property that $c(W^i) = W^i$ for $i = 0,1$, then W is called a **Real $\mathbb{Z}_2$-graded module** for $C\ell(\mathbb{R}^{r,s})$.

We denote by $\mathfrak{M}R_{r,s}$ and $\widehat{\mathfrak{M}R}_{r,s}$ the Grothendieck groups of equivalence classes of Real modules (and Real $\mathbb{Z}_2$-graded respectively) for $C\ell(\mathbb{R}^{r,s})$.

There is an important basic relationship between these Real algebras and the algebras $C\ell_{r,s}$ defined in §3. To begin, consider the complexification $\mathbb{C}\ell(\mathbb{R}^{r,s})$ of $C\ell(\mathbb{R}^{r,s})$ and note that the involution c has a unique extension to a $\mathbb{C}$-antilinear involution on $\mathbb{C}\ell(\mathbb{R}^{r,s})$. Any Real $C\ell(\mathbb{R}^{r,s})$-module is naturally a $\mathbb{C}\ell(\mathbb{R}^{r,s})$-module by extension, and the condition (10.12) continues to hold. Considering these algebras and their modules is equivalent to considering those above.

Now the main claim is that $\mathbb{C}\ell(\mathbb{R}^{r,s})$ is the appropriate complexification of $C\ell_{r,s}$. Under the ordinary complexification, all the algebras $C\ell_{r,s}$ with the same value of $r + s$ become isomorphic. However, if we also introduce the involution, this is not so.

Recall that $C\ell_{r,s}$ is the Clifford algebra generated by $\mathbb{R}^r \oplus \mathbb{R}^s$ with the quadratic form $q_{r,s}(x,y) = \|x\|^2 - \|y\|^2$. Let $\mathfrak{M}_{r,s}$ and $\hat{\mathfrak{M}}_{r,s}$ denote the Grothendieck groups of $\mathbb{R}$-modules (and $\mathbb{Z}_2$-graded $\mathbb{R}$-modules respectively) for the Clifford algebra $C\ell_{r,s}$.

Proposition 10.9. *There is a natural equivalence*

$$\mathfrak{M}_{r,s} \xrightarrow{\approx} \mathfrak{M}R_{r,s} \tag{10.13}$$

for each (r,s) defined by assigning to a real $C\ell_{r,s}$-module W, the complex module $W \otimes_{\mathbb{R}} \mathbb{C}$ with involution given by complex conjugation and with the $C\ell(\mathbb{R}^{r,s})$-multiplication engendered by setting

$$(x,y) \cdot w \equiv xw + iyw \qquad \text{for } (x,y) \in \mathbb{R}^r \oplus \mathbb{R}^s. \tag{10.14}$$

Furthermore, under the natural inclusion $i: \mathbb{R}^r \oplus \mathbb{R}^s \hookrightarrow \mathbb{R}^{r+1} \oplus \mathbb{R}^s$ given by $i(x_1, \ldots, x_r, y_1, \ldots, y_s) = (x_1, \ldots, x_r, 0, y_1, \ldots, y_s)$ the diagram

$$\begin{array}{ccc} \mathfrak{M}_{r+1,s} & \xrightarrow{\approx} & \mathfrak{M}R_{r+1,s} \\ \downarrow{\scriptstyle i^*} & & \downarrow{\scriptstyle i^*} \\ \mathfrak{M}_{r,s} & \xrightarrow{\approx} & \mathfrak{M}R_{r,s} \end{array}$$

commutes. Hence, there are natural graded isomorphisms:

$$\mathfrak{M}_{*,*}/i^*\mathfrak{M}_{*+1,*} \xrightarrow{\approx} \mathfrak{M}R_{*,*}/i^*\mathfrak{M}R_{*+1,*}. \tag{10.15}$$

All the analogous statements hold in the $\mathbb{Z}_2$-graded case.

Proof. The multiplication given in (10.14) has the property: $(x,y) \cdot (x,y) \cdot w = -(\|x\|^2 + \|y\|^2)w$ and therefore extends to $C\ell(\mathbb{R}^{r,s})$. Furthermore, we see that $c[(x,y) \cdot w] = \overline{(x + iy)w} = (x - iy)\bar{w} = c(x,y) \cdot c(w)$ where both $(\overline{\ \ })$ and c denote complex conjugation. Hence, the map (10.13) is well defined. The inverse is given by taking the real module to be the fixed-point set of c and replacing multiplication by $(x,y) \in \mathbb{R}^r \oplus \mathbb{R}^s$ with multiplication by $x - iy$. The remaining details are left as an exercise for the reader. ∎

REMARK 10.10. As in previous cases the graded tensor product of modules makes $\widehat{\mathfrak{M}R}_{*,*}$ and $\widehat{\mathfrak{M}}_{*,*}$ into graded rings. The multiplication is preserved by the equivalence (10.13) in the graded case. In particular, the map

$$\widehat{\mathfrak{M}}_{*,*}/i^*\widehat{\mathfrak{M}}_{*+1,*} \longrightarrow \widehat{\mathfrak{M}R}_{*,*}/i^*\widehat{\mathfrak{M}R}_{*+1,*} \tag{10.16}$$

is a *ring isomorphism.*

The advantage of the groups $\widehat{\mathfrak{M}R}_{r,s}$ is that they are natural for extending the Atiyah-Bott-Shapiro construction given in §9. On the other hand, the groups $\widehat{\mathfrak{M}}_{r,s}$ and the quotients $\widehat{\mathfrak{M}}_{r,s}/i^*\widehat{\mathfrak{M}}_{r+1,s}$ are particularly easy to compute by using the results of §§4 and 5.

Suppose that $W = W^0 \oplus W^1$ is a Real $\mathbb{Z}_2$-graded module for $C\ell(\mathbb{R}^{r,s})$, and consider the associated element $\varphi(W) \in KR_{\text{cpt}}(\mathbb{R}^{r,s}) \cong KR^{r,s}(\text{pt})$ given by

$$\varphi(W) = [E_0, E_1; \mu]$$

where $E_k = \mathbb{R}^{r,s} \times W^k$ for $k = 0,1$, are the trivial product bundles and where $\mu: E_0 \to E_1$ is defined by

$$\mu(z,w) = (z, z \cdot w)$$

for $z = (x,y) \in \mathbb{R}^{r,s}$ and for $w \in W^0$. Since $z \cdot z \cdot w = -\|z\|^2 w$, the map μ is a bundle isomorphism outside the origin and so $[E_0, E_1; \mu]$ is an element of $KR_{\text{cpt}}(\mathbb{R}^{r,s})$ (cf. Remark 10.7). This gives a graded ring homomorphism

$$\varphi: \widehat{\mathfrak{M}R}_{*,*} \longrightarrow KR^{*,*}(\text{pt}). \tag{10.17}$$

Arguing as in §9 one can show that $\varphi \equiv 0$ on $i^*\widehat{\mathfrak{M}R}_{*+1,*}$ and so φ descends to the quotient. The arguments of Atiyah-Bott-Shapiro [1] then carry through to prove the following (see Atiyah [2]).

Theorem 10.11. *The map defined above,*

$$\varphi: \widehat{\mathfrak{M}R}_{*,*}/i^*\widehat{\mathfrak{M}R}_{*+1,*} \longrightarrow KR^{*,*}(\text{pt}),$$

is a graded ring isomorphism.

Via the isomorphism (10.16) this relates the groups $KR^{r,s}(\text{pt})$ to the algebras $C\ell_{r,s}$. In particular, the (1,1)-Periodicity Theorem is reflected in the (1,1)-periodicity for these algebras (cf. (4.3) of Theorem 4.1). Furthermore, since $KR^{s,0}(\text{pt}) = KR^{-s}(\text{pt}) = KO^{-s}(\text{pt})$ and $C\ell_{s,0} \cong C\ell_s$ we recapture here the real ABS-isomorphism of Theorem 9.27.

From (1,1)-periodicity we see that

$$KR^{0,0}(\text{pt}) \cong KR^{1,1}(\text{pt}) \cong KR^{2,2}(\text{pt}) \cong \cdots \cong \mathbb{Z}.$$

For our later discussion of the $C\ell_n$-linear Atiyah-Singer operator (in III.16) it will be useful to have certain explicit generators for these groups, which we shall now present. Recall that as a Real space $\mathbb{R}^{n,n}$ is just $\mathbb{C}^n$ with involution given by complex conjugation, and therefore we can write $KR^{n,n}(\text{pt}) \cong KR_{\text{cpt}}(\mathbb{C}^n)$. We will give an explicit generator for $KR_{\text{cpt}}(\mathbb{C}^n)$ by using the Clifford algebra $\mathbb{C}\ell_n = \mathbb{C}\ell(\mathbb{C}^n)$ and its natural $\mathbb{Z}_2$-grading $\mathbb{C}\ell_n = \mathbb{C}\ell_n^0 \oplus \mathbb{C}\ell_n^1$. For $\varphi \in \mathbb{C}\ell_n$, let $R_\varphi: \mathbb{C}\ell_n \to \mathbb{C}\ell_n$ and $L_\varphi: \mathbb{C}\ell_n \to \mathbb{C}\ell_n$ denote right and left multiplication by φ respectively. For $x + iy \in \mathbb{C}^n$, we consider the map

$$(R_x + iL_y): \mathbb{C}\ell_n^0 \longrightarrow \mathbb{C}\ell_n^1 \tag{10.18}$$

which, since $(R_x + iL_y)(R_x - iL_y) = -(\|x\|^2 + \|y\|^2)\text{Id}$, is invertible when $x + iy \neq 0$. This map (10.18) is clearly a Real endomorphism, i.e., $\overline{(R_x + iL_y)\varphi} = (R_x + iL_{-y})\bar{\varphi}$.

Proposition 10.12. *The element*

$$\varepsilon_n \equiv [\mathbb{C}\ell_n^0, \mathbb{C}\ell_n^1; R_x + iL_y] \in KR_{\text{cpt}}(\mathbb{C}^n)$$

is a generator of the group $KR_{\text{cpt}}(\mathbb{C}^n) = KR^{n,n}(\text{pt}) \cong \mathbb{Z}$ for all $n \geq 1$.

Proof. Because of the isomorphism of graded algebras $\mathbb{C}\ell_n = \mathbb{C}\ell_1 \hat{\otimes} \cdots \hat{\otimes} \mathbb{C}\ell_1$ (which follows from Proposition 3.2) it suffices to consider the case $n = 1$. The element ε_1 can be shown to coincide with the generator of the image of φ in Theorem 10.11 as follows. To begin note that since $\mathbb{C}\ell_1^0 = \mathbb{C} \cdot 1$, right and left multiplication coincide on $\mathbb{C}\ell_1^0$. Consequently, $\varepsilon_1 \cong [\mathbb{C},\mathbb{C}; m]$ where $m_{(x,y)}: \mathbb{C} \to \mathbb{C}$ is given by scalar multiplication: $m_{(x,y)}(w) = (x + iy)w$.

On the other hand we have $\widehat{\mathfrak{M}R}_{1,1} \cong \mathbb{Z} \oplus \mathbb{Z}$ with generators given as follows. For $(x,y) \in \mathbb{R}^{1,1}$, define a $\mathbb{C}$-linear map $\mathbb{C} \oplus \mathbb{C} \to \mathbb{C} \oplus \mathbb{C}$ by the matrix

$$\begin{pmatrix} 0 & -\bar{m} \\ m & 0 \end{pmatrix} \tag{10.19}$$

At (x,y), the square is $-(\|x\|^2 + \|y\|^2)\text{Id}$. Hence, this action extends to make $\mathbb{C} \oplus \mathbb{C}$ (with involution given by complex conjugation) into a complex $\mathbb{Z}_2$-graded $\mathbb{C}\ell(\mathbb{R}^{1,1})$-module. This is the first generator. The second generator is obtained by interchanging factors. That these are independent generators is easily verified by working through the natural isomorphisms

$$\widehat{\mathfrak{M}R}_{1,1} \cong \hat{\mathfrak{M}}_{1,1} \cong \mathfrak{M}_{0,1}$$

(cf. Proposition 10.9 and Theorem 3.7). These isomorphisms commute with i^*, and working out the details for the simple case $\mathfrak{M}_{0,1}/i^*\mathfrak{M}_{1,1}$ shows that either of our distinguished generators becomes a generator of the quotient: $\widehat{\mathfrak{M}R}_{1,1}/i^*\widehat{\mathfrak{M}R}_{2,1} \cong KR^{1,1}(\text{pt})$. Clearly under the isomorphism φ of Theorem 10.11 our module above with multiplication 10.18 becomes the element $[\mathbb{C},\mathbb{C}; m]$ as claimed.

CHAPTER II

Spin Geometry and the Dirac Operators

In this chapter one finds the soul of the book. It is here that we examine the structures and concepts that are central to differential geometry—manifolds, vector bundles, connections, curvature, etc. If one also introduces metrics (in the sense of Riemann), then it is unavoidable that Clifford algebras and spin groups will enter the discussion. This is for the following reason.

There is a principle by which the natural operations on vector spaces such as direct sum, tensor product, exterior power, etc., carry over canonically to vector bundles. In the same fashion, the natural operations for vector spaces with quadratic forms carry over to vector bundles with metrics. In particular, suppose that $\pi: E \to X$ is a riemannian vector bundle. Then in each fibre, $E_x = \pi^{-1}(x)$, the quadratic form $\|v\|^2 = \langle v, v\rangle$ can be used to construct a Clifford algebra $C\ell(E_x)$. The result is a bundle $C\ell(E) \to X$ of algebras over X called the Clifford bundle of E. It carries all the natural properties of Clifford algebras such as the $\mathbb{Z}_2$-grading, the transpose endomorphism and the *L*-operator. This rich structure is basic for the study of E itself.

In the light of Chapter I it is natural to ask whether one can also find a vector bundle $\$(E) \to X$ with the property that each fibre $\$(E_x)$ is an *irreducible* module over $C\ell(E_x)$. The answer is in general no. We shall discuss the obstruction involved. By taking $E = TX$ this will lead to the notion of a spin manifold and spin cobordism. The bundle $\$(TX)$, when it exists, plays a central role in the study of the global geometry and topology of X.

In general, given any bundle of modules S over $C\ell(TX)$, furnished with a suitable metric and connection, one can associate to S a self-adjoint, first-order elliptic operator $D: \Gamma(S) \to \Gamma(S)$ called the Dirac operator for S. (This is done in §5.) From the algebraic operations of Chapter I one can often construct natural splittings $S = S^+ \oplus S^-$ with respect to which the Dirac operator has the form

$$D = \begin{pmatrix} 0 & D^- \\ D^+ & 0 \end{pmatrix}$$

where $D^+: \Gamma(S^+) \to \Gamma(S^-)$ and $D^-: \Gamma(S^-) \to \Gamma(S^+)$ are formal adjoints of one another. This gives a unified procedure for constructing all the "classical" elliptic operators on a manifold, namely: the Euler characteristic operator, the signature operator, and the Atiyah-Singer operator. Details of this are given in §6.

In §7 we introduce the notion of a $C\ell_k$-linear operator and discuss in detail some of the basic examples. Over a compact manifold any $C\ell_k$-linear Dirac operator D has an analytic index $\text{ind}_k(D) \in KO^{-k}(\text{pt})$ defined by applying the Atiyah-Bott-Shapiro isomorphism to the residue class of the Clifford module [ker D] (see I.9).

Roughly speaking, every Dirac operator represents the square root of a Laplace operator. In euclidean space this statement is unambiguously true. Over general manifolds, however, the difference $D^2 - \nabla^*\nabla$ between the square of the Dirac operator and the standard "connection laplacian" is a certain universal expression involving curvature and Clifford multiplication. Deriving such formulas and using them to draw global conclusions about curvature and topology is referred to as "Bochner's method." We shall show how our universal formula specializes to give the classical Bochner formulas on exterior differential forms, as well as the Lichnerowicz formula for spinors. Using curvature identities derived in §5, we shall then give a succinct proof of the theorem of Gallot and Meyer concerning curvature and homology spheres. We shall also derive formulas for the Atiyah-Singer operator with coefficients in a bundle. This will be quite useful later in studying manifolds of positive scalar curvature.

§1. Spin Structures on Vector Bundles

Let $\pi : E \to X$ be a real n-dimensional vector bundle over a manifold X. We assume this bundle is equipped with a **riemannian** structure, that is, a positive definite inner product continuously defined in the fibres. Such a structure always exists.

We assume also that the bundle is **oriented,** i.e., that there is an orientation continuously defined on the fibres. This structure does not always exist. To analyze the situation, we consider the bundle $P_O(E)$ of orthonormal frames in E. This is the principal O_n-bundle whose fibre at a point $x \in X$ is the set of orthonormal bases of $E_x \equiv \pi^{-1}(x)$. The bundle of orientations in E is then just the quotient $\text{Or}(E) = P_O(E)/SO_n$, where two bases of E_x are identified if the orthogonal matrix transforming one to the other has determinant $+1$. Note that $\text{Or}(E)$ is a 2-sheeted covering space of X and that E is orientable if and only if this covering space is the trivial one. We recall now the following elementary fact:

Lemma 1.1. *Let* $\text{Cov}_2(X)$ *be the set of equivalence classes of 2-sheeted covering spaces of* X. *Then there is a natural isomorphism*

$$\text{Cov}_2(X) \approx H^1(X; \mathbb{Z}_2). \tag{1.1}$$

Note. This is a special case of the isomorphism: $H^1(X;G) \approx$ {equivalence classes of principal G-bundles on X}, which is proved in Appendix A.

Proof. Assuming X is connected, we can decompose the isomorphism (1.1) as follows: $H^1(X;\mathbb{Z}_2) \xrightarrow{\approx} \text{Hom}(H_1(X),\mathbb{Z}_2) \xrightarrow{\approx} \text{Hom}(\pi_1(X),\mathbb{Z}_2) \xrightarrow{\approx} \text{Cov}_2(X)$. The second isomorphism follows from the fact that $H_1(X)$ is the abelianization of $\pi_1(X)$. The third isomorphism is a restatement of the elementary fact that 2-sheeted coverings of X are in one-to-one correspondence with subgroups of index 2 in $\pi_1(X)$. The case where X is not connected now follows immediately. ∎

From (1.1) we now see that, for each vector bundle E over X, the 2-sheeted covering space $\text{Or}(E)$ determines an element $w_1(E) \in H^1(X;\mathbb{Z}_2)$, called the **first Stiefel-Whitney class** of E. Directly from the definitions we have the following fact.

Theorem 1.2. *A vector bundle E over X is orientable if and only if $w_1(E) = 0$. Furthermore if $w_1(E) = 0$, then the distinct orientations on E are in one-to-one correspondence with elements of $H^0(X;\mathbb{Z}_2)$.*

The second statement simply says that there are two possible orientations of E over each connected component of X.

The definition of $w_1(E)$ given above is in accord with the one given via classifying spaces (see App. B). To prove this it suffices to show the following:

(i) This definition of w_1 is **natural**, i.e., $w_1(f^*E) = f^*w_1(E)$ for any bundle E over X and any continuous map $f: X' \to X$.
(ii) This definition of $w_1(\mathbb{E})$ gives the non-zero element in $H^1(\text{BO}_n;\mathbb{Z}_2) \cong \mathbb{Z}_2$ when $\mathbb{E}$ is the universal n-plane bundle over the classifying space BO_n.

Fact (i) is obvious since $P_O(f^*E) = f^*P_O(E)$ and, therefore, $\text{Or}(f^*E) = f^*\text{Or}(E)$. Fact (ii) is true since otherwise every n-plane bundle would be orientable.

By establishing properties (i) and (ii) it is possible to prove the equivalence of a number of quite different definitions of $w_1(E)$. For example, suppose X is connected. Then from the fibration $\text{O}_n \to P_O(E) \to X$, there is an exact sequence.

$$0 \longrightarrow H^0(X;\mathbb{Z}_2) \longrightarrow H^0(P_O(E);\mathbb{Z}_2) \longrightarrow H^0(\text{O}_n;\mathbb{Z}_2) \xrightarrow{w_E} H^1(X;\mathbb{Z}_2). \tag{1.2}$$

We can define $w_1(E) = w_E(g_1)$ where g_1 is the generator of $H^0(\text{O}_n;\mathbb{Z}_2)$. This definition of $w_1(E)$ has property (i) since the sequence (1.2) is natural and property (ii) since the sequence (1.2) is exact and for the universal bundle $\mathbb{E}$, $P_O(\mathbb{E}) = EO_n$ is contractible. From the exactness of (1.2) it is clear that $w_1(E) = 0$ iff $P_O(E)$ is disconnected, i.e., iff E is orientable.

We shall examine other equivalent definitions of w_1 later in this section. At the moment we pass on to the next possible simplification of the structure group of a bundle E. Note that if E is orientable, then choosing an orientation is equivalent to choosing a principal SO_n-bundle $P_{\mathrm{SO}}(E) \subset P_O(E)$. This embedding is, of course, compatible with the action of $\mathrm{SO}_n \subset \mathrm{O}_n$. Having thereby made the structure group of E 0-connected, one might ask whether it is possible to make the structure group 1-connected. This leads us to the concept of a spin structure.

Let E be an oriented n-dimensional riemannian vector bundle over a manifold X, and let $P_{\mathrm{SO}}(E)$ be its bundle of oriented orthonormal frames. Recall that for $n \geqq 3$ we have the universal covering homomorphism $\xi_0 : \mathrm{Spin}_n \to \mathrm{SO}_n$ with kernel $\{1, -1\} \cong \mathbb{Z}_2$.

DEFINITION 1.3. Suppose $n \geqq 3$. Then a **spin structure** on E is a principal Spin_n-bundle $P_{\mathrm{Spin}}(E)$ together with a 2-sheeted covering

$$\xi : P_{\mathrm{Spin}}(E) \longrightarrow P_{\mathrm{SO}}(E)$$

such that $\xi(pg) = \xi(p)\xi_0(g)$ for all $p \in P_{\mathrm{Spin}}(E)$ and all $g \in \mathrm{Spin}_n$.

When $n = 2$, a spin structure on E is defined analogously, with Spin_n replaced by SO_2 and $\xi_0 : \mathrm{SO}_2 \to \mathrm{SO}_2$ the connected 2-fold covering. When $n = 1$, $P_{\mathrm{SO}}(E) \cong X$ and a spin structure is simply defined to be a 2-fold covering of X.

Note that the diagram

$$\begin{array}{ccc} P_{\mathrm{Spin}}(E) & \xrightarrow{\ \xi\ } & P_{\mathrm{SO}}(E) \\ & {\scriptstyle \pi'}\searrow \quad \swarrow{\scriptstyle \pi} & \\ & X & \end{array} \tag{1.3}$$

where π and π' are the bundle projections, is commutative. Note also that ξ restricted to the fibres corresponds to the covering ξ_0. The diagram of fibrations is

$$\begin{array}{cccc} & \mathrm{Spin}_n & \xrightarrow{\ \xi_0\ } & \mathrm{SO}_n \\ \nearrow & \downarrow & & \downarrow \\ \mathbb{Z}_2 & & & \\ \searrow & \downarrow & & \downarrow \\ & P_{\mathrm{Spin}_n}(E) & \xrightarrow{\ \xi\ } & P_{\mathrm{SO}}(E) \\ & {\scriptstyle \pi'}\searrow & & \swarrow{\scriptstyle \pi} \\ & & X & \end{array}$$

On the other hand, suppose $\xi : P_{\mathrm{Spin}}(E) \to P_{\mathrm{SO}}(E)$ is a 2-sheeted covering which is non-trivial on the fibres of X, i.e., so that the diagram

$$\begin{array}{ccccc} & & \mathrm{Spin}_n & \xrightarrow{\xi_0} & \mathrm{SO}_n \\ & \nearrow & \downarrow & & \downarrow \\ \mathbb{Z}_2 & & & & \\ & \searrow & & & \\ & & P_{\mathrm{Spin}}(E) & \xrightarrow{\xi} & P_{\mathrm{SO}}(E) \end{array}$$

commutes. Then setting $\pi' = \pi \circ \xi$ makes $P_{\mathrm{Spin}}(E)$ a fibre bundle over X. To make this a principal Spin_n-bundle we must lift the action of SO_n on $P_{\mathrm{SO}}(E)$ to a compatible action of Spin_n on $P_{\mathrm{Spin}}(E)$. The proof that this lifting exists is a straightforward application of elementary covering space theory. We conclude the following (even for $n = 1$ and 2).

Theorem 1.4. *The spin structures on E are in natural one-to-one correspondence with 2-sheeted coverings of $P_{\mathrm{SO}}(E)$ which are non-trivial on the fibres of π.*

This can be reinterpreted via Lemma 1.1 as follows:

Corollary 1.5. *Suppose X is connected. Then the spin structures on E are in natural one-to-one correspondence with elements of $H^1(P_{\mathrm{SO}}(E); \mathbb{Z}_2)$ whose restriction to the fibre of $P_{\mathrm{SO}}(E)$ is non-zero.*

We are now in a position to discuss the question of existence and uniqueness of spin structures. Associated to the fibration $\mathrm{SO}_n \xrightarrow{i} P_{\mathrm{SO}}(E) \xrightarrow{\pi} X$ there is an exact sequence

$$0 \longrightarrow H^1(X; \mathbb{Z}_2) \xrightarrow{\pi^*} H^1(P_{\mathrm{SO}}(E); \mathbb{Z}_2) \xrightarrow{i^*} \mathrm{H}^1(\mathrm{SO}_n; \mathbb{Z}_2) \xrightarrow{w_E} H^2(X; \mathbb{Z}_2) \tag{1.4}$$

which can be deduced from the Serre spectral sequence. In analogy with the definition of w_1 via the sequence (1.2) we make the following definition:

DEFINITION 1.6. The image $w_2(E) = w_E(g_2) \in H^2(X; \mathbb{Z}_2)$ of the generator g_2 of $H^1(\mathrm{SO}_n; \mathbb{Z}_2) \cong \mathbb{Z}_2$ is the **second Stiefel-Whitney class** of the oriented bundle E.

To prove that this (or any other) definition agrees with the one given via classifying spaces, it again suffices to show (cf. (i) and (ii) above):

(i′) This definition of w_2 is natural.
(ii′) This definition of $w_2(\mathbb{E})$ gives the non-zero element in $H^2(\mathrm{BSO}_n; \mathbb{Z}_2) \cong \mathbb{Z}_2$ when $\mathbb{E}$ is the universal oriented n-plane bundle over the classifying space BSO_n.

Property (i′) follows from the naturality of the sequence (1.4). Property (ii′) follows from the exactness of (1.4) and the fact that $P_{\mathrm{SO}}(\mathbb{E}) = \mathrm{ESO}_n$ is contractible.

From Corollary 1.5 and the exactness of the sequence (1.4) we immediately conclude the following.

Theorem 1.7. *Let E be an oriented vector bundle over a manifold X. Then there exists a spin structure on E if and only if the second Stiefel-Whitney class of E is zero.*

Furthermore, if $w_2(E) = 0$, then the distinct spin structures on E are in one-to-one correspondence with the elements of $H^1(X; \mathbb{Z}_2)$.

Note that this result holds for all n. When $n = 2$, the class w_2 is just the mod 2 reduction of the Euler class. More generally, the following is true.

REMARK 1.8 (cf. Milnor-Stasheff [1]). If E is the real $2n$-dimensional bundle underlying a complex n-dimensional vector bundle $\mathbf{E}$, then

$$w_2(E) \equiv c_1(\mathbf{E})(\text{mod } 2)$$

where $c_1(\mathbf{E})$ is the first Chern class of $\mathbf{E}$. (To see this, it suffices to observe that the map $H^2(\mathrm{BSO}_{2n}; \mathbb{Z}_2) \to H^2(\mathrm{BU}_n; \mathbb{Z}_2)$, induced by the inclusion $\mathrm{U}_n \subset \mathrm{SO}_{2n}$, is an isomorphism.)

REMARK 1.9. Note that the spin structure on a bundle E is independent of the bundle metric on E in the following sense. A spin structure on E uniquely determines a spin structure for any other metric. This follows from Theorem 1.4 and the observation that the inclusion $P_{\mathrm{SO}}(E) \subset P_{\mathrm{GL}^+}(E)$, where $P_{\mathrm{GL}^+}(E)$ is the bundle of all oriented bases in E, is a homotopy equivalence.

REMARK 1.10. To choose an orientation and a spin structure for E is in particular to find structure groups for E which are respectively 0 and 1-connected. Conversely, suppose E is equivalent to a vector bundle with a Lie structure group G. If G is connected, then E is orientable; and if G is simply-connected, then E is spin.

It should be noted that the process of finding successively more highly connected structure groups for E terminates at the spin level. This is because for any simply-connected Lie group G, $\pi_2(G) = 0$, and if $\pi_3(G) = 0$, then G is contractible.

The conditions $w_1 = 0$ and $w_2 = 0$ can be interpreted geometrically as follows:

Proposition 1.11. *Let E be a vector bundle over a manifold X. Then E is orientable if and only if the restriction of E to any circle embedded in X is trivial.*

The proof is obvious. There is an analogue for w_2.

Proposition 1.12. *Let E be an oriented vector bundle of dimension $\geqq 3$ over X. Then E is spin if and only if for any compact surface Σ and any continuous map $f:\Sigma \to X$, the bundle f^*E is trivial.*

Suppose, furthermore, that X is simply-connected and of dimension > 4. Then E is spin if and only if the restriction of E to any 2-sphere embedded in X is trivial.

Proof. The group $H_2(X;\mathbb{Z}_2)$ is generated by maps $f:\Sigma \to X$ of compact surfaces. Hence, $w_2(E) = 0 \Leftrightarrow f^*w_2(E) = w_2(f^*E) = 0$ for all such $f \Leftrightarrow f^*E$ is trivial for all such f (since an oriented bundle of dimension $\geqq 3$ over a surface is trivial if and only if $w_2 = 0$). This proves the first statement. For the second statement it suffices to note that when $\pi_1 X = 0$ and $\dim(X) > 4$, the group $H_2(X;\mathbb{Z}_2)$ is generated by embedded 2-spheres. ∎

REMARK 1.13. This second statement can be refined somewhat. If $\pi_1 X = 0$, then $H_2(X;\mathbb{Z}_2)$ is generated by *immersions* of S^2, and if $\dim X > 4$, the immersions can be deformed to embeddings.

We now consider some alternative definitions of the classes w_1 and w_2. Recall that (cf. App. A) for any topological group G, we can view the equivalence classes of principal G-bundles on X as elements in a "Čech cohomology space" $H^1(X;G)$. As noted in Milnor [7] the short exact sequence of topological groups

$$1 \longrightarrow \mathrm{SO}_n \xrightarrow{\ \iota\ } \mathrm{O}_n \xrightarrow{\ \rho\ } \mathbb{Z}_2 \longrightarrow 0 \tag{1.5}$$

gives an exact sequence

$$H^1(X;\mathrm{SO}_n) \xrightarrow{\ \iota_*\ } H^1(X;\mathrm{O}_n) \xrightarrow{\ \rho_*\ } H^1(X;\mathbb{Z}_2). \tag{1.6}$$

Given an n-dimensional bundle E on X, we define the first Stiefel-Whitney class by setting $w_1(E) = \rho_*([P_{\mathrm{O}}(E)])$. This definition is easily seen to have properties (i) and (ii) and it therefore agrees with our previous definitions.

In a similar way the short exact sequence

$$0 \longrightarrow \mathbb{Z}_2 \longrightarrow \mathrm{Spin}_n \xrightarrow{\ \xi\ } \mathrm{SO}_n \longrightarrow 1 \tag{1.7}$$

gives an exact sequence

$$\begin{aligned} H^0(X;\mathrm{SO}_n) \xrightarrow{\ \delta^0\ } H^1(X;\mathbb{Z}_2) \longrightarrow H^1(X;\mathrm{Spin}_n) \xrightarrow{\ \xi_*\ } \\ H^1(X;\mathrm{SO}_n) \xrightarrow{\ \delta\ } H^2(X;\mathbb{Z}_2) \end{aligned} \tag{1.8}$$

(cf. Hirzebruch [1]). Thus, for an oriented bundle E we define the second Stiefel-Whitney class by setting $w_2(E) = \delta([P_{\mathrm{SO}}(E)])$. This definition has properties (i′) and (ii′) and therefore agrees with our previous ones.

With this definition it is transparent that $w_2(E) = 0$ if and only if $P_{\mathrm{SO}}(E)$ is equivalent to the $\mathbb{Z}_2$-quotient of a principal Spin_n-bundle on X.

REMARK 1.14. This is a convenient time to inject a minor word of caution. It was pointed out by Milnor that *the* Spin_n*-bundles associated to distinct spin structures on E may be equivalent as abstract principal bundles.* Recall that spin-structures on E are in one-to-one correspondence with elements of $H^1(X;\mathbb{Z}_2)$. However, by (1.8) we see that the equivalence classes of principal Spin_n-bundles with $\mathbb{Z}_2$-quotient equal to $[P_{\text{SO}}(E)]$ are in one-to-one correspondence with elements of $H^1(X;\mathbb{Z}_2)/\delta^0(H^0(X;\text{SO}_n))$. Now by definition we have that $H^0(X;\text{SO}_n) = C(X,\text{SO}_n)$, the space of continuous maps from X to SO_n. The map

$$\delta^0 : C(X,\text{SO}_n) \longrightarrow H^1(X;\mathbb{Z}_2) \tag{1.9}$$

is given as follows. For a map $f:X \to SO_n$, set $\delta^0(f) = f^*(1)$ where 1 is the generator of $H^1(\text{SO}_n;\mathbb{Z}_2)$.

The map (1.9) is often surjective. For example, any class which is the mod 2 reduction of an integral class lies in the image. To see this let $f_0:X \to S^1$ represent the integral class and define $f = i \circ f_0 : X \to \text{SO}_n$ where $i: S^1 \hookrightarrow \text{SO}_n$ is not homotopic to zero. Thus, if $H^1(X;\mathbb{Z}_2) \cong H^1(X;\mathbb{Z}) \otimes \mathbb{Z}_2$, then δ^0 is surjective. Similarly if $\dim X < n$, then δ^0 is also surjective. To see this note that every class in $H^1(X;\mathbb{Z}_2)$ can be induced by a map to $\mathbb{P}^n(\mathbb{R}) \subset \text{SO}_n$. Thus, in either of these cases, *all* the principal Spin_n-bundles associated to distinct spin structures on E are abstractly equivalent.

As an example of this phenomenon consider an oriented bundle E over the circle S^1.

Since $H^1(S^1;\mathbb{Z}_2) \cong \mathbb{Z}_2$, there are two spin structures on E. However, any principal Spin_n-bundle over S^1 is equivalent to the trivial one since it admits a cross-section (as does any fibre bundle over S^1 with connected fibre).

Finally we mention a direct definition of w_1 and w_2 via homotopy theory. From the sequence (1.4) there is a fibration

$$\text{BSO}_n \longrightarrow \text{BO}_n \xrightarrow{w} \text{B}\mathbb{Z}_2 = K(\mathbb{Z}_2,1).$$

A map $f_E : X \to \text{BO}_n$ (classifying a bundle E) has a lifting to BSO_n iff $w \circ f_E$ is homotopic to zero. Recall that $[X,K(\mathbb{Z}_2,1)] \cong H^1(X;\mathbb{Z}_2)$. The class $w \circ f_E$ is the first Stiefel-Whitney class $w_1(E)$.

From the sequence (1.6) we have a fibration $K(\mathbb{Z}_2,1) \to \text{BSpin}_n \xrightarrow{B\xi} \text{BSO}_n$. It follows that the cofiber of $B\xi$ is $K(\mathbb{Z}_2,2)$, i.e., there is a fibration

$$\text{BSpin}_n \xrightarrow{B\xi} \text{BSO}_n \xrightarrow{w} K(\mathbb{Z}_2,2).$$

We can now define $w_2(E)$ as we defined $w_1(E)$ above.

We complete this section with an observation concerning Whitney sums.

Proposition 1.15. *Given three vector bundles E', E'' and $E \cong E' \oplus E''$ over a manifold X, a choice of orientation on any two of them uniquely determines* *an orientation on the third. Similarly, a choice of spin-structure on any two of them uniquely determines a spin structure on the third.*

Proof. The statement concerning orientations is obvious since for finite dimensional vector spaces $V = V' \oplus V''$, an orientation on any two canonically determines an orientation on the third.

Suppose now that E, E' and E'' are orientable. Then $w_2(E) = w_2(E') + w_2(E'')$. Hence, if any two are spin, so is the third.

Suppose under the correspondence of Corollary 1.5 $a' \in H^1(P_{SO}(E'); \mathbb{Z}_2)$ and $a'' \in H^1(P_{SO}(E''); \mathbb{Z}_2)$ represent spin structures on E' and E'' respectively. Now consider the cartesian product bundle $E' \times E'' \to X \times X$. We let

$$\Delta : P_{SO}(E' \oplus E'') \longrightarrow P_{SO}(E' \times E'')$$

denote the diagonal map. The class in $H^1(P_{SO}(E' \oplus E''); \mathbb{Z}_2)$ which will represent the spin structure determined by a' and a'' will be $\Delta^* b$ where $b \in H^1(P_{SO}(E' \times E''); \mathbb{Z}_2)$ will be the unique class which extends $a' \times 1 + 1 \times a'' \in H^1(P_{SO}(E') \times P_{SO}(E''); \mathbb{Z}_2)$ under the inclusion $P_{SO}(E') \times P_{SO}(E'') \subset P_{SO}(E' \times E'')$. That there is a unique b can be seen from the following diagram:

$$\begin{array}{ccccccccc}
0 \to H^1(X \times X; \mathbb{Z}_2) & \longrightarrow & H^1(P_{SO}(E' \times E''); \mathbb{Z}_2) & \longrightarrow & H^1(SO_{n+n'}; \mathbb{Z}_2) & \longrightarrow & H^2(X \times X; \mathbb{Z}_2) \\
\downarrow \approx & & \downarrow \approx & & \downarrow \approx & & \downarrow \approx \\
0 \to H^1(X \times X; \mathbb{Z}_2) & \to & H^1(P_{SO}(E') \times P_{SO}(E''); \mathbb{Z}_2) & \to & H^1(SO_n \times SO_{n'}; \mathbb{Z}_2) & \to & H^2(X \times X; \mathbb{Z}_2)
\end{array}$$

That any two of the classes a', a'' and $\Delta^* b$ determines the third is now easy to see. Let $\pi_V : P_{SO}(V) \to X$ be the projection map where $V = E$, E' or E''. Then any spin structure γ' (under the correspondence of Corollary 1.5) on E' may be written $a' + \pi_{E'}^* u'$ and any spin structure γ'' on E'' may be written $a'' + \pi_{E''}^* u''$ for $u', u'' \in H^1(X; \mathbb{Z}_2)$. Then following the above prescription γ' and γ'' determine $\Delta^* b + \pi_{E' \oplus E''}^*(u' + u'')$. Clearly any two of these determine the third. ∎

§2. Spin Manifolds and Spin Cobordism

We are now in a position to discuss the notion of spin structures on manifolds. For convenience all manifolds are assumed to be of class C^∞.

DEFINITION. A **spin manifold** is an oriented riemannian manifold with a spin structure on its tangent bundle.

The Stiefel-Whitney classes $w_i(X)$ of a manifold X are defined to be the Stiefel-Whitney classes of its tangent bundle TX. Hence, by Theorem 1.7 we have the following.

Theorem 2.1. *An oriented riemannian manifold X admits a spin structure if and only if its second Stiefel-Whitney class is zero. Furthermore, if* $w_2(X) = 0$*, then the spin structures on X are in one-to-one correspondence with elements of* $H^1(X;\mathbb{Z}_2)$.

Recall (cf. Remark 1.9) that the choice of spin structure for one riemannian metric on X canonically determines a spin structure for any other riemannian metric on X. Here we are using the fact that, for any metric, the inclusion $P_{SO}(X) \subset P_{GL^+}(X)$ of the bundle of oriented orthonormal tangent frames into the bundle of all oriented tangent frames, is a homotopy equivalence.

Consider now a diffeomorphism $f: X \xrightarrow{\approx} X$ of a spin manifold X. If f is orientation preserving, then there is an induced diffeomorphism $df: P_{GL^+}(X) \to P_{GL^+}(X)$ of the bundle of oriented tangent frames. This map carries fibres to fibres and therefore induces a permutation of the possible spin structures on X (considered as 2-fold coverings of $P_{GL^+}(X)$.) If the given spin structure remains fixed, then f is called a **spin structure-preserving diffeomorphism.**

It is a classical result of W.-T. Wu [1] that the Stiefel-Whitney classes of a compact manifold X depend only of the homotopy type of X. Therefore, the property of having or not having a spin structure is a homotopy invariant, and in particular, it remains the same under changes of the differentiable structure on X. Wu proves his result by giving a homotopy-theoretic formula for the total Stiefel-Whitney class $w = 1 + w_1 + w_2 + \ldots$ of X. Define $v = 1 + v_1 + v_2 + \ldots \in H^*(X;\mathbb{Z}_2)$ by the requirement that

$$(v \cup a)[X] = \mathrm{Sq}(a)[X]$$

for all $a \in H^*(X;\mathbb{Z}_2)$ where $\mathrm{Sq} = 1 + \mathrm{Sq}^1 + \mathrm{Sq}^2 + \ldots$ is the Steenrod square automorphism, and where $[X] \in H_n(X;\mathbb{Z}_2)$, $n = \dim(X)$, denotes the fundamental class. Wu's formula states that

$$w = \mathrm{Sq}(v)$$

(see Milnor-Stasheff [1] for details). Since $\mathrm{Sq}^i(a) = 0$ if $i > \deg(a)$, we see that $v_k = 0$ for $k > [\dim(X)/2]$. Thus, for example, if $\dim(X) = 3$, we see that $v = 1 + v_1 = 1 + w_1$, and it follows that $w_2 = w_1^2$. In particular, if X is orientable, then it is automatically spin. Suppose now that $\dim(X) = 4$ and $w_1 = 0$. Then $v = 1 + v_2 = 1 + w_2$, and we see that w_2 is characterized by the fact that $(w_2 \cup a)[X] = (a \cup a)[X]$ for all $a \in H^2(X;\mathbb{Z}_2)$.

We will now examine some examples of spin manifolds. For convenience we shall use the expression X **is spin** to mean that $w_1(X) = w_2(X) = 0$, (i.e., that X carries at least one spin structure for any riemannian metric on X). Recall that any complex manifold is canonically oriented. Furthermore, by Remark 1.8 we know the following:

REMARK 2.2. A complex manifold X is spin if and only if its first Chern class satisfies $c_1(X) \equiv 0 \pmod 2$.

EXAMPLE 2.3 (the trivial examples). It follows immediately from Theorem 2.1 that any 2-connected manifold carries a unique spin structure. The obvious examples of this type include homotopy spheres, Stiefel manifolds, and simply-connected Lie groups. Of course, any manifold whose tangent bundle is stably parallelizable is spin. This includes, for example, the inverse image of any regular value of a smooth map $f: \mathbb{R}^{n+p} \to \mathbb{R}^p$. It also includes any Lie group and any orientable manifold of dimension $\leqq 3$.

EXAMPLE 2.4. Let $\mathbb{P}^n(k)$ denote the n-dimensional projective space over the (skew) field k. Then

$$\mathbb{P}^n(\mathbb{R}) \text{ is spin iff } n \equiv 3 \pmod 4$$

$$\mathbb{P}^n(\mathbb{C}) \text{ is spin iff } n \text{ is odd}$$

$$\mathbb{P}^n(\mathbb{H}) \text{ is spin for all } n.$$

To see this recall that the total Stiefel-Whitney class of $\mathbb{P}^n(K)$ is $w = 1 + w_1 + w_2 + \ldots = (1+g)^{n+1}$ where g, the generator of the cohomology ring, has dimension 1, 2 and 4 for $K = \mathbb{R}$, $\mathbb{C}$ and $\mathbb{H}$ respectively. When $K = \mathbb{R}$, the conditions $w_1 = 0$, $w_2 = 0$ are equivalent to $(n+1) \equiv n(n+1)/2 \equiv 0 \pmod 2$. The cases $K = \mathbb{C}$ and $K = \mathbb{H}$ are obvious.

EXAMPLE 2.5. The manifold SO_n has two distinct spin structures given as follows: $P(\mathrm{SO}_n) = \mathrm{SO}_n \times \mathrm{SO}_N$ where $N = n(n-1)/2$. The two coverings are:

$$\tilde{P}_1(\mathrm{SO}_n) = \mathrm{SO}_n \times \mathrm{Spin}_N$$

$$\tilde{P}_2(\mathrm{SO}_n) = (\mathrm{Spin}_n \times \mathrm{Spin}_N)/\mathbb{Z}_2$$

where $\mathbb{Z}_2$ acts on $\mathrm{Spin}_n \times \mathrm{Spin}_N$ by the map $(g, h) \mapsto (-g, -h)$.

EXAMPLE 2.6. Let Σ_g be a compact Riemann surface of genus g. As observed in 2.4 above, this surface is spin. (The Euler characteristic $c_1(\Sigma_g)$ is even.) There are exactly 2^{2g} distinct spin structures on X which can be constructed as follows. Let $H^1(\Sigma_g;\mathcal{O}^*)$ denote the equivalence classes of holomorphic complex line bundles on Σ_g. This is an abelian group under the operation of tensor product. There is a short exact sequence

$$0 \longrightarrow J \longrightarrow H^1(\Sigma_g;\mathcal{O}^*) \xrightarrow{c_1} H^2(\Sigma_g;\mathbb{Z}) \longrightarrow 0$$

where $J \cong H^1(\Sigma_g;\mathbb{R})/H^1(\Sigma_g;\mathbb{Z}) \cong T^{2g}$. Let $\tau_0 \in H^1(\Sigma_g;\mathcal{O}^*)$ denote the tangent bundle of Σ_g and note that there exist exactly 2^{2g} elements $\tau \in H^1(\Sigma_g;\mathcal{O}^*)$ such that $\tau^2 = \tau_0$. Recall that for any complex line bundle τ the natural map $\tau \to \tau^2$ is of the form $z \to z^2$ in the fibres. Hence, each bundle τ with $\tau^2 = \tau_0$ determines a 2-fold covering of $P_{SO}(\Sigma_g) \subset \tau_0$ which is non-trivial on the fibres. These coverings realize all the distinct spin structures on Σ_g.

This construction is a special case of the general fact (cf. Hitchin [1] and App. D) that for any compact Kähler manifold X with $c_1(X) \equiv 0$ (mod 2), the spin structures on X are in one-to-one correspondence with holomorphic square roots of the canonical bundle of X.

EXAMPLE 2.7. Let $V^n(d)$ denote the non-singular complex hypersurface of degree d in $\mathbb{P}^{n+1}(\mathbb{C})$. That is, $V^n(d)$ is given in homogeneous coordinates $[Z_0, \ldots, Z_{n+1}]$ for $\mathbb{P}^{n+1}(\mathbb{C})$ as the zeros of a homogeneous polynomial $p(Z_0, \ldots, Z_{n+1})$ of degree d which satisfies the condition $(\nabla p)(Z) \neq 0$ when $Z \neq 0$ and $p(Z) = 0$. The diffeomorphism class of $V^n(d)$ is uniquely determined by the integers n and d.

The first Chern class of $V^n(d)$ for $n > 1$ is

$$c_1 = (n + 2 - d) \cdot g$$

where g is the canonical generator of $H^2(V^n(d); \mathbb{Z})$ (the Kähler form induced from $\mathbb{P}^{n+1}(\mathbb{C})$). It follows that

$$V^n(d) \text{ is spin} \iff n + d \text{ is even.}$$

EXAMPLE 2.8. Let $V^n(d_1, \ldots, d_k)$ be the transverse intersection of hypersurfaces $V^{n+k=1}(d_1), \ldots, V^{n+k+1}(d_k)$ in $\mathbb{P}^{n+k}(\mathbb{C})$. Then for $n > k$

$$c_1 = (n + k + 1 - d_1 - \ldots - d_k) \cdot g$$

where g is the canonical generator of $H^2(V^n(d_1, \ldots, d_k); Z) \cong \mathbb{Z}$. Hence,

$$V^n(d_1, \ldots, d_k) \text{ is spin} \iff n + k + 1 + \sum_{i=1}^{k} d_i \text{ is even.}$$

EXAMPLE 2.9 (N. Hitchin [1]). Let $f: M \to \mathbb{P}^2(\mathbb{C})$ be a p-fold ramified cover, branched over a non-singular curve of degree pq. Then M is spin iff p is even and q is odd.

As we observed above, every oriented manifold of dimension $\leqq 3$ is spin. In higher dimensions the spin condition has a nice geometric interpretation. We shall restrict our attention to manifolds X with the property that the homomorphism $H_2(X;\mathbb{Z}) \to H_2(X;\mathbb{Z}_2)$, given by reduction mod 2, is surjective. This holds whenever X is simply-connected.

Theorem 2.10. *Let X be an oriented n-dimensional manifold as above. If $n \geqq 5$, then X is spin if and only if every compact orientable surface embedded in X has trivial normal bundle. If $n = 4$, then X is spin if and only if the normal bundle to every compact orientable surface embedded in X has even Euler class.*

Proof. Since $H_2(X;\mathbb{Z}_2) = H_2(X;\mathbb{Z}) \otimes \mathbb{Z}_2$, the group $H_2(X;\mathbb{Z}_2)$ is generated by maps of compact orientable surfaces. By the classical theorems of Whitney, every map of a surface into X is homotopic to an embedding if $n \geqq 5$, and to a self-transversal immersion if $n = 4$. In this latter case, we can remove small disks at each self-intesection point and attach an embedded handle, thereby producing an embedded surface in the same homology class. Consequently, $H_2(X;\mathbb{Z}_2)$ is generated by smooth embeddings of compact orientable surfaces.

Let $i: \Sigma \hookrightarrow X$ be such an embedding. Then $i^*w_2(X) = i^*w_2(TX) = w_2(i^*TX) = w_2(T\Sigma \oplus N\Sigma) = w_2(T\Sigma) + w_2(N\Sigma) = w_2(N\Sigma)$. Evaluating on the fundamental class, we have $(w_2(X), i_*[\Sigma]) = (i^*w_2(X), [\Sigma]) = (w_2(N\Sigma), [\Sigma])$. Consequently, if $w_2(X) = 0$, then $w_2(N\Sigma) = 0$. On the other hand, $H_2(X;\mathbb{Z}_2)$ is generated by such surfaces. We conclude that $w_2(X) = 0$ if and only if $w_2(N\Sigma) = 0$ for every compact orientable surface embedded in X.

The normal bundle $N\Sigma$ to Σ is orientable. Therefore, when $\dim(N\Sigma) \geqq 3$ (i.e., when $n \geqq 5$), we have $w_2(N\Sigma) = 0$ if and only if $N\Sigma$ is trivial. When $\dim(N\Sigma) = 2$ (i.e., when $n = 4$), we have $w_2(N\Sigma) \equiv \chi(N\Sigma) \pmod 2$. This completes the proof. ∎

Corollary 2.11. *Let X be a simply-connected manifold of dimension $\geqq 5$. Then X is spin if and only if every 2-sphere embedded in X has trivial normal bundle.*

Proof. Since $H_2(X;\mathbb{Z}) \cong \pi_2(X)$ by the Hurewicz Theorem, we have that $H_2(X;\mathbb{Z}_2)$ is generated by embedded 2-spheres. ∎

Corollary 2.12. *Let X be a compact, simply-connected 4-manifold. Then X is spin if and only if* $(y \cup y)[X] \equiv 0 \pmod 2$ *for each* $y \in H^2(X;\mathbb{Z})$ (*i.e., the intersection form is "even"*).

Proof. Let $y \in H^2(X;\mathbb{Z}) \cong [X, \mathbb{P}^\infty(\mathbb{C})]$ be given by a smooth map $f: X \to \mathbb{P}^3(\mathbb{C})$. Make f transversal to a hyperplane $\mathbb{P}^2(\mathbb{C}) \subset \mathbb{P}^3(\mathbb{C})$ and let $\Sigma = f^{-1}(\mathbb{P}^2(\mathbb{C}))$. Then Σ represents the Poincare dual of y, and $(y \cup y)[x]$ is the self-intersection number of Σ, i.e., the Euler number of $N\Sigma$. ∎

From the classification of quadratic forms (e.g., Husemoller-Milnor [1]) we conclude that the signature of X must be a multiple of 8. This result holds, in fact, for any topological 4-manifold with $w_1 = w_2 = 0$. For smooth manifolds there is the following deeper result (see Chap. IV).

Theorem 2.13 (Rochlin). *The signature of a smooth compact spin 4-manifold is a multiple of 16.*

EXAMPLE 2.14. The complex hypersurface $V^2(4) = \{(Z_0,Z_1,Z_2,Z_3): Z_0^4 + Z_1^4 + Z_2^4 + Z_3^4 = 0\} \subset \mathbb{P}^3(\mathbb{C})$ is a spin manifold with signature 16 (see Example 2.7 above). This is the so-called **Kummer (or K3) surface.**

REMARK. The statement of Corollary 2.12 fails when $\pi_1(X) \neq 0$. However, the following is true. For a compact orientable 4-manifold X, the intersection form is even if and only if $w_2(X)$ is the mod 2 reduction of a torsion class in $H^2(X;\mathbb{Z})$. This was proved by N. Habegger [1] who also showed the following. Let $X = V^2(4)/\mathbb{Z}_2$ be the quotient of the Kummer surface by the involution $(Z_0, Z_1,Z_2,Z_3) \to (\bar{Z}_1,-\bar{Z}_0,\bar{Z}_3,-\bar{Z}_2)$. Then X is orientable and has even intersection form; however, $\text{sig}(X) = 8$. Of course, X cannot be a spin manifold. In fact, $w_2(X)$ is the mod 2 reduction of the unique torsion class in $H^2(X;\mathbb{Z})$.

The remainder of this section will be devoted to a discussion of spin cobordism. We begin with two observations which follow immediately from Proposition 1.15.

Proposition 2.15. *The cartesian product of two spin manifolds is canonically a spin manifold. Any submanifold of a spin manifold with a spin structure on its normal bundle is canonically a spin manifold.*

In particular, if Y is a compact manifold with boundary ∂Y, then any spin structure on Y induces a spin structure on ∂Y as follows. Let ν be the field of interior unit normal vectors along ∂Y. Using ν, one obtains an embedding $P_{SO}(\partial Y) \subset P_{SO}(Y)$, by completing each tangent frame to ∂Y with the given normal vector. The spin structure, considered for example as a 2-sheeted covering, can now be restricted to $P_{SO}(\partial Y)$.

DEFINITION 2.16. Two spin manifolds are said to be **differentiably equivalent** if there is a diffeomorphism between them preserving orientations and spin structures. A compact (not necessarily connected) spin manifold is said to be **spin cobordant to zero** if it is differentiably equivalent to the boundary of a compact spin manifold Y with (orientation and) spin structure induced from Y as above.

Let Ω_n^{Spin} denote the free abelian group generated by equivalence classes of compact connected n-dimensional spin manifolds, modulo the subgroup generated by elements $[X_1] + \dots + [X_k]$ where $X_1 \amalg \dots \amalg X_k$ is spin cobordant to zero. Ω_n^{Spin} is called the ***n*-dimensional spin cobordism group.**

From Proposition 2.15 we know that the product of two spin manifolds has a uniquely determined spin structure. This multiplication makes $\Omega_*^{\text{Spin}} = \bigoplus_{n=0}^{\infty} \Omega_n^{\text{Spin}}$ into a graded ring, called the **spin cobordism ring.**

Note that the equivalence class (and therefore also the cobordism class) of a spin manifold X is independent of the choice of riemannian metric on X.

REMARK 2.17. Given two spin n-manifolds X_1 and X_2, we can form their connected sum $X_1 \# X_2$ and equip it with a spin structure so that $X_1 \# X_2$ and $X_1 \amalg X_2$ are spin cobordant. Thus every spin cobordism class is represented by a connected manifold (see Milnor [7]).

The connected sum operation is a special case of the general procedure of doing surgery on spin manifolds (see Milnor [5] and Kervaire-Milnor [1]). In particular one can show that for $n \geqq 3$ every spin cobordism class is represented by a simply-connected manifold. For $n \geqq 5$, every spin cobordism class is represented by a 2-connected manifold (see Corollary 2.11). Thus by Poincaré duality and the h-cobordism theorem $\Omega_5^{\text{Spin}} = 0$.

For the special case $n = 1$, a spin structure is defined to be a 2-fold covering of $P_{\text{SO}}(S^1) = S^1$. Consider S^1 as the boundary of the 2-disk with its unique spin structure. Then the spin structure induced on S^1 is the **connected** 2-fold covering. Interestingly, the disconnected 2-fold covering is not cobordant to zero, although two copies of it clearly is (a good exercise). Thus, $\Omega_1^{\text{Spin}} \cong \mathbb{Z}_2$.

It is also true that the square of the circle with the bad spin structure is not zero in $\Omega_2^{\text{Spin}} \cong \mathbb{Z}_2$ (another good exercise).

We now briefly review the Thom construction. Let X^n be a compact n-dimensional spin manifold and choose a smooth embedding $X^n \to S^{n+k}$, $k > n$. The spin structures on X^n and S^{n+k} determine a unique spin structure on the normal bundle $N(X^n)$ of X^n (see Proposition 1.15). Hence, there is a bundle map

$$\begin{array}{ccc} N(X^n) & \xrightarrow{\tilde{f}} & \mathbb{E}_k \\ \downarrow & & \downarrow \\ X^n & \xrightarrow{f} & \text{BSpin}_k \end{array}$$

classifying $N(X^n)$, where $\mathbb{E}_k$ is the universal k-plane bundle. The map $\tilde{f}$ descends to a map of Thom spaces $\tilde{f}: \tau(N(X^n)) \to \tau(\mathbb{E}_k) \equiv \text{MSpin}_k$. If we identify $N(X^n)$ with a tubular neighborhood of X^n in S^{n+k}, then $\tau(N(X^n))$ is the space obtained by collapsing the complement of this tubular neighborhood to a point. Thus we get a map $\pi: S^{n+k} \to \tau(N(X^n))$, and the composition $\tilde{f} \cdot \pi: S^{n+k} \to \text{MSpin}_k$ determines an element

$$\langle X^n \rangle \in \pi_n^S(\text{MSpin}) \equiv \lim_{k \to \infty} \pi_{n+k}(\text{MSpin}_k).$$

The classic result of Thom states that this map induces an isomorphism

$$\Omega_n^{\text{Spin}} \xrightarrow{\approx} \pi_n^S(\text{MSpin}). \tag{2.1}$$

We observe now that there is a natural ring homomorphism

$$\rho_*: \Omega_*^{\text{Spin}} \longrightarrow \Omega_*^{\text{SO}} \tag{2.2}$$

where Ω_*^{SO} is the oriented cobordism ring. In fact, in low dimensions we have the following (cf. Milnor-Stasheff [1; §17] and Milnor [7]):

n	Ω_n^{Spin}	Ω_n^{SO}
0	$\mathbb{Z}$	$\mathbb{Z}$
1	$\mathbb{Z}_2$	0
2	$\mathbb{Z}_2$	0
3	0	0
4	$\mathbb{Z}$ (generated by the Kummer surface surface $V^2(4)$ in Example 2.14)	$\mathbb{Z}$ (generated by $\mathbb{P}^2(\mathbb{C})$)
5	0	$\mathbb{Z}_2$
6	0	0
7	0	0
8	$\mathbb{Z} \oplus \mathbb{Z}$ (generated by $\mathbb{P}^2(\mathbb{H})$ and a manifold L^8 such that $4L^8$ is spin cobordant to $V^2(4) \times V^2(4)$)	$\mathbb{Z} \oplus \mathbb{Z}$ (generated by $\mathbb{P}^4(\mathbb{C})$ and $\mathbb{P}^2(\mathbb{C}) \times \mathbb{P}^2(\mathbb{C})$)

Of course the non-zero element $x \in \Omega_1^{\mathrm{Spin}}$ given by the circle with the "bad" spin structure, goes to zero in Ω_1^{SO}. Hence, we have $x \cdot \Omega_*^{\mathrm{Spin}} \subset \ker \rho_*$. In fact, it can be proved that $x \cdot \Omega_*^{\mathrm{Spin}} = \ker \rho_*$ (see Anderson-Brown-Peterson [1]).

The cokernel of ρ_* is not so simple to describe. For example, the signature gives an isomorphism $\Omega_4^{SO} \cong \mathbb{Z}$, and therefore using Theorem 2.13 and Example 2.14 we get a short exact sequence

$$0 \longrightarrow \Omega_4^{\mathrm{Spin}} \xrightarrow{\rho_4} \Omega_4^{SO} \longrightarrow \mathbb{Z}_{16} \longrightarrow 0.$$

Similarly,

$$0 \longrightarrow \Omega_8^{\mathrm{Spin}} \xrightarrow{\rho_8} \Omega_8^{SO} \longrightarrow \mathbb{Z}_{2^7} \longrightarrow 0$$

is exact. The map ρ_* tensored with the rational numbers is an isomorphism. In fact, so is ρ_* tensored with $\mathbb{Z}[\frac{1}{2}]$ (cf. Milnor [5]).

It is known that the oriented cobordism class of a manifold is determined by its Pontryagin and Stiefel-Whitney numbers. Similarly, it has been proved that the spin cobordism class of a manifold is completely determined by its Stiefel-Whitney and KO-characteristic numbers (Anderson-Brown-Peterson [1]). A fundamental such KO-invariant is the ring homomorphism

$$\hat{\mathscr{A}}_* : \Omega_*^{\mathrm{Spin}} \longrightarrow KO^{-*}(\mathrm{pt}) \tag{2.3}$$

which we describe in §3 of this chapter. Recall that

n (mod 8)	0	1	2	3	4	5	6	7
$KO^{-n}(\mathrm{pt})$	$\mathbb{Z}$	$\mathbb{Z}_2$	$\mathbb{Z}_2$	0	$\mathbb{Z}$	0	0	0

The homomorphism $\hat{\mathscr{A}}_n$ is an isomorphism for $n \leqq 7$.

One of the remarkable and very useful consequences of the Atiyah-Singer Index Theory is that this invariant can be computed as the topological index of an elliptic operator naturally defined on any spin manifold in the cobordism class. This will be discussed in Chapter III.

One of the interesting uses of the invariant $\hat{\mathscr{A}}$ is the following. Let Σ^n be an n-dimensional **homotopy sphere,** i.e., a compact differentiable manifold which is homotopy equivalent to the n-sphere S^n. Then Σ^n is cobordant to zero (cf. Kervaire-Milnor [1]), but it is not necessarily spin cobordant to zero. In fact the homotopy n-spheres form a finite abelian group, Θ_n, under the operation of connected sum, and $\hat{\mathscr{A}}_n:\Theta_n \to KO^{-n}(\mathrm{pt})$ is a homomorphism. The following is a consequence of deep results of Adams [3] and Milnor [7].

Theorem 2.18. *For* $n \equiv 1$ or $2 \pmod 8$ *and* $n > 8$, *the homomorphism*

$$\hat{\mathscr{A}}_n:\Theta_n \longrightarrow \mathbb{Z}_2$$

is surjective.

§3. Clifford and Spinor Bundles

We begin this section by briefly describing the **associated bundle construction.** Let $\pi:P \to X$ be a principal G-bundle over a space X, and let Homeo(F) denote the group of homomorphisms of another space F. Give Homeo(F) the compact-open topology. Then to each continuous homomorphism $\rho:G \to \mathrm{Homeo}(F)$, we construct a fibre bundle over X with fibre F as follows. Consider the free left action of G on the product $P \times F$ given by

$$\varphi_g(p,f) = (pg^{-1},\rho(g)f)$$

for $g \in G$ and $(p,f) \in P \times F$. Define $P \times_\rho F$ to be the quotient space (the space of orbits) of this action. One easily sees that the projection $P \times F \to P \xrightarrow{\pi} X$ descends to a mapping

$$\pi_\rho:P \times_\rho F \longrightarrow X$$

which is the fibre bundle over X with fibre F. It is called **the bundle associated to** P *by* ρ.

If X and F are manifolds, G is a Lie group, and P is differentiable, one can consider continuous homomorphisms $\rho:G \to \mathrm{Diff}(F)$, where Diff$(F)$ is the group of diffeomorphisms of F with the usual C^∞ topology. In this case, the associated bundle $\pi_\rho:P \times_\rho F \to X$ is differentiable.

Note that if P is given by transition functions

$$g_{\alpha\beta}:U_\alpha \cap U_\beta \longrightarrow G$$

for $U_\alpha,U_\beta \in \mathscr{U}$, where $\mathscr{U}$ is some open cover of X, then $P \times_\rho F$ is given by the transition functions:

$$\rho \circ g_{\alpha\beta} : U_\alpha \cap U_\beta \longrightarrow \mathrm{Diff}(F).$$

Note also that if $\rho : G \to \mathrm{GL}(V)$ is a linear representation on a vector space V, then $P \times_\rho V$ is a vector bundle over X.

EXAMPLE 3.0. Let X be a manifold and let $p : \tilde{X} \to X$ be its universal covering space. Then $\tilde{X}$ is a principal $\pi_1(X)$-bundle. Choose $\rho \in \mathrm{Hom}(\pi_1 X, \mathbb{Z}_2) \cong H^1(X; \mathbb{Z}_2)$, and consider $\mathbb{Z}_2$ as the group of permutations of the set $\{0,1\}$. Then $\tilde{X} \times_\rho \{0,1\}$ is a 2-sheeted covering of X (see Lemma 1.1).

EXAMPLE 3.1. Let X be a manifold and let $P = P_{\mathrm{GL}}(X)$ be the principal $\mathrm{GL}_n(\mathbb{R})$-bundle of tangent frames. Let $\rho_n : \mathrm{GL}_n(\mathbb{R}) \to \mathrm{GL}(\mathbb{R}^n)$ denote the standard representation, and let ρ_n^* denote the dual representation ($\rho_n^*(g) = \rho_n(g^{-1})^t$). Then

$$TX = P_{\mathrm{GL}}(X) \times_{\rho_n} \mathbb{R}^n \qquad \text{and} \qquad T^*X = P_{\mathrm{GL}}(X) \times_{\rho_n^*} (\mathbb{R}^n)^*$$

where TX and T^*X are the tangent and cotangent bundles of X respectively. Similarly,

$$\Lambda^k TX = P_{\mathrm{GL}}(X) \times_{\Lambda^k \rho_n} \Lambda^k \mathbb{R}^n$$

$$\Lambda^k T^*X = P_{\mathrm{GL}}(X) \times_{\Lambda^k \rho_n^*} (\Lambda^k \mathbb{R}^n)^*$$

$$\bigotimes\nolimits_s^r TX = P_{\mathrm{GL}}(X) \times_{\otimes_s^r \rho_n} \left(\bigotimes\nolimits_s^r \mathbb{R}^n\right)$$

where $\Lambda^k \rho_n$, $\Lambda^k \rho_n^*$ and $\bigotimes_s^r \rho_n$ are the induced exterior power and tensor product representations.

EXAMPLE 3.2. Let X be an oriented riemannian manifold and let $P = P_{\mathrm{SO}}(X)$ be the SO_n-bundle of positively oriented orthonormal frames. If $\rho_n : \mathrm{SO}_n \to \mathrm{SO}(\mathbb{R}^n)$ is the standard representation, then again

$$TX = P_{\mathrm{SO}}(X) \times_{\rho_n} \mathbb{R}^n$$

$$\Lambda^k TX = P_{\mathrm{SO}}(X) \times_{\Lambda^k \rho_n} (\Lambda^k \mathbb{R}^n)$$

$$\bigotimes\nolimits^r TX = P_{\mathrm{SO}}(X) \times_{\otimes^r \rho_n} \left(\bigotimes\nolimits^r \mathbb{R}^n\right)$$

Note that in this case $\rho_n^* = \rho_n$. This corresponds to the canonical isomorphism $TX \cong T^*X$ given by the riemannian metric.

EXAMPLE 3.3. More generally, if E is any oriented riemannian vector bundle over a space X, then

$$E = P_{\mathrm{SO}}(E) \times_{\rho_n} \mathbb{R}^n$$

$$\Lambda^k(E) = P_{\mathrm{SO}}(E) \times_{\Lambda^r \rho_n} (\Lambda^k \mathbb{R}^n)$$

$$\bigotimes\nolimits^r(E) = P_{\mathrm{SO}}(E) \times_{\otimes^r \rho_n} \left(\bigotimes\nolimits^r \mathbb{R}^n\right)$$

Again, since $\rho_n = \rho_n^*$, E and E^* are canonically isomorphic.

These examples suggest the following. Recall that each orthogonal transformation of $\mathbb{R}^n$ induces an orthogonal transformation of $C\ell(\mathbb{R}^n) = C\ell_n$. (It maps the tensor algebra to itself and preserves the ideal.) This induced map on $C\ell_n$ clearly preserves the multiplication. Hence we get a representation:

$$c\ell(\rho_n): SO_n \longrightarrow \mathrm{Aut}(C\ell(\mathbb{R}^n)). \tag{3.1}$$

DEFINITION 3.4. The **Clifford bundle** of the oriented riemannian vector bundle E is the bundle

$$C\ell(E) = P_{SO}(E) \times_{c\ell(\rho_n)} C\ell(\mathbb{R}^n)$$

associated to the representation (3.1).

E is a bundle of vector spaces with inner products, and $C\ell(E)$ is just the associated bundle of Clifford algebras. In fact $C\ell(E)$ could be defined as the quotient bundle:

$$C\ell(E) = \left(\sum_{r=0}^{\infty} \bigotimes^r E\right) \Big/ I(E)$$

where $I(E)$ is the bundle of ideals, i.e., the bundle whose fibre at $x \in X$ is the two-sided $I(E_x)$ in $\sum_{r=0}^{\infty} \bigotimes^r E_x$, generated by elements $v \otimes v + \|v\|^2$ for $v \in E_x$.

It is evident that $C\ell(E)$ is in fact a *bundle of* (*Clifford*) *algebras* over X. The fibrewise multiplication in $C\ell(E)$ gives an algebra structure to the space of sections of $C\ell(E)$.

It is also evident that each of the notions intrinsic to Clifford algebras carries over to Clifford bundles. For example, there is a decomposition

$$C\ell(E) = C\ell^0(E) \oplus C\ell^1(E) \tag{3.2}$$

corresponding to the even-odd decomposition of the algebras. These are the $+1$ and -1 eigenbundles of the bundle automorphism

$$\alpha: C\ell(E) \longrightarrow C\ell(E) \tag{3.3}$$

which extends the map $E \to E$ sending v to $-v$. There is also an intrinsically defined bundle map

$$L: C\ell(E) \longrightarrow C\ell(E) \tag{3.4}$$

which in any fibre $C\ell(E_x)$ is given by

$$L(\varphi) = -\sum_{k=1}^{n} e_k \varphi e_k \tag{3.5}$$

where $\{e_1, \ldots, e_n\}$ is an orthonormal basis of E_x (see Chap. I).

The following is an elementary but important fact:

Proposition 3.5. *There is a canonical vector bundle isometry*

$$\lambda: \Lambda^*(E) \xrightarrow{\approx} C\ell(E) \tag{3.6}$$

under which

$$\lambda(\Lambda^{\text{even}} E) = C\ell^0(E); \qquad \lambda(\Lambda^{\text{odd}} E) = C\ell^1(E) \tag{3.7}$$

and

$$\lambda(\Lambda^p E) = \{\varphi \in C\ell(E) : \alpha \circ L(\varphi) = (n - 2p)\varphi\} \tag{3.8}$$

for $p = 0, \ldots, n$.

Proof. The isomorphism λ follows directly from the canonical isomorphism $\lambda : \Lambda^* \mathbb{R}^n \xrightarrow{\approx} C\ell(\mathbb{R}^n)$ and the fact that $\lambda \circ \Lambda^* \rho_n = c\ell(\rho_n) \circ \lambda$. Equations (3.7) and (3.8) follow from Proposition 3.8 in Chapter 1. ■

As mentioned in the introduction, it is now natural to look for bundles of irreducible modules over the bundle of Clifford algebras $C\ell(E)$. Such bundles can be constructed if $w_2(E) = 0$.

DEFINITION 3.6. Let E be an oriented riemannian vector bundle with a spin structure $\xi : P_{\text{Spin}}(E) \to P_{\text{SO}}(E)$. A **real spinor bundle** of E is a bundle of the form

$$S(E) = P_{\text{Spin}}(E) \times_\mu M,$$

where M is a left module for $C\ell(\mathbb{R}^n)$ and where $\mu : \text{Spin}_n \to \text{SO}(M)$ is the representation given by left multiplication by elements of $\text{Spin}_n \subset C\ell^0(\mathbb{R}^n)$.

Similarly, a **complex spinor bundle** of E is a bundle of the form

$$S_{\mathbb{C}}(E) = P_{\text{Spin}}(E) \times_\mu M_{\mathbb{C}}$$

where $M_{\mathbb{C}}$ is a complex left module for $C\ell(\mathbb{R}^n) \otimes \mathbb{C}$.

If the module M (or $M_{\mathbb{C}}$) is $\mathbb{Z}_2$-graded, the corresponding bundle is said to be **$\mathbb{Z}_2$-graded.**

EXAMPLE 3.7. Consider $C\ell(\mathbb{R}^n)$ as a module over itself by left multiplication ℓ. The corresponding real spinor bundle

$$C\ell_{\text{Spin}}(E) = P_{\text{Spin}}(E) \times_\ell C\ell(\mathbb{R}^n)$$

is a "principal $C\ell(\mathbb{R}^n)$-bundle", i.e., it admits a free action of $C\ell(\mathbb{R}^n)$ on the right. There is a natural embedding $P_{\text{Spin}}(E) \subset C\ell_{\text{Spin}}(E)$ which comes from the embedding $\text{Spin}_n \subset C\ell(\mathbb{R}^n)$. Hence, every real spinor bundle for E can be captured from this one.

A similar remark holds for the complex case.

Of course, the bundle $C\ell_{\text{Spin}}(E)$ differs from the Clifford bundle $C\ell(E)$. They can be compared as follows. Consider the representation

$$\text{Ad} : \text{Spin}_n \longrightarrow \text{Aut}(C\ell(\mathbb{R}^n)) \tag{3.9}$$

given by $\text{Ad}_g(\varphi) = g\varphi g^{-1}$ for $g \in \text{Spin}_n \subset C\ell(\mathbb{R}^n)$. Clearly $\text{Ad}_{-1} = $ identity, and so this representation descends to a representation Ad′ of SO_n. One easily checks that Ad′ is just the representation $c\ell(\rho_n)$ given in (3.1). It follows that

$$C\ell(E) = P_{\mathrm{Spin}}(E) \times_{\mathrm{Ad}} C\ell(\mathbb{R}^n).$$

This leads to the following.

Proposition 3.8. *Let $S(E)$ be a real spinor bundle of E. Then $S(E)$ is a bundle of modules over the bundle of algebras $C\ell(E)$. In particular the sections of the spinor bundle are a module over the sections of the Clifford bundle.*

The corresponding fact holds in the complex and $\mathbb{Z}_2$-graded cases.

Proof. The diagram

$$\begin{array}{ccc} P_{\mathrm{Spin}}(E) \times C\ell(\mathbb{R}^n) \times M & \xrightarrow{\ \mu\ } & P_{\mathrm{Spin}}(E) \times M \\ \Big\downarrow{\rho_g} & & \Big\downarrow{\rho'_g} \\ P_{\mathrm{Spin}}(E) \times C\ell(\mathbb{R}^n) \times M & \xrightarrow{\ \mu\ } & P_{\mathrm{Spin}}(E) \times M \end{array}$$

given by

$$\begin{array}{ccc} (p,\varphi,m) & \longrightarrow & (p,\varphi m) \\ \Big\downarrow & & \Big\downarrow \\ (pg^{-1}, g\varphi g^{-1}, gm) & \longrightarrow & (pg^{-1}, g\varphi m) \end{array}$$

clearly commutes. Therefore, μ descends to a mapping

$$\mu : C\ell(E) \oplus S(E) \longrightarrow S(E), \tag{3.10}$$

which is easily seen to have the desired properties. The corresponding argument goes through in the complex and $\mathbb{Z}_2$-graded cases. ■

We say that two spinor bundles of E are **equivalent** iff they are equivalent as bundles of $C\ell(E)$-modules. A bundle of (real or complex, graded or ungraded) $C\ell(E)$-modules is called **irreducible** if at each x the fibre is irreducible as a (real or complex, graded or ungraded) module over $C\ell(E_x)$.

Recall that every module for $C\ell(\mathbb{R}^n)$ can be written as a direct sum of irreducible ones, and there are at most two equivalence classes of irreducible modules. Consulting §5 of Chapter I we obtain the following:

Proposition 3.9. *Every spinor bundle of E (real or complex, graded or ungraded) can be decomposed into a direct sum of irreducible ones. With the assumption that X is connected, the number N of equivalence classes of irreducible ones depends on the dimension n of E as follows.*

n (mod 8)	*Real* **Ungraded**	*Complex* **Ungraded**	*Real* **Graded**	*Complex* **Graded**
1	1	2	1	1
2	1	1	1	2
3	2	2	1	1
4	1	1	2	2
5	1	2	1	1
6	1	1	1	2
7	2	2	1	1
8	1	1	2	2

Thus, for n even, there is only one irreducible, ungraded spinor bundle of E (over $\mathbb{R}$ or over $\mathbb{C}$). If $n \equiv 6$ or 8 (mod 8), the complex one is just the complexification of the real one. If $n \equiv 2$ or 4 (mod 8), the complexification of the real one splits into two copies of the complex one. This, and the corresponding information for n odd, can be easily deduced from the tables in Chapter I.

We now observe that in certain dimensions any real spinor bundle automatically carries a natural complex or quaternion structure. This is a consequence of the fact that for certain n, each irreducible real $C\ell(\mathbb{R}^n)$-module carries a compatible $\mathbb{C}$ or $\mathbb{H}$ structure, i.e., the module multiplication is $\mathbb{C}$ or $\mathbb{H}$ linear (and hence, the complex or quaternion scalar multiplication in M descends to the quotient $P_{\mathrm{Spin}} \times_\mu M$). The appearance of such structure is periodic in n and can be deduced from Table III in Chapter I. We conclude the following.

Proposition 3.10. *Let E be a real n-dimensional bundle equipped with a spin structure, and let $S(E)$ be any real ungraded spinor bundle of E. If $n \equiv 1$ or 5 (mod 8), then $S(E)$ carries a complex structure such that Clifford multiplication is complex linear in each fibre. If $n \equiv 2$, 3 or 4 (mod 8), then $S(E)$ carries a quaternion structure so that Clifford multiplication is quaternion linear in each fibre.*

Let us now say a word about the $\mathbb{Z}_2$-graded case. There is a natural one-to-one correspondence between classes of bundles of irreducible $\mathbb{Z}_2$-graded modules over $C\ell(E) = C\ell^0(E) \oplus C\ell^1(E)$ and classes of bundles of irreducible modules over $C\ell^0(E)$. Given a bundle $S(E) = S^0(E) \oplus S^1(E)$ of the first kind, $S^0(E)$ is of the second. Given an $S^0(E)$ of the second kind, the bundle

$$S(E) = C\ell(E) \otimes_{C\ell^0(E)} S^0(E)$$

is of the first.

Suppose now that $n = 2m$ and $S_{\mathbb{C}}(E)$ is the irreducible complex spinor bundle of E. We shall show explicitly how to split $S_{\mathbb{C}}(E)$ into a direct sum

$$S_{\mathbb{C}}(E) = S_{\mathbb{C}}^+(E) \oplus S_{\mathbb{C}}^-(E) \tag{3.11}$$

of $C\ell^0(E)$-modules. Interpreting $S^+_{\mathbb{C}}(E)$ as $S^0_{\mathbb{C}}(E)$ and $S^-_{\mathbb{C}}(E)$ as $S^1_{\mathbb{C}}(E)$, or the other way around, gives a $\mathbb{Z}_2$-graded module structure to $S_{\mathbb{C}}(E)$. The two possibilities are the two inequivalent graded modules appearing in the table. The construction is as follows. Consider the global section $\omega_{\mathbb{C}}$ of $C\ell(E) \otimes \mathbb{C}$ which at $x \in X$ is given by

$$\omega_{\mathbb{C}} = i^m e_1 \cdots e_{2m} \tag{3.12}$$

for any positively oriented orthonormal basis $\{e_1, \ldots, e_{2m}\}$ of E_x. Then we have

$$\omega_{\mathbb{C}}^2 = 1 \tag{3.13}$$

$$e\omega_{\mathbb{C}} = -\omega_{\mathbb{C}} e \tag{3.13$'$}$$

for any $e \in C\ell^1(E) \otimes \mathbb{C}$. We then define $S^+_{\mathbb{C}}(E)$ and $S^-_{\mathbb{C}}(E)$ to be the $+1$ and -1 eigenbundles for Clifford multiplication by $\omega_{\mathbb{C}}$. One easily sees from (3.13′) that these bundles have the same dimension and that they form the $\mathbb{Z}_2$-graded modules as stated above.

These bundles can be written as associated bundles in the following way. Let $\Delta^{\mathbb{C}^+}_{2m}$ and $\Delta^{\mathbb{C}^-}_{2m}$ denote the two fundamental complex representations of Spin_{2m}. Then

$$S^\pm_{\mathbb{C}}(E) \cong P_{\mathrm{Spin}}(E) \times_{\Delta^{\mathbb{C}\pm}_{2m}} \mathbb{C}^{2^{m-1}}. \tag{3.14}$$

For $n \equiv 0 \pmod 4$ there is an analogous construction in the real case. Let $S(E)$ be the irreducible real spinor bundle of E and define a global section ω of $C\ell(E)$ by setting

$$\omega = e_1 \cdots e_n \tag{3.15}$$

at $x \in X$ where $\{e_1, \ldots, e_n\}$ is any positively oriented orthonormal basis of E_x. Again, we have that

$$\omega^2 = 1 \tag{3.16}$$

$$\omega e = -e\omega \qquad \text{for all } e \in C\ell^0(E). \tag{3.17}$$

For equations (3.16) and (3.17) it is necessary that n be a multiple of 4. This again determines a decomposition

$$S(E) = S^+(E) \oplus S^-(E) \tag{3.18}$$

into the $+1$ and -1 eigenbundles of the operator given by Clifford multiplication by ω. They make $S(E)$ a $\mathbb{Z}_2$-graded module in two distinct ways, thereby accounting for the 2 in Table 3.1 above.

If $n \equiv 0 \pmod 8$, then $S^\pm(E) \otimes \mathbb{C} \cong S^\pm_{\mathbb{C}}(E)$. This corresponds to the fact that in these dimensions, $\Delta^\pm_n$ are the complexifications of real representations.

If $n \equiv 4 \pmod 8$, then $S^{\mp}(E) \otimes \mathbb{C} \cong S^{\pm}_{\mathbb{C}}(E) \oplus S^{\pm}_{\mathbb{C}}(E)$. In these dimensions $\Delta^{\pm}_n$ are quaternionic.

We now make a fundamental observation concerning these $\mathbb{Z}_2$-graded bundles. Let

$$\mathbb{D}(E) = \{e \in E : \|e\| \leqq 1\}$$

be the unit disk bundle of E with boundary

$$\dot{\mathbb{D}}(E) = \{e \in E : \|e\| = 1\},$$

the unit sphere bundle. Let $\pi : \mathbb{D}(E) \to X$ be the bundle projection. Assume n is even, so that $S^{\pm}_{\mathbb{C}}(E)$ are defined. Then the pull-backs of these bundles over $\mathbb{D}(E)$ are canonically isomorphic on $\dot{\mathbb{D}}(E)$ by the map

$$\mu_e : (\pi^* S^+_{\mathbb{C}})_e \longrightarrow (\pi^* S^-_{\mathbb{C}})_e \tag{3.19}$$

given at $e \in \dot{\mathbb{D}}(E)$ by

$$\mu_e(\sigma) = e \cdot \sigma$$

that is, Clifford multiplication by e itself. Since $e \cdot e = -\|e\|^2 = -1$, each map μ_e is an isomorphism. The pair of bundles $\pi^* S^{\pm}_{\mathbb{C}}$ over $\mathbb{D}(E)$, together with the isomorphism $\mu : \pi^* S^+_{\mathbb{C}} \xrightarrow{\approx} \pi^* S^-_{\mathbb{C}}$ over $\dot{\mathbb{D}}(E)$, given by (3.19), determine a **"difference" element**

$$\eta_{\mathbb{C}}(E) = [\pi^* S^+_{\mathbb{C}}, \pi^* S^-_{\mathbb{C}}; \mu] \in \tilde{K}(\tau(E)) \tag{3.20}$$

where $\tau(E) \equiv \mathbb{D}(E)/\dot{\mathbb{D}}(E)$ is the Thom space of E. Here $\tilde{K}$ denotes reduced complex K-theory (cf. I.9).

If $n \equiv 0 \pmod 4$, the analogous construction clearly goes through in the real case. Here we obtain an element

$$\eta(E) = [\pi^* S^+, \pi^* S^-; \mu] \in \widetilde{KO}(\tau(E)). \tag{3.21}$$

We are now in a position to define the map (2.3) discussed in the last section. Let $\mathbb{E}_{8k}$ be the universal $8k$-plane bundle over BSpin_{8k} with its unique spin structure. ($\mathbb{E}_{8k}$ is the pull-back of the universal $8k$-plane bundle over BSO_{8k} by the map $\mathrm{B}\xi : \mathrm{BSpin}_{8k} \to \mathrm{BSO}_{8k}$.) Let

$$\eta(\mathbb{E}_{8k}) \in \widetilde{KO}(\mathrm{MSpin}_{8k})$$

be the class (3.21) defined above. Now fix n and choose k sufficiently large that we have the isomorphism $\Omega^{\mathrm{Spin}}_n \cong \pi_{n+8k}(\mathrm{MSpin}_{8k})$. Then a cobordism class $[X] \in \Omega^{\mathrm{Spin}}_n$ determines a map $f_X : S^{n+8k} \to \mathrm{MSpin}_{8k}$. We define

$$\hat{\mathscr{A}}_n : \Omega^{\mathrm{Spin}}_n \longrightarrow KO^{-n}(\mathrm{pt}) \tag{3.22}$$

by

$$\hat{\mathscr{A}}_n([X]) = f_X^* \eta(\mathbb{E}_{8k}) \in \widetilde{KO}(S^{8k+n}) \cong \widetilde{KO}(S^n) \equiv KO^{-n}(\mathrm{pt}).$$

§4. Connections on Spinor Bundles

Suppose E is a smooth riemannian vector bundle over a manifold X and that $\xi: P_{\mathrm{Spin}}(E) \to P_{\mathrm{SO}}(E)$ is a spin structure on E. Then, of course, any connection on $P_{\mathrm{SO}}(E)$ can be lifted via ξ to a connection on $P_{\mathrm{Spin}}(E)$, and this, in turn, defines a connection on the associated spinor bundles. In this section we shall give an explicit computation of this spinor connection and its associated curvature tensor in terms of the connection on E.

We begin by briefly recalling some facts from the theory of connections. Let $\pi: P \to X$ be a smooth principal bundle over a manifold X with group G. Then G is a Lie group whose Lie algebra will be denoted by $\mathfrak{g}$. G acts freely from the right on P. Each element $V \in \mathfrak{g}$ determines a vector field $\tilde{V}$ on P by setting

$$\tilde{V}_p = d/dt(p \cdot \exp(tV))\big|_{t=0}.$$

The map $V \to \tilde{V}_p$ gives an isomorphism

$$\mathfrak{g} \cong \mathscr{V}_p \tag{4.1}$$

where $\mathscr{V}_p$ is the tangent space to the orbit through p. The orbits are the fibres of π, and the plane $\mathscr{V}_p$ can be thought of as the "vertical" space through p. A connection is then a choice of an invariant field of complementary "horizontal" spaces.

DEFINITION 4.1. A **connection** on P is a G-invariant field of tangent n-planes τ on P ($n = \dim(X)$) such that the linear map $\pi_*: \tau_p \to T_{\pi p}(X)$ is an isomorphism for all $p \in P$.

At each $p \in P$, τ_p determines a linear projection $T_p(P) \to \mathscr{V}_p$. The canonical isomorphism (4.1) then gives a linear map

$$\omega_p: T_p(P) \longrightarrow \mathfrak{g}. \tag{4.2}$$

This defines a $\mathfrak{g}$-valued 1-form ω on P, called the **connection 1-form.** It has the following properties.

$$\omega(\tilde{V}) \equiv V \qquad \text{for all } V \in \mathfrak{g}. \tag{4.3}$$

$$g^*(\omega) = \mathrm{Ad}_{g^{-1}}\omega \qquad \text{for all } g \in G \text{ acting on the manifold } P. \tag{4.4}$$

Note that given the connection 1-form ω, one can recapture the connection by the relation

$$\tau_p = \ker(\omega_p).$$

The **curvature** of the connection is the $\mathfrak{g}$-valued 2-form Ω given by the equation

$$\Omega = d\omega + [\omega,\omega]. \tag{4.5}$$

(Note that by the skew-symmetry of the Lie bracket, $[\omega,\omega](v,w) \stackrel{\text{def}}{\equiv} [\omega(v),\omega(w)]$ is a $\mathfrak{g}$-valued exterior 2-form.) This form has the following properties.

$$\Omega(\tilde{V},\cdot) \equiv 0 \qquad \text{for all } V \in \mathfrak{g} \tag{4.6}$$

$$g^*\Omega = \text{Ad}_{g^{-1}}(\Omega) \qquad \text{for all } g \in G \tag{4.7}$$

EXAMPLE 4.2 (*orthogonal connections*). Let $P = P_{\text{SO}}(E)$ where E is a smooth, oriented riemannian vector bundle. The Lie algebra of SO_n is the space $\mathfrak{so}_n$ of real, skew-symmetric $n \times n$-matrices. Hence, a connection 1-form ω can be considered as an $n \times n$-matrix of 1-forms $\omega = ((\omega_{ij}))$ where $\omega_{ij} = -\omega_{ji}$. The corresponding curvature is a matrix of 2-forms $\Omega = ((\Omega_{ij}))$ where

$$\Omega_{ij} = d\omega_{ij} + \sum_{k=1}^{n} \omega_{ik} \wedge \omega_{kj}. \tag{4.8}$$

For an orthogonal matrix g, $\text{Ad}_g(\omega) = g\omega g^{-1}$.

Let $e_i \wedge e_j$ denote the elementary skew-symmetric (i,j)-matrix. If $\{e_1, \ldots, e_n\}$ denotes the canonical basis of $\mathbb{R}^n$, this corresponds to the transformation

$$(e_i \wedge e_j)(v) = \langle e_i, v\rangle e_j - \langle e_j, v\rangle e_i. \tag{4.9}$$

The connection and curvature forms can then be written as

$$\omega = -\sum_{i<j} \omega_{ij} e_i \wedge e_j \tag{4.10}$$

$$\Omega = -\sum_{i<j} \Omega_{ij} e_i \wedge e_j \tag{4.11}$$

Given a connection on the bundle $P_{\text{SO}}(E)$ as above, we can define a rule for taking derivatives of sections of E. For any smooth vector bundle E' over X, let $\Gamma(E')$ denote the space of smooth cross-sections of E'.

DEFINITION 4.3. A *covariant derivative* on E is a linear map

$$\nabla : \Gamma(E) \longrightarrow \Gamma(T^*X \otimes E)$$

such that

$$\nabla(fe) = df \otimes e + f\nabla e \tag{4.12}$$

for all $f \in C^\infty(X)$ and all $e \in \Gamma(E)$.

Thus, given a smooth vector field V on X, we obtain a map $\nabla_V : \Gamma(E) \to \Gamma(E)$ called **the covariant derivative with respect to V.** At a given point $x \in X$, $(\nabla_V e)_x$ depends only on V_x and on the values of e in a neighborhood of x.

Proposition 4.4. *Let ω be a connection 1-form on $P_{SO}(E)$ as above. Then ω determines a unique covariant derivative on E by the rule*

$$\nabla e_i = \sum_{j=1}^{n} \tilde{\omega}_{ji} \otimes e_j \tag{4.13}$$

where $\mathscr{E} = (e_1, \ldots, e_n)$ is a local family of pointwise orthonormal sections of E, i.e., a local section of $P_{SO}(E)$, and where $\tilde{\omega} = \mathscr{E}^*\omega$.

This covariant derivative satisfies the rule

$$V\langle e, e'\rangle = \langle \nabla_V e, e'\rangle + \langle e, \nabla_V e'\rangle \tag{4.14}$$

for all $V \in T(X)$ and $e, e' \in \Gamma(E)$, where $\langle \cdot, \cdot \rangle$ *denotes the inner product in E.*

Conversely, any covariant derivative on E satisfying (4.14) *determines a unique connection 1-form by equations* (4.13).

Note. A covariant derivative with property (4.14) will be called **riemannian.**

Proof. Let $\mathscr{E} = (e_1, \ldots, e_n)$ be a frame defined on an open set $U \subseteq X$, and let $((\tilde{\omega}_{ij}))$ be any skew-symmetric $n \times n$-matrix of 1-forms on U. Then equation (4.13) together with property (4.12) defines a unique covariant derivative on $E|_U$. (To see this, note that any section e of E over U can be written uniquely as $e = \sum f_j e_j$ for $f_1, \ldots, f_n \in C^\infty(U)$. Then by (4.12) and (4.13) we have $\nabla e = \sum df_j \otimes e_j + \sum f_j \tilde{\omega}_{kj} \otimes e_k = \sum \{df_k + \sum \tilde{\omega}_{kj} f_j\} \otimes e_k$. This definition of ∇ has properties (4.12) and (4.14) and is therefore a riemannian covariant derivative on U.) Conversely, given $\mathscr{E}$ and a riemannian covariant derivative ∇ on U, we have a skew-symmetric matrix of 1-forms $((\tilde{\omega}_{ij}))$ on U uniquely defined by (4.13).

Consequently, to define a global covariant derivative on E it suffices to assign to each local frame field $\mathscr{E}$ a matrix of local 1-forms $((\tilde{\omega}_{ij}))$ satisfying the following compatibility condition. Suppose $\mathscr{E} = (e_1, \ldots, e_n)$ and $\mathscr{E}' = (e'_1, \ldots, e'_n)$ are two orthonormal frame fields over an open set U, and let $\tilde{\omega} = ((\tilde{\omega}_{ij}))$ and $\tilde{\omega}' = ((\tilde{\omega}'_{ij}))$ be the associated matrices of 1-forms. Then for each $x \in U$, there is a unique orthogonal $n \times n$-matrix $g(x) = ((g_{ij}(x)))$ such that $\mathscr{E}(x) = \mathscr{E}'(x)g(x)$, i.e.,

$$e_i(x) = \sum_{j=1}^{n} e'_j(x) g_{ji}(x).$$

Applying ∇ and using (4.13) we find easily that

$$\tilde{\omega}(x) = g^{-1}(x)\tilde{\omega}'(x)g(x) + g^{-1}(x)dg(x). \tag{4.15}$$

This transformation rule is the required compatibility condition.

Suppose now that $P_{SO}(E)$ is provided with a connection 1-form ω. Then given a local section $\mathscr{E} = (e_1, \ldots, e_n)$ of $P_{SO}(E)$ over an open set U, we get a skew-symmetric matrix of 1-forms $\tilde{\omega} = ((\tilde{\omega}_{ij}))$ on U by setting $\tilde{\omega} = \mathscr{E}^*\omega$. Note that $\mathscr{E}$ determines a local trivialization

$$\varphi : U \times \mathrm{SO}_n \longrightarrow \pi^{-1}(U)$$

of $\pi : P_{\mathrm{SO}}(E) \to X$ by setting $\varphi(x,g) = \mathscr{E}(x)g$. Conversely, φ determines $\mathscr{E}$, since $\mathscr{E}(x) = \varphi(x,e)$. Note that φ is SO_n-equivariant.

We now observe that in this local product determined by φ, the connection 1-form can be written as

$$(\varphi^*\omega)_{x,g} = \mathrm{Ad}_{g^{-1}}(\tilde{\omega}) + g^{-1}dg. \tag{4.16}$$

To see this we first write $\varphi^*\omega = \omega_0 + \omega_1$ where $\omega_0 = \sum a_i(x,g)dx^i$ for local coordinates x^i on X and where $\omega_1 = \sum b_{ij}(x,g)dg_{ij}$. Properties (4.3) and (4.4) imply that $\omega_1 = g^{-1}dg$. By definition we have $\omega_0 = \mathscr{E}^*\omega$ along $U \times \{e\} \subset U \times G$. Property (4.4) then implies that $\omega_0 = \mathrm{Ad}_{g^{-1}}(\tilde{\omega}) = g^{-1} \circ \tilde{\omega} \circ g$ at a general point (x,g).

If we choose a different cross section $\mathscr{E}' = (e_1', \dots, e_n')$ over U, we get a new trivialization $\varphi' : U \times \mathrm{SO}_n \to \pi^{-1}(U)$ by the formula $\varphi'(x,g) = \mathscr{E}'(x)g$. The change of trivializations $\Phi = (\varphi')^{-1} \circ \varphi : U \times \mathrm{SO}_n \to U \times \mathrm{SO}_n$ is given by

$$\Phi(x,g) = (x, g(x)g) \tag{4.17}$$

where $g : U \to \mathrm{SO}_n$ is the change of frames as above, i.e., $\mathscr{E}(x) = \mathscr{E}'(x)g(x)$. Clearly we have that

$$\varphi^*\omega = \Phi^*(\varphi'^*\omega).$$

Using (4.16) and (4.17) we can re-express this as

$$\begin{aligned} \mathrm{Ad}_{g^{-1}}(\tilde{\omega}) + g^{-1}dg &= \mathrm{Ad}_{(g(x)g)^{-1}}(\tilde{\omega}') + (g(x)g)^{-1}d(g(x)g) \\ &= \mathrm{Ad}_{g^{-1}}\{\mathrm{Ad}_{g^{-1}(x)}(\tilde{\omega}') + g^{-1}(x)dg(x)\} + g^{-1}dg. \end{aligned} \tag{4.18}$$

This equation immediately reduces to the compatibility condition (4.15). Consequently, a connection 1-form on $P_{\mathrm{SO}}(E)$ determines a riemannian covariant derivative on E, as claimed. Conversely, given a riemannian covariant derivative, we obtain local 1-forms $\tilde{\omega}$ transforming according to (4.15). This implies that the compatibility condition (4.18) for the existence of a global connection 1-form on $P_{\mathrm{SO}}(E)$ is satisfied. This completes the proof. ■

Given a covariant derivative ∇ on E, it is natural to ask whether the second covariant derivatives commute in an appropriate sense. For this we consider the composition

$$\Gamma(E) \xrightarrow{\nabla} \Gamma(T^* \otimes E) \xrightarrow{\tilde{\nabla}} \Gamma(\Lambda^2 T^* \otimes E)$$

where $\tilde{\nabla}$ is the natural prolongation of ∇ defined on sections of the form $\alpha \otimes e$ by $\tilde{\nabla}(\alpha \otimes e) = d\alpha \otimes e - \alpha \wedge \nabla e$, and we set $R = \tilde{\nabla} \circ \nabla$

Proposition 4.5. *Let ω, $\mathscr{E}$ and ∇ be as in Proposition 4.4, and let Ω be the curvature 2-form of the connection. Then*

$$Re_i = \sum_{j=1}^{n} \tilde{\Omega}_{ji} \otimes e_j \tag{4.19}$$

where $\tilde{\Omega} = \mathscr{E}^*\Omega$.

Proof. From equation (4.8) we have

$$\begin{aligned}\tilde{\nabla}(\nabla e_i) &= \tilde{\nabla}\left(\sum_{j=1}^{n} \tilde{\omega}_{ji} \otimes e_j\right)\\ &= \sum_{j=1}^{n} d\tilde{\omega}_{ji} \otimes e_j + \sum_{j,k=1}^{n} \tilde{\omega}_{kj} \wedge \tilde{\omega}_{ji} \otimes e_k\\ &= \sum_{j=1}^{n} \tilde{\Omega}_{ji} \otimes e_j. \quad \blacksquare\end{aligned}$$

Proposition 4.6. *Let Ω, $\mathscr{E}$ and ∇ be as in Proposition 4.5. Then for local tangent vector fields V and W on X, we have*

$$R_{V,W}e = (\nabla_V\nabla_W - \nabla_W\nabla_V - \nabla_{[V,W]})e. \tag{4.20}$$

Proof. Note that

$$\begin{aligned}\nabla_V\nabla_W e_i &= \nabla_V\left(\sum_j e_j\tilde{\omega}_{ji}(W)\right)\\ &= \sum_{j,k} e_k\tilde{\omega}_{kj}(V)\tilde{\omega}_{ji}(W) + \sum_j e_j V \cdot \tilde{\omega}_{ji}(W)\end{aligned}$$

and therefore

$$\begin{aligned}(\nabla_V\nabla_W - \nabla_W\nabla_V - \nabla_{[V,W]})e_i &= \sum_j e_j\{V \cdot \tilde{\omega}_{ji}(W) - W \cdot \tilde{\omega}_{ji}(V) - \tilde{\omega}_{ji}([V,W])\}\\ &\quad + \sum_{j,k} e_j\{\tilde{\omega}_{jk}(V)\tilde{\omega}_{ki}(W) - \tilde{\omega}_{jk}(W)\tilde{\omega}_{ki}(V)\}\\ &= \sum_j e_j\{d\tilde{\omega}_{ji}(V,W) + \sum_k \tilde{\omega}_{jk} \wedge \tilde{\omega}_{ki}(V,W)\}\\ &= \sum_j e_j\tilde{\Omega}_{ji}(V,W). \quad \blacksquare\end{aligned} \tag{4.21}$$

Notice that from (4.14) we have the relation

$$\langle R_{V,W}e, e'\rangle + \langle e, R_{V,W}e'\rangle = 0. \tag{4.22}$$

Notice also that from the above it follows that the expression $\langle R_{V,W}e, e'\rangle$ is a tensor, that is, at any point $x \in X$, it depends only on the quantities V_x, W_x, e_x, e'_x and not on the local fields V, W, e, e' extending them. Hence, given two tangent vectors V, W at $x \in X$, the curvature gives a well-defined, skew-symmetric endomorphism

$$R_{V,W}: E_x \longrightarrow E_x \tag{4.23}$$

called the **curvature transformation** associated to V and W.

Suppose now that $\pi: P \to X$ is a smooth principal G-bundle over X, that $\rho: G \to \mathrm{SO}_n$ is a representation of G, and that

$$E_\rho = P \times_\rho \mathbb{R}^n$$

is the associated riemannian vector bundle (cf. II.3). Then given a connection on P there is a canonical connection induced on $P(E_\rho)$ as follows. Note that

$$P(E_\rho) = P \times_\rho \mathrm{SO}_n$$

where an element $g \in G$ acts on SO_n via left multiplication by $\rho(g)$. Given a connection τ on P, extend it trivially to $P \times \mathrm{SO}_n$ and then push it forward to $P \times_\rho \mathrm{SO}_n$. One can easily see that this gives a connection τ_ρ on $P(E_\rho)$.

There is a canonical, G-equivariant mapping

$$i: P \longrightarrow P(E_\rho) \tag{4.24}$$

given by

$$p \longrightarrow [(p,e)].$$

(where $[(p,h)]$ denotes the class of $(p,h) \in P \times \mathrm{SO}_n$ in the quotient $P(E_\rho)$.) If the representation ρ is faithful, the mapping (4.24) is an embedding. For simplicity we assume ρ to be faithful.

Proposition 4.7. *Suppose ω and Ω are the connection and curvature forms on P respectively, and let ω_ρ and Ω_ρ denote the corresponding forms for the induced connection on $P(E_\rho)$. Then, considering $P \subset P(E_\rho)$ as above, we have that*

$$\omega_\rho\big|_P = \rho_* \omega$$

and

$$\Omega_\rho\big|_P = \rho_* \Omega$$

where $\rho_*: \mathfrak{g} \to \mathfrak{so}_n$ is the Lie algebra homomorphism associated to $\rho: G \to \mathrm{SO}_n$.

Proof. This follows straightforwardly from the fact that the embedding (4.24) is G-equivariant, i.e., that $i(p \cdot g) = i(p) \cdot \rho(g)$. ∎

We are now in a position to discuss the connections on Clifford and spinor bundles. Let E be an oriented riemannian vector bundle of dimension n, and suppose E is furnished with a riemannian connection, i.e., a connection τ on $P_{SO}(E)$. The Clifford bundle $C\ell(E)$ is associated to E by the representation (3.1):

$$c\ell(\rho_n): SO_n \longrightarrow \text{Aut}(C\ell(\mathbb{R}^n)).$$

Therefore, by the construction above, τ induces a unique connection, τ', on $C\ell(E)$. Note that since $c\ell(\rho_n)$ maps into the automorphisms of $C\ell(\mathbb{R}^n)$, we have that the associated Lie algebra homomorphism is a map

$$c\ell(\rho_n)_*: \mathfrak{so}_n \longrightarrow \text{Der}(C\ell(\mathbb{R}^n))$$

where Der$(\cdot)$ is the Lie algebra of derivations; i.e., $c\ell(\rho_n)_*$ has the property that for each element $A \in \mathfrak{so}_n$

$$\{c\ell(\rho_n)_*A\}(\varphi \cdot \psi) = (\{c\ell(\rho_n)_*A\}\varphi) \cdot \psi + \varphi \cdot (\{c\ell(\rho_n)_*A\}\psi) \tag{4.25}$$

for all $\varphi,\psi \in C\ell(\mathbb{R}^n)$. Recall, furthermore, that under the canonical identification $C\ell(\mathbb{R}^n) \cong \Lambda^*\mathbb{R}^n$, the representation $c\ell(\rho_n)$ becomes $\Lambda^*\rho_n$. Together with Propositions 4.4 and 4.7, this gives the following.

Proposition 4.8. *The covariant derivative ∇ on $C\ell(E)$ acts as a derivation on the algebra of sections, i.e.,*

$$\nabla(\varphi \cdot \psi) = (\nabla\varphi) \cdot \psi + \varphi \cdot (\nabla\psi) \tag{4.26}$$

for any two sections φ and ψ of $C\ell(E)$.

Furthermore, under the canonical identification $C\ell(E) \cong \Lambda^*(E)$, the covariant derivative ∇ preserves the subbundles $\Lambda^p(E)$ and agrees there with the covariant derivative induced by the representation $\Lambda^p\rho_n$ (i.e., the usual covariant derivative).

Corollary 4.9. *The subbundles $C\ell^0(E)$ and $C\ell^1(E)$ are preserved by ∇. Furthermore, the "volume form" $\omega = e_1 \cdots e_n$ is globally parallel; that is,*

$$\nabla\omega = 0.$$

Therefore when $n \equiv 3$ or $4 \pmod 4$, the eigenbundles $C\ell^{\pm}(E) = \{\varphi \in C\ell(E): \omega\varphi = \pm\varphi\}$ are also preserved by ∇.

Proof. The first statement follows from the fact that $C\ell^0(E) \cong \Lambda^{\text{even}}(E)$ and $C\ell^1(E) \cong \Lambda^{\text{odd}}(E)$. The second statement follows from the fact that ω corresponds to the unit section of $\Lambda^n(E)$. ■

In the analogous way, Propositions 4.5 and 4.7 give the following.

Proposition 4.10. *For any pair of tangent vectors V and W at $x \in X$, the curvature transformation*

$$R_{V,W}: C\ell(E_x) \rightarrow C\ell(E_x)$$

is a derivation, i.e.,

$$R_{V,W}(\varphi \cdot \psi) = R_{V,W}(\varphi) \cdot \psi + \varphi \cdot R_{V,W}(\psi) \tag{4.27}$$

for all $\varphi, \psi \in C\ell(E_x)$. *Furthermore,* $R_{V,W}$ *preserves the subspaces* $C\ell^0(E_x)$, $C\ell^1(E_x)$ *and* $C\ell^{\pm}(E_x)$ *as above.*

Suppose now that E carries a spin structure $\xi: P_{\mathrm{Spin}}(E) \to P_{\mathrm{SO}}(E)$ and let $S(E) = P_{\mathrm{Spin}}(E) \times_\mu M$ be an associated real spinor bundle. Here M is a left module over $C\ell(\mathbb{R}^n)$ and $\mu: \mathrm{Spin}_n \to \mathrm{SO}(M)$ is the resulting representation (cf. §3). The connection τ on $P_{\mathrm{SO}}(E)$ lifts via the covering map ξ to a connection $\tilde{\tau}$ on $P_{\mathrm{Spin}}(E)$. This in turn induces a connection and therefore a covariant derivative on $S(E)$. Recall (Proposition 3.8), that the sections of $S(E)$ form a module over the sections of $C\ell(E)$.

Proposition 4.11. *The covariant derivative* ∇ *on* $S(E)$ *acts as a derivative with respect to the module structure over* $C\ell(E)$, *i.e.,*

$$\nabla(\varphi \cdot \sigma) = (\nabla\varphi) \cdot \sigma + \varphi \cdot (\nabla\sigma) \tag{4.28}$$

for any section φ *of* $C\ell(E)$ *and any section* σ *of* $S(E)$.

Corollary 4.12. *If* $n \equiv 3$ *or* $4 \pmod 4$, *the eigenbundles* $S^{\pm}(E) = \{\varphi \in S(E): \omega\varphi = \pm\varphi\}$ *are preserved by* ∇.

Proof of Proposition 4.11. The representations $c\ell(\rho_n)$ ($=$ Ad) and μ preserve the module multiplication, that is, $\mu(g)(\varphi \cdot \sigma) = \{c\ell(\rho_n)(g)\varphi\} \cdot \{\mu(g)\sigma\}$ for all $g \in \mathrm{Spin}_n$, $\varphi \in C\ell(\mathbb{R}^n)$ and $\sigma \in M$ (see the discussion surrounding Proposition 3.8.) Differentiating at the identity, we get that for each element $A \in \mathfrak{so}_n = \mathfrak{spin}_n$

$$\{\mu_* A\}(\varphi \cdot \sigma) = (\{c\ell(\rho_n)_* A\}\varphi) \cdot \sigma + \varphi \cdot (\{\mu_* A\}\sigma). \tag{4.29}$$

The argument is now completed using Propositions 4.4 and 4.7 as before. ∎

We also have the analogous statement for curvature:

Proposition 4.13. *For any pair of tangent vectors* V,W *at* $x \in X$, *the curvature transformation*

$$R_{V,W}: S(E_x) \longrightarrow S(E_x)$$

is a module derivation, i.e.,

$$R_{V,W}(\varphi \cdot \sigma) = R_{V,W}(\varphi) \cdot \sigma + \varphi \cdot R_{V,W}(\sigma) \tag{4.30}$$

for all $\varphi \in C\ell(E_x)$ *and all* $\sigma \in S(E_x)$. *Furthermore,* $R_{V,W}$ *preserves the subspaces* $S^{\pm}(E_x)$ *when they are defined.*

We shall now proceed to explicitly compute the connection and curvature forms on $S(E)$. To do this we need to examine the representation μ.

Recall that $\mathfrak{so}_n$ is generated by the elementary transformations $x \wedge y$; $x,y \in \mathbb{R}^n$, given by the formula

$$(x \wedge y)(v) \equiv \langle x, v\rangle y - \langle y, v\rangle x. \tag{4.31}$$

From Chapter I, §6, we have

$$\mu_*(x \wedge y) = \tfrac{1}{4}[x,y]. \tag{4.32}$$

That is, $\mu_*(x \wedge y)(\sigma) = \frac{1}{4}[x,y] \cdot \sigma$ for all $\sigma \in M$. Recall that on Spin_n, we have $c\ell(\rho_n) = \mathrm{Ad}$. Hence,

$$c\ell(\rho_n)_* x \wedge y = \mathrm{Ad}_* x \wedge y = \mathrm{ad}_{\frac{1}{4}[x,y]}. \tag{4.33}$$

That is, $\mathrm{Ad}_*(x \wedge y)(\varphi) = \frac{1}{4}[[x,y],\varphi]$ for $\varphi \in C\ell(\mathbb{R}^n)$. Note that equation (4.29) follows immediately from (4.32) and (4.33).

Suppose now that $\mathscr{E} = (e_1, \ldots, e_n)$ is an n-tuple of pointwise orthonormal sections of E defined over a contractible open set $U \subseteq X$. $\mathscr{E}$ is just a section of $P_{\mathrm{SO}}(E)$ over U, and it can be lifted to a section $\tilde{\mathscr{E}}$ of $P_{\mathrm{Spin}}(E)$ over U. There are two possible such liftings. They satisfy the relation: $\xi \circ \tilde{\mathscr{E}} = \mathscr{E}$.

The connection 1-form on $P_{\mathrm{Spin}}(E)$ is just the lift $\xi^*\omega$ of the connection 1-form ω on $P_{\mathrm{SO}}(E)$. To obtain a formula of the type given in Proposition 4.4 we want to pull $\xi^*\omega$ down to U by the local section $\tilde{\mathscr{E}}$. This "pull-down" is just

$$\tilde{\omega} = \tilde{\mathscr{E}}^*(\xi^*\omega) = (\xi \circ \tilde{\mathscr{E}})^*(\omega) = \mathscr{E}^*(\omega).$$

Consequently, the scalar 1-forms $\tilde{\omega}_{ij}$ are just the 1-forms we obtained earlier by pulling down the connection ω by the local frame field $\mathscr{E}$.

Now for any spinor bundle $S(E)$ we have a canonical embedding $P_{\mathrm{Spin}}(E) \subset P_{\mathrm{SO}}(S(E))$. Thus, $\tilde{\mathscr{E}}$ can be considered as a section of $P_{\mathrm{SO}}(S(E))$. Let ω^s denote the connection 1-form on $P_{\mathrm{SO}}(S(E))$. Then to apply Proposition 4.4 we want to compute $\tilde{\omega}^s = \tilde{\mathscr{E}}^*(\omega^s)$. However, by Proposition 4.7, we know that ω^s restricted to $P_{\mathrm{Spin}}(E) \subset P_{\mathrm{SO}}(S(E))$ is just $\mu_*(\xi^*\omega)$. Consequently, we have

$$\tilde{\omega}^s = \mu_*\tilde{\omega}.$$

Writing $\tilde{\omega} = -\sum_{i<j} \tilde{\omega}_{ij} e_i \wedge e_j$ (cf. (4.10)) and using (4.32), we can rewrite this as

$$\tilde{\omega}^s = -\tfrac{1}{2} \sum_{i<j} \tilde{\omega}_{ij} e_i e_j. \tag{4.34}$$

(Note that since $e_1, \ldots, e_n$ are orthogonal, $[e_i, e_j] = e_i e_j - e_j e_i = 2e_i e_j$.)

Finally, we note that the embedding $P_{\mathrm{Spin}}(E) \subset P_{\mathrm{SO}}(S(E))$ can be interpreted to mean that every point of $P_{\mathrm{Spin}}(E)$ determines an orthonormal frame in $S(E)$. In particular, the local section $\tilde{\mathscr{E}}$ determines a local section $\mathscr{S} = (\sigma_1, \ldots, \sigma_N)$ of $P_{\mathrm{SO}}(S(E))$. (The other choice of lifting $\tilde{\mathscr{E}}$ just determines the negative frame $-\mathscr{S} = (-\sigma_1, \ldots, -\sigma_N)$.)

Combining the remarks above, we have the following.

Theorem 4.14. *Let ω be the connection 1-form on $P_{\mathrm{SO}}(E)$ and let $S(E)$ be any spinor bundle associated to E. Then the covariant derivative ∇^s on $S(E)$ is given locally by the formula*

$$\nabla^s \sigma_\alpha = \tfrac{1}{2} \sum_{i<j} \tilde{\omega}_{ji} \otimes e_i e_j \cdot \sigma_\alpha \tag{4.35}$$

where $\mathscr{E} = (e_1, \ldots, e_n)$ is a local section of $P_{\mathrm{SO}}(E)$, $\tilde{\omega} = \mathscr{E}^*(\omega)$, and where $\mathscr{S} = (\sigma_1, \ldots, \sigma_N)$ is a local section of $P_{\mathrm{SO}}(S(E))$ determined by $\mathscr{E}$.

Note. The frame field $\mathscr{S} = (\sigma_1, \ldots, \sigma_N)$ is, in fact, only determined up to a constant orthogonal change of framing, that is, up to a choice of orthonormal basis at some point. (This corresponds to a choice of basis in the module M, i.e., on the matrix realization of the representation μ.) This family of framings is characterized by the following property. For any $I = (i_1, \ldots, i_p)$, we have that $e_{i_1} \cdots e_{i_p} \sigma_j = \sum_k C^k_{Ij} \sigma_k$ where the coefficients $\{C^k_{Ij}\}$ are constants.

An analysis similar to the one above can be carried out for the curvature 2-form. Since curvature is a tensor, we do not need to be concerned in this case with the distinguished frame field $\mathscr{S}$ on $S(E)$. Hence, the curvature of $S(E)$ can be expressed in the following very pretty way:

Theorem 4.15. *Let Ω be the curvature 2-form on $P_{\mathrm{SO}}(E)$ and let $S(E)$ be any spinor bundle associated to E. Then the curvature R^s of $S(E)$ is given locally by the formula*

$$R^s \sigma = \tfrac{1}{2} \sum_{i<j} \tilde{\Omega}_{ji} \otimes e_i e_j \cdot \sigma \tag{4.36}$$

where $\mathscr{E} = (e_1, \ldots, e_n)$ is a local section of $P_{\mathrm{SO}}(E)$, $\tilde{\Omega} = \mathscr{E}^*(\Omega)$ and where σ is any section of $S(E)$.

In particular, for any two tangent vectors V and W at $x \in X$, the curvature transformation $R^s_{V,W} : S(E_x) \to S(E_x)$ is given by the formula

$$\boxed{R^s_{V,W}(\sigma) = \tfrac{1}{2} \sum_{i<j} \langle R_{V,W}(e_i), e_j \rangle e_i e_j \sigma} \tag{4.37}$$

where $R_{V,W}$ is the curvature transformation of E_x.

Note that the expression

$$\mathfrak{R}^s_{V,W} \equiv \tfrac{1}{2} \sum_{i<j} \langle R_{V,W}(e_i), e_j \rangle e_i e_j \tag{4.38}$$

is independent of the choice of an orthonormal basis $(e_1, \ldots, e_n)$ for E_x and is therefore invariantly defined. It is skew symmetric in V and W. Consequently, $\mathfrak{R}^s$ can be thought of as a 2-form on X with values in $C\ell(E)$. The formula (4.37) can now be succinctly expressed as

$$R^s_{V,W}(\sigma) = \mathfrak{R}^s_{V,W} \cdot \sigma. \tag{4.39}$$

It is interesting to note that the arguments above can be carried through also for the Clifford bundle $C\ell(E)$ by using the equation (4.33). For tangent vectors V and W at $x \in X$ we define the invariant operator

$$\mathfrak{R}^{c\ell}_{V,W} \equiv \tfrac{1}{2} \sum_{i<j} \langle R_{V,W}(e_i), e_j \rangle \mathrm{ad}_{e_i e_j} \tag{4.40}$$

where, as before, $(e_1, \ldots, e_n)$ is any orthonormal basis of E_x. We then obtain the following expression for the curvature tensor on $C\ell(E) \cong \Lambda^*(E)$ in terms of Clifford multiplication.

Theorem 4.16. *For any two tangent vectors V and W at $x \in X$, the curvature transformation $R^{c\ell}_{V,W}: C\ell(E_x) \to C\ell(E_x)$ is given by the formula*

$$R^{c\ell}_{V,W}(\varphi) = \mathfrak{R}^{c\ell}_{V,W}(\varphi) \tag{4.41}$$

where $\mathfrak{R}^{c\ell}_{V,W}$ is the operator defined above in terms of the curvature transformation $R_{V,W}$ of E_x.

It is an easy exercise to check that the restriction of the operator $\mathfrak{R}^{c\ell}_{V,W}$ to $E_x \subset C\ell(E_x)$ agrees with $R_{V,W}$. One need only verify that $\frac{1}{2}[e_i e_j, e] = (e_i \wedge e_j)(e)$ for $e \in E_x$.

Note that Theorem 4.16 does not depend on the existence of a spin structure on E. In fact, it does not even depend on the existence of an orientation on E.

We conclude this section with some remarks concerning the situation where $E = T(X)$, the tangent bundle of X. In this basic case we abbreviate notation by setting $P_{SO}(X) \equiv P_{SO}(T(X))$, the **(orthonormal) tangent frame bundle of** X, and $C\ell(X) \equiv C\ell(T(X))$, the **Clifford bundle** of X.

Suppose now that $P_{SO}(X)$ is furnished with a connection, and let ∇ denote the corresponding covariant derivative. Then there is an invariantly defined tensor field associated to ∇ as follows. Let V and W be tangent vectors at a point $x \in X$. Extend them to local vector fields (i.e., local sections of $T(X)$) and consider the expression

$$T_{V,W} \equiv \nabla_V W - \nabla_W V - [V,W] \tag{4.42}$$

where $[V,W]$ is the Lie bracket. The value of $T_{V,W}$ at x can be easily shown to be independent of the choice of vector fields extending V_x and W_x (see Helgason [1, Chap. I]). $T_{V,W}$ is clearly bilinear and skew-symmetric. It therefore defines a global 2-form on X with values in $T(X)$ called the **torsion tensor** of the connection. The following result can be found in any basic text in differential geometry.

Theorem 4.17. (The Fundamental Theorem of Riemannian Geometry). *Let $P_{SO}(X)$ denote the tangent frame bundle of a riemannian manifold X. Then there exists a unique connection on $P_{SO}(X)$ with the property that its torsion tensor vanishes identically.*

This connection will be called the **canonical riemannian connection** on X. It induces a canonical connection on $C\ell(X) \cong \Lambda^*(X)$.

Note that if X admits a spin structure $\xi : P_{Spin}(X) \to P_{SO}(X)$, then by lifting, we obtain a canonical riemannian connection on $P_{Spin}(X)$. This, in turn, induces a connection on any spinor bundle associated to $P_{Spin}(X)$.

This situation falls precisely into the general framework developed above. Thus any spinor bundle for X is a bundle of left modules over $C\ell(X)$, and the canonical covariant derivative is a derivation of the module multiplication (see Proposition 4.11).

For most of the basic facts of riemannian geometry, the reader is referred to the general literature which is extensive and quite good. However, there is one fact we use sufficiently often that it is worthwhile to cite it here.

Proposition 4.18. *Let R denote the curvature tensor of a riemannian manifold X, i.e., the curvature tensor of the canonical riemannian connection on the tangent bundle TX. Then R satisfies the following identities:*

$$R_{U,V}W + R_{V,W}U + R_{W,U}V = 0, \tag{4.43}$$

$$\langle R_{U,V}W, Y\rangle = \langle R_{W,Y}U, V\rangle \tag{4.44}$$

for all tangent vectors $U,V,W,Y \in T_xX$, at all points $x \in X$.

§5. The Dirac Operators

Let X be a riemannian manifold with Clifford bundle $C\ell(X)$, and let S be any bundle of left modules over $C\ell(X)$ (i.e., a vector bundle over X such that at each point $x \in X$, the fibre S_x is a left module over the algebra $C\ell(X)_x$.) Assume S is riemannian and is furnished with a remannian connection. Then under these general hypotheses, we can define a canonical first-order differential operator $D : \Gamma(S) \to \Gamma(S)$ called the **Dirac operator** of S, by setting

$$D\sigma \equiv \sum_{j=1}^{n} e_j \cdot \nabla_{e_j}\sigma \tag{5.0}$$

at $x \in X$, where $e_1, \ldots, e_n$ is an orthonormal basis of $T_x(X)$, where ∇ denotes the covariant derivative on S determined by the connection, and where "·" denotes the Clifford module multiplication. The operator D^2 is called the **Dirac laplacian.**

Recall that the **principal symbol** of a differential operator $D: \Gamma(E) \to \Gamma(E)$ is a map which associates to each point $x \in X$ and each cotangent vector $\xi \in T_x^*(X)$, a linear map $\sigma_\xi(D): E_x \to E_x$ defined as follows. If in local coordinates we have

$$D = \sum_{|\alpha| \leqq m} A_\alpha(x) \frac{\partial^{|\alpha|}}{\partial x^\alpha} \quad \text{and} \quad \xi = \sum_k \xi_k dx_k,$$

where m is the order of D, then

$$\sigma_\xi(D) = i^m \sum_{|\alpha| = m} A_\alpha(x)\xi^\alpha.$$

The operator D is **elliptic** if $\sigma_\xi(D)$ is an isomorphism for all $\xi \neq 0$ (see III.1 for further details).

Recall also that the riemannian metric induces a canonical isomorphism between $T(X)$ and $T^*(X)$. Throughout this section we shall consider them as so identified.

Lemma 5.1. *Let D be the Dirac operator of the bundle S defined above. Then for any $\xi \in T^*(X) \cong T(X)$ we have that*

$$\sigma_\xi(D) = i\xi \tag{5.1}$$

$$\sigma_\xi(D^2) = \|\xi\|^2 \tag{5.2}$$

where the symbol on the right denotes Clifford multiplication by the vector ξ in (5.1) and the scalar $\|\xi\|^2$ in (5.2). In particular, both D and D^2 are elliptic operators.

Proof. Fix $x \in X$ and an orthonormal basis $e_1, \ldots, e_n$ of $T_x(X)$. Choose local coordinates $(x_1, \ldots, x_n)$ on X at x such that x corresponds to 0 and e_j corresponds to $(\partial/\partial x_j)_0$ for each j. Under the identification $T_x(X) \cong T_x^*(X)$ we have that e_j also corresponds to $(dx_j)_0$ for each j.

For any local trivialization of S near x, we have that $\nabla_{e_j} = (\partial/\partial x_j)_0 +$ zero-order terms. Hence, at 0 we have that $D = \sum e_j(\partial/\partial x_j)_0 +$ zero-order terms. Consequently, for any cotangent vector $\xi = \sum \xi_j(dx_j)_0$ at 0, we have by definition of the symbol that $\sigma_\xi(D) = i \sum e_j\xi_j = i\xi$. This gives (5.1). Then $\sigma_\xi(D^2) = \sigma_\xi(D) \circ \sigma_\xi(D) = -\xi \cdot \xi = \|\xi\|^2$, and the proof is complete. ■

We now observe that in light of the basic examples it is natural to require that the bundle S have certain additional properties. The first property is that Clifford multiplication by unit vectors in $T(X)$ be orthogonal, i.e., that at each $x \in X$,

$$\langle e\sigma_1, e\sigma_2 \rangle = \langle \sigma_1, \sigma_2 \rangle \tag{5.3}$$

for all $\sigma_1, \sigma_2 \in S_x$ and all unit vectors $e \in T_x(X)$. Since $e^2 = -1$, this is equivalent to the requirement that

$$\langle e\sigma_1, \sigma_2 \rangle + \langle \sigma_1, e\sigma_2 \rangle = 0 \tag{5.3$'$}$$

for all such σ_1, σ_2 and e.

Recall (from the end of §4) that the bundle $C\ell(X)$ carries a canonical riemannian connection, whose associated covariant derivative will be denoted by ∇. Our second requirement is that the covariant derivative on S (which we also denote by ∇) be a module derivation, i.e., that

$$\nabla(\varphi \cdot \sigma) = (\nabla \varphi) \cdot \sigma + \varphi \cdot (\nabla \sigma) \tag{5.4}$$

for all $\varphi \in \Gamma(C\ell(X))$ and all $\sigma \in \Gamma(S)$.

There is a surprisingly large and important collection of bundles with the properties described above. Although these bundles can be quite varied in nature, a substantial part of the theory concerning them can be treated in a uniform way. For this reason we introduce the following concept:

Definition 5.2. A **Dirac bundle** over a riemannian manifold X is a bundle S of left modules over $C\ell(X)$ together with a riemannian metric and connection on S having properties (5.3) and (5.4) above.

Before presenting examples of these bundles, we shall investigate some of their elementary properties. Note that any Dirac bundle S has a canonically associated Dirac operator. Furthermore there is an inner product on $\Gamma(S)$ induced from the pointwise inner product $\langle \cdot, \cdot \rangle$ by setting

$$(\sigma_1, \sigma_2) \equiv \int_X \langle \sigma_1, \sigma_2 \rangle. \tag{5.5}$$

Proposition 5.3. *The Dirac operator of any Dirac bundle over a riemannian manifold is formally self-adjoint, i.e.,*

$$(D\sigma_1, \sigma_2) = (\sigma_1, D\sigma_2)$$

for all compactly supported sections σ_1 and σ_2.

Proof. Fix $x \in X$ and choose an orthonormal tangent frame field $(e_1, \ldots, e_n)$ in a neighborhood of x so that $(\nabla_{e_i} e_j)_x = 0$ for all i,j. This can be done for example, by extending a frame at x by parallel translation along geodesic rays emanating from x. Using properties (5.3)′, (5.4) and (4.14), we have that at x,

$$\begin{aligned}\langle D\sigma_1,\sigma_2\rangle_x &= \sum_j \langle e_j\nabla_{e_j}\sigma_1,\sigma_2\rangle_x = -\sum_j \langle \nabla_{e_j}\sigma_1, e_j\sigma_2\rangle_x \\ &= -\sum_j \{e_j\langle\sigma_1,e_j\sigma_2\rangle - \langle\sigma_1,(\nabla_{e_j}e_j)\sigma_2\rangle - \langle\sigma_1,e_j\nabla_{e_j}\sigma_2\rangle\}_x \\ &= -\sum_j (e_j\langle\sigma_1,e_j\sigma_2\rangle)_x + \langle\sigma_1,D\sigma_2\rangle_x \\ &= \mathrm{div}(V)_x + \langle\sigma_1,D\sigma_2\rangle_x\end{aligned}$$

where V is the vector field defined by the condition that

$$\langle V,W\rangle = -\langle\sigma_1,W\cdot\sigma_2\rangle$$

for all tangent vectors W. The last line in (5.6) is established as follows

$$\begin{aligned}\mathrm{div}(V)_x &\overset{\text{def}}{=} \sum_j \langle\nabla_{e_j}V,e_j\rangle_x \\ &= \sum_j \{e_j\langle V,e_j\rangle - \langle V,\nabla_{e_j}e_j\rangle\}_x \\ &= \sum_j \{e_j\langle V,e_j\rangle\}_x \\ &= -\sum_j \{e_j\langle\sigma_1,e_j\cdot\sigma_2\rangle\}_x.\end{aligned}$$

The first and last expressions in (5.6) are independent of the frame field $(e_1,\dots,e_n)$. Hence, we have established the equation

$$\langle D\sigma_1,\sigma_2\rangle = \mathrm{div}(V) + \langle\sigma_1,D\sigma_2\rangle$$

on X. The proposition follows immediately. ∎

Note that if X is permitted to have a boundary ∂X, then the above argument proves that

$$(D\sigma_1,\sigma_2) - (\sigma_1,D\sigma_2) = \int_{\partial X}\langle\nu\cdot\sigma_1,\sigma_2\rangle \tag{5.7}$$

for compactly supported sections σ_1 and σ_2, where ν denotes the outer unit normal field to ∂X in X.

It is a general consequence of the ellipticity of D that any weak solution to the equation $D\psi = 0$ is of class C^∞ (cf. Theorem 5.2(i) of Chap. III). Because of the formal self-adjointness this may be expressed as follows. Let $\Gamma_{\mathrm{cpt}}(S)$ denote the Frechet space of C^∞ sections of S with compact support. Then any continuous linear functional F on $\Gamma_{\mathrm{cpt}}(S)$ such that $F(D\varphi) = 0$ for all $\varphi\in\Gamma_{\mathrm{cpt}}(S)$, can be represented as $F(\varphi) = (\varphi,\psi)$ where $\psi\in\Gamma(S)$ and $D\psi\equiv 0$. In particular, any locally integrable section ψ of S which is orthogonal to $D\Gamma_{\mathrm{cpt}}(S)$ (i.e., which satisfies the equation $D\psi = 0$ weakly), is of class C^∞ and satisfies the equation $D\psi = 0$ in the usual sense.

Similar comments apply to D^2.

With this in mind, we define $\ker D = \{\varphi \in \Gamma(S) : D\varphi = 0\}$ and $\ker D^2 = \{\varphi \in \Gamma(S) : D^2\varphi = 0\}$, and observe the following:

Theorem 5.4. *Let D be the Dirac operator of any Dirac bundle over a compact riemannian manifold X. Then*

$$\ker D = \ker D^2$$

and this space has finite dimension.

Proof. Finite dimensionality is a direct consequence of elementary elliptic theory (cf. Theorem 5.2(ii) of Chap. III). For the rest, observe that $D^2\varphi = 0 \Rightarrow \|D\varphi\|^2 = (D\varphi, D\varphi) = (\varphi, D^2\varphi) = 0 \Rightarrow D\varphi = 0$. ∎

The results above have an important extension to L^2-sections of a Dirac bundle over any complete manifold. We begin with the following preparatory lemma.

Lemma 5.5. *Let D, S and X be as above. Then for any $f \in C^\infty(X)$ and any $\varphi \in \Gamma(S)$, we have that*

$$D(f\varphi) = (\operatorname{grad} f) \cdot \varphi + fD\varphi. \tag{5.8}$$

Proof. $D(f\varphi) = \sum e_j \cdot \nabla_{e_j}(f\varphi) = \sum e_j\{(e_j f)\varphi + f\nabla_{e_j}\varphi\} = (\sum (e_j f)e_j) \cdot \varphi + fD\varphi) = (\operatorname{grad} f) \cdot \varphi + fD\varphi$. ∎

REMARK 5.6. The relation (5.8) immediately extends to distributional sections Φ of S. To see this, consider $\varphi \in \Gamma_{\text{cpt}}(S)$, and note that from (5.8) and (5.3)$'$, $(fD\Phi, \varphi) \equiv (\Phi, D(f\varphi)) = (\Phi, (\operatorname{grad} f) \cdot \varphi + fD\varphi) = (D(f\Phi) - (\operatorname{grad} f) \cdot \Phi, \varphi)$.

We shall now consider the space $L^2(S)$ of **L^2-sections of S.** This is the natural completion of $\Gamma_{\text{cpt}}(S)$ in the Hilbert space norm introduced above. We now consider D as a symmetric operator on $\Gamma_{\text{cpt}}(S)$ and take its closure (also denoted D), in $L^2(S)$. This gives us an unbounded operator in $L^2(S)$. Its domain $\operatorname{dom}(D)$ consists of all $\varphi \in L^2(S)$ for which there is a sequence $\langle \varphi_n \rangle_{n=1}^\infty \subset \Gamma_{\text{cpt}}(S)$ such that $\varphi_n \to \varphi$ and $D\varphi_n \to \psi =$ "$D\varphi$" in $L^2(S)$.

There is another extension of this operator to $L^2(S)$ which we shall denote D^*. The domain of D^* consists of all $\varphi \in L^2(S)$ such that the distributional image $D\varphi$ is also in $L^2(S)$. Of course, $\operatorname{dom}(D) \subseteq \operatorname{dom}(D^*)$.

We now observe that D^* is simply the adjoint of D. Recall that $\psi \in L^2(S)$ is in the domain of the adjoint D^t if the function $\varphi \mapsto \langle D\varphi, \psi \rangle$ is continuous on $\operatorname{dom}(D)$. Since $\operatorname{dom}(D)$ is dense, there exists a unique element $D^t\psi \in L^2(S)$ such that $\langle D\varphi, \psi \rangle = \langle \varphi, D^t\psi \rangle$ for all $\varphi \in \operatorname{dom}(D)$. Clearly $D^t\psi$ is just the distributional image of ψ under D, and so $D^t = D^*$.

We now prove that on complete manifolds, any Dirac operator is essentially self-adjoint (cf. Wolf [1]).

Theorem 5.7. *Let X be a complete riemannian manifold and let D be the Dirac operator of any Dirac bundle S over X. Then the closure of D in $L^2(S)$ is a self-adjoint operator. Furthermore,*

$$\ker(D) = \ker(D^2)$$

on $L^2(S)$.

Proof. Fix $x_0 \in X$ and let $d(x)$ be a regularization of the distance function from x_0. Choose $\chi \in C^\infty(\mathbb{R})$ so that $0 \leqq \chi \leqq 1$, $\chi(t) \equiv 0$ for $t \geqq 2$, $\chi(t) = 1$ for $t \leqq 1$, and $|\chi'| \leqq 2$. Then set

$$\chi_n(x) = \chi\left(\frac{1}{n} d(x)\right).$$

We want to show that $\operatorname{dom} D = \operatorname{dom}(D^*)$. It suffices to prove that $\operatorname{dom}(D^*) \subseteq \operatorname{dom}(D)$ since the reverse inclusion is obvious. Choose $\varphi \in \operatorname{dom}(D^*)$ and define $\varphi_n = \chi_n \varphi$ for each positive integer n. Then by Remark 5.6 we know that $D\varphi_n = (\operatorname{grad} \chi_n) \cdot \varphi + \chi_n D\varphi$ for each n. Clearly, $\chi_n D\varphi \to D\varphi$ in $L^2(S)$. Furthermore, if we let B_ρ denote the metric ball of radius ρ centered at x_0 in X, then

$$\|(\operatorname{grad} \chi_n) \cdot \varphi\|^2 \leqq \int_{B_{2n} - B_n} \frac{4}{n^2} \|\varphi\|^2 \underset{n\to\infty}{\longrightarrow} 0.$$

Consequently, $D\varphi_n \to D\varphi$ in $L^2(S)$.

Thus, we are reduced to the case where $\varphi \in \operatorname{dom}(D^*)$ has compact support. By a partition of unity over $\operatorname{supp} \varphi$ we can assume that φ has compact support in a local coordinate system. Here, using Fourier transform methods, we can construct a **parametrix** for D, i.e. a pseudo-differential operator P such that

$$DP = 1 - \mathscr{S} \qquad \text{and} \qquad PD = 1 - \mathscr{S}'$$

where $\mathscr{S}$ and $\mathscr{S}'$ are infinitely smoothing operators, and where P, $\mathscr{S}$ and $\mathscr{S}'$ all have Schwartzian kernels supported near the diagonal. Observe now that since $D\varphi \in L^2(S)$, there exists a sequence $\langle\psi_n\rangle_{n=1}^\infty \subset \Gamma_{\text{cpt}}(S)$, with support uniformly bounded in this coordinate neighborhood, such that $\psi_n \to D\varphi$ in $L^2(S)$. We then set $\varphi_n \equiv P\psi_n + \mathscr{S}'\varphi$ and observe that since the Schwartzian kernels of P and $\mathscr{S}'$ are supported near the diagonal, each (smooth) φ_n has compact support. That is, we have $\langle\varphi_n\rangle_{n=1}^\infty \subset \Gamma_{\text{cpt}}(S)$. From the relations above, we see that $\varphi_n \to PD\varphi + \mathscr{S}'\varphi = \varphi$, and that $D\varphi_n = DP\psi_n + D\mathscr{S}'\varphi = \psi_n - \mathscr{S}\psi_n + D\mathscr{S}'\varphi \to D\varphi - \mathscr{S}D\varphi + D\mathscr{S}'\varphi = D\varphi$ (since, clearly, we have $D\mathscr{S}' = \mathscr{S}D$). This completes the proof of the essential self-adjointness of D.

We now prove that $\ker(D^2) \subseteq \ker(D)$. That is, we shall show that any (necessarily smooth) L^2-section φ of S which satisfies the differential equation $D^2\varphi = 0$, also satisfies the equation $D\varphi = 0$. To see this, let χ_n be the sequence of functions above and note that

$$\begin{aligned} 0 &= (D^2\varphi, \chi_n^2\varphi) = (D\varphi, D(\chi_n^2\varphi)) \\ &= (D\varphi, 2\chi_n \operatorname{grad}(\chi_n)\cdot\varphi + \chi_n^2 D\varphi) \\ &= \|\chi_n D\varphi\|^2 + 2(\chi_n D\varphi, \operatorname{grad}(\chi_n)\cdot\varphi). \end{aligned}$$

Consequently, by the Schwartz inequality we have

$$\|\chi_n D\varphi\|^2 \leqq 2\|\chi_n D\varphi\| \, \|\operatorname{grad}(\chi_n)\cdot\varphi\|.$$

Therefore,

$$\|\chi_n D\varphi\| \leqq 2\|\operatorname{grad}(\chi_n)\cdot\varphi\| \leqq \tfrac{4}{n}\|\varphi\|,$$

and we conclude that $\|D\varphi\| = \lim_n \|\chi_n D\varphi\| = 0$. This completes the proof. ∎

Having discussed Dirac bundles in general terms, it is now time to look hard at some important examples. We begin with the basic ones.

Example (an historical case). Let $X = \mathbb{R}^n$, euclidean n-space, and let $S = \mathbb{R}^n \times V$ where V is some finite dimensional module for $C\ell_n$. In this case the Dirac operator is a constant coefficient operator (on V-valued functions) of the form

$$D = \sum_{k=1}^{n} \gamma_k \frac{\partial}{\partial x_k}$$

where each γ_k is a linear map $\gamma_k : V \to V$ and where

$$\gamma_j\gamma_k + \gamma_k\gamma_j = -2\delta_{jk}$$

for all j,k. If we choose a basis for V, these γ_k's will be represented by matrices. The relations above imply that

$$D^2 = \Delta \cdot \mathrm{Id}_V \cong \begin{pmatrix} \Delta & & 0 \\ & \Delta & \\ & & \ddots & \\ 0 & & & \Delta \end{pmatrix}$$

where $\Delta = -\sum \partial^2/\partial x_k^2$ is the positive laplacian in $\mathbb{R}^n$.

This particular operator has historical roots in physics. In the 1920s, the physicist P.A.M. Dirac was searching for a Lorentz-invariant first-order differential operator whose square would be the Klein-Gordon operator. Thus he was essentially led to search for a first order operator D of the form above which satisfied the equation $D^2 = \Delta$. Realizing that the γ_k's must be matrices, he was led immediately by this equation to the above relations, which we recognize now as the generating relations of a representation of $C\ell_n$. See the beautiful book of Dirac [2] for a discussion of the physics.

It is illuminating to consider this euclidean operator in low dimensions.

Let $n = 1$, so that $C\ell_1 = V = \mathbb{C}$. Then we have

$$D = i\frac{\partial}{\partial x_1}$$

the generator of a basic semigroup of unitary operators on L^2.

Let $n = 2$, so that $C\ell_2 = V = \mathbb{H} \cong \mathbb{C} \oplus \mathbb{C}$. The decomposition of $\mathbb{H}$ into $\mathbb{C} \oplus \mathbb{C}$ is natural and corresponds to the $\mathbb{Z}_2$-grading $C\ell_2 = C\ell_2^0 \oplus C\ell_2^1$. With respect to this $\mathbb{Z}_2$-grading, the Dirac operator interchanges even and odd parts. In particular if we identify $C\ell_2^0$ and $C\ell_2^1$ with $\mathbb{C}$ by setting $u + ve_2e_1 \cong u + iv \cong ue_1 + ve_2$, then $D = e_1(\partial/\partial x_1) + e_2(\partial/\partial x_2)$ has the form

$$D = \begin{pmatrix} 0 & -\dfrac{\partial}{\partial z} \\ \dfrac{\partial}{\partial \bar{z}} & 0 \end{pmatrix}$$

where $\partial/\partial\bar{z} \equiv \partial/\partial x_1 + i\partial/\partial x_2$. Thus, the Dirac operator on $\mathbb{R}^2 = \mathbb{C}$, considered as mapping even to odd spinors, is exactly the Cauchy-Riemann operator.

Let $n = 3$, so that $C\ell_3 = \mathbb{H} \oplus \mathbb{H}$ and $V = \mathbb{H}$. $C\ell_3$ has two representations on $\mathbb{H}$ given as follows. Identify $\mathbb{R}^3$ with $\mathrm{Im}(\mathbb{H})$, and let $\{i,j,k\}$ be the standard basis of imaginary quaternions. Then the two representations are generated by letting i,j,k act on either the right or the left in $\mathbb{H}$. Choosing multiplication on the left, we get the following expression for the Dirac operator on $\mathbb{H}$-valued functions

$$D = i\frac{\partial}{\partial x_1} + j\frac{\partial}{\partial x_2} + k\frac{\partial}{\partial x_3}.$$

If we re-express left quaternion multiplication, with respect to the basis $\{1,i,j,k\}$, as 4×4 real matrices, then D becomes

$$D = \begin{pmatrix} 0 & -\partial_1 & -\partial_2 & -\partial_3 \\ \partial_1 & 0 & -\partial_3 & \partial_2 \\ \partial_2 & \partial_3 & 0 & -\partial_1 \\ \partial_3 & -\partial_2 & \partial_1 & 0 \end{pmatrix}$$

where $\partial_k \equiv \partial/\partial x_k$.

Let $n = 4$, so that $C\ell_4 = \mathbb{H}(2)$ and $V = \mathbb{H}^2 = \mathbb{H} \oplus \mathbb{H}$. Here again the splitting corresponds to a $\mathbb{Z}_2$-grading of the module V, and D interchanges parts. To describe the full Dirac operator we consider first the following quaternion analogue of the Cauchy-Riemann operator. Identify $\mathbb{R}^4$ with $\mathbb{H}$ under the standard basis $\{1,i,j,k\}$, and define the following operators on functions from $\mathbb{H}$ to $\mathbb{H}$:

$$\frac{\partial}{\partial \bar{q}} \equiv \frac{\partial}{\partial x_0} + i\frac{\partial}{\partial x_1} + j\frac{\partial}{\partial x_2} + k\frac{\partial}{\partial x^3} \qquad \frac{\partial}{\partial q} \equiv \frac{\partial}{\partial x_0} - i\frac{\partial}{\partial x_1} - j\frac{\partial}{\partial x_2} - k\frac{\partial}{\partial x_3}$$

Then the Dirac operator can be expressed with respect to the splitting $\mathbb{H} \oplus \mathbb{H}$ as

$$D = \begin{pmatrix} 0 & -\dfrac{\partial}{\partial q} \\ \dfrac{\partial}{\partial \bar{q}} & 0 \end{pmatrix}$$

Note the analogy with dimension two.

Note that left quaternion multiplication is always complex linear with respect to the complex structure given by multiplication by i on the right. Thus, left multiplication by i, j and k on $\mathbb{H} = \mathbb{C}^2$ could be represented by complex 2×2-matrices σ_1, σ_2 and σ_3 respectively. Under this convention, the operator $\partial/\partial\bar{q}$ becomes

$$\frac{\partial}{\partial \bar{q}} = \frac{\partial}{\partial x_0} + \sigma_1\frac{\partial}{\partial x_1} + \sigma_2\frac{\partial}{\partial x_2} + \sigma_3\frac{\partial}{\partial x_3}.$$

The matrices σ_k can be chosen to be the classical **Pauli matrices:**

$$\sigma_1 = \begin{pmatrix} i & 0 \\ 0 & -i \end{pmatrix}, \qquad \sigma_2 = \begin{pmatrix} 0 & -1 \\ 1 & 0 \end{pmatrix}, \qquad \sigma_3 = \begin{pmatrix} 0 & i \\ i & 0 \end{pmatrix}$$

Note that these matrices generate the fundamental representation of $C\ell_3$ in complex form.

We could continue this analysis. For general n, one can calculate an enormous $N \times N$-matrix whose entries are linear combinations of $\partial/\partial x_1, \dots, \partial/\partial x_n$. Here N is on the order of 2^n. This matrix will have the property that its square is $\Delta \cdot I$ where I is the $N \times N$ identity matrix. However, this explicit form of D is seldom, if ever, useful. It is always simpler to use the structure of the Clifford module.

It is interesting to note that the concept of D as a generalized Cauchy-Riemann operator is a useful one. Let us fix a dimension n, and let D denote the euclidean Dirac operator acting on functions $f: \mathbb{R}^n \to V$ where V is some fixed $C\ell_n$-module. Recall that the fundamental solution for the

Laplace operator on $\mathbb{R}^n$, for $n > 2$, is $\varphi(x) = c_n/\|x\|^{n-2}$ for an appropriate constant c_n. That is, we have

$$\Delta\varphi = \delta_0$$

where δ_0 is the Dirac δ-function at the origin. Since $D^2 = \Delta \cdot I$, where I is the identity on V, we have that

$$\Phi \equiv D(\varphi I)$$

is the fundamental solution for D. That is,

$$D\Phi = \delta_0 I.$$

From this one can derive very pretty analogues of the Cauchy Integral Formula, and the analysis of D becomes quite accessible.

We now examine Dirac bundles S over more general spaces. Let X be an arbitrary riemannian manifold. Then there are two basic cases.

EXAMPLE 5.8. (the Clifford bundle). Let $S = C\ell(X)$ with its canonical riemannian connection, and view $C\ell(X)$ as a bundle of left modules over itself by left Clifford multiplication. (Note that Property (5.4) was established in Proposition 4.8.) The Dirac operator in this case is a square root of the classical Hodge laplacian. Most of the remainder of this section is devoted to a detailed analysis of this basic case.

EXAMPLE 5.9. (the spinor bundles). Suppose X is a spin manifold with a spin structure on its tangent bundle. Let S be any spinor bundle associated to $T(X)$. Then S is a bundle of modules over $C\ell(X)$, and as shown in §4, S carries a canonical riemannian connection which has property (5.4) (see Proposition 4.11 and the discussion following Theorem 4.17). The Dirac operator in this case was first written down by Atiyah and Singer in their work on the Index Theorem. Finding this operator was a major accomplishment, and for this reason we shall call it the **Atiyah-Singer operator.**

Notation. For spin manifolds X of even dimension we shall denote the (unique) irreducible complex spinor bundle by $\$_{\mathbb{C}}$; and when $\dim(X) \not\equiv 3 \pmod 4$, we denote the irreducible real spinor bundle by $\$$. In both cases the Atiyah-Singer operator will be written $\not{D}$.

These basic examples each generate large families of new examples by the following construction. Let S be a given Dirac bundle with connection ∇^S over a riemannian manifold X, and let E be an arbitrary riemannian vector bundle with connection ∇^E over X. Then the tensor product $S \otimes E$ is again a bundle of left modules over $C\ell(X)$, where for $\varphi \in C\ell(X)$, $\sigma \in S$ and $e \in E$, the module multiplication is given by setting

$$\varphi \cdot (\sigma \otimes e) = (\varphi \cdot \sigma) \otimes e.$$

It is clear that in the tensor product metric on $S \otimes E$ we have that this Clifford multiplication by unit tangent vectors on X is orthogonal (i.e., Property (5.3) is satisfied). Furthermore, we can equip $S \otimes E$ with the canonical **tensor product connection,** $\nabla \equiv \nabla^S \otimes \nabla^E$, which is defined on sections of the form $\sigma \otimes e$ by the formula

$$\nabla(\sigma \otimes e) = (\nabla^S \sigma) \otimes e + \sigma \otimes (\nabla^E e).$$

It is straightforward to verify that with this riemannian connection, the bundle $S \otimes E$ has the derivation property (5.4). Consequently, we have proved the following.

Proposition 5.10. *Let S be any Dirac bundle over a riemannian manifold X, and suppose E is any riemannian bundle with connection. Then the tensor product $S \otimes E$ is again a Dirac bundle over X.*

An interesting example of this construction is the following. Let X be a spin manifold of dimension $8k$, and let $\$$ be the canonical (real) spinor bundle of X. Then under the above construction the bundle $\$ \otimes \$$ is a Dirac bundle over X in two canonical ways (by multiplication in the left or the right factor). It follows from the representation theory (see I.5.18 and IV.10.16–17) that

$$C\ell(X) \cong \$ \otimes \$$$

where the two module structures correspond to multiplication on the left and on the right (by the transpose) in $C\ell(X)$.

Similarly, in all even dimensions we have

$$C\ell(X) \otimes \mathbb{C} \cong \$_{\mathbb{C}} \otimes \$_{\mathbb{C}}^*.$$

REMARK 5.11. The construction above does indeed give rise to a large number of examples. As we shall see later (Remark III.13.11), basically *every elliptic operator* on a spin manifold can be constructed, up to homotopy and degree shift, as a Dirac operator of the form $D: \Gamma(S^0 \otimes E) \to \Gamma(S^1 \otimes E)$ where $S = S^0 \oplus S^1$ is the complex $\mathbb{Z}_2$-graded spinor bundle.

Note that in each of the basic cases above (Examples 5.8 and 5.9), the Dirac bundles and their associated Dirac operators are canonically defined in terms of the riemannian metric on X. Hence, any mathematical objects constructed using these operators are invariants of the riemannian structure on X.

The remainder of this section will be devoted to an analysis of these operators. We begin with the Clifford bundle $C\ell(X)$.

Our first observation is that it is also possible to view $C\ell(X)$ as a bundle of right modules over $C\ell(X)$ (by right Clifford multiplication). Property (5.4) also holds for right multiplication. Hence, we can also define a "right-handed" Dirac operator $\hat{D}$ on $C\ell(M)$ by setting

$$\hat{D}\varphi = \sum_{j=1}^{n} (\nabla_{e_j}\varphi) \cdot e_j. \tag{5.9}$$

This operator is also elliptic and formally self-adjoint. The principal symbol $\sigma_\xi(\hat{D})$ is just right multiplication by $i\xi$.

Recall now that there is a canonical isomorphism $C\ell(X) \cong \Lambda^*(T^*(X)) \equiv \Lambda^*(X)$. The bundle $\Lambda^*(X)$ also has two canonical first order operators, namely the exterior derivative $d: \Lambda^*(X) \to \Lambda^*(X)$ and its formal adjoint $d^*: \Lambda^*(X) \to \Lambda^*(X)$. This adjoint is given by the formula

$$d^* = (-1)^{np+n+1} * d * \tag{5.10}$$

on $\Lambda^p(X)$, where $*: \Lambda^p(X) \to \Lambda^{n-p}(X)$ is the linear map defined by the condition that $\varphi \wedge *\psi = \langle \varphi, \psi \rangle * 1$ where $*1$ is the volume form.

The Dirac operators and the exterior derivative are directly related as follows:

Theorem 5.12. *Under the canonical isomorphism* $C\ell(X) \cong \Lambda^*(X)$, *the Dirac operators of* $C\ell(X)$ *satisfy the following equations:*

$$D \cong d + d^* \tag{5.11}$$

$$\hat{D} \cong (-1)^p(d - d^*) \qquad \text{on } \Lambda^p(X). \tag{5.12}$$

Consequently, since $d^2 = (d^*)^2 = 0$, *they also satisfy*

$$D^2 = \hat{D}^2 = dd^* + d^*d \equiv \Delta \tag{5.13}$$

$$D\hat{D} = \hat{D}D. \tag{5.14}$$

The operator Δ defined in (5.13) is called the **Hodge laplacian.**

Proof. Fix $x \in X$ and choose an orthonormal frame field $(e_1, \dots, e_n)$ in a neighborhood U of x with the property that $(\nabla_{e_j} e_j)_x = 0$. We first observe the following:

Lemma 5.13. *The operators d and d^* are given in U by the formulas*

$$d = \sum_{j=1}^{n} e_j \wedge \nabla_{e_j}$$

$$d^* = -\sum_{j=1}^{n} e_j \llcorner \nabla_{e_j}$$

where "$\llcorner$" denotes contraction in $\Lambda^*(X)$.

Proof. Both expressions are invariantly defined; that is, they are independent of the choice of frame field $(e_1, \dots, e_n)$. To establish the first identity it suffices to show that the operator on the right satisfies the following axioms for d:

(i) $d^2\varphi = 0$
(ii) $d(\varphi \wedge \psi) = d\varphi \wedge \psi + (-1)^p \varphi \wedge d\psi$
(iii) $df = \text{grad}(f)$

for all smooth functions f and all smooth p-forms φ and q-forms ψ.

Property (iii) is obvious. Property (ii) is seen as follows:

$$d(\varphi \wedge \psi) = \sum_{j=1}^{n} e_j \wedge [(\nabla_{e_j}\varphi) \wedge \psi + \varphi \wedge (\nabla_{e_j}\psi)]$$
$$= d\varphi \wedge \psi + (-1)^p \varphi \wedge d\psi.$$

(Here we use the fact that ∇ acts as a derivation on $\Lambda^*(X)$.) To prove property (i) we first note that from the independence of the choice of frame field, it suffices to verify this at the point x where we have $(\nabla_{e_i} e_j)_x = 0$. Furthermore, by linearity it suffices to consider a φ of the form $\varphi = ae_1 \wedge \ldots \wedge e_p$ where a is some smooth function in U. Then one easily sees that at x,

$$d^2\varphi = \sum_{p<j<k} ([e_j,e_k]a)e_j \wedge e_k \wedge e_1 \wedge \ldots \wedge e_p.$$

Since $[e_j,e_k]_x = (\nabla_{e_j}e_k - \nabla_{e_k}e_j)_x = 0$, property (i) is proved, and the first equation is established.

For the second equation, we again consider $\varphi = ae_1 \wedge \ldots \wedge e_p$. Hence, at x we have that

$$d(*\varphi) = \sum_{j=1}^{p} (e_j a)e_j \wedge e_{p+1} \wedge \ldots \wedge e_n.$$

Consequently, at x

$$*(d * \varphi) = \sum_{j=1}^{p} (e_j a) * (e_j \wedge e_{p+1} \wedge \ldots \wedge e_n)$$
$$= \sum_{j=1}^{p} (-1)^{n(p+1)+j-1}(e_j a)e_1 \wedge \ldots \wedge \hat{e}_j \wedge \ldots \wedge e_p$$
$$= (-1)^{n(p+1)} \sum_{j=1}^{n} (e_j a)e_j \llcorner (e_1 \wedge \ldots \wedge e_p)$$
$$= (-1)^{n(p+1)} \sum_{j=1}^{n} e_j \llcorner (\nabla_{e_j}\varphi).$$

This completes the proof of Lemma 5.8. ■

We now recall that under the canonical isomorphism $C\ell(X) \cong \Lambda^*(X)$ we have that

$$e \cdot \varphi \cong e \wedge \varphi - e \llcorner \varphi$$
$$\varphi \cdot e \cong (-1)^p(e \wedge \varphi + e \llcorner \varphi)$$

for all $e \in \Lambda^1(X)$ and all $\varphi \in \Lambda^p(X)$ (see the discussion at the end of §3 in Chap. I). Equations (5.11) and (5.12) now follow directly from Lemma 5.13. This completes the proof of the theorem. ■

This theorem has the following important consequences. On X the **space of harmonic p-forms** is defined to be

$$\mathbf{H} \equiv \bigoplus_{p=0}^{n} \mathbf{H}^p = \ker(\Delta).$$

Corollary 5.14. *If X is compact and without boundary, then*

$$\ker(D) = \ker(\hat{D}) = \ker(\Delta)$$

In particular, under the isomorphism $C\ell(X) \cong \Lambda^*(X)$, the kernels of D and $\hat{D}$ correspond to the space of harmonic forms on X.

Note that by applying Theorem 5.7 to a non-compact complete manifold we conclude that *a differential form which is in L^2 is harmonic if and only if it is both closed and co-closed.*

In the compact case the harmonic forms are related to the topology of X as follows. Consider the so-called **de Rham complex:**

$$0 \longrightarrow \Gamma(\Lambda^0 X) \xrightarrow{d} \Gamma(\Lambda^1 X) \xrightarrow{d} \Gamma(\Lambda^2 X) \xrightarrow{d} \cdots.$$

Since $d^2 = 0$ we can form the quotient: $\mathscr{H}^*(X) \equiv [\ker(d)/\text{image}(d)]^*$. The fundamental theorem of de Rham asserts that for each $p = 0, \ldots, n$, $\mathscr{H}^p(X)$ is isomorphic to $H^p(X;\mathbb{R})$, the pth singular cohomology group of X with real coefficients.

Since we are given a riemannian metric, we can also consider the adjoint sequence:

$$0 \longleftarrow \Gamma(\Lambda^0 X) \xleftarrow{d^*} \Gamma(\Lambda^1 X) \xleftarrow{d^*} \Gamma(\Lambda^2 X) \xleftarrow{d^*} \cdots.$$

The fundamental result of harmonic theory is the following Hodge Decomposition Theorem (Corollary III.5.6):

Theorem 5.15. *Let X be compact, and without boundary. Then there is an orthogonal decomposition*

$$\Gamma(\Lambda^* X) = \mathbf{H} \oplus \operatorname{Im}(d) \oplus \operatorname{Im}(d^*)$$

(where $\operatorname{Im}(\cdot)$ *denotes the image of the operator on* $\Gamma(\Lambda^* X)$*). In particular there is an isomorphism*

$$\mathbf{H}^p \cong H^p(X;\mathbb{R})$$

for each $p = 0, \ldots, n$.

A Note on Orientability. It is clear that orientability is not required for the definition of D, and so the spaces $\mathbf{H}^p$ can be defined for a non-orientable manifold X. Moreover, if X is compact the isomorphism $\mathbf{H}^p \cong H^p(X;\mathbb{R})$ also holds. This is proved as follows. Let $\pi : \tilde{X} \to X$ be the two-fold, orientable covering manifold, and let $\Gamma = \{1,g\} \cong \mathbb{Z}/2\mathbb{Z}$ denote the group of covering transformations of $\tilde{X}$. Then Γ acts naturally on $\Lambda^*(\tilde{X})$. The subspace $\Lambda^*_\Gamma(\tilde{X})$ of Γ-fixed elements is preserved by d, and its cohomology is just the cohomology of X. (Orientability is not required for the de Rham Theorem.) We claim that the inclusion of this cohomology into the cohomology of $\tilde{X}$ is injective. Suppose $\varphi \in \Lambda^p_\Gamma(\tilde{X})$ and $\varphi = d\psi$ for some $\psi \in \Lambda^{p-1}(\tilde{X})$. Set $\psi' = \frac{1}{2}(\psi + g^*\psi)$. Then $\psi' \in \Lambda^{p-1}_\Gamma(\tilde{X})$ and since d commutes with g^*, $\varphi = d\psi'$. This proves the injectivity.

Given a riemannian metric on X, we lift the metric to $\tilde{X}$. The harmonic forms on X lift to Γ-invariant harmonic forms on $\tilde{X}$, and every Γ-invariant harmonic form is such a lift. That is, $\mathbf{H}^*(X) \cong \mathbf{H}^*_\Gamma(\tilde{X})$. We now observe that $\mathbf{H}^*_\Gamma(\tilde{X})$ corresponds to the subspace $H^*_\Gamma(\tilde{X};\mathbb{R}) \subset H^*(X;\mathbb{R})$. To see this we note that if φ is a harmonic form, so is $g^*\varphi$. If φ is cohomologous to a closed Γ-invariant form, then φ and $g^*\varphi$ represent the same cohomology class. Hence $\varphi = g^*\varphi$. We conclude that $\mathbf{H}^*(X) \cong \mathbf{H}^*_\Gamma(\tilde{X}) \cong H^*_\Gamma(X;\mathbb{R}) \cong H^*(X;\mathbb{R})$ as claimed.

The fundamental identities (5.13) and (5.14), established above for the operators D and $\hat{D}$, imply certain important identities for the curvature tensor. In particular we have the following.

Theorem 5.16. *Suppose X is a riemannian manifold, and let R denote the Riemann curvature tensor acting by derivations on the bundle $C\ell(X) \cong \Lambda^*(X)$. Then for any element $\varphi \in C\ell(X)$,*

$$\sum_{i<j} \{e_i e_j R_{e_i,e_j}(\varphi) - R_{e_i,e_j}(\varphi) e_j e_i\} = 0 \tag{5.15}$$

$$\sum_{i,j} e_i R_{e_i,e_j}(\varphi) e_j = 0, \tag{5.16}$$

where $(e_1, \dots, e_n)$ is any orthonormal tangent frame at the point in question. In particular, from (5.15) *we conclude that*

$$\begin{aligned} \sum_{i<j} e_i e_j R_{e_i,e_j}(\varphi) &= -\sum_{i<j} R_{e_i,e_j}(\varphi) e_i e_j \\ &= \frac{1}{2} \sum_{i<j} [e_i e_j, R_{e_i,e_j}(\varphi)] \\ &= \frac{1}{2} \sum_{i<j} \mathrm{ad}_{e_i e_j}(R_{e_i,e_j}(\varphi)). \end{aligned} \tag{5.17}$$

Note. Since $\{e_i \wedge e_j\}_{i<j}$ represents an orthonormal basis of $\Lambda^2(X)$, the formulas above could be easily re-expressed without the use of $(e_1, \dots, e_n)$.

Proof. Let us fix a point $x \in X$ and choose a local orthonormal tangent frame field $(e_1, \dots, e_n)$ such that $(\nabla e_j)_x = 0$ for each j. Then for any section

$\varphi \in \Gamma(C\ell(X))$, we have at x that

$$\begin{aligned} D^2\varphi &= \sum_{i,j} e_i\nabla_{e_i}(e_j\nabla_{e_j}\varphi) \\ &= \sum_{i,j} e_ie_j\nabla_{e_i}\nabla_{e_j}\varphi \\ &= -\sum_i \nabla_{e_i}\nabla_{e_i}\varphi + \sum_{i<j} e_ie_j(\nabla_{e_i}\nabla_{e_j} - \nabla_{e_j}\nabla_{e_i})\varphi \\ &= -\sum_i \nabla_{e_i}\nabla_{e_i}\varphi + \sum_{i<j} e_ie_jR_{e_i,e_j}(\varphi). \end{aligned}$$

Similarly, we conclude that at the point x,

$$\hat{D}^2\varphi = -\sum_i \nabla_{e_i}\nabla_{e_i}\varphi + \sum_{i<j} R_{e_i,e_j}(\varphi)e_je_i.$$

Since $D^2 = \hat{D}^2$ (see (5.13)), we conclude that (5.15) holds at x, and therefore everywhere.

The second equation is proved similarly. We observe that at x,

$$\begin{aligned} D\hat{D}\varphi &= \sum_{i,j} e_i(\nabla_{e_i}\nabla_{e_j}\varphi)e_j \\ \hat{D}D\varphi &= \sum_{i,j} e_i(\nabla_{e_j}\nabla_{e_i}\varphi)e_j. \end{aligned}$$

Subtracting these equations and recalling that $D\hat{D} = \hat{D}D$ completes the proof. ∎

Since equations (5.15) and (5.16) are for Clifford elements, each constitutes 2^n scalar equations. They include the Bianchi identities (consider the Λ^1-component.) They also include a large number of new identities for the curvature transformation of the bundle $\Lambda^*(X)$. These identities will prove useful in the Bochner-type vanishing arguments presented in §8.

We now examine some of the basic operators on $C\ell(X)$ and analyse their relationships with D and $\hat{D}$. Recall that $C\ell(X)$ carries a canonical bundle mapping

$$\alpha : C\ell(X) \longrightarrow C\ell(X) \tag{5.18}$$

which is, on each fibre $C\ell(X)$, the algebra automorphism extending the map -1 on $T_x(X)$. Since $\alpha^2 = 1$, we obtain a decomposition

$$C\ell(X) = C\ell^0(X) \oplus C\ell^1(X) \tag{5.19}$$

where $C\ell^0(X)$ and $C\ell^1(X)$ are the 1 and -1 eigenbundles of α respectively. Under the isomorphism $C\ell(X) \cong \Lambda^*(X)$, we have $C\ell^0(X) \cong \Lambda^{\text{even}}(X)$ and $C\ell^1(X) \cong \Lambda^{\text{odd}}(X)$. For any non-zero vector $e \in T_x(X)$ left (or right) Clifford multiplication gives an isomorphism $e: C\ell^0_x(X) \xrightarrow{\approx} C\ell^1_x(X)$. Hence, if X admits a nowhere vanishing vector field, i.e., if the Euler characteristic of X is zero, then the bundles $C\ell^0(X)$ and $C\ell^1(X)$ are isomorphic.

There is a second canonical bundle mapping

$$L: C\ell(X) \longrightarrow C\ell(X) \tag{5.20}$$

which on $C\ell_x(X)$ is defined by the formula $L(\varphi) = -\sum_{j=1}^n e_j \varphi e_j$ for any orthonormal basis $e_1, \dots, e_n$ of $T_x(X)$. This map is globally diagonalizable and yields the canonical bundle decomposition

$$C\ell(X) = \bigoplus_{p=0}^{n} \Lambda^p(X). \tag{5.21}$$

In particular,

$$L = (-1)^p(n - 2p) \qquad \text{on } \Lambda^p(X) \tag{5.22}$$

for each $p = 0, \dots, n$ (see Chap. I). Clearly α and L commute, and their composition satisfies

$$\alpha \circ L = L \circ \alpha = (n - 2p) \qquad \text{on } \Lambda^p(X) \tag{5.23}$$

for each p.

Finally, we consider the section ω of $\Lambda^n(X) \subset C\ell(X)$ given at each point x by setting

$$\omega = e_1 \cdots e_n \tag{5.24}$$

where $e_1, \dots, e_n$ is any positively oriented orthonormal basis of $T_x(X)$. Since ω is independent of the choice of basis we may for any $x \in X$ choose local fields $e_1, \dots, e_n$ such that $(\nabla e_i)_x = 0$ for each i. This shows that

$$\nabla \omega \equiv 0. \tag{5.25}$$

The section ω satisfies the relations

$$\omega^2 = (-1)^{\frac{n(n+1)}{2}} \tag{5.26}$$

$$\omega e = (-1)^{n-1} e \omega \tag{5.27}$$

for any section e of $T(X) \subset C\ell(X)$.

We now define a canonical bundle map

$$\lambda_\omega: C\ell(X) \longrightarrow C\ell(X) \tag{5.28}$$

by setting

$$\lambda_\omega(\varphi) = \omega \cdot \varphi.$$

If $n \equiv 3$ or $0 \pmod 4$, then $\lambda_\omega^2 = 1$ and we have a decomposition

$$C\ell(X) = C\ell^+(X) \oplus C\ell^-(X) \tag{5.29}$$

where $C\ell^{\pm}(X)$ are the ± 1 eigenbundles of λ_ω. From (5.27) we see that for any non-zero vector $e \in T_x(X)$ at any point x, left Clifford multiplication gives isomorphisms:

$$e : \mathrm{C}\ell_x^{\pm}(X) \longrightarrow \mathrm{C}\ell_x^{\mp}(X) \qquad \text{if } n \equiv 0 \pmod 4 \tag{5.30}$$

$$e : \mathrm{C}\ell_x^{\pm}(X) \longrightarrow \mathrm{C}\ell_x^{\pm}(X) \qquad \text{if } n \equiv 3 \pmod 4. \tag{5.31}$$

The bundles $\mathrm{C}\ell^{\pm}(X)$ can be explicitly written as

$$\mathrm{C}\ell^{\pm}(X) = (1 \pm \omega)\mathrm{C}\ell(X). \tag{5.32}$$

Each such bundle is evidently a submodule under right Clifford multiplication.

The above construction is particularly important since it generalizes immediately to any Dirac bundle S over X. Again we have a bundle isomorphism

$$\lambda_\omega : S \longrightarrow S \tag{5.33}$$

and a corresponding decomposition

$$S = S^+ \oplus S^- \tag{5.34}$$

where $S^{\pm} = (1 \pm \omega) \cdot S$. Moreover, for any non-zero $e \in T_x(X)$, the formulas analogous to (5.30) and (5.31) hold.

Note that for $n \equiv 1$ or $2 \pmod 4$, $\lambda_\omega^2 = -1$. Hence if we complexify S and consider the operator $i\lambda_\omega$ we obtain a splitting $S \otimes C = (S \otimes \mathbb{C})^+ \oplus (S \otimes \mathbb{C})^-$. In these dimensions, λ_ω defines a complex structure in S, and this splitting is the usual $(1,0)$, $(0,1)$-decomposition for the complex structure.

Let us turn our attention back to the Clifford bundle $\mathrm{C}\ell(X)$ and examine some of the elementary properties of the operators defined above.

Lemma 5.17. *The operators α, L and λ_ω satisfy the following relations*

(i) $\alpha L - L\alpha = 0$
(ii) $\alpha\lambda_\omega + (-1)^{n-1}\lambda_\omega\alpha = 0$
(iii) $L\lambda_\omega + (-1)^n\lambda_\omega L = 0.$

Proof. (i) was observed above. For (ii) we note that $\alpha(\omega\varphi) = \alpha(\omega)\alpha(\varphi) - (-1)^n\omega\alpha(\varphi)$. For (iii) we see that by (5.27), $L(\omega\varphi) = -\sum e_j\omega\varphi e_j = -(-1)^{n-1}\sum \omega e_j\varphi e_j = (-1)^{n-1}\omega L(\varphi)$. ∎

It follows from (iii) and (5.22) above that $\lambda_\omega(\Lambda^p(X)) = \Lambda^{n-p}(X)$. In fact λ_ω is related to the $*$-operator (cf. (5.10)) as follows:

$$\left.\begin{aligned} \omega\varphi &= (-1)^{p(n-p)+\frac{1}{2}p(p+1)} *\varphi \\ \varphi\omega &= (-1)^{\frac{1}{2}p(p+1)} *\varphi \end{aligned}\right\} \quad \text{for } \varphi \in \Lambda^p(X). \tag{5.35}$$

To see this it suffices to consider $\varphi = e_1 \cdots e_p$. Then $*\varphi = e_{p+1} \cdots e_n$ and the computation is easy.

Lemma 5.18. *The operators α, L and λ_ω considered as sections of* $\text{Hom}(\mathrm{C}\ell(X),\mathrm{C}\ell(X))$, *are globally parallel. That is,*

$$[\nabla,\alpha] = [\nabla, L] = [\nabla, \lambda_\omega] = 0.$$

Proof. That α is parallel follows from the fact that -1 is parallel in $\text{Hom}(T(X),T(X))$ and ∇ is a derivation. That λ_ω is parallel follows from (5.25). For L we fix a point $x \in X$ and choose a local orthonormal frame field $e_1, \dots, e_n$ such that $(\nabla e_j)_x = 0$ for each j. Then at the point x, $\nabla(L\varphi) = -\nabla \sum e_j\varphi e_j = -\sum \{(\nabla e_j)\varphi e_j + e_j(\nabla\varphi)e_j + e_j\varphi(\nabla e_j)\} = -\sum e_j(\nabla\varphi)e_j = L(\nabla\varphi)$. This completes the proof. ■

Corollary 5.19. *The subbundles $\Lambda^p(X)$ and $\mathrm{C}\ell^{\pm}(X)$ (when defined) are preserved by covariant differentiation.*

We now examine the relationship of these operators to the Dirac operators.

Proposition 5.20. *Let D and $\hat{D}$ be the Dirac operators on $\mathrm{C}\ell(X)$ defined above. Then the following relations hold*

(i) $D\alpha + \alpha D = \hat{D}\alpha + \alpha\hat{D} = 0$
(ii) $D\lambda_\omega + (-1)^n\lambda_\omega D = \hat{D}\lambda_\omega - \lambda_\omega\hat{D} = 0$
(iii) $DL + LD = 2\hat{D}$; $\hat{D}L + L\hat{D} = 2D$.

Corollary 5.21. *The operator $\Delta = D^2 = \hat{D}^2$ satisfies the relations*

$$[\alpha, \Delta] = [\lambda_\omega, \Delta] = [L, \Delta] = 0.$$

In particular we have that $\lambda_\omega : \mathbf{H}^p \xrightarrow{\approx} \mathbf{H}^{n-p}$. This is the Poincaré Duality Isomorphism.

Proof of Proposition 5.20. We shall consider only D. The arguments for $\hat{D}$ are similar. Let φ be a section of $\mathrm{C}\ell(X)$. Then using the lemmas above, we have

$$\text{(i)}\quad D(\alpha\varphi) = \sum_j e_j\nabla_{e_j}(\alpha\varphi) = \sum_j e_j\alpha(\nabla_{e_j}\varphi) = -\alpha\Big(\sum_j e_j\nabla_{e_j}\varphi\Big) = -\alpha(D\varphi)$$

$$\text{(ii)}\quad D(\omega\varphi) = \sum_j e_j\nabla_{e_j}(\omega\varphi) = \sum e_j\omega\nabla_{e_j}\varphi = (-1)^{n-1}\omega\sum e_j\nabla_{e_j}\varphi = (-1)^{n-1}\omega D\varphi$$

$$\begin{aligned}\text{(iii)}\quad D(L\varphi) &= \sum_j e_j\nabla_{e_j}(L\varphi) = \sum_j e_j L(\nabla_{e_j}\varphi) \\ &= -\sum_{j,k} e_j e_k(\nabla_{e_j}\varphi)e_k \\ &= -\sum_{j,k}(-e_k e_j - 2\delta_{kj})(\nabla_{e_j}\varphi)e_k \\ &= -L(D\varphi) + 2\hat{D}(\varphi).\end{aligned}$$

This completes the proof. ∎

The above arguments carry directly over to the following case.

Proposition 5.22. *Let S be any Dirac bundle over X. Then the Dirac operator on S satisfies the relation*

$$D\lambda_\omega = (-1)^{n-1}\lambda_\omega D.$$

We shall conclude this section with some remarks on the variation of the Atiyah-Singer operator under changes of metric.

Notice that all of the standard bundles of riemannian geometry—the tangent bundle, the cotangent bundle, and their tensor products—have structure group $\mathrm{GL}_n(\mathbb{R})$. They exist independently of any metric considerations. In fact, the introduction of a riemannian metric amounts to a (simultaneous) reduction of the structure group of these bundles to SO_n.

When X is a spin manifold, the situation for the canonical spinor bundle of X is completely different. *The spinor bundle itself depends on the choice of riemannian metric.*

This last statement can be made precise as follows. Let $P_{\mathrm{GL}^+}(X)$ be the oriented frame bundle of X and suppose that $\dim(X) = n > 2$. Denote by $\widetilde{\mathrm{GL}}_n^+(\mathbb{R}) \to \mathrm{GL}_n^+(\mathbb{R})$ the 2-fold, universal covering group of $\mathrm{GL}_n^+(\mathbb{R})$. Since X is spin, there exists a principal $\widetilde{\mathrm{GL}}_n^+(\mathbb{R})$-bundle $P_{\widetilde{\mathrm{GL}}^+}(X)$ with a $\widetilde{\mathrm{GL}}_n^+(\mathbb{R})$-equivariant bundle map $P_{\widetilde{\mathrm{GL}}^+}(X) \to P_{\mathrm{GL}^+}(X)$. Introducing a riemannian metric gives a reduction of the structure groups and a commutative diagram

$$\begin{array}{ccc} P_{\mathrm{Spin}}(X) & \longrightarrow & P_{\widetilde{\mathrm{GL}}^+}(X) \\ \big\downarrow & & \big\downarrow \\ P_{\mathrm{SO}}(X) & \longrightarrow & P_{\mathrm{GL}^+}(X) \end{array}$$

Now one could hope for the existence of a finite dimensional representation $\sigma : \widetilde{\mathrm{GL}}_n^+(\mathbb{R}) \to \mathrm{GL}(V)$ whose restriction to Spin_n is an irreducible spinor representation. The associated vector bundle

$$S = P_{\widetilde{\mathrm{GL}}^+}(X) \times_\sigma V$$

would then be the canonical spinor bundle, and choosing a metric on X would induce a metric on S (as in the case of the tensor bundles.). This would give us a fixed vector bundle on which we could consider a family of Atiyah-Singer operators associated to each variation of the riemannian structure on the base. Unfortunately, this hope cannot be fulfilled.

Lemma 5.23. *The Lie group $\widetilde{\mathrm{GL}}_n^+(\mathbb{R})$ $(n > 2)$ has no finite dimensional representations other than those which descend to* $\mathrm{GL}_n^+(\mathbb{R})$.

Proof. Consider the subgroup $\mathrm{SL}_n(\mathbb{R}) \subset \mathrm{GL}_n^+(\mathbb{R})$ and its 2-fold, universal covering group $\widetilde{\mathrm{SL}}_n(\mathbb{R}) \subset \widetilde{\mathrm{GL}}_n^+(\mathbb{R})$. Since $\widetilde{\mathrm{SL}}_n(\mathbb{R})$ contains the kernel of the homomorphism $\widetilde{\mathrm{GL}}_n^+(\mathbb{R}) \to \mathrm{GL}_n^+(\mathbb{R})$, it will clearly suffice to prove the assertion for this subgroup. Let $\varphi: \widetilde{\mathrm{SL}}_n(\mathbb{R}) \to \mathrm{GL}_N(\mathbb{R})$ be any Lie group homomorphism. Let $\varphi_*: \mathfrak{sl}_n(\mathbb{R}) \to \mathfrak{gl}_N(\mathbb{R})$ denote the associated Lie algebra homomorphism, and consider its complexification $\varphi_* \otimes \mathbb{C}: \mathfrak{sl}_n(\mathbb{C}) \to \mathfrak{gl}_N(\mathbb{C})$. Since $\mathrm{SL}_n(\mathbb{C})$ is simply-connected, the elementary theory of Lie groups tells us that $\varphi_* \otimes \mathbb{C}$ is induced by a homomorphism of Lie groups $\Phi: \mathrm{SL}_n(\mathbb{C}) \to \mathrm{GL}_N(\mathbb{C})$. It follows that

$$\varphi = \Phi|_{\mathrm{SL}_n(\mathbb{R})}$$

in a neighborhood of the identity. By the uniqueness of analytic continuation this identity holds everywhere, and so the representation descends to $\mathrm{SL}_n(\mathbb{R})$ as claimed. ∎

Note that the argument just given does not apply to the conformal group $\mathrm{C}_n \equiv \{g \in \mathrm{GL}_n(\mathbb{R}): g = \lambda g_0 \text{ for } \lambda \in \mathbb{R}^+ \text{ and } g_0 \in \mathrm{SO}_n\}$. Indeed, *conformal* changes in the metric on the base manifold can be lifted to a fixed spinor bundle, and one can study there the associated change in the Atiyah-Singer operator. A basic and important fact is that the Atiyah-Singer Dirac operator remains essentially invariant under all conformal changes of the metric.

We now make this last statement precise. Fix a riemannian spin manifold X with metric $\langle \cdot,\cdot \rangle$, and consider the conformally related metric

$$\langle \cdot,\cdot \rangle' = e^{2u} \langle \cdot,\cdot \rangle$$

where u is some smooth function on X. Let X' denote this riemannian manifold with metric $\langle \cdot,\cdot \rangle'$. To each orthonormal tangent frame $\mathscr{E} = \{e_1, \ldots, e_n\}$ on X we can associate the orthonormal frame $\psi(\mathscr{E}) = \{e_1', \ldots, e_n'\}$ on X', where $e_j' = \exp(-u)e_j$ for each j. This gives us an SO_n-equivariant map $\psi: P_{\mathrm{SO}}(X) \to P_{\mathrm{SO}}(X')$ which lifts to a Spin_n-equivariant map

$$\psi : P_{\mathrm{Spin}}(X) \longrightarrow P_{\mathrm{Spin}}(X') \tag{5.36}$$

between principal Spin_n-bundles. (We have chosen the spin structure on X' which is topologically equivalent to the one given on X.) For any fixed spinor representation $\mu : \mathrm{Spin}_n \to \mathrm{SO}(M)$ we have associated spinor bundles $S = P_{\mathrm{Spin}}(X) \times_\mu M$ and $S' = P_{\mathrm{Spin}}(X') \times_\mu M$, and the map (5.36) gives us a bundle isometry

$$\psi_\mu : S \longrightarrow S'. \tag{5.37}$$

We now modify this isometry by setting

$$\Psi = e^{-\frac{n-1}{2}u} \psi_\mu \tag{5.38}$$

The resulting map $\Psi : S \to S'$ is a bundle isomorphism which is conformal on each fibre. The basic result is the following:

Theorem 5.24. *Let $\not{D} : \Gamma(S) \to \Gamma(S)$ and $\not{D}' : \Gamma(S') \to \Gamma(S')$ be the canonical Atiyah-Singer Dirac operators defined over the conformally related riemannian manifolds X and X' respectively. Then*

$$\not{D}' = \Psi \circ \not{D} \circ \Psi^{-1} \tag{5.39}$$

Corollary 5.25. *Let $\not{D}$ and $\not{D}'$ be as in 5.24. Then*

$$\dim(\ker \not{D}) = \dim(\ker \not{D}').$$

In other words the dimension of the space of harmonic spinors remains constant under pointwise conformal changes of the riemannian metric.

Note. Suppose we have a decomposition $S = S^+ \oplus S^-$ defined by a canonical volume element as in the next chapter. Then Corollary 5.25 applies to each of the operators $\not{D}^+$ and $\not{D}^-$, that is, $\dim(\ker \not{D}^\pm) = \dim(\ker(\not{D}')^\pm)$.

Proof of Theorem 5.24. We have two metrics $\langle \cdot,\cdot \rangle$ and $\langle \cdot,\cdot \rangle'$ defined on the same vector bundle $T = TX = TX'$. Let ∇ and ∇' respectively denote the associated canonical riemannian connections. It is straightfoward to verify that for vector fields V and W we have

$$\nabla'_V W = \nabla_V W + (V \cdot u)W + (W \cdot u)V - \langle V,W \rangle \mathrm{grad}(u)$$

where the gradient is taken in the metric $\langle \cdot,\cdot \rangle$. (Check the axioms.) Suppose now that $\mathscr{E} = \{e_1, \ldots, e_n\}$ and $\psi(\mathscr{E}) = \{e'_1, \ldots, e'_n\}$ are local orthonormal frame fields for $\langle \cdot,\cdot \rangle$ and $\langle \cdot,\cdot \rangle'$ respectively, and let $\omega_{ji} = \langle \nabla e_i, e_j \rangle$ and $\omega'_{ji} = \langle \nabla' e'_i, e'_j \rangle'$ be the associated 1-forms. One easily finds that for any tangent vector V,

$$\omega'_{ji}(V) = \omega_{ji}(V) + (e_i \cdot u)\langle V, e_j\rangle - (e_j \cdot u)\langle V, e_i\rangle.$$

The local tangent frame field $\mathscr{E} = \{e_1, \dots, e_n\}$ determines a local frame field $\mathscr{S} = \{\sigma_1, \dots, \sigma_N\}$ for S. Similarly, $\mathscr{E}' = \{e'_1, \dots, e'_n\}$ determines a frame field $\mathscr{S}' = \{\sigma'_1, \dots, \sigma'_N\}$ for S' where $\sigma'_j = \psi_\mu(\sigma_j)$ for each j. From (4.35) we see that the induced connections on S and S' are related as follows:

$$\begin{aligned}
\nabla^{S'}_V \sigma'_\alpha &= \tfrac{1}{4} \sum_{i,j} \omega'_{ji}(V) e'_i e'_j \sigma'_\alpha \\
&= \tfrac{1}{4} \psi_\mu \left\{ \sum_{i,j} \omega'_{ji}(V) e_i e_j \sigma_\alpha \right\} \\
&= \tfrac{1}{4} \psi_\mu \left\{ \sum_{i,j} (\omega_{ji}(V) + (e_i u)\langle V, e_j\rangle - (e_j u)\langle V, e_i\rangle) e_i e_j \sigma_\alpha \right\} \\
&= \psi_\mu \left\{ \nabla^S_V \sigma_\alpha + \tfrac{1}{4} (\operatorname{grad}(u) \cdot V - V \cdot \operatorname{grad}(u)) \sigma_\alpha \right\}.
\end{aligned}$$

Since $\operatorname{grad}(u) \cdot V = -V \cdot \operatorname{grad}(u) - 2\langle \operatorname{grad}(u), V\rangle$, we conclude the following.

Lemma 5.27. *Let ∇^S and $\nabla^{S'}$ denote the riemannian connections on S and S' respectively. Then*

$$\nabla^{S'} = \psi_\mu \circ \left\{ \nabla^S - \frac{1}{2} V \cdot \operatorname{grad}(u) - \frac{1}{2}(V \cdot u) \right\} \circ \psi_\mu^{-1}.$$

Corollary 5.28.

$$D\!\!\!/' = \psi_\mu \circ \left\{ D\!\!\!/ + \frac{1}{2}(n-1)\operatorname{grad}(u) \right\} \circ \psi_\mu^{-1}$$

We now observe that for any constant, α,

$$\begin{aligned}
D\!\!\!/(e^{\alpha u}\sigma) &= e^{\alpha u}\left(D\!\!\!/\sigma + \alpha \sum_j (e_j u) e_j \sigma \right) \\
&= e^{\alpha u}(D\!\!\!/\sigma + \alpha \operatorname{grad}(u) \cdot \sigma),
\end{aligned}$$

and therefore

$$\begin{aligned}
\Psi \circ D\!\!\!/ \circ \Psi^{-1} &= e^{-\frac{n-1}{2}u} \psi_\mu \circ D\!\!\!/ \circ \left(e^{\frac{n-1}{2}u} \psi_\mu^{-1} \right) \\
&= \psi_\mu \circ \left(D\!\!\!/ + \frac{1}{2}(n-1)\operatorname{grad}(u) \right) \circ \psi_\mu^{-1} = D\!\!\!/'. \quad \blacksquare
\end{aligned}$$

§6. The Fundamental Elliptic Operators

In this section we shall use the Clifford bundle and its modules to systematically derive the fundamental elliptic operators in riemannian geometry, that is, the Euler characteristic operator, the signature operator, and the Atiyah-Singer $\hat{A}$-operator.

The basic construction is the following. Let S be a Dirac bundle with Dirac operator D over a riemannian manifold X, and suppose that S is **$\mathbb{Z}_2$-graded.** This means that there is a parallel decomposition

$$S = S^0 \oplus S^1 \tag{6.1}$$

so that $C\ell^i(X) \cdot S^j \subseteq S^{i+j}$ for all $i,j \in \mathbb{Z}_2$. From the definition (5.1) of the Dirac operator it is clear that D is of the form

$$D = \begin{pmatrix} 0 & D^1 \\ D^0 & 0 \end{pmatrix} \tag{6.2}$$

where

$$D^0 : \Gamma(S^0) \longrightarrow \Gamma(S^1) \quad \text{and} \quad D^1 : \Gamma(S^1) \longrightarrow \Gamma(S^0). \tag{6.3}$$

Since D is self-adjoint, we see that D^0 and D^1 are adjoints of one another.

The ellipticity of D established in §5 implies that each operator D^k is also elliptic. In fact, the principal symbol of D^k at a cotangent vector ξ is simply Clifford multiplication by $i\xi$, that is,

$$\sigma_\xi(D^k) = i\xi \cdot : S^k \longrightarrow S^{k+1} \qquad \text{for } k \in \mathbb{Z}_2 \tag{6.4}$$

(see Lemma 5.1). Since $\xi \cdot \xi = -\|\xi\|^2$, we see that $\sigma_\xi(D^k)$ is an isomorphism for $\xi \neq 0$. It is a fact that over a compact manifold, the kernel and cokernel of an elliptic operator P are of finite dimension, and a basic invariant of P is its **index** which is defined as

$$\text{ind } P = \dim(\ker P) - \dim(\text{coker } P). \tag{6.5}$$

Since D^1 is the adjoint of D^0, there is an isomorphism $\ker(D^1) \cong \text{coker}(D^0)$, and so we have that

$$\text{ind } D^0 = \dim(\ker D^0) - \dim(\ker D^1). \tag{6.6}$$

From (6.2) it is clear that

$$\ker D = \ker D^0 \oplus \ker D^1. \tag{6.7}$$

In particular, if D is injective, then $\text{ind } D^0 = 0$.

Whenever X is oriented and even-dimensional, we recall (cf. (3.11)ff) that there is an important method for introducing a $\mathbb{Z}_2$-grading on any Dirac bundle S. Let $\omega_{\mathbb{C}}$ be the **complex volume element,** given in terms of a positively oriented orthonormal tangent frame $(e_1, \ldots, e_{2m})$ by

$$\omega_{\mathbb{C}} \equiv i^m e_1 \cdots e_{2m} \tag{6.8}$$

where $2m = \dim(X)$. This is a globally defined section of $\mathbb{C}\ell(X) = C\ell(X) \otimes \mathbb{C}$ with the following properties.

$$\nabla \omega_{\mathbb{C}} = 0, \tag{6.9}$$

$$\omega_{\mathbb{C}}^2 = 1, \tag{6.10}$$

$$\omega_{\mathbb{C}} e = -e\omega_{\mathbb{C}} \qquad \text{for any } e \in TX. \tag{6.11}$$

Property (6.9) follows easily from the derivation property (4.26) of ∇. (Fix $x \in X$ and choose e_j's with $(\nabla e_j)_x = 0$.). For the rest, see Proposition 3.3 of Chapter I.

Suppose now that S is a Dirac bundle over X (complex if m is odd). Then S has a decomposition

$$S = S^+ \oplus S^- \tag{6.12}$$

into the $+1$ and -1 eigenbundles for multiplication by $\omega_{\mathbb{C}}$. These bundles can be simply expressed as

$$S^{\pm} = (1 \pm \omega_{\mathbb{C}})S. \tag{6.13}$$

From (6.9) we see that this decomposition is parallel, and from (6.11) we see that for any $e \in TX$,

$$e \cdot S^{\pm} \subseteqq S^{\mp}. \tag{6.14}$$

This means that after identifying S^0 with S^+ and S^1 with S^-, the decomposition (6.12) gives a $\mathbb{Z}_2$-grading on S.

We now examine some important examples of this construction.

Example 6.1 (**the Euler characteristic operator**). Let X be a compact riemannian manifold and consider the basic case where

$$S \equiv C\ell(X) = C\ell^0(X) \oplus C\ell^1(X)$$

(cf. (3.2)). By Theorem 5.7 we see that under the canonical isomorphism $C\ell(X) \cong \Lambda^*(X)$, the operator $D^0 : \Gamma(C\ell^0(X)) \to \Gamma(C\ell^1(X))$ corresponds to the operator

$$d + d^* : \Gamma(\Lambda^{\text{even}}(X)) \longrightarrow \Gamma(\Lambda^{\text{odd}}(X)).$$

Consequently, we have

$$\begin{aligned} \operatorname{ind} D^0 &= \dim \mathbf{H}^{\text{even}} - \dim \mathbf{H}^{\text{odd}} \\ &= \text{the Euler characteristic of } X. \end{aligned}$$

Example 6.2 (**the signature operator**). Let X be a compact, oriented riemannian manifold of dimension $4k$ and consider again the basic case

$$S \equiv C\ell(X) = C\ell^+(X) \oplus C\ell^-(X)$$

where the $\mathbb{Z}_2$-grading is now given as above by the complex volume element $\omega_{\mathbb{C}} = (-1)^k\omega$ (see (6.12) and also (5.26)ff). There is a corresponding decomposition

$$\ker D = \ker D^+ \oplus \ker D^-.$$

Since $\omega_{\mathbb{C}}$ is parallel, it preserves $\ker D$ and the subspaces $\ker D^\pm$ are just the ± 1 eigenspaces under multiplication by $\omega_{\mathbb{C}}$ on $\ker D$. That is,

$$\ker D^\pm = (1 \pm \omega_{\mathbb{C}})\ker D.$$

Now, under the canonical isomorphism $C\ell(X) \cong \Lambda^*(X)$ we know that $\ker D \cong \mathbf{H} = \mathbf{H}^0 \oplus \cdots \oplus \mathbf{H}^{4k}$, the space of harmonic forms (see Corollary 5.9). Furthermore, under this isomorphism, left multiplication by $\omega_{\mathbb{C}}$ corresponds to the Hodge $*$-operator, that is, for $\varphi \in \Lambda^p(X)$,

$$\omega_{\mathbb{C}} \cdot \varphi = (-1)^{k + \frac{p(p-1)}{2}} * \varphi \tag{6.15}$$

(see (5.35)). Consequently, for each $p = 0, \ldots, 2k$ we have an isomorphism

$$\omega_{\mathbb{C}} : \mathbf{H}^p \xrightarrow{\approx} \mathbf{H}^{4k-p}.$$

This, in turn, implies that the space $\mathbf{H}(p) \equiv \mathbf{H}^p \oplus \mathbf{H}^{4k-p}$, for $p < 2k$, has a decomposition

$$\mathbf{H}(p) = \mathbf{H}^+(p) \oplus \mathbf{H}^-(p)$$

where the subspaces $\mathbf{H}^\pm(p) \equiv (1 \pm \omega_{\mathbb{C}})\mathbf{H}(p)$ are of the *same dimension*. Since $\ker D^\pm = \mathbf{H}^\pm = \mathbf{H}^\pm(0) \oplus \cdots \oplus \mathbf{H}^\pm(2k-1) \oplus (\mathbf{H}^{2k})^\pm$, where $(\mathbf{H}^{2k})^\pm = (1 \pm \omega_{\mathbb{C}})\mathbf{H}^{2k}$, we conclude that

$$\begin{aligned} \operatorname{ind} D^+ &= \dim(\mathbf{H}^{2k})^+ - \dim(\mathbf{H}^{2k})^- \\ &= \operatorname{sig}(X). \end{aligned}$$

For this last statement we recall that the **signature** of X, denoted $\operatorname{sig}(X)$, is defined to be the signature of the quadratic form

$$Q(\varphi,\psi) \equiv \int_X \varphi \wedge \psi \cong ([\varphi] \cup [\psi])([X])$$

on $\mathbf{H}^{2k} \cong H^{2k}(X;\mathbb{R})$. Since $* \cong \omega_{\mathbb{C}}$ in dimension $2k$ and since

$$\int_X \varphi \wedge *\varphi = \|\varphi\|^2,$$

we see that this signature is just the difference of the dimensions of the $+1$ and -1 eigenspaces of $*$ on $\mathbf{H}^{2k}$.

EXAMPLE 6.3 (**the Atiyah-Singer $\hat{A}$-operator**). Let X be a compact riemannian spin manifold of dimension $4k$ and consider the complex spinor bundle $\$_{\mathbb{C}}$ with Dirac operator $\not{D}$. We split $\$_{\mathbb{C}} \cong \$_{\mathbb{C}}^+ \oplus \$_{\mathbb{C}}^-$ under the complex volume element as above. Then it is a consequence of the Atiyah-Singer Index Theorem in Chapter III that

$$\operatorname{ind}(\not{D}^+) = \hat{A}(X)$$

where $\hat{A}(X)$ is a rational Pontryagin number of X called the **$\hat{A}$-genus.** We shall examine this invariant in detail in §11 of Chapter III; however, some discussion of it is in order here.

The $\hat{A}$-genus has an important multiplicative property like that of the signature. If X and Y are compact oriented manifolds. Then

$$\hat{A}(X \times Y) = \hat{A}(X) \times \hat{A}(Y). \tag{6.16}$$

The $\hat{A}$-genus is, in general, *not* an integer. For example, for a compact 4-manifold X, it is a fact that

$$-\hat{A}(X) = \frac{1}{8}\operatorname{sig}(X) = \frac{1}{24}p_1(X) \tag{6.17}$$

where $p_1(X)$ is the first Pontryagin number of X. In particular, for the complex projective plane $\mathbb{P}^2(\mathbb{C})$, we have $H^2(\mathbb{P}^2(\mathbb{C})) \cong \mathbb{Z}$ and so the signature is 1. It follows that $\hat{A}(\mathbb{P}^2(\mathbb{C})) \cong -1/8$.

The index of an elliptic operator is, of course, an integer. Hence, we conclude the following result (cf. Atiyah-Hirzebruch [1]).

The $\hat{A}$-genus of a compact spin manifold is an integer. (6.18)

Note also that from (6.17) we retrieve the basic fact that the signature of a spin 4-manifold must be a multiple of 8 (see Corollary 2.12 and the following discussion).

An important set of spin manifolds with non-zero $\hat{A}$-genus is provided by the hypersurfaces $V^{2n}(d)$ of complex projective space $\mathbb{P}^{2n+1}(\mathbb{C})$ (see Example 2.7). Recall that the manifold $V^{2n}(d)$ is spin if and only if the degree d is even. It is a fact that (cf. Lawson-Michelsohn [1])

$$\hat{A}(V^{2n}(d)) = \frac{2^{-2n}d}{(2n+1)!}\prod_{k=1}^{n}(d^2-(2k)^2). \tag{6.19}$$

Thus, each of the spin manifolds $V^{2n}(2d)$, for $d > n$, has non-zero $\hat{A}$-genus. Taking products of these gives further examples by (6.16).

Each of the fundamental examples above gives rise to a family of associated operators by the process of taking "coefficients in a bundle." This works as follows. Let $S = S^0 \oplus S^1$ be a $\mathbb{Z}_2$-graded Dirac bundle as before, and let E be any riemannian bundle with connection over X. Then the bundle

$$S \otimes E = (S^0 \otimes E) \oplus (S^1 \otimes E)$$

is again a $\mathbb{Z}_2$-graded Dirac bundle over X (see Proposition 5.10).

EXAMPLE 6.4. Let $S = C\ell(X) = C\ell^+(X) \oplus C\ell^-(X)$ be as in Example 6.2. Then for any bundle E over X we can construct the **twisted signature operator** $D_E^+ : \Gamma(C\ell^+(X) \otimes E) \to \Gamma(C\ell^-(X) \otimes E)$ whose index is

$$\operatorname{sig}(X;E) \equiv \{\operatorname{ch}_2 E \cdot \mathbf{L}(X)\}[X].$$

This formula is explained in detail in Chapter III.

EXAMPLE 6.5. Let $\mathcal{S}_{\mathbf{C}} = \mathcal{S}_{\mathbf{C}}^+ \oplus \mathcal{S}_{\mathbf{C}}^-$ be the complex spinor bundle of Example 6.3. Then for any bundle E over X we can construct the **twisted Atiyah-Singer operator** $\not{D}_E^+ : \Gamma(S_{\mathbf{C}}^+ \otimes E) \to \Gamma(S_{\mathbf{C}}^- \otimes E)$ whose index is

$$\hat{A}(X;E) \equiv \{\operatorname{ch} E \cdot \hat{\mathbf{A}}(X)\}[X].$$

Again see Chapter III for details.

§7. $C\ell_k$-Linear Dirac Operators

There is a variation of the constructions given above, due to Atiyah and Singer, which has proved to be very important in riemannian geometry. To introduce the concept we return to a point mentioned earlier (in Example 3.7). Let $P_{\mathrm{Spin}}(X)$ be the principal Spin_n-bundle of an n-dimensional spin manifold X. Then from the representation $\ell : \mathrm{Spin}_n \to \mathrm{Hom}(C\ell_n, C\ell_n)$ given by left multiplication, we have the associated vector bundle

$$\mathfrak{S}(X) \equiv P_{\mathrm{Spin}} \times_\ell C\ell_n. \tag{7.1}$$

Since right multiplication commutes with ℓ, we see that there is a right action of the algebra $C\ell_n$ on the bundle $\mathfrak{S}(X)$ which preserves the fibres. This action makes $\mathfrak{S}(X)$ a bundle of rank-1 $C\ell_n$-modules.

The idea now is to construct elliptic operators and an appropriate index theory which take into account this action of $C\ell_n$.

We shall begin with the construction of such an operator in the basic case of $\mathfrak{S}(X)$. Note first that the action of $C\ell_n$ on $\mathfrak{S}(X)$ clearly commutes with Clifford multiplication by elements of $C\ell(X)$.

Since $\mathfrak{S}(X)$ is associated to $P_{\mathrm{Spin}}(X)$, it carries the canonical riemannian connection, and as such it is clearly a Dirac bundle over X. In fact, *as a vector bundle $\mathfrak{S}(X)$ is simply a direct sum of irreducible (real) spinor bundles of X*. (This comes from the decomposition of $C\ell_n$ into irreducible modules under left-multiplication.)

The right action of $C\ell_n$ on $\mathfrak{S}(X)$ is **parallel** in the riemannian connection, i.e., for any section $\sigma \in \Gamma(\mathfrak{S}(X))$ and any element $\varphi \in C\ell_n$, we have $\nabla(\sigma \cdot \varphi) = (\nabla\sigma) \cdot \varphi$. (This is evident since the holonomy in $\mathfrak{S}(X)$ is left multiplication by elements of Spin_n. It can also be seen directly from the methods of §4.)

From the definition (7.1) we see that the decomposition $C\ell_n = C\ell_n^0 \oplus C\ell_n^1$ gives rise to a parallel decomposition

$$\mathfrak{S}(X) = \mathfrak{S}^0(X) \oplus \mathfrak{S}^1(X) \tag{7.2}$$

which is not only a $\mathbb{Z}_2$-grading over the bundle $C\ell(X)$ but also over the free $C\ell_n$-action. That is, we have

$$\mathfrak{S}^i(X) \cdot C\ell_n^j \subseteq \mathfrak{S}^{i+j}(X) \tag{7.3}$$

for all $i, j \in \mathbb{Z}_2$.

Since $\mathfrak{S}(X)$ is a Dirac bundle it carries a canonical Dirac operator $\mathfrak{D}$, which, with respect to the decomposition (7.2) is of the form

$$\mathfrak{D} = \begin{pmatrix} 0 & \mathfrak{D}^1 \\ \mathfrak{D}^0 & 0 \end{pmatrix}. \tag{7.4}$$

Furthermore, *this operator commutes with the action of* $C\ell_n$. To see this, note that $\mathfrak{D}(\sigma\varphi) = \sum e_j \nabla_{e_j}(\sigma\varphi) = \sum (e_j \nabla_{e_j}\sigma)\varphi = \mathfrak{D}(\sigma)\varphi$, since multiplication by φ is parallel and commutes with multiplication by elements from $C\ell(X)$. This operator $\mathfrak{D}$ is called the **$C\ell_n$-linear Atiyah-Singer operator** of X.

We now proceed as above and consider the restricted operator

$$\mathfrak{D}^0 : \Gamma(\mathfrak{S}^0(X)) \longrightarrow \Gamma(\mathfrak{S}^1(X)).$$

From (7.3) and the paragraph above we conclude the following:

Lemma 7.1. *The operator $\mathfrak{D}^0$ is a real, elliptic first-order operator which commutes with the action of $C\ell_n^0 \cong C\ell_{n-1}$ on $\mathfrak{S}(X) = \mathfrak{S}^0(X) \oplus \mathfrak{S}^1(X)$.*

This construction is sufficiently important that we shall axiomatize it.

DEFINITION 7.2. By a **$C\ell_k$-Dirac bundle** over a riemannian manifold X we mean a real Dirac bundle $\mathfrak{S}$ over X, together with a right action $C\ell_k \hookrightarrow \text{Aut}(\mathfrak{S})$ which is parallel and commutes with multiplication by elements of $C\ell(X)$.

This can be thought of as a Dirac bundle which carries "scalar multiplication" by $C\ell_k$. Notice that a $C\ell_1$- or Cl_2-Dirac bundle is just a complex or quaternionic Dirac bundle respectively.

DEFINITION 7.3. A $C\ell_k$-Dirac bundle $\mathfrak{S}$ is said to be **$\mathbb{Z}_2$-graded** if it carries a $\mathbb{Z}_2$-grading $\mathfrak{S} = \mathfrak{S}^0 \oplus \mathfrak{S}^1$ as a Dirac bundle, which is simultaneously a $\mathbb{Z}_2$-grading for the $C\ell_k$-action (that is, (7.3) is satisfied).

Any $C\ell_k$-Dirac bundle $\mathfrak{S}$ has a canonically associated Dirac operator $\mathfrak{D}$ which commutes with the $C\ell_k$-action. If $\mathfrak{S}$ is $\mathbb{Z}_2$-graded, then $\mathfrak{D}$ decomposes as in (7.4), and we get an elliptic operator

$$\mathfrak{D}^0 : \Gamma(\mathfrak{S}^0) \longrightarrow \Gamma(\mathfrak{S}^1) \tag{7.5}$$

which commutes with the action of $C\ell_k^0 \cong C\ell_{k-1}$.

When X is compact, we can directly define an analytic index for such operators as follows. Since $\mathfrak{D}^0$ commutes with $C\ell_k^0 \cong C\ell_{k-1}$, the kernel of $\mathfrak{D}^0$ is a finite dimensional $C\ell_{k-1}$-module, and thereby ker $\mathfrak{D}^0$ determines an element in the Grothendieck group $\mathfrak{M}_{k-1}$ of such modules. Consider now the residue class of this element in $\mathfrak{M}_{k-1}/i^*\mathfrak{M}_k$, where i^* is induced by the homomorphism $i_* : C\ell_{k-1} \to C\ell_k$ determined by the inclusion map $i : \mathbb{R}^{k-1} \hookrightarrow \mathbb{R}^k$. Recall (from Chap. I, §9) that these quotient groups are naturally isomorphic to the KO-groups of a point.

DEFINITION 7.4. Let $\mathfrak{S} = \mathfrak{S}^0 \oplus \mathfrak{S}^1$ be a $\mathbb{Z}_2$-graded $C\ell_k$-Dirac bundle over a compact manifold. Then the **analytic index** of the Dirac operator $\mathfrak{D}^0 : \Gamma(\mathfrak{S}^0) \to \Gamma(\mathfrak{S}^1)$, denoted by $\text{ind}_k(\mathfrak{D}^0)$, is the residue class

$$[\ker \mathfrak{D}^0] \in \mathfrak{M}_{k-1}/i^*\mathfrak{M}_k \cong KO^{-k}(\text{pt}).$$

We recall that the groups $KO^{-*}(\text{pt})$ are the same as the stable homotopy groups of the orthogonal group, that is,

$$KO^{-k}(\text{pt}) = \begin{cases} \mathbb{Z} & k \equiv 0 \pmod 4 \\ \mathbb{Z}_2 & k \equiv 1 \text{ or } 2 \pmod 8 \\ 0 & \text{otherwise.} \end{cases} \tag{7.6}$$

A standard argument shows that this index is constant under deformations of the operator. One of the deepest aspects of the work of Atiyah and Singer is the computation of this index topologically (see III.16). The applications of this result are among the most far-reaching in all of differential geometry. For this reason we devote the remainder of this section to a detailed examination of such operators particularly in the cases of fundamental interest in geometry.

We begin with a remark which will be useful when studying the multiplicative properties of the operator. Recall from Chapter I (Proposition 5.20) that there is a natural equivalence between the category of (ungraded) modules over $C\ell_{k-1}$ and the category of $\mathbb{Z}_2$-graded modules over $C\ell_k$. This equivalence induces a natural isomorphism

$$\mathfrak{M}_{k-1}/i^*\mathfrak{M}_k \cong \hat{\mathfrak{M}}_k/i^*\hat{\mathfrak{M}}_{k+1} \tag{7.7}$$

where as before $\hat{\mathfrak{M}}_k$ denotes the Grothendieck group of finite dimensional $\mathbb{Z}_2$-graded $\mathbb{R}$-modules over $C\ell_k$. The map from the graded to the ungraded case is given by taking the even part.

Given now a Dirac operator $\mathfrak{D} : \Gamma(\mathfrak{S}) \to \Gamma(\mathfrak{S})$ as above, we see that $\ker(\mathfrak{D})$ is a $\mathbb{Z}_2$-graded $C\ell_k$-module. The even part of $\ker(\mathfrak{D})$ is exactly

$\ker(\mathfrak{D}^0)$. This gives the following:

ALTERNATIVE DEFINITION 7.4. Let $\mathfrak{S} = \mathfrak{S}^0 \oplus \mathfrak{S}^1$ be a $\mathbb{Z}_2$-graded $C\ell_k$-Dirac bundle over a compact manifold. The **analytic index** of the Dirac operator $\mathfrak{D}$ of $\mathfrak{S}$ is the residue class

$$[\ker \mathfrak{D}] \in \hat{\mathfrak{M}}_k/i^*\hat{\mathfrak{M}}_{k+1} \cong KO^{-k}(\mathrm{pt}).$$

Via the isomorphism (7.7) this index coincides with the index $[\ker \mathfrak{D}^0]$ given in Definition 7.4.

This second definition is actually more natural and is important for understanding the multiplicative properties of the index. With this second definition we see that the Clifford index generalizes the classical one in that $\mathrm{ind}_0(\cdot) = \mathrm{ind}(\cdot) = \dim_{\mathbb{R}}(\ker\cdot) - \dim_{\mathbb{R}}(\mathrm{coker}\cdot)$. To see this, note first that $C\ell_0 = \mathbb{R}$ and $C\ell_1 = \mathbb{C}$. A $\mathbb{Z}_2$-graded $C\ell_0$-module is just a pair of real vector spaces $V^0 \oplus V^1$. Now $[V \oplus 0] = -[0 \oplus V]$ in $\hat{\mathfrak{M}}_0/i^*\hat{\mathfrak{M}}_1$ because $V \oplus V \cong V \otimes \mathbb{C}$ extends to be a graded $C\ell_1$-module. Consequently, $\mathrm{ind}_0(\mathfrak{D}) = [\ker \mathfrak{D}^0 \oplus \ker \mathfrak{D}^1] \cong [\ker \mathfrak{D}^0 \oplus 0] - [\ker \mathfrak{D}^1 \oplus 0]$ as claimed.

REMARK 7.5. All of these constructions could be carried out in the complex category. One could consider complex $C\ell_k$-Dirac bundles, etc. Here the index will be valued in

$$K^{-k}(\mathrm{pt}) = \begin{cases} \mathbb{Z} & \text{if } k \text{ is even} \\ 0 & \text{if } k \text{ is odd.} \end{cases}$$

Unfortunately, this leads to nothing essentially new, so we have concentrated our attention on the real case.

Some examples are in order. The first and most illuminating one comes by taking the "$C\ell_k$-ification" of an ordinary elliptic operator.

EXAMPLE 7.6. Let $S = S^0 \oplus S^1$ be an ordinary real $\mathbb{Z}_2$-graded Dirac bundle over a compact manifold X, and let

$$D = \begin{pmatrix} 0 & D^1 \\ D^0 & 0 \end{pmatrix}$$

be its Dirac operator. We now consider an irreducible real $\mathbb{Z}_2$-graded module $V = V^0 \oplus V^1$ over the Clifford algebra $C\ell_k$, and take the tensor product

$$\mathfrak{S} = S \otimes V$$

where V is here considered as the trivialized bundle $V \times X \to X$. This bundle is, in a natural way, a $\mathbb{Z}_2$-graded $C\ell_k$-Dirac bundle. The grading $\mathfrak{S} = \mathfrak{S}^0 \oplus \mathfrak{S}^1$ is given by

$$\mathfrak{S}^0 = (S^0 \otimes V^0) \oplus (S^1 \otimes V^1) \qquad \text{and} \qquad \mathfrak{S}^1 = (S^0 \otimes V^1) \oplus (S^1 \otimes V^0).$$

Of course, multiplication by $C\ell_k$ takes place on the V-factor. The associated Dirac operator $\mathfrak{D}$ on $\mathfrak{S}$ is simply the extension of D, i.e.,

$$\mathfrak{S} = D \otimes \mathrm{Id}_V.$$

Consequently, we have that

$$\ker \mathfrak{D}^0 = ((\ker D^0) \otimes V^0) \oplus ((\ker D^1) \otimes V^1),$$

and therefore

$$\{\ker \mathfrak{D}^0\} = b_0\{V^0\} + b_1\{V^1\} \in \mathfrak{M}_{k-1}$$

where $b_j \equiv \dim_{\mathbb{R}}(\ker D^j)$. In passing to the quotient $\mathfrak{M}_{k-1}/i^*\mathfrak{M}_k$ we see that $[V^0] = -[V^1]$ since $V^0 \oplus V^1$ is a $C\ell_k$-module. (*Caution:* one must show that $V^0 \oplus V^1$ can be made a $C\ell_k$-module in such a way that V^0 and V^1 are invariant subspaces under the subalgebra $C\ell_{k-1} \subset C\ell_k$. This requires a case by case check modulo 8.). Hence,

$$\mathrm{ind}_k(\mathfrak{D}^0) = (b^0 - b^1)[V^0] = \mathrm{ind}(D^0)[V^0] \in KO^{-k}(\mathrm{pt}),$$

and since $[V_0]$ generates this group for each k (see I.9) we conclude that

$$\mathrm{ind}_k\mathfrak{D}^0 = \begin{cases} \mathrm{ind}\, D^0 & \text{if } k \equiv 0 \pmod 4 \\ \mathrm{ind}\, D^0 \pmod 2 & \text{if } k \equiv 1 \text{ or } 2 \pmod 8 \\ 0 & \text{otherwise.} \end{cases} \tag{7.8}$$

This formula shows that, as one would expect, nothing essentially new can be found by this trivial construction. The interesting examples are those where the $C\ell_k$-structure is more intrinsic to the geometry. This is the case in the following:

EXAMPLE 7.7 (the Kervaire Semicharacteristic). Let X be a compact oriented (riemannian) manifold, and take $\mathfrak{S}$ to be the Clifford bundle $C\ell(X) = C\ell^0(X) \oplus C\ell^1(X)$ (considered as a $\mathbb{Z}_2$-graded Dirac bundle as in Example 6.1). If X has dimension $4\ell + 1$, then $\mathfrak{S}$ is naturally a $C\ell_1$-Dirac bundle as follows. Consider the oriented volume form $\omega = e_1 \cdots e_{4\ell+1}$ of X. This form is parallel and satisfies $\omega^2 = -1$. Hence, right multiplication by ω in $C\ell(X)$ makes $C\ell(X)$ a $C\ell_1$-Dirac bundle. Since ω is of odd degree, right multiplication by ω interchanges $C\ell^0(X)$ and $C\ell^1(X)$. Hence, $C\ell(X)$ is $\mathbb{Z}_2$-graded as a $C\ell_1$-Dirac bundle. Hence, the operator $\mathfrak{D}^0 = D^0 \cong (d + d^*)|_{\Lambda^{\mathrm{even}}}$ has an index in $KO^{-1}(\mathrm{pt}) \cong \mathbb{Z}_2$. To compute this index we want to find the residue class of $\ker \mathfrak{D}^0$ if $\mathfrak{M}_0/i^*\mathfrak{M}_1$. Since $C\ell_1 \cong \mathbb{C}$ and $C\ell_0 \cong \mathbb{R}$, we easily identify $\mathfrak{M}_0$ as the Grothendieck group of equivalence classes of real finite dimensional vector spaces, and $\mathfrak{M}_1$ as the complex analogue. Hence $i^*\mathfrak{M}_1$ is given by spaces of even dimension, and we clearly have

$$\begin{aligned} \operatorname{ind}_1(\mathfrak{D}^0) &\equiv \dim_{\mathbb{R}}(\ker D^0) \pmod 2 \\ &\equiv \dim_{\mathbb{R}}(\mathbf{H}^{\text{even}}) \pmod 2 \\ &\equiv \sum_{j=0}^{2\ell} b_{2j}(X) \pmod 2 \end{aligned}$$

where $b_i(X)$ denotes the ith Betti number of X. In other words, $\operatorname{ind}_1(\mathfrak{D}^0)$ is exactly **the Kervaire semicharacteristic** of X, a basic cobordism invariant of an oriented $(4\ell + 1)$-manifold.

Note that in order that the volume element ω have square -1 and be of odd degree, one is restricted to dimensions $4\ell + 1$.

It should be pointed out that an exactly analogous construction fails for the signature complex (Example 6.2), since the subbundles $C\ell^{\pm}(X) = (1 \pm \omega_{\mathbb{C}})C\ell(X)$ are each invariant under right multiplication by any Clifford element. Hence, this is never a grading for right multiplication.

We now pass to an example which illustrates this construction.

THE FUNDAMENTAL CASE (**the Atiyah-Milnor-Singer Invariant**). Let X be a compact spin manifold of dimension n, and let $\mathfrak{S}(X) = \mathfrak{S}^0(X) \oplus \mathfrak{S}^1(X)$ be the $\mathbb{Z}_2$-graded $C\ell_n$-Dirac bundle given by (7.1). It is a consequence of the Atiyah-Singer theorem that $\operatorname{ind}_n \equiv \operatorname{ind}_n(\mathfrak{D}^0)$ is a spin cobordism invariant of X, and in fact gives a graded ring homomorphism

$$\Omega_*^{\text{Spin}} \xrightarrow{\operatorname{ind}_*} KO^{-*}(\text{pt}). \tag{7.9}$$

This homomorphism coincides with the one defined homotopy-theoretically in (3.22). We shall return to this in Chapter III.

Because of their fundamental nature, it is useful to examine these $C\ell_n$-spinor bundles $\mathfrak{S}(X)$ in some detail. We begin with the simplest case where $n = 8k$.

Consider an irreducible real left module V_{8k} for the Clifford algebra $C\ell_{8k}$. Let $\tilde{V}_{8k}$ denote the right $C\ell_{8k}$-module obtained from V_{8k} by simply multiplying by the transposed element, i.e., by setting $v \cdot \varphi \equiv \varphi^t \cdot v$ for $\varphi \in C\ell_{8k}$ and $v \in V_{8k}$. Then there is an isomorphism of bimodules

$$C\ell_{8k} \cong V_{8k} \otimes \tilde{V}_{8k}. \tag{7.10}$$

To see this we recall from Proposition 5.18 of Chapter I, that in all even dimensions we have the complex bimodule isomorphism

$$\mathbb{C}\ell_{2k} \cong V_{2k}^{\mathbb{C}} \otimes \tilde{V}_{2k}^{\mathbb{C}} \tag{7.11}$$

where $V_{2k}^{\mathbb{C}}$ denotes the irreducible complex $C\ell_{2k}$-module and where $\mathbb{C}\ell_{2k} \cong C\ell_{2k} \otimes_{\mathbb{R}} \mathbb{C}$. The assertion (7.10) follows from (7.11) since in dimensions $8k$ the complex case is simply the complexification of the real one.

We conclude from (7.1) that in these dimensions

$$\begin{aligned} \mathfrak{S}(X) &= P_{\mathrm{Spin}} \times_{\ell} C\ell_{8k} = P_{\mathrm{Spin}} \times_{\Delta_{8k}\otimes 1} (V_{8k} \otimes \tilde{V}_{8k}) \\ &= (P_{\mathrm{Spin}} \times_{\Delta_{8k}} V_{8k}) \otimes \tilde{V}_{8k} \\ &= \$(X) \otimes \tilde{V}_{8k} \end{aligned} \tag{7.12}$$

where $\$(X)$ denotes the real spinor bundle of X, and where $\tilde{V}_{8k}$ now becomes the constant (trivialized) $\tilde{V}_{8k}$ bundle.

Recall now that V_{8k} has two inequivalent $\mathbb{Z}_2$-gradings, obtained one from the other by interchanging the factors. Either one gives the same $\mathbb{Z}_2$-grading on the tensor product $V_{8k} \otimes \tilde{V}_{8k}$, and the bimodule isomorphism (7.10) is a $\mathbb{Z}_2$-graded one.

We are now exactly in the situation of Example 7.6. That is, $\mathfrak{D} = D\!\!\!/ \otimes \mathrm{Id}_V$, where $D\!\!\!/$ is the Atiyah-Singer operator on spinors. We can therefore read off from (7.8) that

$$\mathrm{ind}_{8k}(\mathfrak{D}^0) = \mathrm{ind}(D\!\!\!/^0) = \hat{A}(X). \tag{7.13}$$

The situation in dimensions $8k + 4$ is very much similar. The principal difference is that in these dimensions the irreducible real module V_{8k+4} for $C\ell_{8k+4}$ is also the irreducible complex module; that is,

$$V_{8k+4} = [V_{8k+4}^{\mathbb{C}}]_{\mathbb{R}}. \tag{7.14}$$

(Recall that $C\ell_{8k+4}$ is a quaternion, and hence also a complex, matrix algebra.) From (7.11) we know that $V_{8k+4}^{\mathbb{C}} \otimes_{\mathbb{C}} \tilde{V}_{8k+4}^{\mathbb{C}} \cong \mathbb{C}\ell_{8k+4} \cong C\ell_{8k+4} \otimes_{\mathbb{R}} \mathbb{C}$, which yields the real bundle isomorphism

$$2\, C\ell_{8k+4} \cong V_{8k+4}^{\mathbb{C}} \otimes_{\mathbb{C}} \tilde{V}_{8k+4}^{\mathbb{C}}. \tag{7.15}$$

This implies, as above, that there is an isomorphism of real $\mathbb{Z}_2$-graded $C\ell_{8k+4}$-bundles

$$2\mathfrak{S}(X) \cong \$_{\mathbb{C}}(X) \otimes_{\mathbb{C}} \tilde{V}_{8k+4}^{\mathbb{C}} \tag{7.16}$$

where $\$_{\mathbb{C}}(X)$ is the complex spinor bundle of X and $V_{8k+4}^{\mathbb{C}}$ is the trivialized bundle. Hence, we have that $2\mathfrak{D} = D\!\!\!/_{\mathbb{C}} \otimes_{\mathbb{C}} \mathrm{Id}_{\tilde{V}^{\mathbb{C}}}$. Using (7.14) and the arguments above, it is not difficult to see that

$$\mathrm{ind}_{8k+4}(\mathfrak{D}^0) = \tfrac{1}{2}\hat{A}(X). \tag{7.17}$$

Notice that implicit in equation (7.17) is the fact that in these dimensions the $\hat{A}$-genus of a spin manifold is an *even* integer. A direct proof of this can be given by observing that the Atiyah-Singer operator in these dimensions is quaternion linear, and so the complex dimensions of its kernel and cokernel are even. This fact, in dimension four, is just Rochlin's Theorem 2.13.

We now examine this index in the interesting cases where it is an integer modulo 2. We begin with dimension $8k + 1$ and the observation that from the classification of Clifford algebras we have

$$C\ell_{8k+1} \cong C\ell_{8k} \otimes \mathbb{C} \cong C\ell_{8k} \oplus i\, C\ell_{8k}. \tag{7.18}$$

In fact, the element i in this algebra corresponds exactly to the volume form $\omega = e_1 \cdots e_{8k+1}$, which is clearly central and has $\omega^2 = -1$. Thus, the decomposition in (7.18) gives the $\mathbb{Z}_2$-grading on $C\ell_{8k+1}$. In particular, we have

$$C\ell^0_{8k+1} \cong C\ell_{8k} \qquad \text{and} \qquad C\ell^1_{8k+1} = \omega\, C\ell^0_{8k+1} \cong i\, C\ell_{8k}. \tag{7.19}$$

These isomorphisms show that

$$\begin{gathered} V_{8k+1} \cong V_{8k} \otimes \mathbb{C} \cong V_{8k} \oplus iV_{8k} \\ V^0_{8k+1} \cong V_{8k} \qquad \text{and} \qquad V^1_{8k+1} = \omega V^0_{8k+1} \cong iV_{8k}. \end{gathered} \tag{7.20}$$

Combining this with (7.11), we find that

$$\begin{aligned} C\ell_{8k+1} = \mathbb{C}\ell_{8k} &\cong V^{\mathbb{C}}_{8k} \otimes_{\mathbb{C}} \tilde{V}^{\mathbb{C}}_{8k} \\ &= (V_{8k} \otimes \mathbb{C}) \otimes_{\mathbb{C}} (\tilde{V}_{8k} \otimes \mathbb{C}) \\ &= (V^0_{8k+1} \otimes \mathbb{C}) \otimes_{\mathbb{C}} (\tilde{V}^0_{8k+1} \otimes \mathbb{C}) \\ &= (V^0_{8k+1} \otimes \mathbb{C}) \otimes_{\mathbb{R}} \tilde{V}^0_{8k+1} \\ &= V_{8k+1} \otimes_{\mathbb{R}} \tilde{V}^0_{8k+1}. \end{aligned}$$

This implies as above that

$$\mathbf{\$}(X) \cong \$(X) \otimes_{\mathbb{R}} \tilde{V}^0_{8k+1} \tag{7.22}$$

and that $\mathbf{D} \cong D\!\!\!/ \otimes \mathrm{Id}_{\tilde{V}^0}$. It follows immediately that $\ker(\mathbf{D}^0) = \mathbf{H}^0 \otimes \tilde{V}^0_{8k+1}$ where $\mathbf{H}^0 \equiv \ker(D\!\!\!/^0)$. Thus we have that

$$\mathrm{ind}_{8k+1}(\mathbf{D}^0) \equiv \dim_{\mathbb{R}} \mathbf{H}^0 \qquad (\mathrm{mod}\ 2). \tag{7.23}$$

This index can be reinterpreted in more elementary terms as follows. Observe that from the discussion above, we have in these dimensions that the spinor bundle is the complexification of a real bundle, i.e., $\$(X) = \$^0(X) \oplus \omega\, \$^0(X)$. We can construct from the Dirac operator $D\!\!\!/^0 : \Gamma(\$^0(X)) \to \Gamma(\$^1(X))$, an operator

$$P : \Gamma(\$^0(X)) \longrightarrow \Gamma(\$^0(X)) \tag{7.24}$$

by setting

$$P \equiv \omega \cdot D\!\!\!/^0. \tag{7.25}$$

This operator is elliptic and *skew-adjoint*. To prove the skew adjointness, we note that since ω is parallel, commutes with $D\!\!\!/$ and satisfies $\omega^2 = -1$, we have $(P\sigma, \tau) = (\omega \not{D}^0\sigma, \tau) = (\omega^2 \not{D}^0\sigma, \omega\tau) = -(\not{D}^0\sigma, \omega\tau) = -(\sigma, \not{D}^0\omega\tau) = -(\sigma, \omega\not{D}^0\tau) = -(\sigma, P\tau)$.

It is an elementary fact that the parity of a real skew-adjoint Fredholm operator is conserved under deformations (cf. III.10). Thus for a skew-adjoint elliptic operator P on a compact manifold, the **mod-2 index**

$$\mathrm{ind}_{\mathbb{Z}_2}(P) \equiv \dim(\ker P) \quad (\mathrm{mod}\ 2) \tag{7.26}$$

is well defined. The result (7.23) can be reexpressed as

$$\mathrm{ind}_{8k+1}(\mathfrak{D}^0) = \mathrm{ind}_{\mathbb{Z}_2}(P) \tag{7.27}$$

where P is given by (7.25).

In the final case of dimension $8k + 2$ there is a strongly analogous situation. Here we have from the classification of Clifford algebras that

$$C\ell_{8k+2} \cong C\ell_{8k} \otimes \mathbb{H} \quad \text{and} \quad C\ell^0_{8k+2} \cong C\ell_{8k} \otimes \mathbb{C}, \tag{7.28}$$

$$V_{8k+2} \cong V_{8k} \otimes \mathbb{H} \quad \text{and} \quad V^0_{8k+2} \cong V_{8k} \otimes \mathbb{C}. \tag{7.29}$$

Using (7.11) one can deduce from here that

$$\begin{aligned} C\ell_{8k+2} &\cong V_{8k+2} \otimes_{\mathbb{H}} \tilde{V}_{8k+2} \\ &\cong V_{8k+2} \otimes_{\mathbb{C}} \tilde{V}^0_{8k+2}. \end{aligned}$$

It follows as before that

$$\mathfrak{S}(X) \cong \$(X) \otimes_{\mathbb{C}} \tilde{V}^0_{8k+2} \tag{7.31}$$

with respect to which $\mathfrak{D} \cong \not{D} \otimes \mathrm{Id}_{\tilde{V}}$. Consequently, $\ker(\mathfrak{D}^0) \cong \ker(\not{D}^0) \otimes_{\mathbb{C}} \tilde{V}^0_{8k+2}$, and so

$$\mathrm{ind}_{8k+2}(\mathfrak{D}^0) \equiv \dim_{\mathbb{C}}(\ker \not{D}^0) \quad (\mathrm{mod}\ 2). \tag{7.32}$$

Now since the representation V_{8k+2} is $\mathbb{H}$-linear, the bundle $\$(X)$ carries a parallel quaternion structure, i.e., there are parallel endomorphisms I, J, K of $\$(X)$ which satisfy the standard quaternion relations: $I^2 = J^2 = K^2 = -1$, $IJ + JI = IK + KI = JK + KJ = 0$. The $\mathbb{Z}_2$-grading on $\$(X)$ can be written in terms of these as

$$\$(X) = \$^0(X) \oplus J(\$^0(X)). \tag{7.33}$$

We can then define a skew-hermitian operator

$$P = J \circ \not{D}^0$$

on the bundle $\$^0(X)$, and as before we have

$$\mathrm{ind}_{8k+2}(\mathfrak{D}^0) \equiv \mathrm{ind}_{\mathbb{Z}_2}(P) \tag{7.34}$$

where $\mathrm{ind}_{\mathbb{Z}_2}(P) \equiv \dim_{\mathbb{C}}(\ker P)\ (\mathrm{mod}\ 2)$.

We shall now examine this mod 2 index of the Atiyah-Singer operator in low dimensions. The computations here are of some importance since, as we shall see, the map $\text{ind}_*:\Omega_*^{\text{Spin}} \to KO^{-*}(\text{pt})$ is a ring homomorphism, and the non-zero element $\eta \in KO^{-1}(\text{pt})$ has the property that: ηx and $\eta^2 x$ are not zero whenever x is an odd multiple of the generator in degree $8k$.

EXAMPLE 7.8 (the circle as a spin manifold). Consider the circle S^1 with a riemannian metric. The oriented orthonormal frame bundle is canonically diffeomorphic to the circle itself, i.e., $P_{\text{SO}}(S^1) \cong S^1$, since there is exactly one oriented unit tangent vector at each point. A spin structure on S^1 is a 2-fold covering $P_{\text{Spin}}(S^1) \to P_{\text{SO}}(S^1)$ of the circle. As we noted in Chapter I, there are two such coverings, one connected and one with two components; and it is the non-connected covering $S^1 \times \mathbb{Z}_2 \to S^1$ which is not spin-cobordant to zero. We shall refer to this as the **interesting spin structure** on S^1.

Notice that since $C\ell_1 \cong \mathbb{C}$ we have that

$$\mathfrak{S}(S^1) = S^1 \times \mathbb{C}$$

where the product structure gives the connection. Of course $C\ell_1^0 \cong \mathbb{R}$ and the $\mathbb{Z}_2$-grading on $C\ell_1$ is the standard decomposition $\mathbb{C} = \mathbb{R} \oplus i\mathbb{R}$. Thus, $\mathfrak{S}^0(S^1) = S^1 \times \mathbb{R}$ and $\mathfrak{S}^1(S^1) = S^1 \times i\mathbb{R}$. Sections of $\mathfrak{S}$ are just complex-valued functions $f(s)$ on the circle, and the Dirac operator is simply

$$\mathfrak{D} = i\frac{d}{ds} \tag{7.35}$$

where s is arc-length on S^1.

The kernel of $\mathfrak{D}^0:\Gamma(\mathfrak{S}^0) \to \Gamma(\mathfrak{S}^1)$ is the set of real-valued constant functions on S^1. The dimension of this space as a $C\ell^0 \cong \mathbb{R}$ module is one. Hence, we have that for the interesting spin structure on S^1,

$$\text{ind}_1(S^1) \neq 0, \tag{7.36}$$

i.e., $\text{ind}_1(S^1)$ is the generator of $KO^{-1}(\text{pt}) \cong \mathbb{Z}_2$.

EXERCISE. Show directly that $\text{ind}_1 = 0$ for the "uninteresting" spin structure on S^1. Show also that the connected sum of the intersecting spin structure with itself is the uninteresting one.

EXAMPLE 7.9 (the torus as a spin manifold). Let $T = S^1 \times S^1$ be the flat square torus. The oriented orthonormal frame bundle is canonically trivialized

$$P_{\text{SO}}(T) = T \times S^1.$$

The spin structure on T given by squaring the interesting spin structure on the circle is the covering:

$$P_{\mathrm{Spin}}(T) = T \times S^1 \xrightarrow{\mathrm{Id} \times z^2} T \times S^1 = P_{\mathrm{SO}}(T).$$

Since $C\ell_2 \cong \mathbb{H}$ and $C\ell_2^0 \cong \mathbb{C}$, we see as before that $\mathfrak{S}(T) = T \times \mathbb{H}$ and $\mathfrak{S}^0(T) = T \times \mathbb{C}$. The kernel of $\mathfrak{D}^0$ is just the complex-valued constant functions, and we conclude that

$$\mathrm{ind}_2(S^1 \times S^1) \neq 0 \tag{7.37}$$

where S^1 carries the interesting spin structure.

EXERCISE. Compute ind_2 for the remaining spin-structures on T.

EXERCISE. Prove directly that $\mathrm{ind}_2(S^2) = 0$.

It is a good time to summarize what we have established so far. To each riemannian spin manifold X, we have associated the canonical bundle $\mathfrak{S}(X)$, given in (7.1). This is a $\mathbb{Z}_2$-graded $C\ell_n$-Dirac bundle and its Dirac operator has an index, which we denote $\mathrm{ind}_k(X)$ in $KO^{-k}(\mathrm{pt})$. We have proved the following.

Theorem 7.10. *Let X be a compact spin manifold of dimension n. When $n \equiv 1$ or $2 \pmod 8$, let $\mathbf{H} = \ker(\not{D})$ denote the space of real harmonic spinors, that is, the kernel of the Atiyah-Singer operator on the irreducible real spinor bundle of X. Then*

$$\mathrm{ind}_n(X) = \begin{cases} \dim_{\mathbb{C}} \mathbf{H} \pmod 2 & \text{if } n \equiv 1 \pmod 8 \\ \dim_{\mathbb{H}} \mathbf{H} \pmod 2 & \text{if } n \equiv 2 \pmod 8 \\ \tfrac{1}{2}\hat{A}(X) & \text{if } n \equiv 4 \pmod 8 \\ \hat{A}(X) & \text{if } n \equiv 0 \pmod 8 \end{cases}$$

Furthermore, for the interesting spin structure on S^1 and for its square on $S^1 \times S^1$, this invariant is non-zero.

We shall now investigate the multiplicative properties of ind_*. To begin we recall the ring structure on $KO^{-*}(\mathrm{pt})$ which was discussed in Chapter I, §9. We set

$$\left.\begin{aligned} \eta &= \text{the generator of } KO^{-1}(\mathrm{pt}) \cong \mathbb{Z}_2 \\ y &= \text{the generator of } KO^{-4}(\mathrm{pt}) \cong \mathbb{Z} \\ x &= \text{the generator of } KO^{-8}(\mathrm{pt}) \cong \mathbb{Z} \end{aligned}\right\} \tag{7.38}$$

Then the multiplicative structure on $KO^{-*}(\mathrm{pt})$ is given by the following (cf. I.9):

$$KO^{-*}(\mathrm{pt}) \cong \mathbb{Z}[\eta, y, x]/\langle 2\eta, \eta^3, \eta y, y^2 - 4x\rangle \tag{7.39}$$

where the grading is fixed by the requirement that $\deg(\eta) = 1$, $\deg(y) = 4$ and $\deg(x) = 8$.

Our next result is that ind_* is, in fact, a ring homomorphism. More specifically, we mean the following. The homotopy invariance of the index shows that $\text{ind}_n(X)$ is independent of the choice of riemannian metric on X. (Any two metrics are smoothly homotopic.) We consider then the free abelian group $\mathbf{M}_n^{\text{Spin}}$ generated by spin-structure preserving diffeomorphism classes of connected spin manifolds. The sum $\mathbf{M}_*^{\text{Spin}} = \bigoplus_n \mathbf{M}_n^{\text{Spin}}$ is naturally a graded ring under the direct product of manifolds.

Theorem 7.11. *The mapping*

$$\text{ind}_* : \mathbf{M}_*^{\text{Spin}} \longrightarrow KO^{-*}(\text{pt})$$

is a surjective graded-ring homomorphism.

Proof. The map ind_* is additive by definition. To prove that it is multiplicative, we use the Alternative Definition 7.4 of the index. Let X_1 and X_2 be compact riemannian spin manifolds of dimensions n_1 and n_2 respectively. Let $\mathfrak{D}_k : \Gamma(\mathfrak{S}_k) \to \Gamma(\mathfrak{S}_k)$ be the Atiyah-Singer operator for the $C\ell_{n_k}$-Dirac bundle $\mathfrak{S}_k = \mathfrak{S}(X_k)$ defined in (7.1), for $k = 1,2$. Let $\mathfrak{D}$: $\Gamma(\mathfrak{S}) \to \Gamma(\mathfrak{S})$ be the corresponding object for $X \equiv X_1 \times X_2$ with the product riemannian and spin structure. For this structure we have

$$P_{\text{Spin}}(X_1 \times X_2) \supset P_{\text{Spin}}(X_1) \times_{\mathbb{Z}_2} P_{\text{Spin}}(X_2)$$

where $\mathbb{Z}_2$ acts by $(-1,-1)$ on the product. From here one sees easily, using (7.1), that

$$\mathfrak{S} = \mathfrak{S}_1 \hat{\otimes} \mathfrak{S}_2 \tag{7.40}$$

the exterior $\mathbb{Z}_2$-graded tensor product. This is a $C\ell_{n_1+n_2} = (C\ell_{n_1} \hat{\otimes} C\ell_{n_2})$-Dirac bundle, and one can straightforwardly verify that with respect to (7.40), the Atiyah-Singer operator of $\mathfrak{S}$ can be written as

$$\mathfrak{D} = \mathfrak{D}_1 \otimes \text{Id}_2 + \alpha_1 \otimes \mathfrak{D}_2$$

where $\alpha_1 : C\ell_{n_1} \to C\ell_{n_1}$ is the automorphism extending $-\text{Id}$ on $\mathbb{R}^{n_1}$. Since $\alpha_1\mathfrak{D}_1 + \mathfrak{D}_1\alpha_1 = 0$, it follows that

$$\mathfrak{D}^2 = \mathfrak{D}_1^2 \otimes \text{Id}_2 + \text{Id}_1 \otimes \mathfrak{D}_2^2.$$

Each of the operators $\mathfrak{D}^2$, $\mathfrak{D}_1^2$ and $\mathfrak{D}_2^2$ is non-negative, self-adjoint and elliptic over a compact manifold. It follows from standard spectral theory that $\ker(\mathfrak{D}^2) = \ker(\mathfrak{D}_1^2) \otimes \ker(\mathfrak{D}_2^2)$. Since $\ker(\mathfrak{D}_j) = \ker(\mathfrak{D}_j^2)$, we conclude that

$$\ker(\mathfrak{D}) = \ker(\mathfrak{D}_1) \hat{\otimes} \ker(\mathfrak{D}_2) \tag{7.41}$$

where the tensor product in (7.41) inherits the structure of a $\mathbb{Z}_2$-graded tensor product of $\mathbb{Z}_2$-graded Clifford modules from the structure of (7.40). This graded tensor product is exactly the multiplication in $KO^{-*} \cong \hat{\mathfrak{M}}_*/i^*\hat{\mathfrak{M}}_{*+1}$. Thus, we have established that

$$\text{ind}_{n_1+n_2}(\mathfrak{D}) = \text{ind}_{n_1}(\mathfrak{D}_1)\,\text{ind}_{n_2}(\mathfrak{D}_2)$$

and so ind_* is a ring homomorphism.

To see that ind_* is surjective we need only show that it maps onto the set of multiplicative generators $\{\eta, y, x\}$ given in (7.38). Example 7.8 shows that the circle with interesting spin structure maps onto η. The $K3$-surface Y given in Example 2.14 is a compact spin 4-manifold whose $\hat{A}$-genus is 2. Hence, $\text{ind}_4(Y) = \frac{1}{2}\hat{A}(Y) = y$. Finally, to produce a spin 8-manifold X with $\text{ind}_8(X) = x$, we must find one such that $\hat{A}(X) = 1$. For this one could take $X = M_0^8$, the almost parallelizable 8-manifold of index 224 constructed in Kervaire-Milnor [1]. Alternatively, one could take the 4-disk bundle E over S^4 with $\chi(E) = 1$ and $p_1(E)^2 = 900$. Then $\partial E = S^7$, and the manifold $X = E \cup_{S^7} D^8$, obtained by attaching an 8-disk along this boundary, is a closed spin 8-manifold with $\hat{A}(X) = 1$ (see Milnor [7]). This completes the proof of Theorem 7.11. ∎

REMARK 7.12. To prove the index theorem for this fundamental case, it remains only to prove that ind_* is a spin-cobordism invariant. It will then descend to a ring-homomorphism $\text{ind}_*: \Omega_*^{\text{Spin}} \to KO^{-*}(\text{pt})$, which can be seen to agree with the map (3.22) directly. (For the torsion part, this requires some argument.)

It is worth noting that all of the above could be carried out with coefficients in a vector bundle. Let X be a compact riemannian spin manifold of dimension n, and let E be a real vector bundle over X with an orthogonal connection. Then the bundle $\mathfrak{S}(X) \otimes E$ is naturally a $\mathbb{Z}_2$-graded $C\ell_n$-Dirac bundle with a Dirac operator which we shall denote by $\mathfrak{D}_E$. We shall denote

$$\text{ind}_E(X) = \text{ind}_n(\mathfrak{D}_E) \in KO^{-n}(\text{pt}). \tag{7.42}$$

Now we also have the fundamental real spinor bundle $\$$ over X and we can take the tensor product $\$ \otimes E$ in the spirit of Example 6.5 above. We denote the Dirac operator on $\$ \otimes E$ by $D\!\!\!\!/_E$. Then arguing exactly as we did above proves the following.

Theorem 7.13. *Let X be a compact spin manifold of dimension n, and let E be a real vector bundle over X with an orthogonal connection. When $n \equiv 1$ or 2 (mod 8), let $\mathbf{H}_E \equiv \ker(D\!\!\!\!/_E)$ denote the space of real harmonic E-valued* *spinors. Then*

$$\operatorname{ind}_E(X) = \begin{cases} \dim_{\mathbf{C}} \mathbf{H}_E \pmod 2 & \text{if } n \equiv 1 \pmod 8 \\ \dim_{\mathbf{H}} \mathbf{H}_E \pmod 2 & \text{if } n \equiv 2 \pmod 8 \\ \frac{1}{2}\{\operatorname{ch} E \cdot \hat{\mathbf{A}}(X)\}[X] & \text{if } n \equiv 4 \pmod 8 \\ \{\operatorname{ch} E \cdot \hat{\mathbf{A}}(X)\}[X] & \text{if } n \equiv 0 \pmod 8 \end{cases}$$

A topological computation of this index can be given as follows. Embed $X^n \subset S^{n+8k}$ for k sufficiently large, and identify the normal bundle ν of X with a tubular neighborhood of the embedding. The spin structure on X determines a unique spin structure on ν (via Proposition 1.15 and the uniqueness of the spin structure on S^{n+8k}). Let $\$^+(\nu)$ and $\$^-(\nu)$ denote the canonical real spinor bundles of ν (whose dimension is $8k$). Lift $\$^{\pm}(\nu)$ to the total space of ν by the projection $\pi: \nu \to X$, and at each non-zero vector $n \in \nu$ consider the isomorphism $\mu_n: \pi^*\$^+(\nu) \xrightarrow{\approx} \pi^*\$^-(\nu)$ given by Clifford multiplication by n. Then the difference element

$$\tau_\nu \equiv [\pi^*\$^+(\nu), \pi^*\$^-(\nu); \mu]$$

represents a class in the relative KO-group $KO(\nu, \nu - X)$, where $X \subset \nu$ is the zero-section. This is the KO-theory **Thom class** of ν.

Given a real bundle E over X, we can consider the class

$$\tau_\nu(E) = \tau_\nu \cdot [\pi^*E] \in KO(\nu, \nu - X).$$

Since ν is embedded as a domain $\nu \subset S^{n+8k}$, we have an excision isomorphism $j: KO(\nu, \nu - X) \cong KO(S^{n+8k}, S^{n+8k} - X)$. Composing this with the natural map $i: KO(S^{n+8k}, S^{n+8k} - X) \to \widetilde{KO}(S^{n+8k})$ and applying Bott Periodicity $\beta: \widetilde{KO}(S^{n+8k}) \cong \widetilde{KO}(S^n) \equiv KO^{-n}(\text{pt})$, we obtain a class

$$\hat{\mathscr{A}}_E(X) = \beta \circ i \circ j(\tau_\nu(E)) \in KO^{-n}(\text{pt})$$

One assertion of the Atiyah-Singer $C\ell_k$-Index Theorem, proved in III.16, is that

Theorem 7.14.

$$\operatorname{ind}_E(X) = \hat{\mathscr{A}}_E(X).$$

The "cobordism invariance" in this case asserts that this map ind $= \hat{\mathscr{A}}$ determines a transformation

$$\hat{\mathscr{A}}_*: \Omega_*^{\text{Spin}}(\text{BO}) \longrightarrow KO^{-*}(\text{pt})$$

where $\Omega_*^{\text{Spin}}(\text{BO})$ denotes the spin-bordism of the space BO.

REMARK 7.15. It is interesting to note that in dimensions 1 and 2 (mod 8), the index can be changed even by twisting with a **flat** bundle. (This is not true of the usual index.) For example consider the non-trivial real flat line bundle ℓ over the circle (the Möbius band). Using the analysis of Example 7.8, one easily finds that $\mathfrak{S}(S^1) \otimes \ell \cong \ell \oplus i\ell$. Since the kernel of $\mathfrak{D}_\ell \cong i(d/ds)$ consists of locally constant sections, we see that ker $(\mathfrak{D}_\ell) = \{0\}$ and

$$\mathrm{ind}_\ell(S^1) = 0.$$

Recall that $\mathrm{ind}_1(S^1) \neq 0$.

This sensitivity to flat bundles is a reflection of the interesting fact that these (mod 2)-invariants are *not local*, i.e., they cannot be computed by universal formulas involving only the local data of the operator, as in the Gauss-Bonnet and Chern-Weil theorems. Another indication is that the (mod 2)-invariants are not multiplicative under coverings (see Atiyah-Singer [4]).

§8. Vanishing Theorems and Some Applications

Suppose X is a compact riemannian manifold and let D^2 be the Dirac laplacian on the bundle $C\ell(X)$. One of the major results in riemannian geometry was the following discovery of S. Bochner. There exists a second, naturally defined laplacian $\nabla^*\nabla$ on $C\ell(X)$. It is self-adjoint, non-negative and has the same symbol as D^2. The difference, $D^2 - \nabla^*\nabla$, which is necessarily an operator of order ≤ 1, is, in fact, of order zero and can be expressed in terms of the curvature tensor of X. Using harmonic theory, Bochner was thereby able to conclude the vanishing of certain Betti numbers of X under appropriate positivity assumptions on the curvature tensor.

Following Bochner's original paper, arguments of this kind have appeared repeatedly in real and complex geometry. They are known generically as "Bochner's method."

In this section we shall give a systematic derivation of Bochner-type formulas on general Dirac bundles. This will include the classical formulas on $C\ell(X)$, although even here the algebra is vastly simplified by using Clifford multiplication. It will also include the Lichnerowicz formula for the Dirac laplacian on spinor bundles, and generalizations of this to spinors with coefficients in an arbitrary vector bundle. This latter result is very useful in the study of manifolds of positive scalar curvature (cf. Gromov-Lawson [1], [2], [3]). Everything will follow from a single elegant formula which applies to *any* Dirac bundle.

We begin with the definition of the operator $\nabla^*\nabla$. Let E be any riemannian vector bundle over X, and assume E has a riemannian connection with covariant derivative ∇. Then to any pair of tangent vector fields V and W on X, we associate an **invariant second derivative** $\nabla^2_{V,W}$: $\Gamma(E) \to \Gamma(E)$ by setting

$$\nabla^2_{V,W}\varphi \equiv \nabla_V \nabla_W \varphi - \nabla_{\nabla_V W}\varphi. \tag{8.1}$$

(Here, $\nabla_V W$ is the riemannian covariant derivative on X.) At any point $x \in X$, the operator $\nabla^2_{V,W}$ depends only on the values V_x and W_x, i.e., it is tensorial in these variables. This is evidently true for V_x since it is a general property of the covariant derivative. That it is true for W_x follows from the identity

$$\nabla^2_{V,W} - \nabla^2_{W,V} = R_{V,W} \tag{8.2}$$

where R is the curvature tensor of E. Equation (8.2) is an immediate consequence of the fact that $\nabla_V W - \nabla_W V = [V,W]$.

Now given a smooth section φ of E, we see that $\nabla^2_{\cdot,\cdot}\varphi$ is a section of $T^* \otimes T^* \otimes E$; that is, at each point it defines a bilinear form on the tangent space with values in E. The **connection laplacian**

$$\nabla^*\nabla : \Gamma(E) \longrightarrow \Gamma(E)$$

is defined by taking the trace, i.e.,

$$\nabla^*\nabla\varphi \equiv -\operatorname{trace}(\nabla^2_{\cdot,\cdot}\varphi). \tag{8.3}$$

In terms of a local orthonormal tangent frame field $(e_1, \ldots, e_n)$ on X,

$$\nabla^*\nabla\varphi = -\sum_{j=1}^{n} \nabla^2_{e_j,e_j}\varphi.$$

It is easy to see that the symbol of $\nabla^*\nabla$ at a cotangent vector ξ is

$$\sigma_\xi(\nabla^*\nabla) = \|\xi\|^2, \tag{8.4}$$

and so $\nabla^*\nabla$ is elliptic. We shall now show that it is also symmetric and $\geqq 0$.

Recall that the inner product $(\cdot,\cdot)$ on $\Gamma(E)$ is defined by integration: $(\varphi, \psi) = \int_X \langle \varphi, \psi \rangle$. Similarly, we define

$$(\nabla\varphi, \nabla\psi) = \int_X \langle \nabla\varphi, \nabla\psi \rangle \tag{8.5}$$

where $\langle \nabla\varphi, \nabla\psi \rangle$ is defined in terms of local orthonormal tangent frames $(e_1, \ldots, e_n)$ by the expression $\langle \nabla\varphi, \nabla\psi \rangle = \sum_j \langle \nabla_{e_j}\varphi, \nabla_{e_j}\psi \rangle$.

Proposition 8.1. *The operator $\nabla^*\nabla : \Gamma(E) \to \Gamma(E)$ is non-negative and formally self-adjoint. In particular,*

$$(\nabla^*\nabla\varphi, \psi) = (\nabla\varphi, \nabla\psi) \tag{8.6}$$

for all $\varphi, \psi \in \Gamma(E)$ provided that one of φ or ψ has compact support.

If X is compact, then $\nabla^*\nabla\varphi = 0$ if and only if $\nabla\varphi \equiv 0$ i.e., if and only if φ is globally parallel.

Proof. Fix $x \in X$ and choose a local orthonormal tangent frame field $(e_1, \ldots, e_n)$ with the property that $(\nabla e_j)_x = 0$ for all j. Then we have at the point x that

$$\begin{aligned}\langle \nabla^*\nabla\varphi, \psi\rangle &= -\sum_j \langle \nabla_{e_j}\nabla_{e_j}\varphi, \psi\rangle \\ &= -\sum_j \{e_j\langle \nabla_{e_j}\varphi, \psi\rangle - \langle \nabla_{e_j}\varphi, \nabla_{e_j}\psi\rangle\} \\ &= -\operatorname{div}(V) + \langle \nabla\varphi, \nabla\psi\rangle\end{aligned} \tag{8.7}$$

where V is the tangent vector field on X defined by the condition that $\langle V, W\rangle = \langle \nabla_W\varphi, \psi\rangle$ for all $W \in T(X)$. The last line of (8.7) is proved as follows. At x,

$$\begin{aligned}\operatorname{div}(V) &\equiv \sum_j \langle \nabla_{e_j}V, e_j\rangle \\ &= \sum_j e_j\langle V, e_j\rangle \\ &= \sum_j e_j\langle \nabla_{e_j}\varphi, \psi\rangle.\end{aligned}$$

Equation (8.6) now follows by integration of (8.7). The remainder of the proposition is a consequence of formula (8.6). ■

Using arguments similar to those given for Theorem 5.7, one can show that *on a complete riemannian manifold the operator $\nabla^*\nabla$ is essentially self-adjoint*, i.e., it has a unique self-adjoint closed extension on $L^2(S)$. Furthermore, *the kernel of $\nabla^*\nabla$ on $L^2(S)$ consists of the parallel sections of S*, i.e., those which satisfy $\nabla\sigma = 0$. If X has infinite volume, then no such sections exist except $\sigma = 0$, since $\|\sigma\|$ is constant.

We suppose now that S is any Dirac bundle over X, and we define a canonical section $\mathfrak{R}$ of Hom(S,S) by the formula

$$\mathfrak{R}(\varphi) \equiv \tfrac{1}{2} \sum_{j,k=1}^{n} e_j \cdot e_k \cdot R_{e_j,e_k}(\varphi) \tag{8.8}$$

where $(e_1, \ldots, e_n)$ is any orthonormal tangent frame at the point in question, where $R_{V,W}$ is the curvature transformation of S, and where the dot "·" denotes Clifford multiplication.

Theorem 8.2 (the general Bochner Identity). *Let D be the Dirac operator and $\nabla^*\nabla$ the connection laplacian for any Dirac bundle S. Then*

$$\boxed{D^2 = \nabla^*\nabla + \mathfrak{R}}\,. \tag{8.9}$$

Proof. Fix $x \in X$ and choose a local orthonormal frame field $(e_1, \ldots, e_n)$ such that $(\nabla e_j)_x = 0$ for all j. Then using (8.2) we have at x that

$$\begin{aligned}
D^2 &= \sum_{j,k} e_j \cdot \nabla_{e_j}(e_k \cdot \nabla_{e_k}) \\
&= \sum_{j,k} e_j \cdot e_k \cdot \nabla_{e_j} \nabla_{e_k} \\
&= \sum_{j,k} e_j \cdot e_k \cdot \nabla^2_{e_j, e_k} \\
&= -\sum_j \nabla^2_{e_j, e_j} + \sum_{j<k} e_j \cdot e_k \cdot (\nabla^2_{e_j, e_k} - \nabla^2_{e_k, e_j}) \\
&= \nabla^* \nabla + \mathfrak{R}. \qquad \blacksquare
\end{aligned}$$

Recall that the **Ricci transformation** of $T(X) = \Lambda^1(X)$ is defined by the formula

$$\operatorname{Ric}(\varphi) \equiv -\sum_{j=1}^{n} R_{e_j, \varphi}(e_j) \tag{8.10}$$

where R is the curvature transformation of $T(X)$. This determines a bilinear form, called the **Ricci curvature form,** by setting

$$\operatorname{Ric}(\varphi, \psi) \equiv \langle \operatorname{Ric}(\varphi), \psi \rangle. \tag{8.11}$$

From the fundamental identity (4.44) for the Riemann curvature tensor, the form Ric is seen to be symmetric.

A consequence of Theorem 8.2 applied to the bundle $S = C\ell(X)$ is the following

Corollary 8.3. *Let Δ be the Hodge laplacian and $\nabla^*\nabla$ the connection laplacian of the tangent bundle $T(X)$. Then*

$$\boxed{\Delta = \nabla^* \nabla + \operatorname{Ric}}. \tag{8.12}$$

Proof. Consider $T(X) = \Lambda^1(X) \subset C\ell(X)$. Since $D^2 = \Delta$ it suffices to compute the right hand side of (8.9) for vectors $\varphi \in \Lambda^1(X)$. Note that since $[L, \nabla] = [L, \Delta] = 0$ (cf. Lemma 5.18 and Corollary 5.21), both of the operators $\nabla^*\nabla$ and Δ preserve the subbundle $\Lambda^1(X)$. Hence, so does the operator $\mathfrak{R}$. Therefore, using the identities (4.43) and (4.44), we have

$$\begin{aligned}
\mathfrak{R}(\varphi) &= \tfrac{1}{2} \sum_{i,j} e_i e_j R_{e_i, e_j}(\varphi) \\
&= \tfrac{1}{2} \sum_{i,j,k} e_i e_j \langle R_{e_i, e_j}(\varphi), e_k \rangle e_k
\end{aligned}$$

$$= \tfrac{1}{6} \sum_{i \neq j \neq k \neq i} \langle R_{e_i,e_j}(e_k) + R_{e_k,e_i}(e_j) + R_{e_j,e_k}(e_i), \varphi \rangle e_i e_j e_k$$
$$+ \tfrac{1}{2} \sum_{i,j} e_i e_j \langle R_{e_i,e_j}(\varphi), e_j \rangle e_j + \tfrac{1}{2} \sum_{i,j} e_i e_j \langle R_{e_i,e_j}(\varphi), e_i \rangle e_i$$
$$= -\sum_{i,j} \langle R_{e_i,e_j}(\varphi), e_j \rangle e_i = -\sum_{i,j} \langle R_{\varphi,e_j}(e_i), e_j \rangle e_i$$
$$= -\sum_{i,j} \langle R_{e_j,\varphi}(e_j), e_i \rangle e_i = -\sum_j R_{e_j,\varphi}(e_j)$$
$$= \mathrm{Ric}(\varphi). \quad \blacksquare$$

Note. As pointed out above, we know from (8.9) that $\mathfrak{R}(\Lambda^1(X)) \subset \Lambda^1(X)$. However, $e_i e_j e_k \in \Lambda^3(X)$ if $i \neq j \neq k \neq i$. Hence, the third line of the computation above constitutes a proof of the first Bianchi identity (4.43). In Theorem 5.16 we saw more sophisticated examples of curvature identities that also followed from operator identities and Clifford multiplication.

Corollary 8.3 has the following important consequence:

Theorem 8.4 (Bochner). *Let X be a compact riemannian manifold without boundary. If* Ric > 0, *then the first Betti number $b_1(X)$ is zero. The conclusion also holds if* Ric $\geqq 0$ *and* >0 *at one point.*

Proof. Suppose $b_1(X) \equiv \dim H^1(X;\mathbb{R}) > 0$. Then there exists a non-zero harmonic 1-form $\varphi \in \mathbf{H}^1 \cong H^1(X;\mathbb{R})$ by Theorem 5.15. By Corollary 8.3 and Proposition 8.1 we have that

$$\int_X \mathrm{Ric}(\varphi,\varphi) = -(\nabla^*\nabla\varphi, \varphi) = -\|\nabla\varphi\|^2. \tag{8.13}$$

If Ric $\geqq 0$, we conclude that $\nabla\varphi \equiv 0$, i.e., φ is parallel. In particular, $\|\varphi\|$ is constant. Hence, if at some point we have Ric > 0, then $\int \mathrm{Ric}(\varphi,\varphi) > 0$ and we have a contradiction. ■

Note that this argument also proves that when Ric $\geqq 0$, *every harmonic 1-form is parallel.* Under the metric correspondence $T^*X \cong TX$, the parallel 1-forms become parallel vector fields. Thus we can conclude the following

Theorem 8.5. *Let X be a compact riemannian manifold of non-negative Ricci curvature. Then $b_1(X)$ equals the dimension of the space of parallel vector fields on X. In particular,*

$$b_1(X) \leqq \dim(X)$$

with equality if and only if X is a flat torus.

Proof. Let $k = b_1(X)$. Then by the argument above, there are k linearly independent parallel vector fields on X. Parallel vector fields are linearly independent if and only if they are linearly independent at each point. Hence, $k \leqq \dim(X)$, and $k = \dim(X)$ if and only if X has a globally parallel framing. That X must then be a flat torus can be seen as follows. We may choose parallel vector fields $E_1, \ldots, E_k$ which are pointwise orthonormal. Since $[E_i, E_j] = \nabla_{E_i} E_j - \nabla_{E_j} E_i = 0$ for all i,j, these vector fields generate a locally free $\mathbb{R}^k$-action. Since $k = \dim(X)$, we see that X is an orbit of this action, i.e., $X \cong \mathbb{R}^k/\Lambda$, where Λ is a lattice in $\mathbb{R}^k$. The metric on X clearly agrees with the usual one on $\mathbb{R}^k$. ∎

Any parallel vector field generates an isometric flow, and its integral curves are geodesics. Thus even when $k = b_1(X) < \dim(X)$ (and Ric $\geqq 0$), we get a locally free action of $\mathbb{R}^k$ by isometries on X with totally geodesic orbits. We can also consider the dual basis $\varphi_1, \ldots, \varphi_k$ of parallel 1-forms. Integration of $\varphi = (\varphi_1, \ldots, \varphi_k)$ gives a riemannian submersion $J: X \to T^k \equiv \mathbb{R}^k/\Lambda$ where Λ is the lattice in $\mathbb{R}^k$ generated by the "periods." That is, $\Lambda \equiv \{\lambda \in \mathbb{R}^k : \lambda = \int_\gamma \varphi$ for some closed curve γ in $X\}$. By construction, the map J is a covering map on each orbit. With a little more work one can show that the universal covering space $\tilde{X}$ of X splits as a riemannian product $\tilde{X} = \mathbb{R}^k \times X_0$ where X_0 is compact. The original manifold X is a (possibly twisted) riemannian product of T^k with X_0.

REMARK. The first statement in Theorem 8.4 was considerably improved by Myers [1]. Much stronger results for non-compact complete manifolds with Ric $\geqq 0$ were proved by Cheeger and Gromoll [2]. It is a deep theorem of Gromov [1], that there exist explicit a priori bounds, depending only on dimension, for all the Betti numbers of a compact manifold of non-negative sectional curvature. Such bounds do not exist under the weaker hypothesis *Ric* $\geqq 0$ (except for degree 1) by examples of Sha and Yang [1].

As one might guess, theorems similar to 8.4 can be proved for the higher Betti numbers. For each p, there is a positivity assumption on the curvature tensor which guarantees that $b_p(X) = 0$. For detailed statements of these results the reader is referred to Bochner and Yano [1] and to Goldberg [1]. However, one of the more quotable results of this type can be proved rather easily using the Clifford formalism. We present this now.

Recall that the curvature tensor $\langle R_{V_1,V_2} V_3, V_4 \rangle$ is antisymmetric in (V_1, V_2) and in (V_3, V_4) and symmetric under the interchange of these pairs (see (4.44)). Hence, R can be considered as a symmetric endomorphism $\mathbf{R}: \Lambda^2(X) \to \Lambda^2(X)$. We call this transformation the **curvature operator** and say that it is **positive** (or **non-negative**) if all of its eigenvalues are < 0 (or $\leqq 0$ respectively).

Theorem 8.6 (Gallot and Meyer [1]). *Let X be a compact riemannian n-manifold (without boundary) with the property that its curvature operator* *is positive at every point. Then all the Betti numbers, $b_p(X)$ are zero for $p = 1, \ldots, n-1$, i.e., X is a real homology sphere. The same conclusion holds if the curvature operator is $\geqq 0$ and >0 at some point.*

Proof. It will suffice to prove that positivity of the curvature operator implies that

$$\langle \mathfrak{R}(\varphi),\varphi \rangle > 0 \qquad \text{for all non-zero } \varphi \in \Lambda^p(X),\ p = 1,\ldots,n-1. \tag{8.14}$$

The argument then proceeds as in the proof of Theorem 8.4. From the curvature identity (5.17) we have that

$$\begin{aligned}\langle \mathfrak{R}(\varphi),\varphi \rangle &= \sum_{i<j} \langle e_i e_j R_{e_i,e_j}(\varphi), \varphi \rangle \\ &= \tfrac{1}{2} \sum_{i<j} \langle [e_i e_j, R_{e_i e_j}(\varphi)], \varphi \rangle \\ &= -\tfrac{1}{2} \sum_{i<j} \langle R_{e_i,e_j}(\varphi), [e_i e_j, \varphi] \rangle.\end{aligned}$$

From Theorem 4.16 we know that the curvature transformation $R_{V,W}$: $C\ell(X) \to C\ell(X)$ can be written as $R_{V,W} = \frac{1}{2}\sum_{i<j} \langle R_{V,W}(e_i), e_j \rangle \mathrm{ad}_{e_i e_j}$. Hence,

$$\begin{aligned}\langle \mathfrak{R}(\varphi), \varphi \rangle &= -\tfrac{1}{4} \sum_{\substack{i<j \\ k<\ell}} \langle R_{e_i,e_j}(e_k), e_\ell \rangle \langle \mathrm{ad}_{e_i e_j}(\varphi), \mathrm{ad}_{e_k e_\ell}(\varphi) \rangle \qquad (8.15) \\ &= -\tfrac{1}{4} \sum_{\substack{i<j \\ k<\ell}} \langle \mathbf{R}(e_i e_j), e_k e_\ell \rangle \langle \mathrm{ad}_{e_i e_j}(\varphi), \mathrm{ad}_{e_k e_\ell}(\varphi) \rangle.\end{aligned}$$

Observe now that the elements $\{e_i e_j\}_{i<j}$ form an orthonormal basis of $\Lambda^2(X) \subset C\ell(X)$. The last expression in (8.15) is clearly independent of the choice of orthonormal basis. That is, we can write

$$\langle \mathfrak{R}(\varphi), \varphi \rangle = -\tfrac{1}{4} \sum_{\alpha,\beta} \langle \mathbf{R}(\xi_\alpha), \xi_\beta \rangle \langle \mathrm{ad}_{\xi_\alpha}(\varphi), \mathrm{ad}_{\xi_\beta}(\varphi) \rangle \tag{8.16}$$

where $\{\xi_\alpha\}_\alpha$ is any orthonormal basis of $\Lambda^2(X) \subset C\ell(X)$. We choose a basis that diagonalizes $\mathbf{R}$. Let $\{\lambda'_\alpha\}$ be the eigenvalues of $\mathbf{R}$ and set $\lambda_\alpha = -\frac{1}{4}\lambda'_\alpha$. Then (8.16) becomes

$$\langle \mathfrak{R}(\varphi), \varphi \rangle = \sum_\alpha \lambda_\alpha \|\mathrm{ad}_{\xi_\alpha}(\varphi)\|^2$$

where $\lambda_\alpha > 0$ for all α. This proves that $\langle \mathfrak{R}(\varphi), \varphi \rangle \geqq 0$ and equality holds iff $\mathrm{ad}_\xi(\varphi) = 0$ for all $\xi \in \Lambda^2(X)$. Thus, it remains only to prove the following:

Lemma 8.7. *Consider a form $\varphi \in \Lambda^p(\mathbb{R}^n) \subset C\ell(\mathbb{R}^n)$ for $1 \leqq p \leqq n-1$. If $\mathrm{ad}_\xi(\varphi) = 0$ for all $\xi \in \Lambda^2(\mathbb{R}^n)$, then $\varphi = 0$.*

Proof. Recall that the representation ad_ξ on $\Lambda^p(\mathbb{R}^n)$ is just the standard representation of the Lie algebra $\mathfrak{so}(n) \cong \Lambda^2(\mathbb{R}^n)$. This lemma is therefore equivalent to the well-known fact that these representations have no fixed vectors for $p \neq 0$ or n. However, since an elementary proof is possible we shall give it.

Let $e_1, \ldots, e_n$ be the standard basis of $\mathbb{R}^n$ and write $\varphi = \sum_{|I|=p} a_I e_I$. By assumption, $[e_i e_j, \varphi] = 0$ for all $i < j$. One can see easily that

$$[e_i e_j, e_I] = \begin{cases} 0 & \text{if both } i,j \notin I \\ 0 & \text{if both } i,j \in I \\ 2e_i e_j e_I & \text{otherwise.} \end{cases}$$

If $i \notin I$ and $j \in I$, then $e_i e_j e_I = \pm e_{I \cup \{i\} - \{j\}}$. Therefore, $[e_i e_j, \varphi] = 0$ implies that $a_I = 0$ whenever $i \in I$ or $j \in I$ but not both $i,j \in I$. Applying this for all $i < j$ shows that $\varphi = 0$, provided that $p \neq 0$ or n. This completes the proof of the lemma and the theorem. ∎

We now take up the case of spinor bundles. From now on we assume that X is a compact spin manifold with a fixed spin structure on its tangent bundle. Let S be any spinor bundle for $T(X)$ endowed with its canonical riemannian connection.

Before stating the vanishing theorem in this case we recall one of the simplest invariants of the riemannian curvature tensor, the so-called **scalar curvature.** This is a function $\kappa : X \to \mathbb{R}$ defined by setting

$$\kappa \equiv \text{trace}(\text{Ric}) = -2\,\text{trace}(\mathbf{R}).$$

In terms of an orthonormal tangent frame $(e_1, \ldots, e_n)$ at a point $x \in X$,

$$\kappa = -\sum_{i,j=1}^{n} \langle R_{e_i, e_j}(e_i), e_j \rangle. \tag{8.17}$$

When X has dimension two, κ coincides with the classical Gauss curvature function.

Theorem 8.8 (A. Lichnerowicz [1]). *Let X be a spin manifold and suppose S is any bundle of spinors over X endowed with the canonical riemannian connection. Let $D\!\!\!/$ denote the Atiyah-Singer operator and $\nabla^*\nabla$ the connection laplacian on S. Then*

$$\boxed{D\!\!\!/^2 = \nabla^*\nabla + \tfrac{1}{4}\kappa}. \tag{8.18}$$

This formula has the following striking consequence. We say that a spin manifold X **has no harmonic spinors,** if $\ker D\!\!\!/ = 0$ for any spinor bundle associated to $T(X)$.

Corollary 8.9. *Any compact spin manifold of positive scalar curvature admits no harmonic spinors. In fact, the same conclusion holds if the scalar curvature is $\geqq 0$ and > 0 at some point.*

Proof of Corollary 8.9. This follows from formula (8.18) as before. Suppose $\sigma \in \Gamma(S)$ satisfies $D\!\!\!/\sigma = 0$. Then by integration of (8.18) we find

$$\int_X \kappa\|\sigma\|^2 = -(\nabla^*\nabla\sigma,\sigma) = -\|\nabla\sigma\|^2.$$

If $\kappa \geqq 0$, then we must have $\nabla\sigma = 0$. Hence $\|\sigma\|$ is constant, and if $\kappa(x) > 0$ for any point x, then $\int \kappa\|\sigma\|^2 > 0$ and we have a contradiction. ∎

Note that we have also proved the following:

Corollary 8.10. *On a compact spin manifold with* $\kappa \equiv 0$, *every harmonic spinor is globally parallel.*

Proof of Theorem 8.8. We need to compute the curvature term in equation (8.9) for the canonical spinor connection. In Theorem 4.15 we established that for all $V,W \in T_x(X)$, the curvature transformation $R^s_{V,W}\colon S_x \to S_x$ is given by the formula $R^s_{V,W} = \frac{1}{4}\sum_{k,\ell} \langle R_{V,W}(e_k),e_\ell\rangle e_k e_\ell$, where $R_{V,W}\colon T_x(X) \to T_x(X)$ is the curvature transformation of X and where $(e_1,\dots,e_n)$ is any orthornomal basis of $T_x(X)$. Consequently, using the identities (4.43) and (4.44) we see that:

$$\begin{aligned}
\mathfrak{R} &= \tfrac{1}{2}\sum_{i,j} e_i e_j R^s_{e_i,e_j} \\
&= \tfrac{1}{8}\sum_{i,j,k,\ell} \langle R_{e_i,e_j}(e_k), e_\ell\rangle e_i e_j e_k e_\ell \\
&= \tfrac{1}{8}\sum_{\ell}\Bigg\{\frac{1}{3}\sum_{\substack{i,j,k\\ \text{distinct}}} \langle R_{e_ie_j}(e_k) + R_{e_ke_i}(e_j) + R_{e_je_k}(e_i), e_\ell\rangle e_i e_j e_k \\
&\qquad + \sum_{i,j} \langle R_{e_ie_j}(e_i), e_\ell\rangle e_i e_j e_i + \sum_{i,j} \langle R_{e_ie_j}(e_j), e_\ell\rangle e_i e_j e_j\Bigg\} e_\ell \\
&= \tfrac{1}{4}\sum_{i,j,\ell} \langle R_{e_ie_j}(e_i), e_\ell\rangle e_j e_\ell \\
&= -\tfrac{1}{4}\sum_{j,\ell} \operatorname{Ric}(e_j,e_\ell) e_j e_\ell \\
&= \tfrac{1}{4}\kappa. \quad ∎
\end{aligned}$$

As a consequence of the Atiyah-Singer Index Theorem applied to the fundamental spin complex (Example 6.3), we have the following:

Theorem 8.11. *Let X be a compact spin manifold of dimension* $4k$. *If X admits a metric of positive scalar curvature, then* $\hat{A}(X) = 0$.

More generally, applying the same argument to the Dirac operator $\mathfrak{D}$ of the bundle (7.1), we find that in any dimension, positive scalar curvature implies that the analytic index $\operatorname{ind}_n(\mathfrak{D}^0)$ must vanish (see Definition 7.4).

From the $C\ell_k$-Index Theorem, this implies the following:

Theorem 8.12. *Let X be a compact spin manifold. If X admits a metric of positive scalar curvature, then $\hat{\mathscr{A}}(X) = 0$.*

By Theorem 7.10 we know that this result implies Theorem 8.11 above. However, it gives new information in dimensions one and two (mod 8).

Recall that $\hat{\mathscr{A}}(X)$ is an invariant of the spin-cobordism class of X. The above results therefore explicitly present large classes of manifolds which cannot carry metrics of positive scalar curvature. In particular, from Theorem 2.8 we have the following striking result:

Theorem 8.13 (N. Hitchin [1]). *In every dimension $n \equiv 1$ or 2 (mod 8) where $n > 8$, there exist compact differentiable manifolds which are homeomorphic to the n-sphere but which do not admit any riemannian metric with positive scalar curvature. In fact, they admit no metrics such that $\kappa \geqq 0$ but $\not\equiv 0$.*

Such spheres are hardly Platonic. This theorem can be perhaps best appreciated in light of the current positive results. The standard metric on the standard n-sphere $S^n = \{x \in \mathbb{R}^{n+1} : \|x\| = 1\}$ is of course the most uniformly positively curved manifold. Its curvature transformation is given by the formula $-R_{V,W} = V \wedge W$, i.e.,

$$-R_{V,W}(U) = \langle V, U\rangle W - \langle W, U\rangle V \tag{8.19}$$

for all $U,V,W \in T_x(S^n)$. Hence, $\text{Ric}(V) = (n-1)V$ and $\kappa \equiv n(n-1)$. The most symmetric exotic spheres (see Hsiang and Hsiang [1]) are the **Kervaire spheres** which have dimension $4k+1$. They can be constructed by taking the boundary of the manifold obtained by plumbing together two copies of the tangent disk bundle of S^{2k+1}:

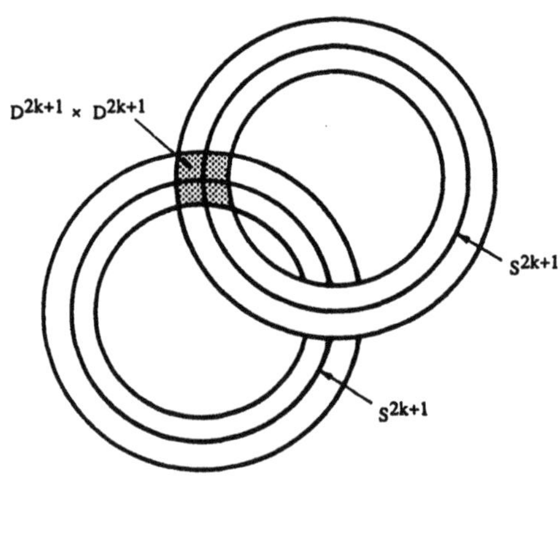

However, Brieskorn has proved that these manifolds can be described algebraically as follows (see Hirzebruch and Mayer [1]). For each integer d, consider the complex polynomial

$$p_d(z_0, \ldots, z_n) = z_0^d + z_1^2 + z_2^2 + \ldots + z_{n+1}^2.$$

Let $V(d) = \{z \in \mathbb{C}^{n+2} : p_d(z) = 0\}$; $S^{2n+3} = \{z \in \mathbb{C}^{n+2} : \|z\| = 1\}$; and set

$$M^{2n+1}(d) = V(d) \cap S^{2n+3}. \tag{8.20}$$

For $n = 2k$ and $d \equiv 3$ or $5 \pmod 8$, $M^{2n+1}(d)$ is a Kervaire sphere.

In this thesis Hernandez [1] proved that the Brieskorn manifolds $M^{2n+1}(d)$ carry metrics of positive Ricci curvature. Moreover, Gromoll and Meyer [2] have proved that a certain exotic 7-sphere carries non-negative *sectional* curvature which is strictly positive on an open subset. It is therefore something of a surprise that in dimension nine there exist exotic spheres which carry no metric of even positive *scalar* curvature!

Manifolds which carry positive scalar curvature are not hard to find. Any homogeneous space G/H, where G is a compact Lie group, carries a metric with $\kappa \geqq 0$. Furthermore, it carries $\kappa > 0$ unless G/H is a torus. More generally one has the following:

Theorem 8.14 (Lawson and Yau [1]). *Let X be a compact manifold which admits an effective differentiable action by a compact, connected, non-abelian Lie group. Then X admits a metric of positive scalar curvature.*

Corollary 8.15. *Let X be a compact spin manifold such that $\hat{\mathscr{A}}(X) \neq 0$. Then the only compact, connected Lie transformation groups of X are tori. In particular, this conclusion holds for any exotic sphere which does not bound a spin manifold.*

In dimensions $4k$, it is a result of Atiyah and Hirzebruch [3] that a compact connected spin manifold X with $\hat{A}(X) \neq 0$ does not even admit an S^1-action! (See IV.3.)

REMARK 8.16. We point out that the results above definitely require that X be a spin manifold. We know from (6.19) that the complex projective spaces $\mathbb{P}^{2k}(\mathbb{C})$ have

$$\hat{A}(\mathbb{P}^{2k}(\mathbb{C})) = (-1)^k 2^{-4k} \binom{2k}{k}. \tag{8.21}$$

These spaces are all homogeneous; $\mathbb{P}^n(\mathbb{C}) = \mathrm{U}(n+1)/\mathrm{U}(1) \times \mathrm{U}(n-1)$. Hence, they have metrics with $\kappa > 0$ and large non-abelian Lie transformation groups. However, as we saw in §2, $\mathbb{P}^{2k}(\mathbb{C})$ is not a spin manifold.

Compact manifolds of positive scalar curvature are now rather well understood. A thorough discussion will be given in Chapter IV.

We conclude this section with an important generalization of the Lichnerowicz formula (Theorem 8.8) to the case of "twisted" spinor bundles. This result is also quite useful in applications as we shall see in Chapter IV.

Let X be a compact riemannian spin manifold, and let S be a spinor bundle for X with the canonical riemannian connection. Let E be any vector bundle over X equipped with an arbitrary orthogonal connection. Then the bundle $S \otimes E$, equipped with the tensor product connection, is again a Dirac bundle over X (see Proposition 5.10). Here the Clifford multiplication takes place in the "S-factor". We now define a smooth, symmetric bundle endomorphism

$$\mathfrak{R}^E : S \otimes E \longrightarrow S \otimes E$$

by the formula

$$\mathfrak{R}^E(\sigma \otimes \varepsilon) \equiv \tfrac{1}{2} \sum_{j,k=1}^{n} (e_j e_k \sigma) \otimes (R^E_{e_j,e_k} \varepsilon) \tag{8.22}$$

on vectors $\sigma \otimes \varepsilon$ of simple type. Here R^E denotes the curvature of the bundle E, and, as usual, $(e_1, \ldots, e_n)$ denotes an orthonormal tangent frame to X at the point in question. The sum in (8.22) is essentially the trace of a bilinear object defined on $\Lambda^2 TX$.

Note. In the case that S is a complex spinor bundle, we may assume E to be complex and endowed with a unitary connection. The tensor product of S with E can then be taken over the complex numbers.

Theorem 8.17. *Let X be a riemannian spin manifold with scalar curvature κ, and let $S \otimes E$ be any twisted spinor bundle over X as above. Then the Dirac operator $D\!\!\!/_E$ and the connection laplacian $\nabla^*\nabla$ of $S \otimes E$ satisfy the identity:*

$$\boxed{D\!\!\!/_E^2 = \nabla^*\nabla + \tfrac{1}{4}\kappa + \mathfrak{R}^E}\,. \tag{8.23}$$

REMARK 8.18. Note that the operator $\mathfrak{R}^E$ depends linearly and universally on the components of the curvature tensor R^E of E. Thus, if E is flat, $\mathfrak{R}^E = 0$, and if R^E is small, then $\mathfrak{R}^E$ is correspondingly small by an estimate depending only on dimension.

Proof. Recall that the covariant derivative of the tensor product connection on $S \otimes E$ acts as a derivation, i.e.,

$$\nabla(\sigma \otimes \varepsilon) = (\nabla^S \sigma) \otimes \varepsilon + \sigma \otimes (\nabla^E \varepsilon)$$

where ∇^S, ∇^E denote the covariant derivatives on S and E respectively. The commutator of two derivations is again a derivation. This fact (or direct verification) shows that the curvature transformation of $S \otimes E$ is also a derivation, i.e.,

$$R(\sigma \otimes \varepsilon) = (R^S\sigma) \otimes \varepsilon + \sigma \otimes (R^E\varepsilon)$$

where R^S and R^E denote the curvature transformations of S and E respectively.

We now wish to compute the curvature term $\mathfrak{R}$ in the general Bochner Identity (8.9). It is given by

$$\begin{aligned}
\mathfrak{R}(\sigma \otimes \varepsilon) &= \tfrac{1}{2} \sum_{j,k=1}^{n} e_j e_k R_{e_j,e_k}(\sigma \otimes \varepsilon) \\
&= \tfrac{1}{2} \sum_{j,k=1}^{n} e_j e_k \{(R^S_{e_j,e_k}\sigma) \otimes \varepsilon + \sigma \otimes (R^E_{e_j,e_k}\varepsilon)\} \\
&= \left(\tfrac{1}{2} \sum_{j,k=1}^{n} e_j e_k R^S_{e_j,e_k}\sigma\right) \otimes \varepsilon + \tfrac{1}{2} \sum_{j,k=1}^{n} (e_j e_k \sigma) \otimes (R^E_{e_j,e_k}\varepsilon).
\end{aligned}$$

Now from the Lichnerowicz calculation (see the proof of Theorem 8.8), we know that the first term in the last line above is just $\frac{1}{4}\kappa$. Hence from (8.22) we conclude that

$$\mathfrak{R}(\sigma \otimes \varepsilon) = \tfrac{1}{4}\kappa(\sigma \otimes \varepsilon) + \mathfrak{R}^E(\sigma \otimes \varepsilon),$$

and the proof is complete. ∎

CHAPTER III

Index Theorems

In this chapter we shall present the analytic underpinnings of the subject of spin geometry. In particular we shall formulate and prove various forms of the Atiyah-Singer Index Theorem. This will include the classical theorem and its consequent cohomological formula for the index. It will also include the Index Theorem for G-Operators, the Index Theorem for Families, and the Index Theorem for $C\ell_k$-Linear Operators. This last result is one of the deepest in the theory. It involves indices in KO-theory which are in general not locally computable on the manifold. Our exposition of this result differs somewhat from that which currently appears in the literature.

There are now in existence many beautiful and illuminating proofs of the more classical index theorems which use the asymptotics of the heat kernel. This is a method that was pioneered by Gilkey and Patodi in the late 1960s. We have elected to present here instead the arguments which originally appeared in Atiyah-Singer [1]–[5]. This is in part because the Atiyah-Singer methods lead to the non-local results just mentioned (and these results are not accessible by heat equation techniques). It is also, however, because their arguments, which proceed in the spirit of Grothendieck, are really quite beautiful and simple. The essential idea is this. One observes that the index is an insensitive object, unperturbed by rather brutal changes in the analytic data. Furthermore, the index is a "functorial" object, which transforms nicely with respect to global operations such as the embedding of one manifold into another, the addition and multiplication of operators, etc. Using an appropriate form of K-theory, one then formulates a topologically defined index possessing the same transformation properties as the analytic index. By performing manipulations allowed in the theory, everything can be reduced to the trivial case where the manifold is a point. Here the analytic and the topological indices are easily seen to coincide, and it follows that they must coincide in general.

Our ambition has been to make the presentation in this chapter reasonably self-contained. All the requisite material on pseudodifferential operators is developed assuming only a knowledge of elementary Fourier analysis. Along the way a proof of the generalized Hodge Decomposition Theorem is given. Most of the material necessary for the derivation and computation of the cohomological formulas is also presented in detail.

The exposition in this chapter owes much to the writings of Atiyah and Singer and also to Gilkey and Nirenberg. The reader is encouraged to consult the excellent literature on the subject of pseudodifferential operators and index theory which has appeared over the past twenty-five years.

§1. Differential Operators

This section presents the basic notion of a linear elliptic differential operator over a manifold X. We begin by fixing notation. For an n-tuple of non-negative integers $\alpha = (\alpha_1, \dots, \alpha_n)$, we set $|\alpha| = \sum_k \alpha_k$, and for each $\xi \in \mathbb{R}^n$ we set $\xi^\alpha = \xi_1^{\alpha_1}\xi_2^{\alpha_2} \cdots \xi_n^{\alpha_n}$. In local coordinates $(x_1, \dots, x_n)$ on X we define the differentiation operators D^α by $i^{|\alpha|}D^\alpha \equiv \partial^{|\alpha|}/\partial x^\alpha \equiv \partial^{|\alpha|}/\partial x_1^{\alpha_1}\partial x_2^{\alpha_2} \cdots \partial x_n^{\alpha_n}$. Recall that for a smooth vector bundle E on X, the symbol $\Gamma(E)$ denotes the space of smooth (i.e., C^∞) cross-sections of E.

Definition 1.1. A **differential operator of order m** on X is a linear map $P:\Gamma(E) \to \Gamma(F)$, where E and F are smooth complex vector bundles over X, with the following property. Each point of X has a neighborhood U with local coordinates $(x_1, \dots, x_n)$ and local trivializations: $E|_U \xrightarrow{\approx} U \times \mathbb{C}^p$ and $F|_U \xrightarrow{\approx} U \times \mathbb{C}^q$, in which P can be written in the form:

$$P = \sum_{|\alpha| \leq m} A^\alpha(x) \frac{\partial^{|\alpha|}}{\partial x^\alpha} \tag{1.1}$$

where each $A^\alpha(x)$ is a $q \times p$-matrix of smooth complex-valued functions and where $A^\alpha \neq 0$ for some α with $|\alpha| = m$.

A **real differential operator of order m** is defined similarly with $\mathbb{C}$ replaced by $\mathbb{R}$.

Observe that if we make a change of the local trivializations of $E|_U$ and $F|_U$ by smooth maps $g_E: U \to GL_p(\mathbb{C})$ and $g_F: U \to GL_q(\mathbb{C})$ respectively, then in these new trivializations P has the form

$$\begin{aligned} P &= g_F\left(\sum_{|\alpha| \leq m} A^\alpha \frac{\partial^{|\alpha|}}{\partial x^\alpha}\right)g_E^{-1} \\ &= \sum_{|\alpha| \leq m} \underset{\sim}{A}^\alpha \frac{\partial^{|\alpha|}}{\partial x^\alpha} \end{aligned}$$

where the $\underset{\sim}{A}^\alpha$'s are again $p \times q$-matrices of smooth functions of x and where

$$\underset{\sim}{A}^\alpha = g_F A^\alpha g_E^{-1} \qquad \text{for } |\alpha| = m. \tag{1.2}$$

If we make a change of local coordinates $\tilde{x} = \tilde{x}(x)$ on U, then using the fact that

$$\frac{\partial}{\partial x_j} = \sum_{k=1}^n \frac{\partial \tilde{x}_k}{\partial x_j}\frac{\partial}{\partial \tilde{x}_k} \qquad \text{for each } j,$$

we find that P again takes the form

$$P = \sum_{|\alpha| \leq m} \tilde{A}^\alpha(\tilde{x}) \frac{\partial^\alpha}{\partial \tilde{x}^\alpha}$$

where

$$\tilde{A}^\alpha = \sum_{|\beta| = m} A^\beta \left[\frac{\partial \tilde{x}}{\partial x}\right]_\beta^\alpha \qquad \text{for } |\alpha| = m \tag{1.3}$$

and where $[\partial\tilde{x}/\partial x]_*^*$ denotes the symmetrization of the mth tensor power of the Jacobian matrix $((\partial\tilde{x}_k/\partial x_j))$.

Equations (1.2) and (1.3) together imply that the coefficients $\{i^m A^\alpha\}_{|\alpha|=m}$ represent a well-defined section $\sigma(P)$ of the bundle $(\odot^m TX) \otimes \text{Hom}(E,F)$ where $\odot$ denotes symmetric tensor product.

DEFINITION 1.2. The section $\sigma(P) \in \Gamma((\odot^m TX) \otimes \text{Hom}(E,F))$ is called the **principal symbol** of the differential operator P.

Recall that for a vector space V, the space $\odot^m V$ is canonically isomorphic to the space of homogeneous polynomial functions of degree m on V^*. Hence, for each cotangent vector $\xi \in T_x^*X$, the principal symbol gives an element

$$\sigma_\xi(P) : E_x \longrightarrow F_x. \tag{1.4}$$

If we fix local coordinates and trivializations as in Definition 1.1, we find that for $\xi = \sum \xi_k \, dx_k$

$$\sigma_\xi(P) = i^m \sum_{|\alpha| = m} A^\alpha(x) \xi^\alpha. \tag{1.5}$$

It is now possible to present one of the fundamental concepts of this chapter.

DEFINITION 1.3. Let P be a differential operator of order m over a manifold X. Then P is **elliptic** if for each non-zero cotangent vector $\xi \in T^*X$, the principal symbol $\sigma_\xi(P) : E_x \to F_x$ is invertible.

EXAMPLE 1.4. Let $E = F$ be the trivialized line bundle and consider the **Laplace-Beltrami operator** $\Delta : C^\infty(X) \to C^\infty(X)$ of a given riemannian metric on X. In local coordinates $(x_1, \ldots, x_n)$ we have

$$\begin{aligned}\Delta f &= \frac{1}{\sqrt{g}} \sum_{j,k=1}^n \frac{\partial}{\partial x_j}\left(\sqrt{g}\, g^{jk} \frac{\partial f}{\partial x^k}\right) \\ &= \sum_{j,k=1}^n g^{jk} \frac{\partial^2 f}{\partial x_j \partial x_k} + \text{lower order terms}\end{aligned}$$

where $\sum g_{jk}\,dx_j\,dx_k$ is the metric tensor and where $((g^{jk})) = ((g_{jk}))^{-1}$. For a given cotangent vector $\xi = \sum \xi_k\,dx_k$ we find that

$$\sigma_\xi(P) = -\sum g^{jk}\xi_j\xi_k = -\|\xi\|^2$$

which is certainly invertible (as a linear map $\mathbb{C} \to \mathbb{C}$) for $\xi \neq 0$.

EXAMPLE 1.5. Let S be a Dirac bundle over a riemannian manifold X (see II.5.2), and consider the associated Dirac operator $D:\Gamma(S) \to \Gamma(S)$. It is straightforward to show that

$$\sigma_\xi(D) = i\xi\cdot$$

where "$\xi\cdot$" denotes "Clifford multiplication by ξ." Since $\xi \cdot \xi\cdot = -\|\xi\|^2\mathrm{Id}$, this map is certainly invertible for $\xi \neq 0$.

EXAMPLE 1.6. Let S be as in Example 1.5 and consider the Dirac Laplacian $D^2:\Gamma(S) \to \Gamma(S)$. Then one has that

$$\sigma_\xi(D^2) = \|\xi\|^2\mathrm{Id}.$$

This shows, from the discussion in Chapter II, §5, that the Hodge Laplacian on exterior p-forms is an elliptic operator. The proof of the following statement is an easy exercise.

Proposition 1.7. *Let $P:\Gamma(E) \to \Gamma(F)$, $P':\Gamma(E) \to \Gamma(F)$ and $Q:\Gamma(F) \to \Gamma(L)$ be differential operators over X where P and P' have the same order. Then for all $\xi \in T^*X$ and for all $t,t' \in \mathbb{R}$, one has that*

$$\sigma_\xi(tP + t'P') = t\sigma_\xi(P) + t'\sigma_\xi(P')$$

and

$$\sigma_\xi(Q \circ P) = \sigma_\xi(Q) \circ \sigma_\xi(P).$$

This proposition says that the symbol is a rather nice object when considered to live on the cotangent bundle. Given $P:\Gamma(E) \to \Gamma(F)$, we pull back the bundles to T^*X via the projection $\pi:T^*X \to X$ and consider the principal symbol as a bundle map

$$\sigma(P):\pi^*E \longrightarrow \pi^*F. \tag{1.6}$$

If P is elliptic, this map is an isomorphism away from the zero section, and we can assign topological data to the operator as follows. Fix a metric on X and set $DX = \{\xi \in T^*X : \|\xi\| \leq 1\}$. Then via the construction given in Chapter I, §9, the symbol of P defines a class

$$i(P) \equiv [\pi^*E,\pi^*F;\sigma(P)] \in K(DX,\partial DX). \tag{1.7}$$

IMPORTANT FACT 1.8. Suppose X is a spin manifold of even dimension and that P is the Atiyah-Singer operator on complex spinors. Then the class $i(P)$, when restricted to any fibre, becomes the element η which generates the Bott periodicity mapping in $K(D^{2k}, S^{2k-1}) \cong \tilde{K}(S^{2k}) \cong K(\mathrm{pt})$ (see I.9). This means that on an even-dimensional spin manifold, *the principal symbol of the Atiyah-Singer operator gives a K-theory orientation on the cotangent bundle*, i.e., a generator of the Thom-isomorphism (see Appendix C).

Similarly, on a spin manifold of dimension $8k$, *the principal symbol of the real Atiyah-Singer operator gives a KO-theory orientation on the cotangent bundle.*

§2. Sobolev Spaces and Sobolev Theorems

Let E be a hermitian vector bundle with connection ∇ on a compact riemannian manifold X. Given $u \in \Gamma(E)$ we have $\nabla u \in \Gamma(T^*X \otimes E)$, and using the tensor product connection on $T^*X \otimes E$, we have $\nabla\nabla u \in \Gamma(T^*X \otimes T^*X \otimes E)$. This process continues, and for any k we can define the norm

$$\|u\|_k^2 \equiv \sum_{j=0}^{k} \int_X |\underbrace{\nabla\nabla \cdots \nabla}_{j \text{ times}} u|^2, \tag{2.1}$$

called the **basic Sobolev *k*-norm** on $\Gamma(E)$. An easy exercise shows the equivalence class of this norm to be independent of the choice of metrics and connection. The completion of $\Gamma(E)$ in this norm is the **Sobolev space** $L_k^2(E)$. It is straightforward to verify the following:

Proposition 2.1. *A differential operator $P:\Gamma(E) \to \Gamma(F)$ of order m extends to a bounded linear map $P:L_k^2(E) \to L_{k-m}^2(F)$ for all $k \geqq m$.*

Ultimately we shall see that if P is elliptic then these extensions have finite dimensional kernel and "cokernel" which consist of smooth sections and are independent of k.

Our aim at present is to establish some analytical tools. This is best done using Fourier transform methods. To this end we select a good system of trivializations of our bundle E. To start we choose a finite covering of X by *closed* coordinate balls $y_\beta : U_\beta \to \bar{B}^n = \{y \in \mathbb{R}^n : |y| \leqq 1\}$, $\beta = 1, \ldots, N$. Over each ball U_β, we choose a smooth trivialization of E

$$E|_{U_\beta} \xrightarrow{\approx} U_\beta \times \mathbb{C}^p$$

which possesses a smooth extension to an open neighborhood of U_α. We further assume that the open balls of radius $1/\sqrt{2}$ cover X, i.e., $X = \bigcup_{\alpha=1}^N B_\beta$ where $B_\beta \equiv \{p \in U_\beta : |y_\beta(p)|^2 < \frac{1}{2}\}$.

We now change each coordinate y_β to a local coordinate x_β by setting

$$x_\beta = \frac{1}{\sqrt{1 - |y_\beta|^2}}\, y_\beta.$$

Note that $x_\beta : U^0_\beta \xrightarrow{\approx} \mathbb{R}^n$ and that $x_\beta(B_\beta) = B^n = \{x \in \mathbb{R}^n : |x| < 1\}$. Furthermore, under the given trivialization over U_β, any smooth section of E restricts to become a bounded function $u : \mathbb{R}^n \to \mathbb{C}^p$. In fact the function $|D^\alpha u(x)|(1 + |x|)^{|\alpha|}$ is bounded for any α.

Let's now choose a smooth partition of unity $\{\chi_\beta\}_{\beta=1}^N$ subordinate to the covering $\{B_\beta\}_{\beta=1}^N$. Any section $u \in \Gamma(E)$ can now be written as $u = \sum u_\beta$ where $u_\beta \equiv \chi_\beta u$. In our system of coordinate trivializations, each u_β becomes a smooth function with compact support in the unit ball B^n.

DEFINITION 2.2. Any system of local coordinates for X and local trivializations for E, together with a partition of unity, all chosen as above, will be called a **good presentation** of E. Good presentations of each of a family of vector bundles over X having the same local coordinates and the same partition of unity, will be called a **good presentation of the family.**

Using a good presentation of E, we can reduce the study of $\Gamma(E)$ to the study of smooth $\mathbb{C}^p$-valued functions with compact support in B^n. Here we can apply the classical **Fourier transform**

$$\hat{u}(\xi) \equiv (2\pi)^{-n/2} \int_{\mathbb{R}^n} e^{-i\langle x,\xi\rangle} u(x)\, dx \tag{2.2}$$

whose elementary properties we summarize (see Taylor [2] as a basic reference). We assume all functions to be $\mathbb{C}^p$-valued, and define the **Schwartz space** $\mathcal{S} \equiv \{u \in C^\infty(\mathbb{R}^n) : \forall \alpha, k, \exists C_{\alpha,k}$ such that $|D^\alpha u(x)| \le C_{\alpha,k}(1 + |x|)^{-k}$ on $\mathbb{R}^n\}$. (Recall that $D^\alpha \equiv i^{-|\alpha|}\partial^{|\alpha|}/\partial x^\alpha$.)

The Fourier transform defines an isomorphism $(\cdot)\hat{}\, : \mathcal{S} \to \mathcal{S}$ whose inverse is given by the following "Inversion Formula"

$$u(x) = (2\pi)^{-n/2} \int_{\mathbb{R}^n} e^{i\langle x,\xi\rangle} \hat{u}(\xi)\, d\xi. \tag{2.3}$$

$$\widehat{D^\alpha_x u}(\xi) = \xi^\alpha \hat{u}(\xi). \tag{2.4}$$

$$\widehat{x^\alpha u}(\xi) = D^\alpha_\xi \hat{u}(\xi). \tag{2.5}$$

$$(u,v)_{L^2} = (\hat{u},\hat{v})_{L^2} \qquad \text{(Plancherel's Formula)} \tag{2.6}$$

where $(u,v)_{L^2} \equiv \int \langle u,v\rangle$ is the usual L^2 inner product.

DEFINITION 2.3. For $s \in \mathbb{R}$ and $u \in \mathcal{S}$, the **Sobolev s-norm** is the norm $\|u\|_s$ given by the formula

$$\|u\|_s^2 \equiv \int (1 + |\xi|)^{2s} |\hat{u}(\xi)|^2\, d\xi. \tag{2.7}$$

The completion of $\mathcal{S}$ in this norm is the **Sobolev space L^2_s.**

REMARK 2.4. Let s be a positive integer. Then there are constants c_1 and c_2 so that $c_1(1 + |\xi|)^{2s} \leqq 1 + |\xi|^2 + \ldots + |\xi|^{2s} \leqq c_2(1 + |\xi|)^{2s}$. It follows from formulas (2.4) and (2.6) that there are constants C_1 and C_2 so that

$$C_1\|u\|_s^2 \leqq \sum_{|\alpha| \leqq s} \int |D^\alpha u(x)|^2\, dx \leqq C_2\|u\|_s^2. \tag{2.8}$$

This means that the norm (2.1) (for the trivialized $\mathbb{C}^p$-bundle with the trivial connection) is equivalent to the norm (2.7).

For any integer $k \geqq 0$, let C^k denote the space of k-times continuously differentiable functions on $\mathbb{R}^n$ equipped with the **uniform C^k-norm,** defined for $u \in C^k$ by

$$\|u\|_{C^k}^2 \equiv \sup_{\mathbb{R}^n} \sum_{|\alpha| \leqq k} |D^\alpha u|^2. \tag{2.9}$$

Our first main result is the following:

Theorem 2.5 (The Sobolev Embedding Theorem). *For each real number $s > (n/2) + k$, there is a constant K_s such that*

$$\|u\|_{C^k} \leqq K_s\|u\|_s \tag{2.10}$$

for all $u \in \mathscr{S}$. Consequently there is a continuous embedding

$$L_s^2 \subset C^k \tag{2.11}$$

for each such s.

Proof. We begin with $k = 0$. For $x \in \mathbb{R}^n$, formula (2.3) gives

$$\begin{aligned} |u(x)| &\leqq (2\pi)^{-n/2} \int |\hat{u}(\xi)|\, d\xi \\ &= (2\pi)^{-n/2} \int (1 + |\xi|)^{-2s}(1 + |\xi|)^{2s}|\hat{u}(\xi)|\, d\xi. \end{aligned}$$

Since $(1 + |\xi|)^{-2s}$ is integrable if $2s > n$ we have by the Schwarz inequality that

$$\begin{aligned} |u(x)|^2 &\leqq (2\pi)^{-n} \int (1 + |\xi|)^{-2s}\, d\xi \int (1 + |\xi|)^{2s}|\hat{u}(\xi)|^2\, d\xi \\ &= K_s^2\|u\|_s^2. \end{aligned}$$

Repeating this argument for each derivative $D^\alpha u$ with $|\alpha| < s - (n/2)$ and then summing, proves the result. ∎

Observe that since $(1 + |\xi|)^{2s'} \leqq (1 + |\xi|)^{2s}$ if $s' < s$ we have that

$$\|u\|_{s'} \leqq \|u\|_s \qquad \forall\, s' < s.$$

Hence there is a continuous inclusion $L_s^2 \subset L_{s'}^2$ for all $s' < s$. When restricted to functions with support in a fixed compact set, this inclusion is "compact."

Theorem 2.6 (The Rellich Lemma). *Let $\{u_j\}_{j=1}^{\infty}$ be a sequence of functions with support in B^n such that $\|u_j\|_s \leqq C$ for all j. Then for any $s' < s$ there is a subsequence which is Cauchy in the norm $\|\cdot\|_s$ and therefore converges in $L^2_{s'}$.*

Proof. We begin by recalling another elementary fact concerning the Fourier transform. Let φ be a smooth $\mathbb{C}$-valued function with compact support in $\mathbb{R}^n$. Then for all integrable functions u,

$$\widehat{\varphi u} = \hat{\varphi}*\hat{u} \quad \text{and} \quad \widehat{\varphi * u} = \hat{\varphi}\hat{u} \tag{2.12}$$

where "$*$" denotes the **convolution product** given by

$$\varphi*u(x) \equiv \int \varphi(x-y)u(y)\,dy. \tag{2.13}$$

Suppose now that $\operatorname{supp} u \subset B^n$ and $\varphi \equiv 1$ on B^n. Then $u = \varphi u$ and so $\hat{u} = \hat{\varphi}*\hat{u}$. Taking derivatives gives the formula

$$D^\alpha \hat{u}(\xi) = \int (D^\alpha \hat{\varphi})(\xi - \eta)\hat{u}(\eta)\,d\eta,$$

and applying the Schwarz Inequality then gives

$$\begin{aligned} |D^\alpha \hat{u}(\xi)|^2 &\leqq \int (1 + |\eta|)^{-2s}|D^\alpha \hat{\varphi}|^2(\xi - \eta)\,d\eta \int (1 + |\eta|)^{2s}|\hat{u}(\eta)|^2\,d\eta \\ &\equiv K_\alpha(\xi)\|u\|_s^2 \end{aligned}$$

where $K_\alpha(\xi)$ is the continuous function defined by the first integral.

Applying (2.14) to the given sequence $\{u_j\}_{j=1}^{\infty}$ shows that the sequence $\{D^\alpha \hat{u}_j\}_{j=1}^{\infty}$ is uniformly bounded on compact subsets of $\mathbb{R}^n$ for any α. In particular the sequence $\{\hat{u}_j\}_{j=1}^{\infty}$ is uniformly equicontinuous on compact subsets, and by the Arzela-Ascoli Theorem there is a subsequence which is uniformly Cauchy on compact subsets.

Fix $r > 0$ and split the integral

$$\begin{aligned} \|u_j - u_k\|_{s'}^2 = &\int_{|\xi| > r} (1 + |\xi|)^{2s'}|\hat{u}_j(\xi) - \hat{u}_k(\xi)|^2\,d\xi \\ &+ \int_{|\xi| \leqq r} (1 + |\xi|)^{2s'}|\hat{u}_j(\xi) - \hat{u}_k(\xi)|^2\,d\xi \end{aligned}$$

For $|\xi| > r$, we have that $(1 + |\xi|)^{2s'} \leqq (1 + r)^{-2(s-s')}(1 + |\xi|)^{2s}$, and so the first integral is $\leqq \|u_j - u_k\|_s^2/(1 + r)^{2(s-s')} \leqq 2C/r^{2(s-s')}$. Hence, for any given $\varepsilon > 0$ we can make the first integral less than $\varepsilon/2$ for all j and k by choosing r sufficiently large. The second integral is then bounded above by a constant multiple of

$$\sup_{|\xi| \leqq r} |\hat{u}_j(\xi) - \hat{u}_k(\xi)|^2.$$

Hence, by the previous paragraph there is a J so that the second integral is $\leqq \varepsilon/2$ for all $j,k \geqq J$, and the sequence $\{u_j\}_{j=1}^{\infty}$ is Cauchy in $L^2_{s'}$ as claimed. ■

Combining the theorems above gives the following:

Corollary 2.7. *Let $\{u_j\}_{j=1}^{\infty}$ be a sequence in L_s^2 with $\operatorname{supp} u_j \subset B^n$ and $\|u_j\|_s \leqq c$ for all j. If $s > (n/2) + k$, then there is a subsequence which converges to a function $u \in C_0^k$ in the uniform C^k-norm.*

We denote by $z \cdot w = \sum_{j=1}^{p} z_j w_j$ the standard $\mathbb{C}$-bilinear pairing on $\mathbb{C}^p$.

Theorem 2.8. *The bilinear function*

$$(u,v) \equiv \int \hat{u}(\xi) \cdot \hat{v}(\xi)\, d\xi$$

is a perfect pairing on $L_s^2 \times L_{-s}^2$, that is, it identifies L_{-s}^2 with the dual of L_s^2, for any $s \in \mathbb{R}$.

Proof. For $u,v \in \mathscr{S}$, we have

$$(u,v) = \int \hat{u}(\xi)(1 + |\xi|)^s \cdot \hat{v}(\xi)(1 + |\xi|)^{-s}\, d\xi,$$

and so by the Schwarz Inequality,

$$|(u,v)| \leqq \|u\|_s \|v\|_{-s}.$$

Hence, the bilinear function has a continuous extension to $L_s^2 \times L_{-s}^2$. In particular, for any $v \in L_{-s}^2$, we have that

$$\sup_{\|u\|_s = 1} |(u,v)| \leqq \|v\|_{-s}. \tag{2.15}$$

It remains to establish equality. For this we choose u so that $\hat{u}(\xi) = \overline{\hat{v}(\xi)}(1 + |\xi|)^{-2s}$. Then we find $|\hat{u}|^2(1 + |\xi|)^{2s} = |\hat{v}|^2(1 + |\xi|)^{-2s}$ so that $\|u\|_s = \|v\|_{-s}$. Furthermore,

$$(u,v) = \int |\hat{v}|^2(1 + |\xi|)^{-2s}\, d\xi = \|v\|_{-s}^2.$$

Hence $(u,v)/\|u\|_s = \|v\|_{-s}$ and (2.15) can be made an equality. We have established the isomorphism $L_{-s}^2 = (L_s^2)^*$. ■

Corollary 2.9. *Let $T: \mathscr{S} \to \mathscr{S}$ and $T^*: \mathscr{S} \to \mathscr{S}$ be linear maps such that $(Tu, v) = (u, T^*v)$ for all $u,v \in \mathscr{S}$. If for some $s \in \mathbb{R}$ and some constant c, the map T satisfies the condition*

$$\|Tu\|_s \leqq c\|u\|_s \qquad \textit{for all } u \in \mathscr{S}, \tag{2.16}$$

then T^* satisfies the condition

$$\|T^*v\|_{-s} = c\|v\|_{-s} \qquad \textit{for all } v \in \mathscr{S}.$$

In particular, if T extends to a bounded linear map $T: L_k^2 \to L_k^2$ for all positive integers k, then T^* extends to a bounded map $T^*: L_{-k}^2 \to L_{-k}^2$ for all negative integers $-k$.

Proof. Given $u,v \in \mathscr{S}$ we have $|(T^*v,u)| = |(v,Tu)| \leqq \|v\|_{-s}\|Tu\|_s \leqq c\|v\|_{-s}\|u\|_s$. Hence, by Theorem 2.8 we find $\|T^*v\|_{-s} = \sup\{|(T^*v,u)| : \|u\|_s = 1\} \leqq c\|v\|_{-s}$. ∎

There is a stronger conclusion possible here which is proved by the interpolation methods of Calderon. Since we shall only need $\|\cdot\|_k$ for $k \in \mathbb{Z}$ in our work here, we simply state the result.

Theorem 2.10. *Let $T: \mathscr{S} \to \mathscr{S}$ be a linear map such that (2.16) is satisfied for $s = s_1$ and $s = s_2$. Then T also satisfies (2.16) for all values of s between s_1 and s_2.*

Corollary 2.9 gives the following key to transferring our local results to global results on a compact manifold.

Proposition 2.11. *Let A be a smooth matrix-valued function on $\mathbb{R}^n$ so that $|D^\alpha A|$ is bounded for all α. Then the map $T: \mathscr{S} \to \mathscr{S}$ given by $Tu = Au$ extends to a bounded linear map $T: L_s^2 \to L_s^2$ for all $s \in \mathbb{R}$.*

Proof. By formula (2.6) we see that $T^*u = A^t u$ where $A^t(x)$ denotes the transpose of the matrix $A(x)$. For any integer $k \geqq 0$, the maps T and T^* are clearly bounded with respect to the classical norm

$$\|u\|_k' = \left\{ \sum_{|\alpha| \leq k} \int |D^\alpha u|^2 \right\}^{\frac{1}{2}}$$

which is equivalent to $\|u\|_k$ (see 2.4). The result now follows immediately from 2.9 and 2.10. ∎

For any open set $\Omega \subset \mathbb{R}^n$, let $L_{s,\Omega}^2$ denote the $\|\cdot\|_s$-closure of $C_0^\infty(\Omega) \equiv \{u \in \mathscr{S} : \operatorname{supp} u \subset \Omega\}$.

Proposition 2.12. *Let Ω, Ω' be bounded open sets with smooth boundary in $\mathbb{R}^n$, and let $\Phi: \bar{\Omega} \to \bar{\Omega}'$ be a diffeomorphism. Then the map $T: C_0^\infty(\Omega') \to C_0^\infty(\Omega)$ given by $Tu \equiv u \circ \Phi$, extends to a bounded linear map $T: L_{s,\Omega'}^2 \to L_{s,\Omega}^2$ for all s.*

Suppose similarly that $\Phi: \mathbb{R}^n \to \mathbb{R}^n$ is a diffeomorphism which is linear outside a compact subset. Then the map $T: \mathscr{S} \to \mathscr{S}$ given by $Tu = u \circ \Phi$ extends to a bounded map $T: L_s^2 \to L_s^2$ for all s.

Proof. We begin with the second statement. Note that $(Tu,v) = (u,T^*v)$ where $T^*v = j(\Phi)u \circ \Phi^{-1}$ where $j(\Phi)$ denotes the Jacobian determinant of Φ^{-1}. As above we easily see that T and T^* are bounded for the norm

$\| \ \|_k' \simeq \| \ \|_k$ for any integer $k \geq 0$. Hence, 2.9 and 2.10 apply to give the result.

The first statement is proved similarly using the straightforward extensions of 2.8, 2.9 and 2.10 to the case of $\Omega \subset \mathbb{R}^n$. ∎

Proposition 2.13. *Let $P = \sum_{|\alpha| \leq m} A^\alpha(x) D^\alpha$ be a differential operator of order m on $\mathbb{R}^n$ whose coefficients are bounded as in Proposition 2.11. Then $P : \mathscr{S} \to \mathscr{S}$ extends to a bounded linear map $P : L_s^2 \to L_{s-m}^2$ for all s.*

Proof. Consider $P = D^\alpha$. Since $|\widehat{D^\alpha u}|^2 = |\xi^\alpha|^2 |\hat{u}|^2$, we find immediately that $\|D^\alpha u\|_s^2 = \int (1 + |\xi|)^{2s} |\xi^\alpha|^2 |\hat{u}(\xi)|^2 \, d\xi \leq \|u\|_{s+|\alpha|}^2$ for any α and s. Applying the triangle inequality and (2.11) completes the proof. ∎

We are now in a position to globalize. Let E be a smooth vector bundle over a compact manifold X and fix a good presentation for E with coordinates $x_\beta : U_\beta \to \mathbb{R}^n$, $\beta = 1, \ldots, N$, and with partition of unity $\{\chi_\beta\}$ subordinate to the covering $\{B_\beta = x_\beta^{-1}(B^n)\}$ (see 2.2).

Since $\sum \chi_\beta = 1$, any $u \in \Gamma(E)$ can be written as $u = \sum u_\beta$ where $u_\beta = \chi_\beta u$. For $s \in \mathbb{R}$, a Sobolev s-norm can be defined on $\Gamma(E)$ by setting

$$\|u\|_s \equiv \sum_{\beta=1}^{N} \|u_\beta\|_s \tag{2.17}$$

where $\|u_\beta\|_s$ is defined in terms of the presentation: $E|_{U_\beta} \cong \mathbb{R}^n \times \mathbb{C}^p$. The following is a direct consequence of Propositions 2.11 and 2.12 which assert the bounded effect of changes of coordinates and trivializations.

Proposition 2.14. *For any $s \in \mathbb{R}$ and any smooth vector bundle E over a compact manifold, the equivalence class of the norm $\|\cdot\|_s$ is independent of the good presentation chosen to define it. Furthermore, when $s = k$ is a non-negative integer, the equivalence class of $\|\cdot\|_k$ is the same as that defined by (2.1) for any choice of metric or connection on E.*

This justifies our use of the symbol $\|\cdot\|_s$ without subscripts to indicate the presentation chosen. It is now straightforward to see that our main theorems, 2.5–2.8 and 2.13, can be globalized. We summarize the result here.

Theorem 2.15. *Let E and F be smooth vector bundles over a compact manifold X of dimension n.*

(1) *For each integer $k \geq 0$ and each $s > (n/2) + k$, there is a continuous inclusion $L_s^2(E) \subset C^k(E)$. Furthermore, every sequence $\{u_j\}_{j=1}^\infty$ which is bounded in the $\|\cdot\|_s$-norm, has a subsequence which converges in the uniform C^k-norm.*

(2) *For any riemannian volume measure μ on X, the bilinear map on $\Gamma(E) \times \Gamma(E^*)$ given by setting*

$$(u,u^*) = \int_X u^*(u)\, d\mu$$

extends to a perfect pairing $L_s^2(E) \times L_{-s}^2(E^*)$ for all s.

(3) *Multiplication $T_A u \equiv Au$ by any element $A \in \Gamma(\mathrm{Hom}(E,F))$ extends to a bounded linear map $T_A: L_s^2(E) \to L_s^2(F)$ for all s.*

(4) *Any differential operator $P:\Gamma(E) \to \Gamma(F)$ of order m extends to a bounded linear map $P: L_s^2(E) \to L_{s-m}^2(F)$ for all s.*

§3. Pseudodifferential Operators

The concept of a pseudodifferential operator has its roots in the following observation. Let $P = \sum A^\alpha(x)D^\alpha$ be a differential operator on $\mathbb{R}^n$ acting on functions u with, say, compact support. By Fourier Inversion (2.3) any such u can be written as

$$u(x) = (2\pi)^{-n/2} \int e^{i\langle x,\xi\rangle}\hat{u}(\xi)\, d\xi.$$

Applying P we find that

$$Pu(x) = (2\pi)^{-n/2} \int e^{i\langle x,\xi\rangle} p(x,\xi)\hat{u}(\xi)\, d\xi \tag{3.1}$$

where

$$p(x,\xi) = \sum_{|\alpha| \leqq m} A^\alpha(x)\xi^\alpha \tag{3.2}$$

is the (total) **symbol of P.** Replacing p by a more general function of x and ξ defines a pseudodifferential operator. Note that in (3.2) the order of P corresponds to the degree of p as a polynomial in ξ. In the general case one must be careful with growth in the ξ-variable.

DEFINITION 3.1. Fix $m \in \mathbb{R}$. A smooth (matrix-valued) function $p(x,\xi)$ on $\mathbb{R}^n \times \mathbb{R}^n$ is said to be a **symbol of order m** if for each α,α' there is a constant $C_{\alpha\alpha'}$ such that

$$|D_x^\alpha D_\xi^{\alpha'} p(x,\xi)| \leq C_{\alpha\alpha'}(1 + |\xi|)^{m-|\alpha'|} \tag{3.3}$$

for all x,ξ. Let Sym^m denote the space of these symbols.

Proposition 3.2. *To each $p \in \mathrm{Sym}^m$ the formula* (3.1) *defines a linear operator $P:\mathscr{S} \to \mathscr{S}$. If p has compact x-support, this operator has a continuous extension $P: L_{s+m}^2 \to L_s^2$ for all s.*

Proof. If $u \in \mathscr{S}$, then $\hat{u} \in \mathscr{S}$. For any integer $N > 0$, we have

$$\begin{aligned} |x|^{2N} Pu(x) &= (-1)^N \int [\Delta_\xi^N e^{i\langle x,\xi\rangle}] p(x,\xi)\hat{u}(\xi)\, d\xi \\ &= (-1)^N \int e^{i\langle x,\xi\rangle} \Delta_\xi^N [p(x,\xi)\hat{u}(\xi)]\, d\xi. \end{aligned}$$

The second integral is bounded by (3.3) and growth properties of $\hat{u}$. Hence, $Pu \in \mathscr{S}$.

To prove the second part note first that integration by parts gives

$$\zeta^\alpha \int e^{i\langle x,\zeta\rangle} p(x,\xi)\, dx = \int e^{i\langle x,\zeta\rangle} D_x^\alpha p(x,\xi)\, dx.$$

Since p has compact x-support, (3.3) then implies that

$$\left| \int e^{i\langle x,\zeta\rangle} p(x,\xi)\, dx \right| \le C_t (1 + |\xi|)^m (1 + |\zeta|)^{-t}$$

for each $t \in \mathbb{Z}^+$. It follows that

$$\begin{aligned} \Psi(\xi,\eta) &\overset{\text{def}}{=} \left| \int e^{i\langle x,\xi-\eta\rangle} p(x,\xi)\, dx \right| (1 + |\xi|)^{-s-m} (1 + |\eta|)^s \\ &\le C_t (1 + |\xi|)^{-s} (1 + |\eta|)^s (1 + |\xi - \eta|)^{-t} \\ &\le \tilde{C}_t (1 + |\xi - \eta|)^{-t+|s|} \end{aligned}$$

where C_t and $\tilde{C}_t$ are constants that depend on t. In particular there exists a constant C such that

$$\int \Psi(\xi,\eta)\, d\xi < C \qquad \text{and} \qquad \int \Psi(\xi,\eta)\, d\eta < C$$

for all ξ and η. Using the pairing from (2.8) we now compute that

$$\begin{aligned} (Pu,v) &= \int \widehat{Pu}(\eta) \cdot \hat{v}(\eta)\, d\eta \\ &= \int \left\{ \int e^{-i\langle x,\eta\rangle} Pu(x)\, dx \right\} \cdot \hat{v}(\eta)\, d\eta \\ &= \iint \left\{ \int e^{i\langle x,\xi-\eta\rangle} p(x,\xi)\, dx \right\} \hat{u}(\xi) \cdot \hat{v}(\eta)\, d\xi\, d\eta. \end{aligned}$$

Setting $U(\xi) = \hat{u}(\xi)(1 + |\xi|)^{s+m}$ and $V(\eta) = \hat{v}(\eta)(1 + |\eta|)^{-s}$, we find that

$$\begin{aligned} |(Pu,v)| &\le \iint \Psi(\xi,\eta) U(\xi) \cdot V(\eta)\, d\xi\, d\eta \\ &\le \left\{ \iint \Psi(\xi,\eta) U^2(\xi)\, d\xi\, d\eta \right\}^{\frac{1}{2}} \left\{ \iint \Psi(\xi,\eta) V^2(\eta)\, d\xi\, d\eta \right\}^{\frac{1}{2}} \\ &\le C \|u\|_{s+m} \|v\|_{-s}. \end{aligned}$$

Applying duality and interpolation completes the proof. ∎

The operators P given in Proposition 3.2 are called **pseudodifferential operators of order m on $\mathbb{R}^n$**, and the space of all such is denoted by ΨDO_m.

We shall study some of the properties of these "local" operators before globalizing them to bundles on manifolds.

Note that a pseudodifferential operator can have order $-m < 0$. Such an operator is said to be **smoothing of order** $\boldsymbol{m}$. A linear map $\tau : \mathscr{S} \to \mathscr{S}$ which extends to a bounded linear map $\tau : L_s \to L_{s+m}$ for all s and m is called an **infinitely smoothing operator.** Note that by the Sobolev Embedding Theorem 2.5, we have $\tau(L_s) \subset C^\infty$ for all s.

Two pseudodifferential operators P and P' will be called **equivalent** if $P - P'$ is an infinitely smoothing operator.

We want now to examine what happens to the symbols of pseudodifferential operators under the operations of composition, of taking adjoints and of changing coordinates. This is referred to as the "symbol calculus." To this end we make the following definition.

DEFINITION 3.3. Let P be a pseudodifferential operator with symbol p. Then p is said to have a **formal development**

$$p \sim \sum_{j=1}^{\infty} p_j$$

(each $p_j \in \mathrm{Sym}^{m_j}$ for some m_j) if for each integer m, there is a K so that $p - \sum_{j=1}^{k} p_j \in \mathrm{Sym}^{-m}$ for all $k \geqq K$. (Hence, the corresponding operator is m-smoothing for all $k \geqq K$.)

The utility of such formal developments is evident from the following result.

Proposition 3.4. *Any formal series $\sum_{j=1}^{\infty} p_j$, where $p_j \in \mathrm{Sym}^{m_j}$ and $m_j \to -\infty$, is the formal development of a pseudodifferential operator. This operator is unique up to equivalence.*

Proof. By grouping terms we can assume $m_{j+1} < m_j$ for all j. Fix a smooth function $\varphi : \mathbb{R}^+ \to [0,1]$ such that $\varphi(t) = 0$ for $t \leq 1$ and $\varphi(t) = 1$ for $t \geq 2$. For any sequence of radii $\{r_j\}_{j=1}^{\infty}$ with $\lim r_j = \infty$ the symbol

$$p(x,\xi) = \sum_{j=1}^{\infty} \varphi(|\xi|/r_j)p_j(x,\xi)$$

is well defined since the sum is finite for each (x,ξ). Set $\tilde{\varphi}(\xi) = \varphi(|\xi|)$ for $\xi \in \mathbb{R}^n$, and for each j define $m_j = (j+1)^n\|\tilde{\varphi}\|_{C^j}$. Recall that for each α,α' and j there is a constant $C_{\alpha\alpha' j}$ such that $|D_x^\alpha D_\xi^{\alpha'} p_j| \leqq C_{\alpha\alpha' j}(1 + |\xi|)^{m_j - |\alpha'|}$. Let $C_j = \max\{C_{\alpha\alpha' j} : |\alpha| \leq j \text{ and } |\alpha'| \leq j\}$ and choose $r_j > m_j 2^j C_j$. Then for any $k > j$ and for $|\alpha| \leq j$, $|\alpha'| \leq j$, we have

$$\begin{aligned} |D_x^\alpha D_\xi^{\alpha'} p_j| &\leqq C_j(1 + |\xi|)^{m_j - |\alpha'|} \\ &\leqq C_j(1 + |\xi|)^{-1}(1 + |\xi|)^{m_k - |\alpha'|} \\ &\leqq \frac{1}{m_j 2^j}(1 + |\xi|)^{m_k - |\alpha'|} \end{aligned}$$

for all $|\xi| \geqq r_j$. Setting $\varphi_j(\xi) = \varphi(|\xi|/r_j)$, we see that

$$|D_x^\alpha D_\xi^{\alpha'} \varphi_j p_j| \leqq \frac{1}{2^j}(1 + |\xi|)^{m_k - |\alpha'|}$$

for all $|\alpha|,|\alpha'| \leqq j < k$ and for all ξ. It follows easily that $p \equiv \sum \varphi_j p_j \in \mathrm{Sym}^{m_1}$ and, moreover, that for all k

$$p - \sum_{j=1}^{k} p_j \in \mathrm{Sym}^{m_{k+1}}.$$

This proves the existence of the operator. Its uniqueness up to infinitely smoothing operators is obvious. ∎

Before beginning the symbol calculus it is useful to note that up to equivalence any pseudodifferential operator can be made "local" in the sense that Pu always has support in a neighborhood of supp u. For this and many other basic calculations we shall need the following:

Workhorse Theorem 3.5. *Let $a(x,y,\xi)$ be a smooth matrix-valued function on $\mathbb{R}^n \times \mathbb{R}^n \times \mathbb{R}^n$ with compact x- and y-support. Fix $m \in \mathbb{R}$ and assume that for each α,β,γ, there is a constant $C_{\alpha\beta\gamma}$ such that $|D_x^\alpha D_y^\beta D_\xi^\gamma a| \leqq C_{\alpha\beta\gamma}(1 + |\xi|)^{m-|\gamma|}$. Then the operator $K: \mathscr{S} \to \mathscr{S}$ given by*

$$(Ku)(x) \equiv (2\pi)^{-n} \iint e^{i\langle x-y,\xi\rangle} a(x,y,\xi) u(y)\, dy\, d\xi \tag{3.4}$$

is a pseudodifferential operator whose symbol k has asymptotic development

$$k(x,\xi) \sim \sum_\alpha \frac{i^{|\alpha|}}{\alpha!} (D_\xi^\alpha D_y^\alpha a)(x,x,\xi). \tag{3.5}$$

Proof. Note that the y-integral in (3.4) is a Fourier transform. Using the rule $\widehat{uv} = \hat{u} * \hat{v}$ (and neglecting constants $(2\pi)^{-n/2}$ which will take care of themselves), we find that

$$\begin{aligned}(Ku)(x) &= \iint e^{i\langle x,\xi\rangle} \hat{a}(x,\xi-\eta,\xi)\hat{u}(\eta)\, d\eta\, d\xi \\ &= \int e^{i\langle x,\eta\rangle} \int e^{i\langle x,\xi-\eta\rangle} \hat{a}(x,\xi-\eta,\xi)\, d\xi\, \hat{u}(\eta)\, d\eta \\ &\stackrel{\text{def}}{=} \int e^{i\langle x,\eta\rangle} k(x,\eta)\hat{u}(\eta)\, d\eta\end{aligned}$$

where $\hat{a}$ denotes Fourier transform in the second variable. The interchange of integration is allowed since for each integer ℓ we have

$$|\hat{a}(x,\xi-\eta,\xi)|\,|\hat{u}(\eta)| \leq C_\ell(1 + |\xi|)^m(1 + |\xi - \eta|)^{-\ell}(1 + |\eta|)^{-\ell},$$

and the right hand side is integrable for ℓ sufficiently large. Formula (3.6) shows that K is pseudodifferential with symbol

$$k(x,\eta) = \int e^{i\langle x,\xi-\eta\rangle}\hat{a}(x,\xi-\eta,\xi)\,d\xi = \int e^{i\langle x,\zeta\rangle}\hat{a}(x,\zeta,\zeta+\eta)\,d\zeta.$$

For each integer ℓ we have the Taylor expansion in the third variable:

$$\hat{a}(x,\zeta,\zeta+\eta) = \sum_{|\alpha|\le\ell}\frac{i^{|\alpha|}}{\alpha!}(D_\eta^\alpha\hat{a})(x,\zeta,\eta)\zeta^\alpha + R_\ell(x,\zeta,\zeta+\eta)$$

where the remainder is given by the formula

$$R_\ell(x,\zeta,\zeta+\eta) = (\ell+1)i^{\ell+1}\sum_{|\mu|=\ell+1}\frac{1}{\mu!}\int_0^1(D_\eta^\mu\hat{a})(x,\zeta,t\zeta+\eta)\zeta^\mu(1-t)^\ell\,dt$$

Recalling that $(\cdot)^\wedge$ denotes the Fourier transform in the middle variable, we find that

$$\begin{aligned}\int e^{i\langle x,\zeta\rangle}(D_\eta^\alpha\hat{a})(x,\zeta,\eta)\zeta^\alpha\,d\zeta &= \int e^{i\langle x,\zeta\rangle}\zeta^\alpha\widehat{(D_\eta^\alpha a)}(x,\zeta,\eta)\,d\zeta\\ &= \int e^{i\langle x,\zeta\rangle}\widehat{(D_y^\alpha D_\eta^\alpha a)}(x,\zeta,\eta)\,d\zeta\\ &= (D_y^\alpha D_\eta^\alpha a)(x,x,\eta).\end{aligned}$$

Consequently, the symbol of K can be written in the form

$$k(x,\eta) = \sum_{|\alpha|\le\ell}\frac{i^{|\alpha|}}{\alpha!}(D_y^\alpha D_\eta^\alpha a)(x,x,\eta) + r_\ell(x,\eta)$$

and by Proposition 3.4 it will suffice to show that

$$r_\ell(x,\eta) \overset{\text{def}}{=} \int e^{i\langle x,\zeta\rangle}R_\ell(x,\zeta,\zeta+\eta)\,d\zeta \in \mathrm{Sym}^{m-(\ell+1)}$$

for each ℓ. To prove this we first show that for each α, β, and k there is a constant $C_{\alpha\beta k}$ such that the inequality

$$|D_x^\alpha D_\eta^\beta\hat{a}(x,\zeta,\zeta+\eta)| \le C_{\alpha\beta k}(1+|\zeta+\eta|)^{m-|\beta|}(1+|\zeta|)^{-k}$$

is satisfied for all x,ζ,η. To establish this we first note that from our basic assumption on a we have

$$\begin{aligned}|\zeta^\gamma D_x^\alpha D_\eta^\beta\hat{a}(x,\zeta,\tilde{\zeta}+\eta)| &= \left|\int\zeta^\gamma(D_x^\alpha D_\eta^\beta a)(x,y,\tilde{\zeta}+\eta)e^{-i\langle y,\zeta\rangle}\,dy\right|\\ &= \left|\int(D_x^\alpha D_y^\gamma D_\eta^\beta a)(x,y,\tilde{\zeta}+\eta)e^{-i\langle y,\zeta\rangle}\,dy\right|\\ &\le C_{\alpha\beta\gamma}\left(\int_{(y\text{-supp})(a)}dy\right)(1+|\tilde{\zeta}+\eta|)^{m-|\beta|}\\ &= C_{\alpha\beta\gamma}(1+|\tilde{\zeta}+\eta|)^{m-|\beta|}\end{aligned}$$

After setting $\tilde{\zeta} = \zeta$, the inequality follows easily. Using this inequality and the above formula for R_ℓ, we calculate that

$$\begin{aligned}
&|D_x^\alpha D_\eta^\beta R_\ell(x,\zeta,\zeta+\eta)| \\
&= \left|(\ell+1) \sum_{|\mu|=\ell+1} \frac{1}{\mu!} \int_0^1 (D_x^\alpha D_\eta^{\beta+\mu}\hat{a})(x,\zeta,t\zeta+\eta)\zeta^\mu(1-t)^\ell\, dt\right| \\
&\leqq c_{\alpha\beta\ell k} \int_0^1 (1+|t\zeta+\eta|)^{m-(\ell+1)-|\beta|}(1+|\zeta|)^{-k}|\zeta|^{\ell+1}(1-t)^\ell\, dt \\
&\leqq C_{\alpha\beta\ell k}(1+|\eta|)^{m-(\ell+1)-|\beta|}(1+|\zeta|)^{\ell+1-k}.
\end{aligned}$$

From this inequality it follows that there are constants $C_{\alpha\beta}$ so that

$$|D_x^\alpha D_\eta^\beta r_\ell(x,\eta)| \leqq C_{\alpha\beta}(1+|\eta|)^{m-(\ell+1)-|\beta|}.$$

Hence, $r_\ell \in \mathrm{Sym}^{m-(\ell+1)}$ and the theorem is proved. ∎

Theorem 3.5 has the following easy consequences:

Observation 3.6. *If the function $a(x,y,\xi)$ of Theorem 3.5 vanishes for all (x,y) in a neighborhood of the diagonal, then the corresponding operator K given by* (3.4) *is infinitely smoothing.*

Proof. By (3.5) we have $k \sim 0$. ∎

For $A \subset \mathbb{R}^n$ and $\varepsilon > 0$, we set $A_\varepsilon \equiv \{x \in \mathbb{R}^n : \mathrm{distance}(x,A) \leqq \varepsilon\}$. An operator $P: \mathscr{S} \to \mathscr{S}$ is said to be **ε-local** if for all $u \in C_0^\infty$

$$\operatorname{supp} Pu \subset (\operatorname{supp} u)_\varepsilon. \tag{3.7}$$

Corollary 3.7. *Given $P \in \Psi DO_m$ whose symbol has compact x-support, and given any $\varepsilon > 0$, there exists $P_\varepsilon \in \Psi DO_m$ which is equivalent to P and is ε-local.*

Proof. Choose a smooth real-valued function ψ on $\mathbb{R}^n \times \mathbb{R}^n$ such that $\psi(x,y) \equiv 1$ in a neighborhood of the diagonal, and $\psi(x,y) = 0$ if $|x-y| \geqq \varepsilon$. Let p be the symbol of P. Then the operator

$$(P_\varepsilon u)(x) \equiv (2\pi)^{-n} \iint e^{i\langle x-y,\xi\rangle}\psi(x,y)p(x,\xi)u(y)\, dy\, d\xi \tag{3.8}$$

is clearly ε-local. By Theorem 3.5 P_ε is pseudodifferential with symbol $p_\varepsilon \sim p$. Hence, P_ε is equivalent to P. ∎

Corollary 3.8. *Let $\chi = (\chi_1,\chi_2)$ be a pair of real-valued functions with compact support on $\mathbb{R}^n$. Then for any $P \in \Psi DO_m$ the operator P^χ given by*

$$P^\chi(u) \equiv \chi_1 P(\chi_2 u) \tag{3.9}$$

is also in ΨDO_m.

Proof. The operator P^χ can be expressed in the form (3.4) with $a(x,y,\xi) = \chi_1(x)\chi_2(y)p(x,\xi)$. ∎

Theorem 3.9. *Let P be a pseudodifferential operator and u a function in its domain (in L_s^2 for some s, say). Then for any open set $U \subset \mathbb{R}^n$,*

$$u\big|_U \in C^\infty \quad\Longrightarrow\quad Pu\big|_U \in C^\infty.$$

Proof. Suppose $u\big|_U$ is smooth and fix $x \in U$. Choose $\chi = (\chi_1,\chi_2)$ as above so that: $x \in \operatorname{supp}\chi_1 \subset \operatorname{supp}\chi_2$; $\chi_1 \equiv 1$ near x; and $\chi_2 \equiv 1$ in a neighborhood of supp χ_1. Since $\chi_2 u \in C_0^\infty$, we have $\chi_1 P(\chi_2 u) \in C^\infty$. Furthermore, by 3.6 we have that $\chi_1 Pu - \chi_1 P(\chi_2 u) = \chi_1 P((1-\chi_2)u) \in C^\infty$. Consequently, $\chi_1 Pu \in C^\infty$ and so Pu is smooth near x. ∎

The reader can probably see the potential usefulness of the above corollaries in trying to define and study pseudodifferential operators on general manifolds. These considerations motivate the following definition.

Definition. An operator $P \in \Psi DO_m$ is said to have **support in a compact set *K*,** if $\operatorname{supp}(Pu) \subset K$ for all $u \in C_0^\infty$, and if $Pu = 0$ whenever $\operatorname{supp} u \cap K = \varnothing$. The linear space of such operators is denoted $\Psi DO_{K,m}$.

We now begin the local symbol calculus.

Theorem 3.10. *Given $P \in \Psi DO_{K,\ell}$ and $Q \in \Psi DO_{K,m}$ with symbols p and q respectively, the composition $P \circ Q \in \Psi DO_{K,\ell+m}$ has symbol with formal development*

$$\operatorname{sym}(P \circ Q) \sim \sum_\alpha \frac{i^{|\alpha|}}{\alpha!}(D_\xi^\alpha p)(D_x^\alpha q). \tag{3.10}$$

Given $P \in \Psi DO_{K,m}$, we define its formal adjoint P^* by setting

$$(Pu,v)_{L^2} = (u,P^*v)_{L^2} \tag{3.11}$$

for all $u,v \in \mathscr{S}$ with support in K.

Theorem 3.11. *Given $P \in \Psi DO_{K,m}$ with symbol p, its formal adjoint $P^* \in \Psi DO_{K,m}$ has symbol p^* with formal development*

$$p^* \sim \sum \frac{i^{|\alpha|}}{\alpha!} D_\xi^\alpha D_x^\alpha \bar{p}^t \tag{3.12}$$

where $(\cdot)^t$ denotes the transposed matrix.

Theorem 3.12. *Let $\phi: U \to V$ be a diffeomorphism between open subsets of $\mathbb{R}^n$. Then for each compact subset $K \subset U$, ϕ induces a map $\phi_*: \Psi DO_{K,m} \to$*

$\Psi DO_{\phi K,m}$ by setting

$$(\phi_* P)(u) \equiv P(u \circ \phi) \circ \phi^{-1}. \tag{3.13}$$

Proof of Theorem 3.11. Choose $u,v \in C^\infty$ with support in K, and note that

$$\begin{aligned}(Pu, v)_{L^2} &= \iint e^{i\langle x,\xi\rangle}\langle p(x,\xi)\hat{u}(\xi),\ v(x)\rangle\, d\xi\, dx \\ &= \iiint e^{i\langle x-y,\xi\rangle}\langle p(x,\xi)u(y), v(x)\rangle\, dy\, d\xi\, dx \\ &= \iiint \langle u(y), e^{-i\langle x-y,\xi\rangle}\overline{p(x,\xi)}^t v(x)\rangle\, dx\, d\xi\, dy \\ &= (u, P^* v)_{L^2}\end{aligned}$$

Fix a real-valued function $\phi \in C_0^\infty$ such that $\phi \equiv 1$ on K, and note that since $\phi u = u$, we can write

$$(P^* v)(y) = \iint e^{i\langle y-x,\xi\rangle}\phi(y)\overline{p(x,\xi)}^t v(x)\, dx\, d\xi. \tag{3.14}$$

This operator satisfies the conditions of Theorem 3.5 and therefore has a symbol p^* with formal development

$$\begin{aligned}p^*(x,\xi) &\sim \sum_\alpha \frac{i^{|\alpha|}}{\alpha!} D_\xi^\alpha D_y^\alpha \phi(x)\overline{p(y,\xi)}^t \Big|_{x=y} \\ &= \sum_\alpha \frac{i^{|\alpha|}}{\alpha!} D_\xi^\alpha D_x^\alpha \overline{p(x,\xi)}^t\end{aligned}$$

where we use the fact that $\phi\bar{p}^t = \bar{p}^t$ because p has x-support contained in K. ■

Proof of Theorem 3.10. Note that

$$(PQu)(x) = \int e^{i\langle x,\xi\rangle}p(x,\xi)\widehat{Qu}(\xi)\, d\xi$$

and so we need a reasonable expression for $\widehat{Qu}(\xi)$. To find this, note that $Q = (Q^*)^*$ and so from equation (3.14)

$$(Qu)(x) = \iint e^{i\langle x-y,\xi\rangle}\overline{q^*(y,\xi)}^t u(y)\, dy\, d\xi \tag{3.15}$$

for $x \in K$, where $q^* = \mathrm{sym}(Q^*)$. Now (3.15) is just an inverse Fourier transform. Hence,

$$\widehat{Qu}(\xi) = \int e^{-i\langle y,\xi\rangle}r(y,\xi)u(y)\, dy$$

where $r \equiv \overline{(q^*)}^t$. It follows that

$$(PQu)(x) = \iint e^{i\langle x-y,\xi\rangle}p(x,\xi)r(y,\xi)u(y)\, dy\, d\xi$$

to which Theorem 3.5 applies. We conclude that PQ is pseudodifferential with formal development:

$$\begin{aligned}
\operatorname{sym}(PQ) &\sim \sum_\alpha \frac{i^{|\alpha|}}{\alpha!} D_\xi^\alpha D_y^\alpha p(x,\xi) r(y,\xi)\bigg|_{x=y} \\
&= \sum_\alpha \frac{i^{|\alpha|}}{\alpha!} \sum_{\beta+\gamma=\alpha} \frac{\alpha!}{\beta!\gamma!} (D_\xi^\beta p)(D_\xi^\gamma D_x^\alpha r) \\
&= \sum_\alpha \sum_{\beta+\gamma=\alpha} \frac{i^{|\beta|+|\gamma|}}{\beta!\gamma!} (D_\xi^\beta p)(D_\xi^\gamma D_x^\beta D_x^\gamma r) \\
&= \sum_\beta \frac{i^{|\beta|}}{\beta!} (D_\xi^\beta p) D_x^\beta \left(\sum_\gamma \frac{i^{|\gamma|}}{\gamma!} D_\xi^\gamma D_x^\gamma r \right) \\
&\sim \sum_\beta \frac{i^{|\beta|}}{\beta!} (D_\xi^\beta p)(D_x^\beta q)
\end{aligned}$$

where the last line is a consequence of Theorem 3.11. ■

Proof of Theorem 3.12. Write $x = \phi(\tilde{x})$ and $\tilde{x} = \phi^{-1}(x) \equiv \psi(x)$. We note that

$$\begin{aligned}
\tilde{x} - \tilde{y} = \psi(x) - \psi(y) &= \int_0^1 \frac{d}{dt} \psi(tx + (1-t)y)\, dt \\
&= \Psi(x, y) \cdot (x - y)
\end{aligned}$$

where $\Psi(x,y)$ is a smooth matrix-valued function. Since $\Psi(x,x) = (\partial\psi/\partial x)_x$ and ψ is a diffeomorphism, the matrix $\Psi(x,y)$ is invertible for (x,y) in a neighborhood $\mathcal{O}$ of the diagonal. We choose $\chi \in C_0^\infty(\mathcal{O})$ such that $\chi \equiv 1$ in a (smaller) neighborhood of the diagonal.

Let $J = \det(\partial\psi/\partial x)$ denote the Jacobian determinant of ψ, and note that

$$\begin{aligned}
[(\phi_* P)u](x) &= [P(u \circ \phi)](\tilde{x}) \\
&= \iint e^{i\langle \tilde{x} - \tilde{y}, \xi\rangle} p(\tilde{x},\xi) u(\phi \tilde{y})\, d\tilde{y}\, d\xi \\
&= \iint e^{i\langle x - y, \Psi^t(x,y)\xi\rangle} p(\psi(x),\xi) u(y) J(y)\, dy d\xi.
\end{aligned}$$

Let $\mathcal{I}$ denote the integrand in this last integral, and write $\mathcal{I} = \chi\mathcal{I} + (1-\chi)\mathcal{I}$. By 3.6, the integral of $(1-\chi)\mathcal{I}$ represents an infinitely smoothing operator. In the integral of $\chi\mathcal{I}$ we can make a change of coordinates $\xi = [\Psi^t(x,y)]^{-1}\zeta \equiv \Theta(x,y) \cdot \zeta$ and find that, modulo infinitely smoothing operators, we have

$$[(\phi_* P)u](x) \cong \iint e^{i\langle x - y, \zeta\rangle} a(x,y,\zeta) u(y)\, dy d\zeta$$

where

$$a(x,y,\zeta) = \chi(x,y) J(y) |\det \Theta| p(\psi(x), \Theta(x,y)\zeta).$$

By the workhorse Theorem 3.5 we conclude that $\phi_* P$ is pseudodifferential. ■

Applying formula (3.5) and recalling that $\chi \equiv 1$ near the diagonal, we find that

$$\begin{aligned}\operatorname{sym}(\phi_* P) &\sim \sum_\alpha \frac{i^{|\alpha|}}{\alpha!} D_\zeta^\alpha D_y^\alpha J(y)|\det \Theta| p(\psi(x),\Theta\zeta)\Big|_{x=y} \\ &\equiv p\left(\tilde{x}(x), \left(\frac{\partial x}{\partial \tilde{x}}\right)^t \zeta\right) \quad (\text{mod } \operatorname{Sym}^{m-1})\end{aligned}$$

since $\Theta(x,x) = [(\partial\psi/\partial x)^t]^{-1} = (\partial x/\partial\tilde{x})^t$ and $|\det \Theta(x,x)| = J^{-1}(x)$. Equation (3.16) states exactly that, modulo symbols of lower order, the symbol of a pseudodifferential operator transforms like a function on the cotangent bundle. More precisely, let the coordinates for the Fourier transform $(\tilde{x},\xi)$ be considered as standard coordinates for 1-forms $\omega = \sum \xi_j \, d\tilde{x}_j$ on $\tilde{x}$-space.

DEFINITION 3.13. Let $P \in \Psi DO_m$ have symbol $p \in \operatorname{Sym}^m$. Then the **principal symbol** of P is the residue class $\sigma(P) = [p] \in \operatorname{Sym}^m/\operatorname{Sym}^{m-1}$. Our discussion above shows the following:

Corollary 3.14. *The principal symbol $\sigma(P)$ transforms under diffeomorphisms like a function on the cotangent bundle of $\mathbb{R}^n$.*

We are now in a position to consider global questions. Let X be a compact n-dimensional manifold, and let E and F be smooth complex vector bundles over X. We say that a linear map $P : \Gamma(E) \to \Gamma(F)$ is **infinitely smoothing** if it extends to a bounded linear map $P : L_s^2(E) \to L_{s+m}^2(F)$ for all $s,m \in \mathbb{R}$. This implies by (2.15) that $P(L_s^2(E)) \subset \Gamma(F)$ ($= C^\infty$-sections) for all s. It is a straightforward exercise to show that given a riemannian volume measure μ on X, any infinitely smoothing operator can be written as an integral operator:

$$Pu(x) = \int_X K(x,y)u(y)\, d\mu(y) \tag{3.17}$$

where $K(x,y) \in \operatorname{Hom}_{\mathbb{C}}(E_y,F_x)$ varies smoothly on $X \times X$.

DEFINITION 3.15. A linear map $P : \Gamma(E) \to \Gamma(F)$ is called a **pseudodifferential operator of order m** if modulo infinitely smoothing operators P can be written as a finite sum $P = \sum P_\alpha$ where each P_α can be expressed in some system of local coordinates $x_\alpha : U_\alpha \to \mathbb{R}^n$ and smooth bundle trivializations as a pseudodifferential operator of order m with compact support. The linear space of all such operators is denoted $\Psi DO_m(E,F)$. Two such operators are called **equivalent** if they differ by an infinitely smoothing operator.

Any differential operator P of order m is pseudodifferential.

REMARK 3.16. Given a riemannian metric on X and $P \in \Psi DO_m(E,F)$, we see from 3.7 that for any $\varepsilon > 0$ there is an operator $P_\varepsilon \in \Psi DO_m(E,F)$ which is equivalent to P and is ε-local. (Of course, a differential operator is 0-local.)

Using a good presentation of the bundles E and F (cf. 2.2), and patching together local pseudodifferential operators with a partition of unity, one can manufacture interesting elements in $\Psi DO_m(E,F)$ for any $m \in \mathbb{R}$.

Given a riemannian volume measure μ on X, we can associate to any operator $P:\Gamma(E) \to \Gamma(F)$ a **formal adjoint** $P^*:\Gamma(F^*) \to \Gamma(E^*)$ by setting

$$\int_X \langle Pu, v\rangle \, d\mu = \int_X \langle u, P^*v\rangle \, d\mu \tag{3.18}$$

for $u \in \Gamma(E)$ and $v \in \Gamma(F^*)$. The following theorem is an immediate consequence of the previous results of this section.

Theorem 3.17. *Let E,F and G be smooth vector bundles over a compact manifold X and fix operators $P \in \Psi DO_m(E,F)$ and $Q \in \Psi DO_\ell(F,G)$. Then the following statements hold:*

(i) *P extends to a bounded linear map $P:L^2_s(E) \to L^2_{s-m}(F)$ for all s.*
(ii) *For any open set $U \subset X$,*

$$u\big|_U \text{ is } C^\infty \quad \Longrightarrow \quad Pu\big|_U \text{ is } C^\infty.$$

(iii) *$Q \circ P \in \Psi DO_{m+\ell}(E,G)$.*
(iv) *$P^* \in \Psi DO_m(F^*,E^*)$ for any μ.*
(v) *A diffeomorphism $\phi:X \to X$ induces a linear map*

$$\phi_*:\Psi DO_m(\phi^*E,\phi^*F) \longrightarrow \Psi DO_m(E,F)$$

by the formula $\phi^*[(\phi_*P)u] = P(\phi^*u)$.

We now discuss some elements of the symbol calculus in the global setting. Let π^*E and π^*F denote the pull-backs of E and F respectively to the cotangent bundle $\pi:T^*X \to X$.

DEFINITION 3.18. Let p be a smooth cross-section of the bundle $\mathrm{Hom}(\pi^*E, \pi^*F)$ on T^*X. Then p is a **symbol of order m** if in a good presentation of E and F, p defines an element of Sym^m in each local coordinate chart of the presentation (where the variables $(\xi_1,\ldots,\xi_n)$ are canonically identified with the coefficients of cotangent vectors in the basis $\{dx_1,\ldots,dx_n\}$.

It is easy to see that this definition is independent of the good presentation that is used.

We let $\mathrm{Sym}^m(E,F)$ denote the vector space of all such symbols of order m. From Corollary 3.14 and Theorem 3.10 we have the following conclusion.

Theorem 3.19. *Each $P \in \Psi DO_m(E,F)$ has an associated "principal symbol" $\sigma(P)$ defined in the quotient space* $\mathrm{Sym}^m(E,F)/\mathrm{Sym}^{m-1}(E,F)$.

It would be useful to be able to canonically construct for each element $p \in \mathrm{Sym}^m(E,F)$ an associated operator $P \in \Psi DO_m(E,F)$ with the same principal symbol as p. This can be done after introducing a riemannian metric on X and a connection ∇ on E. We proceed as follows. Fix $\rho > 0$ sufficiently small that the exponential map $\exp_x : T_xX \to X$ gives a smooth embedding of the ρ-disk at each x. Fix a smooth "cut-off" function $\psi : [0,\rho] \to [0,1]$ with $\psi \equiv 1$ near 0 and $\psi \equiv 0$ near 1. The riemannian metric determines a Lebesgue measure in each fibre of TX and of T^*X, and we can define a "Fourier Transform" $(\cdot)\hat{} : \Gamma(E) \to \Gamma(\pi^*E)$ as follows. For $\xi \in T_x^*X$, we set

$$\hat{u}(\xi) \equiv \int_{T_xX} e^{-i\langle V,\xi\rangle} U(V)\, dV \tag{3.19}$$

where $U(V) \equiv \psi(|V|)\tilde{u}(\exp_x V)$ and where $\tilde{u}(y)$ denotes the parallel translate of $u(y)$ along the (unique, shortest) geodesic ray joining y to x, when distance $(y,x) < \rho$. For $|V| \geqq \rho$ we set $U(V) = 0$. The function $\hat{u}(\xi)$ lies in the Schwartz class of each fibre, T_x^*X.

Given a symbol $p \in \mathrm{Sym}^m(E,F)$, we now define an operator $P = p(\nabla)$ as follows. For $u \in \Gamma(E)$ we set

$$Pu(x) \equiv (2\pi)^{-n} \int_{T_x^*X} p(\xi)\hat{u}(\xi)\, d\xi. \tag{3.20}$$

We leave as an exercise the proof that $P \in \Psi DO_m(E,F)$. A symbol calculus for operators defined in this way has been worked out in Bokobza-Haggiag [1] and Widom [1].

If we take X to be flat euclidean space, and $E = F$ to be the trivialized line bundles, and if we choose $p(x,\xi) = \sum A^\alpha(x)\xi^\alpha$, then a direct calculation shows that $Pu(x) = \sum A^\alpha(x)D^\alpha u(x)$.

§4. Elliptic Operators and Parametrices

Recall that a differential operator $P : \Gamma(E) \to \Gamma(F)$ over a compact manifold is called elliptic if its principal symbol $\sigma_\xi(P)$ is invertible at all non-zero cotangent vectors ξ. In this section we shall prove the fundamental result that modulo infinitely smoothing operators, an elliptic operator is invertible.

To this end we consider the "local" case of pseudodifferential operators on $\mathbb{R}^n$ which map $\mathbb{C}^k$-valued functions to themselves.

DEFINITION 4.1 An operator $P \in \Psi DO_m$ with symbol p is said to be **elliptic** if there exists a constant $c > 0$ such that for all $|\xi| \geqq c$ the matrix inverse of $p(x,\xi)$ exists and satisfies

$$|p(x,\xi)^{-1}| \leqq c(1+|\xi|)^{-m}. \tag{4.1}$$

For example, an operator P whose symbol is of the form $p(|\xi|)\mathrm{Id}$, where $p(t)$ is a polynomial with constant positive coefficients, is elliptic.

REMARK 4.2. It is straightforward to verify that if $P:\Gamma(E) \to \Gamma(F)$ is an elliptic differential operator, then the local representations of P in a good presentation of E and F are elliptic in the sense of 4.1.

Theorem 4.3. *Let $P \in \Psi DO_m$ be elliptic. Then there exists an operator $Q \in \Psi DO_{-m}$, unique up to equivalence, such that*

$$PQ = \mathrm{Id} - S' \qquad \text{and} \qquad QP = \mathrm{Id} - S \tag{4.2}$$

where S and S' are infinitely smoothing operators.

Proof. Let p be the symbol of P and let c be the constant in Definition 4.1. Set $q_0(x,\xi) = \chi(|\xi|)p(x,\xi)^{-1}$ where $\chi:\mathbb{R}^+ \to [0,1]$ is a smooth function with $\chi(t) = 0$ for $t \leq c$ and $\chi(t) = 1$ for $t \geqq 2c$. ■

Lemma 4.4. $q_0 \in \mathrm{Sym}^{-m}$.

Proof. We must show that for each α and β there is a constant $C_{\alpha\beta}$ so that $|D_x^\alpha D_\xi^\beta q_0| \leqq C_{\alpha\beta}(1+|\xi|)^{-m-|\beta|}$. For $\alpha=\beta=0$, this follows immediately from (4.1). For higher derivatives we first note that $|D_x^\alpha D_\xi^\beta q_0|$ is estimated uniformly in x by derivatives of p^{-1}. Taking derivatives of the equation $pp^{-1} \equiv p^{-1}p \equiv I$ (for $|\xi| \geqq c$) and applying (3.3) and (4.1), we see that $|\partial p^{-1}/\partial\xi_j| = |p^{-1}(\partial p/\partial\xi_j)p^{-1}| \leqq C_j(1+|\xi|)^{-m-1}$ for each j. Taking further derivatives, using (3.3), and applying straightforward induction complete the proof of the lemma. ■

Note that $q_0 p - 1 = pq_0 - 1 = 0$ for $|\xi| \geqq 2c$. Consequently these functions lie in Sym^m for all m and the corresponding operators are infinitely smoothing. Unfortunately, pointwise multiplication of symbols does not give rise to composition of operators. We have instead the complicated formula (3.10):

$$\mathrm{sym}(QP) = \sum_\alpha \frac{i^{|\alpha|}}{\alpha!}(D_\xi^\alpha q)(D_x^\alpha p).$$

Placing q_0 in this formula, we find from the above observation that at least $\mathrm{sym}(Q_0P-1) \in \mathrm{Sym}^{m-1}$. This suggests proceeding inductively to define a formal development

$$q \sim \sum_{k=0}^{\infty} q_k \tag{4.3}$$

where $q_k \in \mathrm{Sym}^{m-k}$ is defined by

$$q_k = -\sum_{j=0}^{k-1}\left[\sum_{|\alpha|+j=k} \frac{i^{|\alpha|}}{\alpha!}(D_\xi^\alpha q_j)(D_x^\alpha p)\right] q_0. \tag{4.4}$$

By Proposition 3.4 there is an operator $Q \in \Psi DO_{-m}$, unique up to equivalence, whose symbol has the formal development (4.3)–(4.4). Since the composition of any pseudodifferential operator with an infinitely smoothing operator is infinitely smoothing we are free to replace P and Q with any operators equivalent to them. In particular, by Corollary 3.7 we may assume P and Q to be 1-local. Theorem 3.10 now applies to prove that $QP - I$ is an infinitely smoothing operator.

A completely analogous argument proves the existence of an operator $Q' \in \Psi DO_{-m}$ such that $PQ' - I$ is infinitely smoothing. Note, however, that $Q \sim Q(PQ') = (QP)Q' \sim Q'$, that is, Q and Q' are equivalent. ∎

The operator Q constructed in Theorem 4.3 is called a **parametrix** for P. Its existence proves many of the basic important facts concerning elliptic operators. Here is an example:

Theorem 4.5. *Let $P \in \Psi DO_m$ be an elliptic operator and choose $u \in L_s^2$, for some s. Then on any open set $U \subseteq \mathbb{R}^n$, it is true that:*

$$Pu \text{ is } C^\infty \text{ on } U \quad \Longrightarrow \quad u \text{ is } C^\infty \text{ on } U. \tag{4.5}$$

Furthermore, if $Pu = \lambda u$ for some $\lambda \in \mathbb{C}$ and if $m > 0$, then u is smooth.

Proof. Choose a parametrix Q with $\mathrm{Id} = QP + S$ as above. By 3.9, if Pu is smooth on U, then $u = QPu + Su$ is smooth on U.

If P is elliptic, so is $P - \lambda\,\mathrm{Id}$ for any $\lambda \in \mathbb{C}$ provided $m > 0$. Hence, by the above, if $(P - \lambda)u = 0$, then u is smooth. ∎

We pass now to the global case. Let E and F be smooth vector bundles over a compact manifold X and consider an elliptic differential operator $P: \Gamma(E) \to \Gamma(F)$ of order m. Choose a good presentation of the bundles E,F with coordinates $x_\beta : U_\beta \to \mathbb{R}^n$ where $\beta = 1, \ldots, N$, and with partition of unity $\{\psi_\beta\}$ subordinate to the covering $\{B_\beta\}$ where $B_\beta = \{p \in U_\beta : |x_\beta(p)| < 1\}$ (see Definition 2.2). In the βth system P defines an elliptic operator $P_\beta \in \Psi DO_m$ for which there is a parametrix $Q_\beta \in \Psi DO_{-m}$ satisfying

$$P_\beta Q_\beta = \mathrm{Id} - S'_\beta \quad \text{and} \quad Q_\beta P_\beta = \mathrm{Id} - S_\beta, \tag{4.6}$$

where S_β and S'_β are infinitely smoothing. By (3.7) we may assume that Q_β and, therefore, also S_β and S'_β are 1-local. Now observe that for any 1-local operator R in the βth coordinate system, the operators $\psi_\beta R$ and

$R\psi_\beta$ have compact support in $2B_\beta \cong \{x_\beta : |x_\beta| < 2\}$, and therefore define global operators in $\Psi DO_*(E,F)$. (In particular, if $u \in \Gamma(E)$, we have $\psi_\beta Ru = \psi_\beta R(\varphi u) + \psi_\beta R(1 - \varphi)u) \equiv \psi_\beta R(\varphi u)$ where φ is any smooth cut-off function with support in U_β and with $\varphi \equiv 1$ on $2B_\beta$.)

We now define global operators $Q,Q' \in \Psi DO_{-m}(E,F)$ and $S,S' \in \Psi DO_{-\infty}(E,F)$ by setting

$$Q = \sum \psi_\beta Q_\beta, \qquad Q' = \sum Q_\beta \psi_\beta,$$
$$S = \sum \psi_\beta S_\beta, \qquad S' = \sum S'_\beta \psi_\beta.$$

It follows immediately from (4.6) that $PQ'u = \sum P_\beta Q_\beta(\psi_\beta u) = \sum \psi_\beta u - \sum S'_\beta(\psi_\beta u) = u - S'u$ since $\sum \psi_\beta \equiv 1$. Similarly, one finds that $QP = \mathrm{Id} - S$, and therefore also that $Q \sim Q(PQ') = (QP)Q' \sim Q'$, that is, Q and Q' are equivalent. We have proved the following main result:

Theorem 4.6. *Let $P:\Gamma(E) \to \Gamma(F)$ be an elliptic differential operator of order m over a compact manifold. Then there is an operator $Q \in \Psi DO_{-m}(F,E)$, unique up to equivalence, such that*

$$PQ = \mathrm{Id} - S' \qquad \text{and} \qquad QP = \mathrm{Id} - S \tag{4.7}$$

where S and S' are infinitely smoothing operators.

The operator Q is called a **parametrix** for P.

Notice that the equations (4.7) imply that $PQP = P - S'P = P - PS$ and $QPQ = Q - QS' = Q - SQ$, and consequently that

$$PS = S'P \quad \text{and} \quad QS' = SQ. \tag{4.8}$$

Furthermore, it is evident that

$$S|_{\ker P} = \mathrm{Id} \qquad \text{and} \qquad S'|_{\ker Q} = \mathrm{Id}. \tag{4.9}$$

REMARK 4.7. Theorem 4.6 carries over to pseudodifferential operators. An operator $P \in \Psi DO_m(E,F)$ is said to be **elliptic** if its principal symbol $\sigma(P) \in \mathrm{Sym}^m(E,F)/\mathrm{Sym}^{m-1}(E,F)$ has a representative p which is pointwise invertible outside a compact set in T^*X and satisfies the estimate $|p(\xi)^{-1}| \leq C(1 + |\xi|)^{-m}$ for some constant C and some riemannian metric on X. A straightforward adaptation of the arguments above shows that *Theorem 4.6 remains valid if the word "differential" is replaced by "pseudo-differential."* This fact will not be used here, so we leave the details to the reader.

REMARK 4.8. Given a differential operator $P:\Gamma(E) \to \Gamma(F)$ of order m and a riemannian volume measure on X, let $P^*:\Gamma(F^*) \to \Gamma(E^*)$ be the formal adjoint given by (3.18). Then P^* is again a differential operator of order m whose principal symbol $\sigma(P^*)$ is the pointwise transpose of $\sigma(P)$. In particular, *P is elliptic if and only if P^* is elliptic.*

§5. Fundamental Results for Elliptic Operators

In this section we shall prove some basic theorems for elliptic operators. These will include the fundamental elliptic estimates, the classical "Hodge Decomposition Theorem," the spectral decomposition for self-adjoint elliptic operators, and estimates for the growth of eigenvalues, used later to establish the strong convergence properties of the heat kernel.

Throughout this section E and F will denote smooth vector bundles over a compact n-dimensional manifold X. We shall prove our results here for differential operators but many of them carry over to pseudo-differential operators of positive order.

To begin we recall some basic concepts concerned with a bounded linear operator $T: H_1 \to H_2$ between Hilbert spaces. The **kernel of T** is the subspace $\ker(T) \equiv \{v \in H_1 : Tv = 0\}$, and the **range of T** is the subspace $\mathrm{Im}(T) \equiv \{Tv \in H_2 : v \in H_1\}$. The **cokernel of T** is the quotient space $\mathrm{coker}(T) \equiv H_2/\overline{\mathrm{Im}(T)}$ by the closure of the range. The operator is called **Fredholm** if its kernel and cokernel are finite dimensional and its range is closed. Its **index** is then defined to be the integer $\mathrm{ind}(T) = \dim(\ker T) - \dim(\mathrm{coker}\, T)$.

At the other extreme from Fredholm operators are compact operators. A bounded operator $T: H_1 \to H_2$ is said to be **compact** if the image of each bounded sequence from H_1 has a subsequence which converges in H_2. By the Sobolev Embedding Theorem, the inclusion $L^2_{s'}(E) \subset L^2_s(E)$ for $s' > s$ is compact. In particular any infinitely smoothing operator $S: L^2_s(E) \to L^2_s(E)$ is compact. The Fredholm operators are exactly those which are invertible modulo the compact ones.

Lemma 5.1. *Let $T: H_1 \to H_2$ and $Q: H_2 \to H_1$ be bounded linear maps such that $QT = \mathrm{Id} - S_1$ and $TQ = \mathrm{Id} - S_2$ where S_1 and S_2 are compact. Then T and Q are Fredholm operators.*

Proof. Since $S_1|_{\ker(T)} = \mathrm{Id}$ and S_1 is compact, $\ker(T)$ must be finite dimensional. Taking adjoints, we find $Q^*T^* = \mathrm{Id} - S_2^*$, and since S_2^* is compact we conclude as above that $\dim(\ker T^*) = \dim(\mathrm{coker}\, T) < \infty$. It remains to prove that $\mathrm{Im}(T)$ is closed. By restricting to $(\ker T)^\perp$ we may assume that T is injective. Let $v_k = Tu_k$, $k = 1,2,\ldots$, be a sequence such that $v_k \to v$ in H_2. We want to show that $v = Tu$ for some $u \in H_1$. We note first that the sequence $\{u_k\}_{k=1}^\infty$ is bounded. Otherwise by passing to a subsequence, we can assume that $\|u_k\| \to \infty$ and so $T(u_k/\|u_k\|) = v_k/\|u_k\| \to 0$. Since $QT = I - S_1$ and S_1 is compact, we may assume by passing to a subsequence that $\lim(u_k/\|u_k\|) = \lim S_1(u_k/\|u_k\|) = w$ where $\|w\| = 1$. However, by continuity $Tw = 0$, and since T is injective, $w = 0$. We conclude that $\{u_k\}_{k=1}^\infty$ must be bounded.

Consider now the convergent sequence $Qv_k = QTu_k = u_k - S_1(u_k) \to Qv$. Since $\{u_k\}_{k=1}^{\infty}$ is bounded and S_1 is compact we may assume, after passing to a subsequence, that $S_1(u_k) \to u_\infty$. Applying T to the line above we find $\lim(Tu_k - TS_1u_k) = v - Tu_\infty = TQv$. Hence, $v \in \mathrm{Im}(T)$ and we have proved that $\mathrm{Im}(T)$ is closed. Therefore T (and by symmetry Q also) is Fredholm. ∎

We now state our first main theorem:

Theorem 5.2. *Let $P:\Gamma(E) \to \Gamma(F)$ be an elliptic operator of order m over a compact manifold X. Then the following is true:*

(i) *For any open set $U \subset X$ and any $u \in L^2_s(E)$,*

$$Pu\big|_U \in C^\infty \quad \Longrightarrow \quad u\big|_U \in C^\infty.$$

(ii) *For each s, P extends to a Fredholm map $P: L^2_s(E) \to L^2_{s-m}(F)$ whose index is independent of s.*

(iii) *For each s there is a constant C_s such that*

$$\|u\|_s \leqq C_s(\|u\|_{s-m} + \|Pu\|_{s-m})$$

for all $u \in L^2_s$. Hence the norms $\|\cdot\|_s$ and $\|\cdot\|_{s-m} + \|P\cdot\|_{s-m}$ on L^2_s are equivalent.

Proof. Part (i) is a restatement of Theorem 4.5. For part (ii), note first that by Proposition 2.13, P extends to a bounded linear map $P: L^2_s(E) \to L^2_{s-m}(F)$. That this extension is Fredholm follows immediately from the existence of the parametrix (Theorem 4.6) and the lemma above. By part (i), ker P consists of smooth sections and its dimension is therefore independent of s. Similarly, the cokernel of P is isomorphic to the kernel of the adjoint

$$\begin{array}{ccc} L^2_{s-m}(F)^* & \xrightarrow{P^*} & L^2_s(E)^* \\ \wr\| & & \wr\| \\ L^2_{-s+m}(F^*) & \xrightarrow{P^*} & L^2_{-s}(E^*) \end{array} \tag{5.1}$$

which is easily seen to be the natural extension of the formal adjoint of P. Since P^* is elliptic (see 4.8), dim(ker P^*) is also independent of s. This proves part (ii).

For part (iii), let $Q \in \Psi DO_{-m}(F,E)$ be a parametrix as in Theorem 4.7. Then $u = QPu + Su$, and since S is infinitely smoothing, $\|u\|_s \leqq \|QPu\|_s + \|Su\|_s \leqq C(\|Pu\|_{s-m} + \|u\|_{s-m})$. ∎

The above proof shows the following. Let $P:\Gamma(E) \to \Gamma(F)$ be an elliptic operator and $P^*: \Gamma(F^*) \to \Gamma(E^*)$ its formal adjoint (defined using any volume form on X). Define the **index** of P to be

$$\operatorname{ind}(P) \equiv \dim(\ker P) - \dim(\ker P^*). \tag{5.2}$$

Corollary 5.3. *The index of an elliptic operator P equals the index of any of its Fredholm extensions $P: L^2_s(E) \to L^2_{s-m}(F)$.*

Part (iii) of Theorem 5.2 is called a "fundamental elliptic estimate." Notice that if we solve the equation $Pu = v$, it allows us to estimate the $\|\cdot\|_s$-norm of u in terms of the $\|\cdot\|_{s-m}$-norms of u and v. It will be useful in later discussions to write down an important local consequence of this result.

Theorem 5.4. *Let P be an elliptic differential operator (of order >0) defined on an open subset Ω of $\mathbb{R}^n$. Then for every compact subset $K \subset \Omega$ and every integer $k \geq 0$, there is a constant $C_{K,k}$ such that for all solutions u of the equation $Pu = 0$, one has that*

$$\|u\|_{K,C^k} \leqq C_{K,k}\|u\|_{\Omega,L^2} \tag{5.3}$$

where $\|\cdot\|_{K,C^k}$ denotes the uniform C^k-norm on K and $\|\cdot\|_{\Omega,L^2}$ denotes the L^2-norm on Ω.

Proof. Choose $\varphi \in C_0^\infty(\Omega)$ with $\varphi \equiv 1$ on K. Observe that $P(\varphi u) = \varphi Pu + \sum a_\alpha(x)(D^\alpha u)(x)$ where the sum is over $|\alpha| < m = \operatorname{order}(P)$ and where the coefficients a_α depend only on P and φ. Assume $Pu = 0$. Then the fundamental elliptic estimate 5.2(iii) applies to give

$$\begin{aligned}\|u\|_{K,L^2_s} < \|\varphi u\|_{\Omega,L^2_s} &\leqq C(\|\varphi u\|_{\Omega,L^2_{s-m}} + \|P\varphi u\|_{\Omega,L^2_{s-m}}) \\ &\leqq C'\|u\|_{\Omega,L^2_{s-1}}\end{aligned}$$

Taking a sequence $K \subset\subset \Omega_1 \subset\subset \Omega_2 \subset\subset \cdots \subset\subset \Omega_N = \Omega$, applying the argument repeatedly, and then using the Sobolev Embedding Theorem 2.15(1) completes the proof. ∎

It is often useful, when studying an operator $P: \Gamma(E) \to \Gamma(F)$ to consider non-negative operators P^*P and PP^* where $P^*: \Gamma(F) \to \Gamma(E)$ is the formal adjoint defined via bundle metrics. For this reason we shall assume from this point on that E and F are equipped with hermitian inner products and unitary connections, and that X is furnished with a riemannian metric. All connections will be denoted by ∇. The Sobolev norms $\|\cdot\|_k$ with $k \in \mathbb{Z}^+$ will be given explicitly by (2.1).

Given an mth order differential operator $P: \Gamma(E) \to \Gamma(F)$ and bundle metrics as above, we define the **formal adjoint** of P to be the map P^*: $\Gamma(F) \to \Gamma(E)$ such that $(Pu,v)_{L^2} = (u,P^*v)_{L^2}$ for all $u \in \Gamma(E)$ and all $v \in \Gamma(F)$. Integration by parts shows that P^* exists and is also a differential operator of order m. Furthermore, $\sigma(P^*) = \sigma(P)^*$ and so P^* is elliptic if and only if P is elliptic.

A differential operator $P:\Gamma(E) \to \Gamma(E)$ is then said to be **self-adjoint** if $P = P^*$. The Dirac operator of a Dirac bundle is always self-adjoint, and so, of course, are its powers (see II.5.3.). An important example for riemannian geometry is the Dirac operator D on the Clifford bundle $C\ell(X)$. Under the canonical isomorphism $C\ell(X) \cong \Lambda^*(X)$ we have $D \cong d + d^*$, and so $D^2 \cong dd^* + d^*d \equiv \Delta$, the Hodge Laplacian (see II.5.12).

Theorem 5.5. *Let $P:\Gamma(E) \to \Gamma(E)$ be an elliptic self-adjoint differential operator over a compact riemannian manifold. Then there is an L^2-orthogonal direct sum decomposition:*

$$\Gamma(E) = \ker P \oplus \operatorname{Im} P \tag{5.4}$$

Proof. With respect to the decomposition $L^2(E) = \ker P \oplus (\ker P)^\perp$ any element $u \in \Gamma(E) \subset L^2(E)$ can be written as $u = u_0 + u_1$ with $Pu_0 = 0$. Since u and u_0 are smooth, so is u_1.

Consider now the Fredholm map $P:L^2(E) \to L^2_{-m}(E)$ whose adjoint $P^*:L^2_m(E) \to L^2(E)$ (cf. 5.1) is the natural extension of P itself. We clearly have that $(\ker P)^\perp = \operatorname{Im} P^* = P(L^2_m(E))$. Hence, we can write $u_1 = P\tilde{u}_1$ for $\tilde{u}_1 \in L^2_m(E)$. Since u_1 is smooth, so is $\tilde{u}_1$ by Theorem 5.2. ■

Consider the special case $E = \Lambda^*X$ and $P = \Delta = dd^* + d^*d$ (see II.5.12 forward). Clearly, $\operatorname{Im} \Delta = \operatorname{Im} d + \operatorname{Im} d^*$. However, since $(d^*\varphi, d\psi)_{L^2} = (\varphi, d^2\psi)_{L^2} = 0$ for all φ,ψ, this sum is L^2-orthogonal. Letting $\mathbf{H}^* \equiv \ker \Delta$ denote the **harmonic** forms we have the following:

Corollary 5.6 (The Hodge Decomposition Theorem). *Let X be a compact riemannian n-manifold. Then there is an L^2-orthogonal direct sum decomposition of the smooth p-forms*

$$\Gamma(\Lambda^p X) = \mathbf{H}^p \oplus \operatorname{Im} d \oplus \operatorname{Im} d^* \tag{5.5}$$

for $p = 0, \ldots, n$.

Theorem 5.5 has another important consequence:

Corollary 5.7. *Let $P:\Gamma(E) \to \Gamma(E)$ be a self-adjoint elliptic operator over a compact riemannian manifold, and let $H:\Gamma(E) \to \ker P$ be the orthogonal projection given by the decomposition (5.4) Then there is an operator G: $\Gamma(E) \to \Gamma(E)$ of degree $-m$, called the "Green's operator," such that*

$$PG = GP = \mathrm{Id} - H.$$

Proof. The map $P:\operatorname{Im} P \to \operatorname{Im} P$ is an algebraic isomorphism and has an inverse P^{-1}. Let $G = P^{-1} \circ (I - H)$. G obviously extends to a bounded map $G: L^2_s(E) \to L^2_{s+m}(E)$ which is a Hilbert space isomorphism on $(\ker P)^\perp$ for all s. ∎

We now consider the spectral theory for a self-adjoint elliptic operator $P: \Gamma(E) \to \Gamma(F)$. For $\lambda \in \mathbb{C}$ we consider the **λ-eigenspace of P** given by

$$E_\lambda \equiv \ker(P - \lambda I), \tag{5.6}$$

and we say that λ is an **eigenvalue of P** if $\dim E_\lambda > 0$.

Theorem 5.8. *Let $P: \Gamma(E) \to \Gamma(E)$ be a self-adjoint elliptic differential operator of order $m > 0$ over a compact riemannian n-manifold. Then each eigenspace of P is finite-dimensional and consists of smooth sections. The eigenvalues of P are real, discrete and tend rapidly to infinity in the following sense. If*

$$d(\Lambda) \equiv \dim\left(\bigoplus_{|\lambda| \leqq \Lambda} E_\lambda\right), \tag{5.6}$$

Then there is a constant c such that

$$d(\Lambda) \leqq c\Lambda^{n(n+2m+2)/2m}. \tag{5.7}$$

Furthermore, the eigenspaces of P furnish complete orthonormal systems for $L^2(E)$, i.e., there is a Hilbert space direct sum decomposition

$$L^2(E) = \bigoplus_\lambda E_\lambda. \tag{5.8}$$

Proof. The first statement follows immediately from the fact that $P - \lambda I$ is elliptic. To see that each eigenvalue is real, note that if $Pu = \lambda u$, then $\lambda\|u\|^2 = (Pu,u)_{L^2} = (u,Pu)_{L^2} = \bar{\lambda}\|u\|^2$.

To prove the estimate (5.7) we proceed as follows. Given $\varepsilon > 0$, we say that a subset $A \subset X$ is **ε-dense** if for each point $x \in X$ there is a point $a \in A$ with $\text{dist}(x,a) < \varepsilon$. For each $\varepsilon > 0$, let $N(\varepsilon)$ be the minimal possible number of elements in an ε-dense subset of X. One sees easily that for some constant C.

$$N(\varepsilon) \leqq C\varepsilon^{-n} \qquad \text{for all } \varepsilon > 0. \tag{5.9}$$

Consider now a function $u \in E_\lambda$ and note that for each integer $k > 0$, $P^k u = \lambda^k u$. Hence by the elliptic estimates 5.2(iii) we see that there is a constant C_k, independent of u, such that

$$\|u\|_{mk} \leqq C_k(\|u\|_0 + \|P^k u\|_0). \tag{5.10}$$

Set $E(\Lambda) = \bigoplus_{|\lambda| \leq \Lambda} E_\lambda$ and write $u \in E(\Lambda)$ as $u = \sum a_\lambda u_\lambda$ where $u_\lambda \in E_\lambda$ has $\|u_\lambda\| = 1$. Then $P^k u = \sum a_\lambda \lambda^k u_\lambda$, and since $u_\lambda \perp u_\mu$ for $\lambda \neq \mu$, we have that $\|Pu\|_0^2 = \sum |a_\lambda|^2 |\lambda|^{2k} \leqq \sum |a_\lambda|^2 \Lambda^{2k} = \Lambda^{2k}\|u\|_0^2$. Consequently, by (5.10) we have that

$$\|u\|_{mk} \leqq C_k(1 + \Lambda^k)\|u\|_0$$

for all $u \in E(\Lambda)$. If we assume that $mk > (n/2) + 1$, the Sobolev Embedding Theorem 2.15 gives us a constant C_k' such that

$$\sup_X |\nabla u| \leqq C_k'(1 + \Lambda^k)\|u\|_0 \tag{5.11}$$

for all $u \in E(\Lambda)$. Suppose now that for a given $\varepsilon > 0$ we have $d(\Lambda) = \dim E(\Lambda) > N(\varepsilon)$. Then there will be an element $u \in E(\Lambda)$ with $\|u\|_0 = 1$ such that $u = 0$ on an ε-dense subset of points in X. It then follows from (5.11) that

$$\sup_X |u| \leqq \varepsilon C_k'(1 + \Lambda^k).$$

However, this is impossible if $\varepsilon C_k'(1 + \Lambda^k)\text{vol}(X)^{\frac{1}{2}} < 1$ since $\int |u|_1^2 = 1$. We conclude that $d(\Lambda) \leqq N(\varepsilon_\Lambda)$ where $\varepsilon_\Lambda^{-1} \equiv 2C_k'(1 + \Lambda^k)\text{vol}(X)^{\frac{1}{2}} \leqq C_k''\Lambda^k$. Consequently, by (5.9) we have $d(\Lambda) \leqq C\varepsilon_\Lambda^{-n} \leqq c_k\Lambda^{nk}$. Choosing $k = [(n + 2m + 2)/2m]$ completes the proof of the estimate (5.7).

It remains to prove (5.8). Let V_s denote the closure in $L_s^2(E)$ of the subspace consisting of finite linear combinations of eigensections of P. Note that by 5.2(ii) the map $P: L_s^2(E) \to L_{s-m}^2(E)$ is an isomorphism on $(\ker P)^\perp$. Hence, $P: V_s^\perp \to V_{s-m}^\perp$ is a Hilbert space isomorphism for all s. We want to show that $V_0^\perp = \{0\}$. Let's assume $V_0^\perp \neq \{0\}$ and consider

$$\mu^2 = \inf\{(P^2u, u) = \|Pu\|_0^2 : u \in V_0^\perp \text{ and } \|u\|_0 = 1\}.$$

Note that $V_{2m}^\perp$ is dense in $V_0^\perp$ and so $\mu^2 < \infty$. (This density follows from the fact that L_{2m}^2-orthogonality implies L_0^2-orthogonality, V_{2m} is dense in V_0 and L_{2m}^2 is dense in L_0^2.)

Choose a sequence $\{u_k\}_{k=1}^\infty \subset V_0^\perp$ with $\|u_k\|_0 = 1$ for all k and $\lim(P^2u_k, u_k) = \mu^2$. Since $\|u_k\|_0^2 + \|Pu_k\|_0^2$ is uniformly bounded we know from Theorem 5.2(iii) and the compactness of the embedding $L_m^2(E) \subset L_0^2(E)$, that there exists a subsequence, also denoted $\{u_k\}_{k=1}^\infty$, such that $u_k \to u$ in $L_0^2(E)$. We claim that

$$((P^2 - \mu^2)u, v)_0 = (u, (P^2 - \mu^2)v)_0 = 0 \tag{5.12}$$

for all $v \in V_{2m}^\perp \cong (V_{-2m})^*$. If not, then there exists $v \in V_{2m}^\perp$ with $\|v\|_0 = 1$ and $((P^2 - \mu^2)u, v) = -\alpha < 0$. For $t \in \mathbb{R}$, consider the vectors $u_k(t) = u_k + tv$ and note that

$$0 \leqq \frac{((P^2 - \mu^2)u_k(t), u_k(t))_0}{\|u_k(t)\|_0^2} \longrightarrow \frac{-2\alpha t + t^2}{1 + 2t(u,v) + t^2} < 0$$

for $|t|$ sufficiently small. This proves (5.12), which in turn proves that $P^2u = \mu^2u$. Since P^2 is elliptic, the μ^2-eigenspace is finite dimensional and P-invariant. Consequently it contains an eigenvector of P in contradiction with the definition of $V_0^\perp$. This completes the proof of Theorem 5.8. ■

A self-adjoint operator P is said to be **positive** if $(Pu,u)_0 \geqq 0$ for all $u \in \Gamma(E)$. Theorem 5.8 can be reformulated in the positive case as follows:

Corollary 5.9. *Let $P:\Gamma(E) \to \Gamma(E)$ be a positive self-adjoint elliptic differential operator of order $m > 0$ over a compact manifold. Then there is a complete orthonormal basis $\{u_k\}_{k=1}^{\infty}$ of $L_0^2(E)$ such that*

$$Lu_k = \lambda_k u_k \qquad \text{for all } k \tag{5.13}$$

where $0 \leq \lambda_1 \leqq \lambda_2 \leqq \ldots \to \infty$. In fact for some constant $c > 0$,

$$\lambda_k \geqq ck^{2m/n(n+2m+2)} \qquad \text{for all } k. \tag{5.14}$$

Proof. The first statement is clear. For the second note that $d(\lambda_k) = k$. ■

§6. The Heat Kernel and the Index

Let $P:\Gamma(E) \to \Gamma(E)$ be a positive self-adjoint elliptic differential operator of order m over a compact riemannian n-manifold X. In this section we shall explicitly construct the **heat operator** $e^{-tP}:\Gamma(E) \to \Gamma(E)$, for $t > 0$, which is an infinitely smoothing operator with the property that if $u_t = e^{-tP}u$, for some $u \in \Gamma(E)$, then u_t satisfies the equation

$$\frac{d}{dt}u_t + Pu_t = 0. \tag{6.1}$$

We shall define e^{-tP} as an integral operator of the form

$$(e^{-tP}u)(x) = \int_X K_t(x,y)u(y)\,dy \tag{6.2}$$

where $K_t(x,y):E_y \to E_x$ is a linear map depending smoothly on x,y and t. (Hence, K is a smooth section of the obvious bundle over $\mathbb{R} \times X \times X$.) This kernel K is called the **heat kernel for P** and is defined as follows. Let $\{u_k\}_{k=1}^{\infty}$ be a complete orthonormal basis of $L^2(E)$ consisting of eigensections of P with $Pu_k = \lambda_k u_k$ and $0 \leqq \lambda_1 \leqq \lambda_2 \leqq \ldots \to \infty$ as guaranteed by Corollary 5.9. We set

$$K_t(x,y) = \sum_{k=1}^{\infty} e^{-\lambda_k t}u_k(x) \otimes u_k^*(y) \tag{6.3}$$

where $v^* \in E_y^*$ denotes the element such that $v^*(u) = \langle u,v\rangle_y$ for all $u \in E_y$.

Lemma 6.1. *For any $r \geqq 0$ and any closed interval $I \subset (0,\infty)$, the series (6.3) converges uniformly in the C^r-topology on $I \times X \times X$.*

Proof. Fix a positive integer s with $ms > (n/2) + r$. By the Sobolev Embedding Theorem 2.15 and the Fundamental Elliptic Estimates 5.2(iii) we have constants c and c', depending only on s, such that each u_k satisfies the inequality:

$$\|u_k\|_{C^r} \leqq c'\|u_k\|_{ms} \leqq c(\|u_k\|_0 + \|P^s u_k\|_0) = c(1 + \lambda_k^s)$$

where $\|\cdot\|_{C^r}$ denotes the function C^r-norm on X. By Corollary 5.9, λ_k satisfies the inequality

$$\lambda_k \geqq k^\gamma$$

where $\gamma = 2m/n(n + 2m + 2)$. This implies that

$$e^{-\lambda_k t}\lambda_k^s \leqq (e^{-k^\gamma t})(k^{s\gamma})$$

for all $k > (s/t)^{1/\gamma}$. The uniform C^r-convergence of (6.3) now follows directly from the convergence of the integral $\int_1^\infty e^{-tx}x^s\,dx$. ■

This lemma has the following immediate consequence:

Theorem 6.2. *For each $t > 0$ the operator $e^{-tP}: \Gamma(E) \to \Gamma(E)$ is infinitely smoothing. Furthermore, given $u \in L_s^2(E)$, for any s, the section $u(t,x) \equiv (e^{-tP}u)(x)$ is C^∞ on $\mathbb{R} \times X$ and satisfies the "heat equation," $\partial u/\partial t = -Pu$.*

Proof. The first two statements are an immediate consequence of the fact that K is C^∞. To see that $\partial u/\partial t = -Pu$, note that

$$\begin{aligned}\frac{\partial}{\partial t} K_t(x, y) &= -\sum e^{-\lambda_k t}\lambda_k u_k(x) \otimes u_k^*(y)\\ &= -P_x K_t(x,y). \quad ■\end{aligned}$$

For a given operator P as above, we introduce the following important concept.

DEFINITION 6.3. The **trace of the heat kernel for P** is the function

$$\begin{aligned}\operatorname{tr}(e^{-tP}) &= \int_X \operatorname{trace}_x[K_t(x,x)]\,dx\\ &= \sum_{k=1}^{\infty} e^{-\lambda_k t}\end{aligned}$$

which is defined and analytic for all $t > 0$.

Note that $\operatorname{trace}_x(u_k(x) \otimes u_k^*(x)) = |u_k(x)|^2$ and so the second equality follows from the fact that $\|u_k\|_0^2 = 1$ for all k.

EXAMPLE. Let X be the flat cubical torus $\mathbb{R}^n/2\pi\mathbb{Z}^n$, where $\mathbb{R}^n$ carries the standard metric, and let $P = -\Delta = -\sum \partial^2/\partial\theta_k^2$ acting on functions. The normalized eigenfunctions are given by $u_N(\theta) = (2\pi)^{-n/2}e^{i\langle N,\theta\rangle}$ for $N = (N_1, \ldots, N_n) \in \mathbb{Z}^n$. Note that $-\Delta u_N = |N|^2 u_N$, and so

$$K_t(\theta, \theta') = \sum_{N \in \mathbb{Z}^n} e^{-|N|^2 t + i\langle N, \theta - \theta'\rangle}$$

and

$$\operatorname{tr}(e^{t\Delta}) = \sum_{k=0}^{\infty} a(k)e^{-kt}$$

where $a(k)$ denotes the number of elements $N \in \mathbb{Z}^n$ with $|N|^2 = k$. For $n = 1$, we find

$$\operatorname{tr}(e^{t\Delta}) = \sum_{-\infty}^{\infty} e^{-k^2 t} \underset{t \searrow 0}{\sim} \sqrt{\pi/t}.$$

Let us return now to the case of a general elliptic differential operator $P:\Gamma(E) \to \Gamma(F)$ over a compact riemannian manifold X. We assume E and F are equipped with bundle metrics, and we consider the "Laplace operators":

$$P^*P:\Gamma(E) \longrightarrow \Gamma(E) \qquad \text{and} \qquad PP^*:\Gamma(F) \longrightarrow \Gamma(F).$$

Since $(P^*Pu, u)_0 = \|Pu\|_0^2$ and $(PP^*v, v)_0 = \|P^*v\|_0^2$, we see that these elliptic operators are *self-adjoint* and *positive.* It is furthermore evident that $\ker P^*P = \ker P$ and $\ker PP^* = \ker P^*$. Consequently, we have (cf.(5.2)) that

$$\operatorname{ind}(P) = \dim(\ker P^*P) - \dim(\ker PP^*) \tag{6.5}$$

On the other hand, *the operators P^*P and PP^* have exactly the same sequence of non-zero eigenvalues.* To see this, set $E_\lambda \equiv \{u \in \Gamma(E) : P^*Pu = \lambda u\}$ and $F_\lambda \equiv \{v \in \Gamma(F) : PP^*v = \lambda v\}$ for $\lambda \in \mathbb{R}$. Observe that if $u \in E_\lambda$, then $(PP^*)(Pu) = P(P^*Pu) = \lambda(Pu)$; that is, $P(E_\lambda) \subset F_\lambda$. Similarly, we find that $P^*(F_\lambda) \subset E_\lambda$. Since $P^*P = \lambda \operatorname{Id}$ on E_λ, we conclude that $P: E_\lambda \xrightarrow{\approx} F_\lambda$ is an isomorphism for all $\lambda \neq 0$.

Let $0 < \lambda_1 \leqq \lambda_2 \leqq \lambda_3 \leqq \ldots \to \infty$ denote the common non-zero eigenvalues of P^*P and PP^* (listed to multiplicity). Then taking the difference of the traces of the heat kernels, we get enormous cancellation(!):

$$\begin{aligned}\operatorname{tr}(e^{-tP^*P}) - \operatorname{tr}(e^{-tPP^*}) &= (\dim E_0 + \textstyle\sum e^{-\lambda_k t}) - (\dim F_0 + \textstyle\sum e^{-\lambda_k t}) \\ &= \dim E_0 - \dim F_0.\end{aligned}$$

This proves the following:

Theorem 6.4. *Let $P:\Gamma(E) \to \Gamma(F)$ be an elliptic differential operator over a compact riemannian manifold, and let $P^*:\Gamma(F) \to \Gamma(E)$ be defined via inner products in E and F. Then*

$$\operatorname{ind}(P) = \operatorname{tr}(e^{-tP^*P}) - \operatorname{tr}(e^{-tPP^*})$$

for all $t > 0$.

Observe that as $t \to \infty$, the operator e^{-tP^*P} converges strongly to the orthogonal projection $H_E: L^2(E) \to \ker P$. Consequently the direct sum

$D_t \equiv e^{-tP^*P} \oplus (-e^{-tPP^*})$ acting on $\Gamma(E \oplus F)$ can be thought of as a "homotopy" of operators which, as $t \to \infty$, converges to $H_E \oplus (-H_F)$, the "difference of harmonic projections." We see that for all t, D_t is of trace class and its trace, $\mathrm{tr}(D_t) = \mathrm{ind}(P)$, is time independent. It is useful to consider the operator D_t at the other end of the "homotopy," as $t \searrow 0$.

When $\deg(P) = 1$, it turns out that as $t \searrow 0$, the heat kernel for P^*P has an asymptotic expansion

$$\mathrm{trace}_x K_t(x,x) \sim \sum_{k=0}^{\infty} \rho_k(x) t^{(k-n)/2} \tag{6.6}$$

Where $\rho_k(x)$ are densities on X which are locally and explicitly computable in terms of the geometry of X and P. Theorem 6.4 says that the index of P depends only on the coefficient $\rho_n(x)$ for the operators P^*P and PP^*. Careful computations of these terms yields a proof of the classical Atiyah-Singer Index Theorem.

§7. The Topological Invariance of the Index

The index of an elliptic operator is a quite stable object. It remains constant during continuous perturbations of the operator and, in fact, it can be shown to depend only on the homotopy class of the principal symbol. A proof of this fact is one of the main objectives of this section. In succeeding sections we shall consider elliptic operators with additional structure, such as operators which are $C\ell_k$-linear for some k or operators which commute with a given action of a compact Lie group. In each case we shall define a refined index and examine its elementary homotopy invariance properties.

The index of an elliptic operator $P:\Gamma(E) \to \Gamma(F)$ is the index of any of its Fredholm extensions $P: L^2_s(E) \to L^2_{s-m}(F)$. For this reason we begin with a discussion of Fredholm operators.

Let H_1 and H_2 be separable Hilbert spaces and let $\mathscr{L} = \mathscr{L}(H_1, H_2)$ denote the Banach space of bounded linear maps from H_1 to H_2 with norm given by $\|T\| = \sup\{\|Tv\| : \|v\| \leq 1\}$. It is elementary to verify that the subset $\mathscr{L}^{\mathrm{iso}} \subset \mathscr{L}$ of linear isomorphisms is open in the norm topology (cf. Palais [1]). We are interested here in the subset $\mathfrak{F} = \mathfrak{F}(H_1, H_2)$ of **Fredholm operators from H_1 to H_2** (see §5). To each $T \in \mathfrak{F}$, we have defined the **index** of T:

$$\mathrm{ind}\, T = \dim(\ker T) - \dim(\mathrm{coker}\, T).$$

Proposition 7.1. *The map* $\mathrm{ind}: \mathfrak{F} \to \mathbb{Z}$ *is locally constant on* $\mathfrak{F}$, *and induces a bijection*

$$\mathrm{ind}: \pi_0(\mathfrak{F}) \xrightarrow{\approx} \mathbb{Z}$$

between the connected components of $\mathfrak{F}$ *and the integers* $\mathbb{Z}$.

Proof. We begin the proof with a lemma which will be useful later on.

Lemma 7.2. *Fix $T_0 \in \mathfrak{F}$ and let $V \subset H_1$ be a closed subspace of finite codimension with $V \cap (\ker T_0) = \{0\}$. Then there is a neighborhood $\mathcal{U}$ of T_0 in $\mathcal{L}$ such that for all T in $\mathcal{U}$ we have:*

(i) $V \cap (\ker T) = \{0\}$,
(ii) *TV is closed in H_2,*
(iii) *the subspace $W \equiv (T_0 V)^\perp \subset H_2$ projects isomorphically:*

$$W \cong H_2/TV.$$

Furthermore, the isomorphisms of (iii) *assemble to give that:*

(iv) *The family $\bigcup_{T \in \mathcal{U}} (H_2/TV) \to \mathcal{U}$, topologized as a quotient of $\mathcal{U} \times H_2$, is equivalent to the trivial bundle $\mathcal{U} \times W \to \mathcal{U}$.*

Proof. To each $T \in \mathcal{L}$ we associate the bounded linear map

$$\tilde{T}: W \oplus V \longrightarrow H_2 \tag{7.1}$$

given by setting $\tilde{T}(w,v) = w + Tv$. This correspondence $T \to \tilde{T}$ defines a continuous map $\mathcal{L} \to \mathcal{L}(W \oplus V, H_2)$ in the norm topology. Since $\tilde{T}_0$ is an isomorphism, so is $\tilde{T}$ for all T in a neighborhood $\mathcal{U}$ of T_0. This establishes (i)–(iv). ∎

Corollary 7.3. *$\mathfrak{F}$ is open in $\mathcal{L}$.*

Proof. Choose $V = (\ker T_0)^\perp$ in Lemma 7.2 and note that $\mathcal{U} \subset \mathfrak{F}$ by (i), (ii), and (iii). ∎

Corollary 7.4. *The index is constant on connected components of $\mathfrak{F}$.*

Proof. Fix $T_0 \in \mathfrak{F}$, set $V = (\ker T_0)^\perp$, and let $\mathcal{U}$ be given by Lemma 7.2. It will suffice to show that $\operatorname{ind} T = \operatorname{ind} T_0$ for all $T \in \mathcal{U}$. Fix $T \in \mathcal{U}$ and consider the (non-orthogonal) direct sum decomposition $H_1 = (\ker T) \oplus Z \oplus V$ where $Z \equiv (\ker T \oplus V)^\perp$ (see 7.2 (i)). By the Open Mapping Theorem, T induces an isomorphism between $Z \oplus V$ and the closed subspace $TZ \oplus TV$. Setting $\operatorname{coker} T \equiv (TH_1)^\perp = (TZ \oplus TV)^\perp$, we obtain the following factoring of T:

$$\begin{array}{ccc} H_1 & \xrightarrow{\quad T \quad} & H_2 \\ \wr\Vert & & \wr\Vert \\ \ker T \oplus Z \oplus V & \longrightarrow & \operatorname{coker} T \oplus TZ \oplus TV \end{array} \tag{7.2}$$

Dividing by $V = (\ker T_0)^\perp$ gives an isomorphism

$$\ker T_0 \cong \ker T \oplus Z. \tag{7.3}$$

Dividing by TV and using 7.2(iii) (with $W = \text{coker } T_0$) gives an isomorphism

$$\text{coker } T_0 \cong \text{coker } T \oplus TZ \cong \text{coker } T \oplus Z. \tag{7.4}$$

It follows immediately that ind T_0 = ind T. ■

We shall examine this argument again when the maps have more structure. The proof of Proposition 7.1 is completed by the following:

Lemma 7.6. *Any two operators $T_0, T_1 \in \mathfrak{F}$ with the same index lie in the same connected component of $\mathfrak{F}$.*

Proof. Since $\text{ind}(T^*) = -\text{ind } T$, it suffices to consider the case where $\text{ind } T_0 = \text{ind } T_1 \geqq 0$. To begin we note that any $T \in \mathfrak{F}$ with $\text{ind } T \geqq 0$ is homotopic in $\mathfrak{F}$ to a surjective map. To see this, choose any linear surjection $L: \ker T \twoheadrightarrow \text{coker } T = (\text{Im } T)^{\perp}$, and consider the homotopy $T + tL$, $t \geqq 0$. We may assume therefore that T_0 and T_1 are both surjective. Set $K_j = \ker(T_j)$ and consider the decomposition $H_1 = K_j \oplus K_j^{\perp}$ for each j. We have the isomorphism $B \equiv T_1^{-1} T_0 : K_0^{\perp} \xrightarrow{\approx} K_1^{\perp}$, and since K_0 and K_1 have the same finite dimension, we can choose an isomorphism $A: K_0 \to K_1$. The direct sum $C = A \oplus B: K_0 \oplus K_0^{\perp} \to K_1 \oplus K_1^{\perp}$ is an automorphism of H_1 such that $T_0 = T_1 C$. Suppose we can find a continuous family of isomorphisms $C_t: H_1 \to H_1$, $0 \leq t \leq 1$, with $C_0 = C$ and $C_1 = \text{Id}$. Then $T_t \equiv T_1 C_t$ connects T_0 to T_1. Therefore we are done when we have proved the following:

Lemma 7.7. *The set $\mathscr{L}^{\times} = \mathscr{L}^{\times}(H,H)$ of isomorphisms of Hilbert space is connected.*

Proof. We fix $C \in \mathscr{L}^{\times}$ and show that C can be connected to the identity. To begin we put C in polar form $C = UA$ where A is the positive square root of the positive self-adjoint operator C^*C, and where (since $U^*U = A^{-1}C^*CA^{-1} = A^{-1}A^2A^{-1} = \text{Id}$), U is unitary. Since the positive bounded self-adjoint operators form a convex set, we see that C is homotopic to U. Now U has a spectral decomposition of the form $U = \int_0^{2\pi} e^{i\lambda}\, d\pi_\lambda$ and can be written as $U = e^{iT}$ where $T = \int_0^{2\pi} \lambda\, d\pi_\lambda$. The homotopy $U_t = e^{itT}$, $0 \leq t \leq 1$, completes the proof of Lemma 7.7 and so also of Lemma 7.6 and Proposition 7.1. ■

Note. When $H_1 = H_2 \equiv H$, the composition of operators makes $\pi_0 \mathfrak{F}$ a semigroup. Interestingly, *the map* $\text{ind}: \pi_0 \mathfrak{F} \to \mathbb{Z}$ *is a group isomorphism.* To prove this we need only observe that

$$\text{ind}(S \circ T) = \text{ind}(S) + \text{ind}(T) \tag{7.5}$$

for all $S,T \in \mathfrak{F}$. To begin let $\pi : H \to (\ker S)^{\perp}$ be orthogonal projection and note that since T is homotopic to $\pi \circ T$ in $\mathfrak{F}$ (by a linear homotopy) we can assume that Image$(T) \subset (\ker S)^{\perp}$. The assertion is now easily checked.

Let us now return to the case of elliptic operators over a compact manifold X. A family $P_t : \Gamma(E) \to \Gamma(F)$, $0 \le t \le 1$, of such operators is said to be **continuous** if in any good presentation of the bundles the coefficients of the local representations $P_t = A^{\alpha}(x,t)D^{\alpha}$ are jointly continuous in x and t. Under this hypothesis the order of the operators must be a constant, say m. Furthermore, for any s the map $[0,1] \to \mathfrak{F}(L^2_s(E), L^2_{s-m}(F))$ given by $t \mapsto P_t$, is continuous in the norm topology. Consequently, ind(P_t) is constant in t by Proposition 7.1.

Recall that two elliptic operators $P_0,P_1 : \Gamma(E) \to \Gamma(F)$ over X are **homotopic** if they can be joined by a continuous family P_t, $0 \le t \le 1$, of elliptic operators. Our remarks above can be restated as follows:

Corollary 7.8. *The index of an elliptic operator on a compact manifold depends only on its homotopy class.*

An immediate consequence is this:

Corollary 7.9. *The index of an elliptic operator on a compact manifold depends only on its principal symbol.*

Proof. If $P_0,P_1 : \Gamma(E) \to \Gamma(F)$ have the same principal symbol, then so does each element in the family $P_t = (1-t)P_0 + tP_1$. ∎

This result suggests making deformations of the principal symbol. Recall that the principal symbol of an elliptic operator is a section $\sigma \in \Gamma[(\odot^m TX) \otimes \text{Hom}(E,F)]$ with the property that

$$\sigma_{\xi} : E \to F \text{ is an isomophism for all } \xi \neq 0. \tag{7.6}$$

We shall say that two such symbols σ_0,σ_1 are **regularly homotopic** if there is a homotopy σ_t, $0 \le t \le 1$, joining them such that $\sigma_{t,\xi}$ satisfies (7.6) for all t.

Theorem 7.10. *The index of an elliptic differential operator on a compact manifold depends only on the regular homotopy class of its principal symbol.*

REMARK 7.11. Using Theorem 3.19 and the subsequent discussion there, one can generalize this result to elliptic pseudodifferential operators.

Proof. In light of 7.8 and 7.9 it will suffice, given σ_t, to construct a family of operators P_t with $\sigma(P_t) = \sigma_t$. This is evidently possible locally given coordinates on X and trivializations of E and F. Patching together with a partition of unity over some good presentation of E and F does the job globally. ∎

The question, manifest at this point, is whether the index of an elliptic operator can be computed directly from its principal symbol. A procedure for doing this was first given by Atiyah and Singer. It will be discussed in detail in §13.

§8. The Index of a Family of Elliptic Operators

In this section we shall present the important concept, due to Atiyah and Singer, of an index for families of elliptic operators. In examining the topological invariance of this index we shall prove that the Fredholm operators on Hilbert space constitute a classifying space for K-Theory.

Let A be a Hausdorff topological space, and let E and F be smooth vector bundles over a compact manifold X. The definition given in §7 of a continuous family $P_t:\Gamma(E)\to\Gamma(F)$, $t\in[0,1]$, of elliptic operators, can be extended directly by replacing $[0,1]$ with A. (We shall call this a **product family.**) However, when A is homotopically non-trivial, it is important to allow the manifold and the operators to twist globally over A (in the spirit of the families of complex analytic objects considered by Kodaira and Spencer).

For a smooth vector bundle $E\to X$, let Diff$(E;X)$ be the set of diffeomorphisms of E which carry fibres to fibres linearly. Note the homomorphism $\beta:\text{Diff}(E;X)\to\text{Diff}(X)$ onto the diffeomorphism group of X. Let $\mathscr{D}\equiv\text{Diff}(E,F;X)$ be the subgroup of Diff$(E\oplus F;X)$ which maps E to E and F to F. This group acts naturally on the space $\text{Op}^m(E,F)$ of all differential operators $P:\Gamma(E)\to\Gamma(F)$ of order $\leqq m$ (by setting $g(P)=g_2\circ P\circ g_1^{-1}$ for $g=(g_1,g_2)$).

DEFINITION 8.1. A continuous *family of smooth vector bundles* over X parameterized by the Hausdorff space A is a fibre bundle $\mathscr{E}\to A$ whose fibre is a smooth vector bundle E over X and whose structure group is Diff$(E;X)$.

From the homomorphism $\beta:\text{Diff}(E;X)\to\text{Diff}(X)$ we get an associated fibre bundle $\mathscr{X}\to A$ whose fibre is X and whose structure group is Diff(X). Note that $\mathscr{E}$ is just a vector bundle over the total space of $\mathscr{X}$ which is smooth on each fibre. Furthermore, this smooth structure is changing continuously over A.

DEFINITION 8.2. By a **continuous pair** of vector bundles over X parameterized by A we mean a bundle $\mathscr{E}\oplus\mathscr{F}\to A$ whose fibre is a split bundle $E\oplus F$ over X and whose structure group is $\mathscr{D}=\text{Diff}(E,F;X)$.

Associated to such a pair is the family of differential operators of order $\leqq m$ from E to F. This is the bundle $\text{Op}^m(\mathscr{E},\mathscr{F})\to A$ associated to the principal $\mathscr{D}$-bundle of $\mathscr{E} \oplus \mathscr{F}$ by the action of $\mathscr{D}$ on $\mathrm{Op}^m(E,F)$. A continuous section P of such a bundle whose fibre is elliptic for all $a \in A$ is called a **family of elliptic operators** parameterized by A.

EXAMPLE 8.3. Let X be a compact manifold and consider its associated Jacobian torus $J_X \equiv H^1(X;\mathbb{R})/H^1(X;\mathbb{Z})$. Recall the isomorphisms: $H^1(X;\mathbb{R}) \cong \mathrm{Hom}(\pi_1 X,\mathbb{R})$ and $H^1(X;\mathbb{Z}) \cong \mathrm{Hom}(\pi_1 X;\mathbb{Z})$. Let $\tilde{X}$ denote the universal covering of X. Then we define a flat complex line bundle L over $X \times J_X$ by taking the quotient

$$L \equiv \tilde{X} \times H^1(X;\mathbb{R}) \times \mathbb{C}/\pi_1(X) \times H^1(X;\mathbb{Z}) \tag{8.1}$$

where the action of $\pi_1 X \times H^1(X;\mathbb{Z})$ is given by

$$\varphi_{(g,h)}(\tilde{x},v,z) \equiv (\tilde{g}x, v+h, e^{2\pi i v(g)}z). \tag{8.2}$$

We think of L as a family of flat line bundles on X parameterized by J_X.

Suppose now that X is even-dimensional and oriented, and let S be any Dirac bundle over X with associated Dirac operator D. (For example, one could choose the Clifford bundle or the spinor bundle for some riemannian metric on X.) Then $S \otimes L = (S^+ \otimes L) \oplus (S^- \otimes L)$ is a continuous pair of vector bundles on X parameterized by J_X. Using the given flat connection on L, the Dirac operator on S extends to a family of Dirac operators

$$D_v^+ : \Gamma(S^+ \otimes L_v) \longrightarrow \Gamma(S^- \otimes L_v) \tag{8.3}$$

for $v \in J_X$. For $S = C\ell(X)$, this family was introduced by G. Lusztig [1] is his proof of the Novikov Conjecture for $\pi_1 \cong \mathbb{Z}^m$.

Suppose now that P is a family of elliptic operators defined over a compact Hausdorff space A. Following Atiyah and Singer [3] we shall define an analytic index $\mathrm{ind}(P)$ in the group $K(A)$. If the dimension of $\ker P_a$ (and therefore also of $\mathrm{coker}\, P_a$) were locally constant on A, then this index would simply be the formal difference of finite dimensional vector bundles:

$$\mathrm{ind}\, P = [\ker P] - [\mathrm{coker}\, P] \in K(A). \tag{8.4}$$

In general these dimensions are not constant, and so we must stabilize the picture. Let $\mathscr{E}$, $\mathscr{F}$ and $\mathscr{X}$ be as above. For $a \in A$, let $\mathscr{E}_a$ (and $\mathscr{F}_a$) denote the fibre of $\mathscr{E}$ (and $\mathscr{F}$ respectively) at a. This is a smooth vector bundle on X and we denote by $\Gamma(\mathscr{E}_a)$ its space of smooth cross-sections.

Lemma 8.4. *There exists a finite set of sections $\{w_1, \dots, w_N\}$ of $\mathscr{F}$ over $\mathscr{X}$ such that for each $a \in A$ the map $\tilde{P}_a : \mathbb{C}^N \oplus \Gamma(\mathscr{E}_a) \to \Gamma(\mathscr{F}_a)$ given by $\tilde{P}_a(t_1, \dots, t_N, \varphi) = \sum t_j w_j|_{\mathscr{X}_a} + P_a(\varphi)$, is surjective for all a. The vector spaces* ker $\tilde{P}_a$ *form a vector bundle over* A, *and the element*

$$[\ker \tilde{P}] - [\mathbb{C}^N] \in K(A) \tag{8.5}$$

depends only on the operator P.

Proof. From the local triviality of fibrations each point $a_0 \in A$ has a neighborhood U in which P_a, $a \in U$, is a product family. For any s, this family extends to a continuous map $P: U \to \mathfrak{F}(L_s^2(E), L_{s-m}^2(F))$, and we can apply Lemma 7.2. For $W = \ker(P_{a_0}^*)$ this lemma shows that the maps $\tilde{P}_a: W \oplus L_s^2(E) \to L_{s-m}^2(F)$ given by $\tilde{P}_a(w,\varphi) = w + P_a(\varphi)$ are all surjective in a neighborhood U' *of* a_0. Theorem 5.2(i) implies that $W \subset \Gamma(F)$ and that the restriction $\tilde{P}_a: W \oplus \Gamma(E) \to \Gamma(F)$ is also surjective for all $a \in U'$. Corollary 7.4 shows that dim(ker $\tilde{P}_a$) is locally constant on U'.

We now globalize this construction. Let $\{w_1, \ldots, w_r\}$ be a basis for W. Each w_j can be considered as a (constant) section of $\mathscr{F}$ over U', or alternatively as a section of the vector bundle $\mathscr{F}$ over the open set $\pi^{-1}(U') \subset \mathscr{X}$ (where $\pi: \mathscr{X} \to A$ is the fibration). Clearly each w_j can be extended to a global section w_j of $\mathscr{F}$ over all of $\mathscr{X}$. Taking a finite covering of A by such neighborhoods, and taking $\{w_1, \ldots, w_N\}$ to be the union of the sections constructed from each neighborhood, we establish the first statement of the lemma. The local constancy of dim(ker P_a) follows by repeating the argument above with W replaced by the direct sum of W's from each neighborhood.

It remains to prove that the class (8.5) is independent of the choice of global sections $w_1, \ldots, w_N$. Let $w'_1, \ldots, w'_{N'}$ be another such choice, and consider the family of maps $\tilde{\tilde{P}}_a: \mathbb{C}^N \oplus \mathbb{C}^{N'} \oplus \Gamma(\mathscr{E}_a) \to \Gamma(\mathscr{F}_a)$ given by $\tilde{\tilde{P}}_a(t,t',\varphi) \equiv \sum t_i w_i(a) + \sum t'_j w'_j(a) + P_a(\varphi)$. We have a commutative diagram

$$\begin{array}{ccccc}
\ker \tilde{P}_a \subset & \mathbb{C}^N \oplus \Gamma(\mathscr{E}_a) & & \xrightarrow{P_a} & \Gamma(\mathscr{F}_a) \\
\cap & \cap & & & \| \\
\ker \tilde{\tilde{P}}_a \subset & \mathbb{C}^N \oplus \mathbb{C}^{N'} \oplus \Gamma(\mathscr{E}_a) & & \xrightarrow{P_a} & \Gamma(\mathscr{F}_a) \\
\downarrow & \downarrow \text{pr} & & & \\
\mathbb{C}^{N'} & = \quad \mathbb{C}^{N'} & & &
\end{array}$$

where the map pr denotes projection. Since the map $\tilde{P}_a$ is surjective one easily sees that the sequence

$$0 \longrightarrow \ker \tilde{P}_a \longrightarrow \ker \tilde{\tilde{P}}_a \longrightarrow \mathbb{C}^{N'} \longrightarrow 0$$

is exact. It follows that $[\ker \tilde{\tilde{P}}_a] - [\mathbb{C}^{N+N'}] = [\ker \tilde{P}_a] - [\mathbb{C}^N]$ in $K(A)$. The argument applies symmetrically to the operators $\tilde{P}'_a: \mathbb{C}^{N'} \oplus \Gamma(\mathscr{E}_a) \to \Gamma(\mathscr{F}_a)$ to prove that $[\ker \tilde{\tilde{P}}_a] - [\mathbb{C}^{N+N'}] = [\ker \tilde{P}'_a] - [\mathbb{C}^{N'}]$, and the proof is complete. ∎

We are now authorized to make the following definition:

DEFINITION 8.5. The **analytic index** of a family of elliptic operators P over a compact Hausdorff space A is the element $\mathrm{ind}(P) \in K(A)$ defined by (8.5).

As one might imagine, this index enjoys invariance properties analogous to those of the ordinary index. It depends only on the principal symbol of the family, and in fact only on the homotopy class of the principal symbol (all taken in the obvious sense; see Atiyah-Singer [3] for details).

This invariance is related to a basic and interesting fact. Let H be a (infinite-dimensional) separable Hilbert space. Recall that any two such spaces are isomorphic. Furthermore, there is **Kuiper's Theorem** which states that the group $\mathscr{L}^\times = \mathscr{L}^\times(H,H)$ of linear isomorphisms of H with the norm topology is contractible. Consider a family of elliptic operators P over A as above. This is a section of the bundle $\mathrm{Op}^m(\mathscr{E};\mathscr{F})$ with structure group $\mathscr{D}$. We can fix s and complete each fibre in the Sobolev norms to get a bundle with fibre $H \cong \mathfrak{F}(L_s^2(E), L_{s-m}^2(F))$. This is the bundle associated to the principle $\mathscr{D}$-bundle by the homomorphism $\mathscr{D} \to \mathscr{L}^\times(H,H)$. Since $\mathscr{L}^\times(H,H)$ is contractible, this bundle is trivial as an $\mathscr{L}^\times(H,H)$-bundle in a homotopically unique fashion. Under the trivialization, the family P becomes a continuous map $P: A \to H = \mathfrak{F}(H_1, H_2)$ where $H_1 = L_s^2(E)$ and $H_2 = L_{s-m}^2(F)$.

Note that the set $\mathfrak{F} = \mathfrak{F}(H,H)$ of Fredholm operators on a Hilbert space H has a continuous associative semi-group structure given by the composition $\mathfrak{F} \times \mathfrak{F} \to \mathfrak{F}$. For any topological space A, this makes the space $[A, \mathfrak{F}]$, of homotopy classes of maps of A into $\mathfrak{F}$, into an associative semigroup.

Theorem 8.6. (cf. Atiyah [4]). *For any compact Hausdorff space A there is a natural isomorphism*

$$\mathrm{ind}: [A, \mathfrak{F}] \longrightarrow K(A) \tag{8.6}$$

It has the property that for any continuous map $f: A' \to A$ between such spaces,

$$\mathrm{ind} \circ f^* = f^* \circ \mathrm{ind} \tag{8.7}$$

Consequently, $\mathfrak{F}$ is a classifying space for K-Theory.

This theorem completely generalizes Proposition 7.1 (where $A = [0,1]$) and is the foundation for proving the homotopy invariance of the index for families.

Proof. Let $T: A \to \mathfrak{F}$ be a continuous map. By Lemma 7.2 and the compactness of A we know that there is a closed subspace $V \subset H$ of finite codimension so that for all $a \in A$

(i) $V \cap (\ker T_a) = \{0\}$,
(ii) $T_a V$ is closed and of finite codimension in H, and
(iii) $H/TV \equiv \bigcup_{a \in A} (H/T_a V)$ is a vector bundle over A.

One then defines

$$\operatorname{ind}(T) \equiv [H/V] - [H/TV] \in K(A), \tag{8.8}$$

where $[H/V] = [A \times (H/V)]$ denotes the trivial bundle. We must show that this element is independent of the choice of V. Let V' be another such choice. Since $V \cap V'$ is also a choice, we may assume $V' \subset V$. There is then an exact sequences of vector bundles: $0 \to V/V' \to H/TV' \to H/TV \to 0$, which shows that $[H/TV] - [H/TV'] = [V/V'] = [H/V] - [H'V']$. Hence, (8.8) depends only on T.

Given $T: A \to \mathfrak{F}$ and a map $f: A' \to A$, the subspace V chosen for T also works for $T \circ f$ and we evidently have $\operatorname{ind}(T \circ f) = f^*(\operatorname{ind} T)$.

Suppose now that $T: A \times I \to \mathfrak{F}$ is a homotopy between T_0 and T_1 where $T_j = T \circ i_j$, and $i_j: A \to A \times I$ is the inclusion $i_j(a) = (a,j)$. By the above paragraph $\operatorname{ind}(T_j) = i_j^*(\operatorname{ind} T)$ for $j = 0,1$, and it is a basic fact that $i_0^* = i_1^*$ in K-theory. This proves the homotopy invariance of ind.

It remains to show that (8.6) is an isomorphism. We show first that it is a homomorphism. Let $T: A \to \mathfrak{F}$ and $T': A \to \mathfrak{F}$ be continuous maps and choose $V \subset H$ for T as above. Note that T' is homotopic to $\mathrm{pr}_v \circ T'$ where $\mathrm{pr}_v: H \to V$ is orthogonal projection. Hence, we may assume $T'H \subseteq V$. Let V' be a choice of subspace for T', and note that V' is also a choice for the composition $T \circ T'$. Therefore we have

$$\operatorname{ind}(T \circ T') = [H/V'] - [H/TT'V']$$
$$\operatorname{ind}(T) = [H/V] - [H/TV]$$
$$\operatorname{ind}(T') = [H/V'] - [H/T'V'].$$

From the exact sequence: $0 \to V/T'V' \xrightarrow{T} H/TT'V' \to H/TV \to 0$ of vector bundles over A, we see that $[H/TT'V'] = [H/TV] + [V/T'V'] = [H/TV] + [H/T'V'] - [H/V]$. Plugging this into the equation above shows that $\operatorname{ind}(T \circ T') = \operatorname{ind}(T) + \operatorname{ind}(T')$ as required.

To prove that ind is surjective we first recall from Chapter I, §9, that every element in $K(A)$ can be written as $[\mathbb{C}^k] - [E]$ where E is a vector bundle on A and $\mathbb{C}^k$ denotes the trivial k-plane bundle. For each integer k we define an operator $S_k \in \mathfrak{F}$ of index k by fixing a complete orthonormal basis $\{e_j\}_{j=1}^{\infty}$ of H and setting

$$S_k(e_j) = \begin{cases} e_{j-k} & \text{if } j - k > 0 \\ 0 & \text{otherwise.} \end{cases}$$

The constant map $T \equiv S_k$ on A has $\operatorname{ind} T = [\mathbb{C}^k]$ for $k \geqq 0$.

Fix a vector bundle E over A and recall from Chapter I, §9, that for some N there is a continuous map $f: E \to \mathbb{C}^N$ which is a linear injection on each fibre. Let $\mathrm{pr}_a: \mathbb{C}^N \to \mathbb{C}^N$ denote orthogonal projection onto $f(E_a)$ and let $\mathrm{pr}_a^\perp = \mathrm{Id} - \mathrm{pr}_a$ denote projection onto the orthogonal complement. Then the map $T_E: A \to \mathfrak{F}(\mathbb{C}^N \otimes H, \mathbb{C}^N \otimes H)$ given by setting $T_E(a) \equiv \mathrm{pr}_a \otimes S_{-1} + \mathrm{pr}_a^\perp \otimes \mathrm{Id}$ has the property that $\mathrm{ind}\, T_E = -[E]$. Choosing an isomorphism $\mathbb{C}^N \otimes H \cong H$ yields an isomorphism $\mathfrak{F}(\mathbb{C}^N \otimes H, \mathbb{C}^N \otimes H) \cong \mathfrak{F}$. Since $\mathrm{ind}(S_k \circ T_E) = [\mathbb{C}^k] - [E]$, we have proved that $\mathrm{ind}: [A, \mathfrak{F}] \to K(A)$ is a surjective homomorphism.

Our next step is the following. Let $\mathfrak{F}^\times$ denote the invertible elements in $\mathfrak{F}$.

Lemma 8.7. *If $T: A \to \mathfrak{F}$ has index zero, then T is homotopic to a map $T': A \to \mathfrak{F}^\times \subset \mathfrak{F}$.*

Proof. Choose a subspace $V \subset H$ as above so that $\mathrm{ind}\, T = [H/V] - [H/TV]$. The hypothesis $\mathrm{ind}\, T = 0$ means that for some k, we have $A \times [(H/V) \oplus \mathbb{C}^k] \cong (H/TV) \oplus \mathbb{C}^k$. This implies that if we replace V by a closed subspace of codimension k in V, we have the bundle isomorphism

$$A \times (H/V) \cong H/TV.$$

It is elementary to verify that there is a continuous map $H/TV \to H$ which carries H/T_aV isomorphically onto $(T_aV)^\perp$ for each $a \in A$. Combining with the isomorphism above, we get a linear map $T^\perp: A \to \mathscr{L}(H/V, H)$ where for each a, $T_a^\perp$ is a linear injection of H/V onto $(T_aV)^\perp$. The direct sum

$$T_a^\times \equiv T_a^\perp \oplus T_a: (H/V) \oplus V \longrightarrow H$$

defines a continuous map $T^\times: A \to \mathfrak{F}^\times \subset \mathfrak{F}$. This map can be connected to T in $\mathfrak{F}$ by the homotopy $T_t = tT^\perp \oplus T$ for $0 \le t \le 1$. This proves the lemma. ∎

Using the theorem of Kuiper [1] that $\mathfrak{F}^\times$ is contractible, we conclude that any $T: A \to \mathfrak{F}$ of index zero is homotopic to the constant map $T \equiv \mathrm{Id}$. Hence ind is injective and Theorem 8.6 is proved. ∎

One of the important results of Atiyah and Singer is the establishment of a topological formula for the index of a family P of elliptic operators. We shall present this formula in §15.

As a final remark we point out that the arguments given above adapt easily to prove the following real analogue of Theorem 8.6. Let $\mathfrak{F}_{\mathbb{R}} = \mathfrak{F}_{\mathbb{R}}(H_{\mathbb{R}})$ denote the space of (real) Fredholm operators acting on real Hilbert space $H_{\mathbb{R}}$.

Theorem 8.8. *For any compact Hausdorff space A there is a natural isomorphism*

$$\mathrm{ind} : [A, \mathfrak{F}_{\mathbb{R}}] \xrightarrow{\approx} KO(A) \tag{8.9}$$

having the functorial property (8.7) *above. Consequently,* $\mathfrak{F}_{\mathbb{R}}$ *is a classifying space for KO-theory.*

An important consequence of 8.6 and 8.8 is the following:

Corollary 8.9. *For all* $k \geqq 0$, *there are isomorphisms:*

$$\pi_k(\mathfrak{F}) \xrightarrow{\approx} \tilde{K}(S^k) \equiv K^{-k}(\mathrm{pt}) \tag{8.10}$$

$$\pi_k(\mathfrak{F}_{\mathbb{R}}) \xrightarrow{\approx} \widetilde{KO}(S^k) \equiv KO^{-k}(\mathrm{pt}). \tag{8.11}$$

§9. The G-Index

In this section we shall study operators which are preserved by the action of a compact Lie group G. To be specific we fix vector bundles E and F over a compact manifold X together with an action of G on the triple (X,E,F). This means a smooth action $\mu: G \times X \to X$ of G on X together with smooth actions of G on both E and F which carry fibres to fibres linearly and which project to μ. In this context we have the following notion:

DEFINITION 9.1. A differential operator $P: \Gamma(E) \to \Gamma(F)$ is called a **G-operator** if $P(g\varphi) = gP(\varphi)$ for all $g \in G$ and all $\varphi \in \Gamma(E)$.

A good example is given by the isometry group G_X of X acting on $C\ell(X) \cong \Lambda^* X$. This action preserves both splittings $C\ell(X) = C\ell^0(X) \oplus C\ell^1(X)$ and $C\ell(X) = C\ell^+(X) \oplus C\ell^-(X)$ and commutes with the Dirac operator. Therefore both D^0 and D^+ are G_X-operators.

Similarly, if X is spin, then either G_X or a two-fold covering of G_X acts on the spinor bundle of X and commutes with the Atiyah-Singer operator.

The basic observation here is that *if P is elliptic and a G-operator, then* ker *P and* coker *P are finite-dimensional representation spaces for G.* This leads us to consider the **representation ring** (or the **ring of virtual representations**) $R(G)$ of G. This can be defined as the free abelian group generated by the equivalence classes of irreducible finite-dimensional complex representations of G. Since every finite-dimensional representation of G can be decomposed uniquely, up to equivalence, into a direct sum of irreducible ones, $R(G)$ can also be defined as the Grothendeick group of all finite-dimensional representations. In other words, each element of $R(G)$ can be expressed as a formal difference $[V] - [W]$, where $[V]$ and $[W]$

are equivalence classes of finite-dimensional representations of G, and where $[V] - [W] \doteq [V'] - [W']$ if and only if $V \oplus W'$ is equivalent to $V' \oplus W$.

The tensor product of representations is distributive over the direct sum operation and makes $R(G)$ into a commutative ring.

EXAMPLE 9.2. Let $G = S^1$ and let t^m stand for the 1-dimensional representation: $\varphi_m(\sigma)z = e^{im\sigma}z$. Then $R(S^1)$ is easily seen to be the ring of Laurent polynomials in $t: R(S^1) \cong \mathbb{Z}[t,t^{-1}]$.

DEFINITION 9.3. Let P be an elliptic G-operator on a compact manifold. Then the **G-index** of P is the element

$$\operatorname{ind}_G(P) = [\ker P] - [\operatorname{coker} P] \in R(G).$$

Note that when $G = \{1\}$, $R(G) \cong \mathbb{Z}$ and we recover the usual index of P.

The G-index can be specialized to individual elements of G. Given a complex representation $\rho: G \to \mathrm{GL}(V)$ we define its associated **character** to be the function $\chi_\rho(g) = \operatorname{trace}(\rho(g))$. This function completely determines the representation up to equivalence, and it has the properties that $\chi_{\rho_1+\rho_2} = \chi_{\rho_1} + \chi_{\rho_2}$, $\chi_{\rho_1 \otimes \rho_2} = \chi_{\rho_1}\chi_{\rho_2}$, etc. (see Adams [1] for details). For this reason $R(G)$ is alternatively called the **character ring** of G. Given a G-operator P as above and given $g \in G$, we can define

$$\operatorname{ind}_g(P) = \operatorname{trace}(g|_{\ker P}) - \operatorname{trace}(g|_{\operatorname{coker} P}). \tag{9.1}$$

This is the difference of the characters of the two representations, $\ker P$ and $\operatorname{coker} P$, evaluated at g. Consequently ind_g is the specialization of ind_G to g as claimed.

Note that for $D \equiv d + d^* : \Lambda^{\text{even}}(X) \to \Lambda^{\text{odd}}(X)$, and for $g \in \operatorname{Isom}(X)$, the number $\operatorname{ind}_g(D)$ is just the classical **Lefschetz number** of g. The general topological formulas for ind_G given by Atiyah, Singer, Bott and Segal represent generalizations of fundamental work of Lefschetz and Hopf.

Our object at the moment is to establish some elementary stability properties of ind_G. As before, the regularity theory of elliptic operators implies that the spaces $\ker P$ and $\ker P^* \cong \operatorname{coker} P$ remain unchanged if we pass to any Sobolev completion $P: L^2_s(E) \to L^2_{s-m}(F)$. Since G is compact, we may choose metrics on X, E and F which are G-invariant. The actions of G on $\Gamma(E)$ and $\Gamma(F)$ then extend to unitary representations on $L^2_s(E)$ and $L^2_s(F)$ which commute with P and P^*. This leads us to examine the following concept.

Let H_1 and H_2 be separable complex Hilbert spaces equipped with unitary representations $G \to U(H_j)$, $j = 1,2$, of a compact Lie group G. Let $\mathfrak{F}_G = \mathfrak{F}_G(H_1, H_2)$ denote the space of all G-equivariant Fredholm maps, i.e., all Fredholm maps $T: H_1 \to H_2$ such that $Tgv = gTv$ for all $g \in G$

and all $v \in H_1$. To each $T \in \mathfrak{F}_G$ we associate the index

$$\operatorname{ind}_G(T) \equiv [\ker T] - [\operatorname{coker} T] \in R(G)$$

Proposition 9.4. *The map* $\operatorname{ind}_G : \mathfrak{F}_G \to R(G)$ *is constant on connected components of* $\mathfrak{F}_G$. *Furthermore, if* H_1 *and* H_2 *are G-isomorphic, then* ind_G *induces an injection*

$$\operatorname{ind}_G : \pi_0(\mathfrak{F}_G) \hookrightarrow R(G). \tag{9.2}$$

Note. The map (9.2) will be a bijection provided that each finite-dimensional irreducible representation of G occurs with infinite multiplicity in $H_1 = H_2$. The map (9.2) is also an additive homomorphism since $\operatorname{ind}_G(T \circ S) = \operatorname{ind}_G(T) + \operatorname{ind}_G(S)$. This is proved exactly as above (see (7.5)).

Proof. The argument for the first statement is identical with the one given for the first part of Proposition 7.1 above. It is only necessary to check that all subspaces are G-invariant and that all maps (such as (7.2), (7.3) and (7.4)) commute with G.

We now assume that $H_1 = H_2 = H$ (as G-spaces). To prove the second statement we must show that any two elements $T_0, T_1 \in \mathfrak{F}_G$ with the same G-index must lie in the same connected component of $\mathfrak{F}_G$. We begin by observing that any $T \in \mathfrak{F}_G$ can be deformed to one which is "relatively prime," that is, one for which no non-zero subspace of ker T is G-isomorphic to a subspace of coker $T = (\operatorname{Im} T)^\perp$. Indeed, if such a G-isomorphism L exists, it can be extended to a map $L : H \to H$ by defining it to be zero on the complementary subspace. The family $T + tL$, $t \geqq 0$, gives the desired deformation.

We may assume now that T_0 and T_1 satisfy this "relatively prime" condition. Consequently, the hypothesis $\operatorname{ind}_G T_0 = \operatorname{ind}_G T_1$ implies that there are G-isomorphisms: $\ker T_0 \cong \ker T_1$ and $\operatorname{coker} T_0 \cong \operatorname{coker} T_1$. We now observe that given two finite dimensional, G-invariant subspaces $V_0, V_1 \subset H$ which are G-isomorphic, there exists a G-isomorphism $C : H \to H$ with $C(V_0) = V_1$. In fact, C may be taken to be the identity on $(V_0 + V_1)^\perp$. The subspace $V_0 \cap V_1$ is G-invariant and its complements in V_0 and V_1 respectively are G-isomorphic; so the existence of C is clear.

It now follows that there exists a G-isomorphism $C : H \to H$ which carries $\operatorname{Im}(T_0)$ isomorphically onto $\operatorname{Im}(T_1)$. We define a second G-isomorphism $C' : H \to H$ by taking the given isomorphism $\ker T_0 \xrightarrow{\approx} \ker T_1$ and extending by the map $T_1^{-1} C T_0 : (\ker T_0)^\perp \xrightarrow{\approx} (\ker T_1)^\perp$. We find that $T_1 = C T_0 (C')^{-1}$, and the existence of a homotopy from T_0 to T_1 is an immediate consequence of the following:

Lemma 9.5. *The set* $\mathscr{L}_G^\times(H, H)$ *of G-isomorphisms of H is connected.*

Proof. The argument here is identical to the one given for Lemma 7.7 above. One needs only to check that the entire construction is G-equivariant. ■

Arguing as in §7 we see that Proposition 9.4 has the following immediate consequence. Let $P:\Gamma(E) \to \Gamma(F)$ be an elliptic G-operator over a compact manifold X.

Corollary 9.6. *The G-index of P depends only on the homotopy class of P in the space of elliptic G-operators. In particular,* $\mathrm{ind}_G(P)$ *depends only on the principal symbol of P.*

Note that the action of G on X, E and F makes $\Sigma \equiv (\odot^m TX) \otimes \mathrm{Hom}(E,F)$ into a G-bundle, i.e., a bundle with a smooth G-action which maps fibres to fibres linearly (and isomorphically). The principal symbol of P is an **invariant** section of Σ, i.e., $\sigma(P) \in \Gamma_G(\Sigma)$ where $\Gamma_G(\Sigma) \equiv \{\sigma \in \Gamma(\Sigma): g\sigma = \sigma$ for all $g \in G\}$. Two principal symbols of elliptic G-operators, $\sigma_0 = \sigma(P_0)$ and $\sigma_1 = \sigma(P_1)$, are said to be **regularly G-homotopic** if there is a regular homotopy σ_t, $0 \le t \le 1$, joining them (see (7.5) forward), such that $\sigma_t \in \Gamma_G(\Sigma)$ for all t.

Theorem 9.7. *The G-index of an elliptic G-operator on a compact manifold depends only on the regular G-homotopy class of its principal symbol.*

Proof. As in the proof of Theorem 7.10 we construct a family of elliptic operators P_t with $\sigma(P_t) = \sigma_t$ for $0 \le t \le 1$. Averaging over G, by integrating

$$\tilde{P}_t\varphi \equiv \int_G (gP_tg^{-1}\varphi)\,dg = \left\{\int_G g(P_t)\,dg\right\}\varphi$$

with respect to Haar measure on G, produces a homotopy of G-operators with $\sigma(\tilde{P}_t) = \sigma_t$ for all t (since σ_t is G-invariant). The result now follows from Corollary 9.6. ■

§10. The Clifford Index

In this section we shall discuss, in general terms, elliptic operators which are $C\ell_k$-linear for some k. Several important examples of such operators have been introduced and discussed in detail in Chapter II, §7. Motivated by these $C\ell_k$-Dirac operators, we make the following definitions.

Definition 10.7. By a **$C\ell_k$-bundle** on a space X we mean a bundle of real, left $C\ell_k$-modules. This is a real vector bundle E over X together with a continuous map $\Psi: C\ell_k \times E \to E$ such that $\Psi_\varphi(\cdot) \equiv \Psi(\varphi,\cdot): E \to E$ is a bundle endomorphism for all $\varphi \in C\ell_k$, and the restriction $C\ell_k \times E_x \to E_x$ makes the fibre into a $C\ell_k$-module for each $x \in X$.

Note that $C\ell_1$-bundles and $C\ell_2$-bundles are simply complex and quaternion bundles respectively. In general, a $C\ell_k$-bundle is thought of as having the algebra $C\ell_k$ as "scalars."

REMARK 10.2. A $C\ell_k$-bundle E will be called **riemannian** if it carries a bundle metric which is preserved under multiplication by each unit vector $e \in \mathbb{R}^k \subset C\ell_k$. Starting with any metric and averaging over the Clifford group, as in I.5.16, makes any $C\ell_k$-bundle riemannian. Note that if E is riemannian, then multiplication by any element $w \in \mathbb{R}^k \subset C\ell_k$ is fibre-wise skew-adjoint. Thus

$$\langle w \cdot u_1, u_2 \rangle + \langle u_1, w \cdot u_2 \rangle \equiv 0 \tag{10.1}$$

for all $u_1, u_2 \in \Gamma(E)$.

Let us fix smooth $C\ell_k$-bundles E and F over a manifold X.

DEFINITION 10.3. A differential operator $P: \Gamma(E) \to \Gamma(F)$ is said to be **$C\ell_k$-linear** (or simply a **$C\ell_k$-operator**) if $\varphi P(u) = P(\varphi u)$ for all $\varphi \in C\ell_k$ and all $u \in \Gamma(E)$.

Tensoring with $C\ell_k$ makes any differential operator trivially into one which is $C\ell_k$-linear. More interesting examples, such as the $C\ell_k$-Dirac operators of II.7, occur when there exists a $\mathbb{Z}_2$-grading. We say that a $C\ell_k$-bundle E is **$\mathbb{Z}_2$-graded** if there is a bundle decomposition $E = E^0 \oplus E^1$ making each fibre into a $\mathbb{Z}_2$-graded $C\ell_k$-module. A $C\ell_k$-linear differential operator $P: \Gamma(E) \to \Gamma(E)$ on such a bundle is called **$\mathbb{Z}_2$-graded** (or simply **graded**) if with respect to the decomposition $E = E^0 \oplus E^1$, it has the form

$$P = \begin{pmatrix} 0 & P^1 \\ P^0 & 0 \end{pmatrix}. \tag{10.2}$$

Note that $P^0: \Gamma(E^0) \to \Gamma(E^1)$ is $C\ell_k^0 \cong C\ell_{k-1}$-linear. When E and X are riemannian, the operator P is (formally) self-adjoint if and only if $P^1 = (P^0)^*$.

We would like to define the analytic index of an elliptic operator of this type. For this we recall the groups

$$\mathfrak{M}_{k-1}/i^*\mathfrak{M}_k \cong \hat{\mathfrak{M}}_k/i^*\hat{\mathfrak{M}}_{k+1} \cong KO^{-k}(\mathrm{pt}) \tag{10.3}$$

where $\mathfrak{M}_k$ and $\hat{\mathfrak{M}}_k$ denote respectively the Grothendieck groups of equivalence classes of $C\ell_k$ and $\mathbb{Z}_2$-graded $C\ell_k$ modules. (For detailed discussions of these, see I.5.20, I.9 and II.7.) Recall also that $KO^{-k}(\mathrm{pt}) \cong KO^{-k+8}(\mathrm{pt})$ for all k, and we have

k	1	2	3	4	5	6	7	8
KO^{-k}	$\mathbb{Z}_2$	$\mathbb{Z}_2$	0	$\mathbb{Z}$	0	0	0	$\mathbb{Z}$

Suppose now that $P:\Gamma(E) \to \Gamma(E)$ is an elliptic self-adjoint graded $C\ell_k$-operator on a compact riemannian manifold X. Then $\ker P$ is a finite-dimensional $\mathbb{Z}_2$-graded $C\ell_k$-module, and we have the following:

DEFINITION 10.4. The analytic **Clifford index of P** is the residue class

$$\operatorname{ind}_k P \cong [\ker P] \in \hat{\mathfrak{M}}_k/i^*\hat{\mathfrak{M}}_{k+1} \cong KO^{-k}(\mathrm{pt}). \tag{10.4}$$

Under the isomorphism (10.3) this element corresponds to the residue class

$$\operatorname{ind}_k P \cong [\ker P^0] \in \mathfrak{M}_{k-1}/i^*\mathfrak{M}_k \tag{10.5}$$

where P^0 is given in (10.2). (The equivalence of these definitions is discussed in detail in Chapter II. See II.7.4.)

The category of self-adjoint graded $C\ell_k$-linear operators is a natural and important one as we have seen by examples in Chapter II. However, there is a twin category which has some advantages when studying stability properties of the Clifford index. This is the category of graded operators which are skew-adjoint and $C\ell_k$-antilinear. A differential operator $P:\Gamma(E) \to \Gamma(F)$ between $C\ell_k$-bundles is said to be **$C\ell_k$-antilinear** if $P(wu) = -wP(u)$ for all $w \in \mathbb{R}^k \subset C\ell_k$ and all $u \in \Gamma(E)$. This is equivalent to the requirement that

$$P(\varphi u) = \alpha(\varphi)P(u) \tag{10.6}$$

for $\varphi \in C\ell_k$ and $u \in \Gamma(E)$, where α denotes the involution of $C\ell_k$ engendered by $w \to -w$.

Suppose now that $E = E^0 \oplus E^1$ is a riemannian $\mathbb{Z}_2$-graded $C\ell_k$-bundle over a compact riemannian manifold X, and let $\tilde{P}:\Gamma(E) \to \Gamma(E)$ be a $\mathbb{Z}_2$-graded elliptic differential operator which is $C\ell_k$-antilinear and formally skew-adjoint. With $\tilde{P}$ written as in (10.2) above, this means that $\tilde{P}^1 = -(\tilde{P}^0)^*$. We define the **analytic Clifford index** of such an operator to be

$$\operatorname{ind}_k \tilde{P} \equiv [\ker \tilde{P}] \in \hat{\mathfrak{M}}_k/i^*\hat{\mathfrak{M}}_{k+1} \cong KO^{-k}(\mathrm{pt}) \tag{10.7}$$

and note as before that this is equivalent to taking $[\ker \tilde{P}^0]$ in $\mathfrak{M}_{k-1}/i^*\mathfrak{M}_k$.

Observation 10.5. *There is a natural transformation between graded elliptic differential operators which are formally self-adjoint and $C\ell_k$-linear, and those which are formally skew-adjoint and $C\ell_k$-antilinear. It is given by associating to*

$$P = \begin{pmatrix} 0 & P^1 \\ P^0 & 0 \end{pmatrix}$$

the operator

$$\tilde{P} = \begin{pmatrix} 0 & -P^1 \\ P^0 & 0 \end{pmatrix}.$$

Evidently, $\operatorname{ind}_k P = \operatorname{ind}_k \tilde{P}$.

We shall now explore the topological invariance of these indices. To begin we recall from 5.9 above that the elliptic self-adjoint operator P above can be diagonalized on $L^2(E)$ with finite-dimensional eigenspace V_j and discrete eigenvalues λ_j with $\lim_j |\lambda_j| = \infty$. Thus, P can be written as $P = \sum \lambda_j \pi_j$ where $\pi_j : L^2(E) \to V_j$ denotes orthogonal projection. In these terms the (essentially) positive operator P^2 is written as $P^2 = \sum \lambda_j^2 \pi_j$. Note that P^2 is $C\ell_k$-linear and preserves the factors of the splitting $L^2(E) = L^2(E^0) \oplus L^2(E^1)$. We shall now define an associated $C\ell_k$-linear operator on $L^2(E)$ which preserves these factors by the formula

$$(1 + P^2)^{-\frac{1}{2}} \equiv \sum (1 + \lambda_j^2)^{-\frac{1}{2}} \pi_j.$$

The operator

$$\mathbf{P} \equiv (1 + P^2)^{-\frac{1}{2}} P \tag{10.8}$$

is a $\mathbb{Z}_2$-graded $C\ell_k$-linear self-adjoint Fredholm operator on $L^2(E) = L^2(E^0) \oplus L^2(E^1)$. We leave as an exercise the verification that this construction associates to a continuous family P_t, $0 \le t \le 1$, of such elliptic operators, a continuous family $\mathbf{P}_t$, $0 \le t \le 1$, of Fredholm operators.

The reader will note that if $\tilde{P}$ is the companion of P in the sense of 10.5, then the operator $\tilde{\mathbf{P}} = (1 + P^2)^{-\frac{1}{2}} \tilde{P} = (1 - \tilde{P}^2)^{-\frac{1}{2}} \tilde{P}$ is a $\mathbb{Z}_2$-graded Fredholm operator which is $C\ell_k$-antilinear and skew-adjoint. These considerations motivate the following definitions.

Let $H = H^0 \oplus H^1$ be an infinite-dimensional separable real Hilbert space which is a graded module for the algebra $C\ell_k$. Assume in addition that for each unit vector $e \in \mathbb{R}^k \subset C\ell_k$, the corresponding map $e : H \to H$ is a skew-adjoint isometry. Such a space will be called a **graded Hilbert module** for $C\ell_k$. Examples are easily constructed by taking the tensor product $H = H' \otimes C\ell_k = (H' \otimes C\ell_k^0) \oplus (H' \otimes C\ell_k^1)$ for some real Hilbert space H'.

Let $\mathfrak{F}_k \subset \mathfrak{F}_{\mathbb{R}}(H, H)$ denote the subset of those Fredholm operators $T : H \to H$ which are $\mathbb{Z}_2$-graded (i.e., $T(H^0) \subseteq H^1$ and $T(H^1) \subseteq H^0$), $C\ell_k$-linear and self-adjoint. Similarly, let $\tilde{\mathfrak{F}}_k \subset \mathfrak{F}_{\mathbb{R}}(H, H)$ denote the subset of graded operators which are $C\ell_k$-antilinear and skew-adjoint. (As should be obvious, an operator $T \in \mathfrak{F}_{\mathbb{R}}(H, H)$ is called **$C\ell_k$-linear** if $T(\varphi u) = \varphi T(u)$ for all $\varphi \in C\ell_k$ and $u \in H$. It is called **$C\ell_k$-antilinear** if $T(\varphi u) = \alpha(\varphi) T(u)$ for all such φ and u.)

As in Observation 10.5 we have a natural homeomorphism $\mathfrak{F}_k \cong \tilde{\mathfrak{F}}_k$. We define the **Clifford index** of an element $T \in \mathfrak{F}_k$ (or $\tilde{\mathfrak{F}}_k$) to be the residue class $\text{ind}_k T = [\ker T] \in \mathfrak{M}_k / i^* \mathfrak{M}_{k+1} \cong KO^{-k}(\text{pt})$. Our first main result is the following:

Proposition 10.6. *The Clifford index*

$$\mathrm{ind}_k : \mathfrak{F}_k(\cong \tilde{\mathfrak{F}}_k) \longrightarrow KO^{-k}(\mathrm{pt})$$

is constant on connected components of $\mathfrak{F}_k$.

Proof. Fix $T \in \mathfrak{F}_k$ and recall that since T is Fredholm, 0 is an isolated point in the spectrum of T. Replacing T by an appropriate scalar multiple we may assume that the non-zero spectrum of T lies outside $[-2, 2]$, i.e., that $T^2 > 2\mathrm{Id}$ on $(\ker T)^\perp$. Choose a neighborhood $\mathfrak{U}$ of T in $\mathfrak{F}_k$ so that for all $S \in \mathfrak{U}$, we have spectrum$(S^2) \subset [0, \frac{1}{2}) \cup (1, \infty)$, and $\|T^2 - S^2\| < 1/2$.

Fix $S \in \mathfrak{U}$ and let $W \subset H$ be the range of the spectral projection of S^2 onto $[0, \frac{1}{2}]$. We claim that orthogonal projection $\mathrm{pr} : H \to \ker T\,(= \ker T^2)$ restricts to give an isomorphism

$$\mathrm{pr} : W \xrightarrow{\approx} \ker T. \tag{10.9}$$

To begin, suppose $v \in W \cap (\ker T)^\perp$ and note that $\langle (T^2 - S^2)v, v\rangle \geqq (2 - \frac{1}{2})\|v\|^2 \geqq \|v\|^2$. Since $\|T^2 - S^2\| < 1/2$ we must have $v = 0$ and so (10.9) is injective. Similarly any $v \in W^\perp \cap \ker T$ satisfies $\langle (S^2 - T^2)v, v\rangle \geqq \|v\|^2$ and is therefore 0. Hence, (10.9) is surjective and the claim is proved.

Observe now that W is a $\mathbb{Z}_2$-graded $C\ell_k$-submodule of H, and furthermore splits into graded submodules $W = \ker S \oplus V$ where $V = (\ker S)^\perp \cap W$ is S-invariant. One easily sees that the projection (10.9) preserves the graded module structure, and so we have a graded $C\ell_k$-module equivalence

$$\ker S \oplus V \cong \ker T.$$

It remains only to enhance the structure of V to that of a graded $C\ell_{k+1}$-module. This is done as follows. Let $S_V : V \to V$ denote the restriction of S to V. This is a symmetric, $\mathbb{Z}_2$-graded $C\ell_k$-linear map, and so also is $J \equiv [S_V^2]^{-\frac{1}{2}} S_V$. Note that $J^2 = \mathrm{Id}$. With respect to the decomposition $V = V^0 \oplus V^1$ we can write J in the form

$$J = \begin{pmatrix} 0 & J^1 \\ J^0 & 0 \end{pmatrix}.$$

The graded endomorphism of V given by

$$\tilde{J} = \begin{pmatrix} 0 & -J^1 \\ J^0 & 0 \end{pmatrix}$$

is $C\ell_k$-antilinear and satisfies $\tilde{J}^2 = -\mathrm{Id}$. It therefore makes V into a graded $C\ell_{k+1}$-module as required. Hence $\mathrm{ind}_k T = \mathrm{ind}_k S$ and the proof is complete. ■

Consider now an elliptic differential operator $P: \Gamma(E) \to \Gamma(E)$ which is $C\ell_k$-linear and graded. Note that its principal symbol is also $C\ell_k$-linear and graded. In this context a regular homotopy of symbols is required to preserve these additional properties. Arguing as in Theorem 7.10 and using the proposition above directly proves the following. Assume, as before, that P is defined over a compact manifold and is self-adjoint.

Theorem 10.7. *The Clifford index of P depends only on the regular homotopy class of its principal symbol.*

Proposition 10.6 implies that ind_k induces a map on the connected components of $\boldsymbol{\mathfrak{F}}_k$. In previous cases this map was a bijection, and the analogous result is almost true here. However, we need some preliminary adjustments. To begin, we shall assume for each k that $H = H^0 \oplus H^1$ is a graded Hilbert module for $C\ell_{k+2}$ and is considered to be a graded $C\ell_k$-module under restriction to $C\ell_k \subset C\ell_{k+2}$. (This algebra inclusion is induced by the euclidean space inclusion $\mathbb{R}^k \subset \mathbb{R}^{k+2}$ as the first k-coordinates.) This assumption assures us, for example, that when $k \equiv 0 \pmod 4$, each of the two distinct irreducible graded $C\ell_k$-modules appear with infinite multiplicity in H.

For each k, we now define a new space $\boldsymbol{\mathfrak{F}}_k$ as follows. If $k \not\equiv -1 \pmod 4$, then $\boldsymbol{\mathfrak{F}}_k \equiv \mathfrak{F}_k \approx \tilde{\mathfrak{F}}_k$. If $k \equiv -1 \pmod 4$, then we consider for each $T \in \mathfrak{F}_k$, the associated operator $w(T) \equiv e_1 \cdots e_k T|_{H^0}$ where $e_1, \ldots, e_k$ is a fixed orthonormal basis of $\mathbb{R}^k \subset C\ell_k$. Note that $w(T): H^0 \to H^0$ is a self-adjoint Fredholm operator. We now decompose $\mathfrak{F}_k$ into three disjoint subsets $\mathfrak{F}_k^+, \mathfrak{F}_k^-$, and $\boldsymbol{\mathfrak{F}}_k$ consisting respectively of those T's such that $w(T)$ is essentially positive, essentially negative, or neither. (A self-adjoint operator is **essentially positive** if it is positive on a closed invariant subspace of finite codimension.) Each of these subsets is open in $\mathfrak{F}_k$ and hence is a union of connected components of $\mathfrak{F}_k$.

We now show that when $k \equiv -1 \pmod 4$, the space $\boldsymbol{\mathfrak{F}}_k$ is not empty. Recall that H is a module for $C\ell_{k+2} \supset C\ell_k$. Let $e_1, \ldots, e_{k+2}$ be an extension of our orthonormal basis above, and write the action of e_{k+1} on $H = H^0 \oplus H^1$ as

$$e_{k+1} = \begin{pmatrix} 0 & -S^* \\ S & 0 \end{pmatrix}.$$

Then the map $\varepsilon_{k+1}: H \to H$ given by

$$\varepsilon_{k+1} = \begin{pmatrix} 0 & S^* \\ S & 0 \end{pmatrix} \tag{10.10}$$

is $C\ell_k$-linear and self-adjoint. Furthermore $w(\varepsilon_{k+1})$ anticommutes with $e_{k+2}\varepsilon_{k+1}$, and so we have $\varepsilon_{k+1} \in \boldsymbol{\mathfrak{F}}_k$.

The following striking result is due to Atiyah and Singer [5]:

Theorem 10.8. *For each k, the Clifford index induces a bijection*

$$\mathrm{ind}_k : \pi_0(\mathfrak{F}_k) \longrightarrow KO^{-k}(\mathrm{pt}). \tag{10.11}$$

Proof. From Proposition 10.6 we know that the map (10.11) is well defined. To show that it is surjective, let $V = V^0 \oplus V^1$ represent an element of $\widehat{\mathfrak{M}}_k/i^*\widehat{\mathfrak{M}}_{k+1} \cong KO^{-k}(\mathrm{pt})$, and let ε_{k+1} be as in (10.10) above. Then the map $\varepsilon_{k+1} \oplus 0$ is an element of $\mathfrak{F}_k(H \oplus V)$ and has kernel V. Since $H \oplus V \cong H$, this shows that the map (10.11) is surjective.

To prove injectivity, it is convenient to first divest ourselves of the grading. Recall that the endomorphisms $e_1^0 \equiv e_k e_1, \ldots, e_{k-1}^0 \equiv e_k e_{k-1}$ make the subspace H^0 into an ungraded module over $C\ell_{k-1} \cong C\ell_k^0$. Let $\mathfrak{F}_k^0$ denote the space of all skew-adjoint, $C\ell_{k-1}$-antilinear Fredholm operators on H^0. There is a natural map

$$(\cdot)^0 : \mathfrak{F}_k \longrightarrow \mathfrak{F}_k^0 \tag{10.12}$$

defined by setting $T^0 \equiv e_k T|_{H^0}$. It is easily checked that this map is a homeomorphism. Passing from graded to ungraded modules as in (10.4) and (10.5), we find that $\mathrm{ind}_k T \cong [\ker T^0]$.

Assume now that we are given two operators $S,T \in \mathfrak{F}_k$ with $\mathrm{ind}_k S = \mathrm{ind}_k T$. It will suffice to show that S^0 and T^0 are homotopic in $\mathfrak{F}_k^0$. Our first step is to observe that we may assume S^0 to be a Hilbert space isometry on $(\ker S^0)^\perp$ (and similarly for T^0). This is accomplished by putting the operator in polar form and deforming away the "radial" part as in the first step of the proof of Lemma 7.7. Properties of skew-adjointness and $C\ell_{k-1}$-antilinearity are preserved. The operator S^0 now satisfies $(S^0)^2 = -\mathrm{Id}$ on $(\ker S^0)^\perp$, and setting $e_k^0 \equiv S^0$ makes $(\ker S^0)$ into a $C\ell_k$-module. The same remarks apply to T^0 of course.

Observe now that our hypothesis $\mathrm{ind}_k S = \mathrm{ind}_k T$ implies that there exist ungraded $C\ell_k$-modules V and W together with a $C\ell_{k-1}$-module isomorphism

$$\ker S^0 \oplus V \cong \ker T^0 \oplus W. \tag{10.13}$$

We now claim that the module V can be realized as an S^0-invariant subspace of $(\ker S^0)^\perp$. This means simply that V is isomorphic to a $C\ell_k$-submodule of $(\ker S^0)^\perp$ where e_k^0 acts by S^0 as above. When $k \not\equiv -1 (\mathrm{mod}\ 4)$, all irreducible $C\ell_k$-modules are equivalent, and our claim is obvious. When $k \equiv -1(\mathrm{mod}\ 4)$, there are two equivalence classes of irreducible $C\ell_k$-modules and we must show that each of them appears with infinite multiplicity in $(\ker S^0)^\perp$. Recall that these two representations are distinguished by whether the central element $\omega = e_1^0 \cdots e_k^0$ acts by 1 or -1. There is a splitting $(\ker S^0)^\perp \equiv M = M^+ \oplus M^-$ where $M^\pm =$

$(1 \pm \omega)M$, and we want to know that $\dim M^+ = \dim M^- = \infty$. However, $\omega = e_1^0 \cdots e_k^0 = e_1^0 \cdots e_{k-1}^0 S^0 = \pm w(S)$ where the sign depends only on k, and so the desired property follows from our assumption that $w(S)$ is neither essentially positive nor essentially negative.

The same argument, of course, shows that W may be realized as a T^0-invariant submodule of $(\ker T^0)^\perp$. We now consider the two orthogonal decompositions:

$$H^0 = (\ker S^0) \oplus V \oplus V' = (\ker T^0) \oplus W \oplus W'.$$

It is clear from the discussion above that V' and W' are isomorphic as $C\ell_k$-modules. This means simply that there is a $C\ell_k$-linear Hilbert space isomorphism $L: V' \xrightarrow{\approx} W'$ such that $S^0 = L^{-1}T^0L$ on V'. By the obvious homotopy we may assume that $S^0 = 0$ on V and $T^0 = 0$ on W. Extending L by the isomorphism (10.13) gives an isomorphism $L: H^0 \to H^0$ such that $S^0 = L^{-1}T^0L$.

The theorem now follows from the fact that the group of $C\ell_k$-linear isometries of H^0 is connected. This fact is proved by suitable adaptation of the argument given for Lemma 7.7 above. ∎

REMARK 10.9. The mapping (10.12) above gives a natural transformation between graded and ungraded operators. All the previous discussion of this section could be so transformed. Consequently we find that *there exists a parallel index theory for real elliptic operators which are skew-adjoint and (ungraded!) $C\ell_{k-1}$-antilinear.*

Let us examine some examples. Let P be the operator. If $k = 1$, we find that $\text{ind}_1 P \equiv \dim_{\mathbb{R}}(\ker P) (\text{mod } 2)$. (Thus, every real skew-adjoint elliptic operator has a well-defined index in $\mathbb{Z}_2$!) If $k = 2$, then $\ker P$ is a $C\ell_1 \cong \mathbb{C}$ module and we find that $\text{ind}_2 P \equiv \dim_{\mathbb{C}}(\ker P)(\text{mod } 2)$. If $k = 4$, then $C\ell_3 \cong \mathbb{H} \oplus \mathbb{H}$ and the bundle E on which P is defined splits as $E = E^+ \oplus E^-$ where $E^\pm = (1 \pm \omega)E$ and $\omega = e_1e_2e_3$. By antilinearity, $P(E^\pm) \subset \Gamma(E^\mp)$ and we split P into P^+ and $P^- = -(P^+)^*$ as before. Note that $\ker P = \ker P^+ \oplus \ker P^-$ and that $\text{ind}_4 P = 0$ if and only if $\dim_{\mathbb{H}}(\ker P^+) = \dim_{\mathbb{H}}(\ker P^-)$. (This is because the $\pm$ spaces are the two distinct $C\ell_3$-modules.) It follows easily that $\text{ind}_4 P = \dim_{\mathbb{H}}(P^+) - \dim_{\mathbb{H}}(P^-) =$ the index of P^+ considered as a quaternionic operator.

We conclude this section with a discussion of a nice result of Atiyah and Singer which states that the spaces $\mathfrak{F}_k$ form a "spectrum" (in the sense of homotopy theory). Together with the periodicity phenomena in Clifford modules and Theorem 8.8, this will give a new proof of Bott periodicity. To begin we notice that for each $k \geqq 1$ there is a natural inclusion

$$\mathfrak{F}_{k+1} \hookrightarrow \mathfrak{F}_k$$

and a distinguished element

$$\varepsilon_{k+1} \in \mathfrak{F}_k$$

given by (10.10). We extend this by convention to $k = 0$ as follows. Let $\mathfrak{F}_0$ denote the space of *all* real Fredholm operators on H^0, and let $\mathfrak{F}_1 \hookrightarrow \mathfrak{F}_0$ be the map which associates to $T \in \mathfrak{F}_1$ the operator $T^0 = e_1 T|_{H^0}$. (This identifies $\mathfrak{F}_1$ with the space of all skew-adjoint operators in $\mathfrak{F}_0$.) Set $\varepsilon_1 = \mathrm{Id} \in \mathfrak{F}_0$.

Let $\Omega\mathfrak{F}_k$ denote the **loop space** of $\mathfrak{F}_k$ defined here to be the set of all continuous paths $\gamma:[0,\pi] \to \mathfrak{F}_k$ with $\gamma(0) = \varepsilon_{k+1}$ and $\gamma(\pi) = -\varepsilon_{k+1}$. This space carries the compact-open topology. It is homotopy equivalent to the space of all paths which both begin and end at ε_{k+1}. The following is the main result of Atiyah-Singer [5] and is presented here without proof.

Theorem 10.10 (Atiyah and Singer). *For each $k \geqq 0$, the map*

$$\mathfrak{F}_{k+1} \longrightarrow \Omega\mathfrak{F}_k$$

which assigns to $T \in \mathfrak{F}_{k+1}$ the path from ε_{k+1} to $-\varepsilon_{k+1}$ given by

$$(\cos t)\varepsilon_{k+1} + (\sin t)T, \qquad 0 \le t \le \pi,$$

is a homotopy equivalence.

This gives us a generalization of Theorem 8.8 above.

Theorem 10.11. *For any compact Hausdorff space A and for any k there is a natural isomorphism*

$$\mathrm{ind}_k : [A, \mathfrak{F}_k] \longrightarrow KO^{-k}(A)$$

with the functorial property (8.7). *Hence, $\mathfrak{F}_k$ is a classifying space for the functor KO^{-k}.*

Proof. The case $k = 0$ is just Theorem 8.8 above. For higher k we have $[A, \mathfrak{F}_k] \cong [A, \Omega^k\mathfrak{F}_0] \cong [\Sigma^k A, \mathfrak{F}_0] \cong KO(\Sigma^k A) \stackrel{\text{def}}{=} KO^{-k}(A)$ where $\Sigma^k A$ denotes the k-fold suspension of A. ∎

Theorem 10.12 (Bott Periodicity). *For each $k \geqq 0$ there is a homeomorphism $\mathfrak{F}_k \cong \mathfrak{F}_{k+8}$. Therefore, by* 10.10 *there is a homotopy equivalence $\mathfrak{F}_k \sim \Omega^8\mathfrak{F}_k$ which implies that*

$$KO^{-k}(A) \cong KO^{-k+8}(A)$$

for any compact Hausdorff space A.

Proof. Let $H = H^0 \oplus H^1$ be a graded Hilbert module for $C\ell_k$ as above, and let $V = V^0 \oplus V^1$ be an irreducible graded (real) module for $C\ell_8$. The graded tensor product $H \hat{\otimes} V$ is a module for $C\ell_k \hat{\otimes} C\ell_8 \cong C\ell_{k+8}$.

There is an isomorphism $V \cong \mathbb{R}^{16}$ which yields an explicit identification $C\ell_8 \cong \mathbb{R}(16)$. The fundamental results of Chapter I, §4, give an isomorphism $C\ell_{k+8} \cong C\ell_k \otimes \mathbb{R}(16)$ (*un*graded tensor product), where elements of the form $1 \otimes \varphi$ act by $\mathrm{Id} \otimes \varphi$ on $H \hat{\otimes} V$.

Let $\mathscr{L}_k(H)$ denote the Banach space of bounded operators on H which are $C\ell_k$-linear and graded. Define a map $\Psi: \mathscr{L}_k(H) \to \mathscr{L}_{k+8}(H \hat{\otimes} V)$ by setting $\Psi(T) = T \otimes \mathrm{Id}$. This map is continuous and injective. Furthermore, we claim that it is surjective. To see this, fix $\tilde{T} \in \mathscr{L}_{k+8}(H \otimes V)$, and for each $x \in H$ consider the map $\tilde{T}_x: V \to H \otimes V$ defined by $\tilde{T}_x(v) = \tilde{T}(x \otimes v)$. Since $\tilde{T}_x(\varphi v) = \varphi \tilde{T}_x(v)$ for all $\varphi \in \mathbb{R}(16)$, we conclude that $\dim(\mathrm{Image}\, \tilde{T}_x)$ is either 0 or 16. It follows that $\mathrm{Image}(\tilde{T}_x) = \ell \otimes V$ for a unique 1-dimensional subspace $\ell \subset H$. Consequently, $\tilde{T}(x \otimes v) = T(x) \otimes v$ for a unique element $T(x) \in H$. The map $T: H \to H$ is easily seen to be bounded, $C\ell_k$-linear and graded. Clearly $\Psi(T) = \tilde{T}$. We have shown that Ψ is surjective and therefore by the Open Mapping Theorem that Ψ is a homeomorphism. It is easy to check that Ψ carries the self-adjoint Fredholm operators in $\mathscr{L}_k(H)$ onto those in $\mathscr{L}_{k+8}(H \otimes V)$. This gives the desired homeomorphism $\mathfrak{F}_k \cong \mathfrak{F}_{k+8}$. ∎

Kuiper's results [1] show that the group of $C\ell_k^0$-linear isometries of H^0 is contractible. This shows that the construction of the homeomorphism above is canonical up to homotopy.

Note that for each $k \geq 0$ we have isomorphisms

$$KO^{-k}(\mathrm{pt}) \xleftarrow[\approx]{\alpha} \pi_0(\mathfrak{F}_k) \xrightarrow[\approx]{\beta} \hat{\mathfrak{M}}_k / i^* \hat{\mathfrak{M}}_{k+1}$$

where α is given by 10.11 and where β is the Clifford index defined in 10.4. It is shown in Atiyah-Singer [5] that the map $\alpha \circ \beta^{-1}$ coincides with the isomorphism given by the Aityah-Bott-Shapiro construction (cf. I.9).

There is a natural ring structure in KO^{-*} where the multiplication is induced by the tensor product of operators. This is defined as follows. Let H_k and H_ℓ be graded Hilbert modules for $C\ell_k$ and $C\ell_\ell$ respectively. The graded tensor product $H_k \hat{\otimes} H_\ell$ is then a module for $C\ell_k \hat{\otimes} C\ell_\ell = C\ell_{k+\ell}$ (see I.5). Representing $\mathfrak{F}_{k+\ell}$ by $\mathfrak{F}_{k+\ell}(H_k \hat{\otimes} H_\ell, H_k \hat{\otimes} H_\ell)$, we define a map

$$\hat{\otimes}: \mathfrak{F}_k \times \mathfrak{F}_\ell \longrightarrow \mathfrak{F}_{k+\ell} \tag{10.14}$$

by requiring that for elements $v \in H_k$ and $w \in H_\ell$ of pure degree with respect to the grading,

$$(S \hat{\otimes} T)(v \otimes w) \equiv (Sv) \otimes w + (-1)^{\deg v} v \otimes (Tw),$$

where $S \in \mathfrak{F}_k$ and $T \in \mathfrak{F}_\ell$. One checks easily that $S \hat{\otimes} T$ is $C\ell_k$-linear and $C\ell_\ell$-linear, and therefore $C\ell_{k+\ell}$-linear. The inner product on $H_k \otimes H_\ell$ is given by $\langle v \otimes w, v' \otimes w' \rangle = \langle v,v' \rangle \langle w,w' \rangle$, and $S \hat{\otimes} T$ is clearly self-adjoint. Since S and T interchange even and odd factors, so does $S \hat{\otimes} T$, and one computes that

$$(S \hat{\otimes} T)^2 = S^2 \otimes \mathrm{Id} + \mathrm{Id} \otimes T^2.$$

From this one sees that $S \hat{\otimes} T$ is Fredholm and that there is an identification

$$\ker(S \hat{\otimes} T) = (\ker S) \hat{\otimes} (\ker T). \tag{10.15}$$

The graded tensor product of modules induces a multiplication in $\hat{\mathfrak{M}}_* \equiv \bigoplus_k \hat{\mathfrak{M}}_k$, which descends to the quotient $KO^{-*}(\mathrm{pt}) \cong \bigoplus(\hat{\mathfrak{M}}_k / i^*\hat{\mathfrak{M}}_{k+1})$. Hence, (10.15) gives the following:

Proposition 10.13. *For $S \in \mathfrak{F}_k$ and $T \in \mathfrak{F}_\ell$, one has that*

$$\mathrm{ind}_{k+\ell}(S \hat{\otimes} T) = \mathrm{ind}_k(S) \hat{\otimes} \mathrm{ind}_\ell(T).$$

It is straightforward to check that the map (10.14) above preserves the subsets $\mathfrak{F}_k$, that is, it restricts to give a continuous mapping

$$\hat{\otimes} : \mathfrak{F}_k \times \mathfrak{F}_\ell \longrightarrow \mathfrak{F}_{k+\ell}.$$

Applying the isomorphism $KO^{-k}(A) \cong [A, \mathfrak{F}_k]$ of 10.11, we get a multiplication

$$KO^{-k}(A) \times KO^{-\ell}(A) \longrightarrow KO^{-k-\ell}(A)$$

defined for any compact Hausdorff space A and for all $k, \ell \geqq 0$. In fact, for any pair of such spaces A and B, we have a transformation

$$KO^{-k}(A) \times KO^{-\ell}(B) \longrightarrow KO^{-k-\ell}(A \times B)$$

given by associating to $f: A \to \mathfrak{F}_k$ and $g: \to \mathfrak{F}_\ell$ the map $(fg)(a,b) = f(a) \hat{\otimes} g(b)$. (When $A = B$ the multiplication is obtained by restricting to the diagonal.) These transformations coincide with the ones conventionally defined in KO-theory (see I.9).

REMARK 10.14. Recall that the homeomorphism (10.12) identifies $\mathfrak{F}_1$ with the space $\mathfrak{F}^{\mathrm{skew}}$ of all skew-adjoint Fredholm operators on real Hilbert space. Consequently, Theorem 10.11 shows that $\mathfrak{F}^{\mathrm{skew}}$ is a classifying space for the functor KO^{-1}.

Similarly, the assignment $T \to \varepsilon_7 T|_{H^0}$ associates to each element $T \in \mathfrak{F}_7$, a $C\ell_6$-linear map of the ungraded Hilbert module H_6 for $C\ell_6$. Note that $C\ell_6 \cong \mathbb{R}(8)$ and we can take H_6 to be the product $H_6 = H \otimes \mathbb{R}^8$ with $\mathbb{R}(8)$ acting on the right-hand factor. Any $\mathbb{R}(8)$-linear map of H_6 is then of the form $A \otimes \mathrm{Id}$. This identifies $\mathfrak{F}_7$ with the space $\mathfrak{F}^{\mathrm{symm}}$ of all self-adjoint Fredholm operators on real Hilbert space, and Theorem 10.11 shows that $\mathfrak{F}^{\mathrm{symm}}$ classifies the functor KO^{-7}.

REMARK 10.15. The entire discussion of this section can be carried through in the complex case. One considers complex Clifford algebras, complex graded modules, complex Hilbert spaces, etc. All of the analogous fundamental results remain true. One essentially recaptures the "classical" index theory discussed in §§7 and 8. However, a new fact which emerges is that the functor K^1 is classified by the skew-adjoint Fredholm operators on complex Hilbert space. The analogue of Theorem 10.10 (also proved in Atiyah-Singer [5]) leads to a proof of Bott Periodicity for the unitary group.

§11. Multiplicative Sequences and the Chern Character

In this section we present some fundamental constructions in K-theory and the theory of characteristic classes. These constructions will be needed later when we discuss the cohomological formula for the index of an elliptic operator.

Throughout the section cohomology groups will be taken with rational coefficients, although much of what we do carries over to more general coefficient rings.

There is a principle underlying much of what we do here. Roughly stated it asserts that for computational purposes *every complex vector bundle is a direct sum of line bundles.* Moreover, if the bundle is the complexification of a real bundle, the non-trivial line bundles occur in complex conjugate pairs. To make this precise we need the following result, often referred to as the "Splitting Principle":

Proposition 11.1. *Let E be a complex vector bundle over a manifold X. Then there exists a manifold $\mathcal{S}_E$ and a smooth, proper fibration $\pi: \mathcal{S}_E \to X$ such that*

(i) *The homomorphism $\pi^*: H^*(X) \to H^*(\mathcal{S}_E)$* is injective.
(ii) *The bundle π^*E splits into a direct sum of complex line bundles:*

$$\pi^*E \cong \ell_1 \oplus \cdots \oplus \ell_n \tag{11.1}$$

Proof. Let $p: \mathbb{P}(E) \to X$ denote the **projectivization** of E, i.e., the bundle whose fibre at x is the projective space $\mathbb{P}(E_x)$ of all complex lines in E_x. The bundle p^*E contains a line bundle defined tautologically at a line $\ell \subset E_x$ to be ℓ itself. Using some fixed hermitian metric in E, this gives us a tautological splitting

$$p^*E = \ell \oplus \ell^{\perp}$$

The homomorphism $p^*: H^*(X; \mathbb{Z}) \to H^*(\mathbb{P}(E); \mathbb{Z})$ is injective by the Leray-Hirsch Theorem C.14 in Appendix C.

We now repeat the process for the bundle $\ell^{\perp}$ and continue inductively to complete the proof. ■

There is a direct analogue of the proposition and its proof for the case of real vector bundles. We shall not state this. However, we do want to signal the following "hybrid" result.

Proposition 11.2. *Let E be an oriented real vector bundle of dimension $2n$ over a manifold X. Then there is a smooth proper fibration $\pi : \mathscr{S}_E \to X$ such that $\pi^* : H^*(X) \to H^*(\mathscr{S}_E)$ is injective and the bundle $\pi^*(E \otimes \mathbb{C})$ splits into complex line bundles:*

$$\pi^*(E \otimes \mathbb{C}) \cong \ell_1 \oplus \bar{\ell}_1 \oplus \cdots \ell_n \oplus \bar{\ell}_n \tag{11.2}$$

where $\bar{\ell}_j$ denotes the "inverse" or "complex conjugate" bundle to ℓ_j (see below). In fact, there is a splitting

$$\pi^* E = E_1 \oplus \cdots \oplus E_n \tag{11.3}$$

into oriented real 2-plane bundles such that $E_k \otimes \mathbb{C} = \ell_k \oplus \bar{\ell}_k$ for each k.

Proof. Suppose, to begin, that $\dim_{\mathbb{R}}(E) = 2$. Fix a metric in E and *let* $J : E \to E$ be the map which rotates each fibre by $\pi/2$ in the positive direction. We let $\mathscr{S}_E = X$ and note that

$$E \otimes \mathbb{C} = \ell \oplus \bar{\ell}$$

where at $x \in X$

$$\ell_x \equiv \{v - iJv : v \in E_x\} \qquad \text{and} \qquad \bar{\ell}_x \equiv \{v + iJ_v : v \in E_x\}, \tag{11.4}$$

are the $+i$ and $-i$ eigenspaces of $J \otimes \mathbb{C}$. Note incidentally that as complex bundles we have the bundle isomorphism

$$E \cong \ell. \tag{11.5}$$

For the general case we fix a metric in E and consider the bundle $p : G(E) \to X$ whose fibre at a point x consists of all oriented 2-dimensional subspaces of E_x. Then there is a canonical splitting

$$p^* E = E_1 \oplus E_1^{\perp}$$

where $E_1 \to G(E)$ is the tautological oriented 2-plane bundle whose fibre at $P \in G(E)$ is P itself. The argument given for Theorem C.14 adapts immediately to prove that the homomorphism $p^* : H^*(X; \mathbb{Z}) \to H^*(G(E); \mathbb{Z})$ is injective. Repeating the process for the bundle $E_1^{\perp}$ and proceeding inductively, we construct the desired splitting bundle $\mathscr{S}_E$. ■

REMARK 11.3. An analogous result holds when $\dim_{\mathbb{R}}(E) = 2n + 1$. One must add a trivial line bundle onto the decompositions (11.2) and (11.3).

For many purposes it is permissible to add a trivial real line bundle to E and work directly in the even-dimensional case.

NOTE 11.4 (conjugate bundles). Recall that if E is a complex vector bundle, then its **conjugate bundle** $\bar{E}$ is obtained from E by redefining scalar multiplication. The new scalar multiplication by $t \in \mathbb{C}$ is the old scalar multiplication by $\bar{t}$. For any given hermitian metric $(\cdot,\cdot)$ in E, the map $v \to \varphi_v(\cdot) \equiv (\cdot, v)$ identifies $\bar{E}$ with $E^* \equiv \mathrm{Hom}_{\mathbb{C}}(E, \mathbb{C})$. If $E = E_0 \otimes \mathbb{C}$ is the complexification of a real vector bundle E_0, then there is a complex bundle isomorphism $E \cong \bar{E}$.

NOTE 11.5 (line bundles). The set $\mathscr{L}(X) \cong H^1(X; S^1)$ of equivalence classes of complex line bundles on a manifold X has a natural commutative multiplication given by the tensor product. Since $\bar{\ell} \otimes \ell \cong \ell^* \otimes \ell \cong$ trivial, the multiplication is invertible, and $\mathscr{L}(X)$ is a group. The first Chern class $c_1 : \mathscr{L}(X) \to H^2(X; \mathbb{Z})$ is a *group isomorphism* (see Appendix A, Example A.5).

The Splitting Principle above can be applied directly to characteristic classes.

OBSERVATION 11.6 (splitting the Euler class). Let E be an oriented real vector bundle of dimension $2n$, and let $\chi(E) \in H^{2n}(X; \mathbb{Q})$ denote its Euler class. If E decomposes into a sum of oriented 2-plane bundles $E = E_1 \oplus \cdots \oplus E_n$, then since $\chi(E \oplus E') = \chi(E)\chi(E')$ we can write

$$\chi(E) = x_1 \cdots x_n \tag{11.6}$$

where $x_k = \chi(E_k)$ for each k. If we now complexify and write $E \otimes \mathbb{C} = \ell_1 \oplus \bar{\ell}_1 \oplus \cdots \oplus \ell_n \oplus \bar{\ell}_n$ then we see from (11.5) that

$$x_k = c_1(\ell_k) \qquad \text{for each } k.$$

Using the Splitting Principle, formula (11.6) can, in fact, be used to define the Euler class.

OBSERVATION 11.7 (splitting the total Chern class). Let E be a complex vector bundle, and denote by

$$c(E) = 1 + c_1(E) + \ldots + c_n(E)$$

the total Chern class of E. Recall that $c(E \oplus E') = c(E)c(E')$. Hence, if E decomposes as a sum of line bundles $E = \ell_1 \oplus \cdots \oplus \ell_n$, then

$$c(E) = \prod_{k=1}^{n} (1 + x_k) \tag{11.7}$$

where $x_k = c_1(\ell_k)$ for $k = 1, \ldots, n$. In particular, we find that

$$c_j(E) = \sigma_j(x_1, \ldots, x_n), \qquad j = 1, \ldots, n \tag{11.8}$$

where σ_j denotes the jth elementary symmetric function of $x_1, \ldots, x_n$.

If E does not split as a sum of line bundles, we may lift it to the space $\mathscr{S}_E$ where it does. The map $H^*(X) \to H^*(\mathscr{S}_E)$ is injective and identifies c_j with $\sigma_j(x_1, \ldots, x_n)$ as in the previous case. Note that any symmetric polynomial expression in the x_j's can be rewritten as a polynomial in $\sigma_1, \ldots, \sigma_n$ (i.e., in $c_1, \ldots, c_n$). This will enable us to define characteristic invariants in a particularly useful way.

As a simple application note that since $c_1(\bar{\ell}) = -c_1(\ell)$, we have

$$c_j(\bar{E}) = (-1)^j c_j(E) \tag{11.9}$$

for all j.

Observation 11.8 (splitting the total rational Pontrjagin class). Let E be a real oriented vector bundle of dimension $2n$, and recall that the rational Pontrjagin classes of E are defined by $p_j(E) = (-1)^j c_{2j}(E \otimes \mathbb{C})$, $j = 1, \ldots, n$. (Since $E \otimes \mathbb{C} \cong \overline{(E \otimes \mathbb{C})}$, the Chern classes of odd degree are zero by (11.9).) The total (rational) Pontrjagin class is defined to be

$$p(E) = 1 + p_1(E) + \ldots + p_n(E).$$

It has the property that $p(E \oplus F) = p(E)p(F)$. If $E \otimes \mathbb{C}$ decomposes as a direct sum, $E \otimes \mathbb{C} = \ell_1 \oplus \bar{\ell}_1 \oplus \cdots \oplus \ell_n \oplus \bar{\ell}_n$, then $c(E \otimes \mathbb{C}) = \prod c(\ell_k)c(\bar{\ell}_k) = \prod (1 - x_k^2)$, and we find that

$$p(E) = \prod_{k=1}^{n} (1 + x_k^2) \tag{11.10}$$

where $x_k = c_1(\ell_k)$ as before. In particular, we have

$$p_j(E) = \sigma_j(x_1^2, \ldots, x_n^2) \tag{11.11}$$

for each j.

Let $\mathbb{Q}[[x]]^\wedge$ denote the set of formal power series in x with rational coefficients and with constant term 1. It is easily seen that $\mathbb{Q}[[x]]^\wedge$ is a group under multiplication. Fix an element $f(x) \in \mathbb{Q}[[x]]$, and for each $n \in \mathbb{Z}^+$ consider the formal power series in n indeterminates given by $f(x_1) \cdots f(x_n)$. This is evidently symmetric in the x_j's, and so it has an expansion of the form

$$f(x_1) \cdots f(x_n) = 1 + F_1(\sigma_1) + F_2(\sigma_1, \sigma_2) + F_3(\sigma_1, \sigma_2, \sigma_3) + \ldots$$

where

$$\sigma_k(x_1, \ldots, x_n) \equiv \sum_{i_1 < \ldots < i_k} x_{i_1} \cdots x_{i_k} \qquad \text{for } 1 \le k \le n$$

denotes the kth elementary symmetric function in $x_1, \ldots, x_n$, and where F_k is **weighted homogeneous** of degree k, i.e.,

$$F_k(t\sigma_1, \ldots, t^k\sigma_k) = t^k F_k(\sigma_1, \ldots, \sigma_k) \qquad \text{for all } t \in \mathbb{Q}.$$

Each of the polynomials $F_k(\sigma_1,\ldots,\sigma_k)$ is well defined and *independent of the number of variables* x_j. This is easily seen by adding more variables and using the obvious fact that $\sigma_k(x_1,\ldots,x_n,0,\ldots,0) = \sigma_k(x_1,\ldots,x_n)$ if $k \leqq n$, and $\sigma_k(x_1,\ldots,x_n,0,\ldots,0) = 0$ if $k > n$.

The sequence of polynomials $\{F_k(\sigma_1,\ldots,\sigma_k)\}_{k=1}^{\infty}$ is called the **multiplicative sequence** determined by the formal power series $f(x)$. It has a universal multiplicative property which we shall now describe.

Let B be a commutative algebra with unit over $\mathbb{Q}$, and assume that B has a direct sum decomposition $B = B^0 \oplus B^1 \oplus B^2 \oplus \cdots$ with the property that $B^k \cdot B^\ell \subseteq B^{k+\ell}$ for all $k,\ell \geqq 0$. For example, B could be $H^{2*}(X;\mathbb{Q})$ or $H^{4*}(X;\mathbb{Q})$ for a space X. It could also be the polynomial ring $\mathbb{Q}[x]$ with $B^k \equiv \mathbb{Q}x^k$.

Given such an algebra B, let $B^\wedge$ denote the set of all formal sums $1 + b_1 + b_2 + \ldots$ where $b_k \in B^k$ for each k. Note that the finite sums (those with only a finite number of non-zero terms) actually belong to B and form a set which is closed under multiplication. This extends to $B^\wedge$ by defining

$$(1+b_1+b_2+\ldots)(1+c_1+c_2+\ldots)=1+(b_1+c_1)+(b_2+b_1c_1+c_2)+\ldots,$$

that is, by defining the nth term of the product to be $\sum b_k c_{n-k}$. Every element of $B^\wedge$ has a multiplicative inverse, and so $B^\wedge$ is an abelian group. As an example, note that if $B = \mathbb{Q}[x]$, then $B^\wedge = \mathbb{Q}[[x]]^\wedge$.

Fix a multiplicative sequence $\{F_k(\sigma_1,\ldots,\sigma_k)\}_{k=1}^{\infty}$. Then to each algebra B as above we associate a map $\mathbf{F}: B^\wedge \to B^\wedge$ by assigning to $b = 1 + b_1 + b_2 + \ldots \in B^\wedge$ the element

$$\mathbf{F}(b) = 1 + F_1(b_1) + F_2(b_1,b_2) + \ldots . \qquad (11.12)$$

Lemma 11.9. *The map $\mathbf{F}: B^\wedge \to B^\wedge$ is a group homomorphism, i.e.,*

$$\mathbf{F}(bc) = \mathbf{F}(b)\mathbf{F}(c)$$

for all $b,c \in B^\wedge$.

Proof. In the polynomial algebra $B \equiv \mathbb{Q}[x_1,\ldots,x_n]$ consider the element $\sigma = (1 + x_1)\cdots(1 + x_n) = 1 + \sigma_1 + \ldots + \sigma_n \in B^\wedge$. By definition of $\{F_k\}$ we have that $\mathbf{F}(\sigma) = f(x_1)\cdots f(x_n) = 1 + F_1(\sigma_1) + F_2(\sigma_1,\sigma_2) + \ldots$. We now increase the number of variables. Let $B \equiv \mathbb{Q}[x_1,\ldots,x_{n+m}]$, and consider the subalgebras $B' = \mathbb{Q}[x_1,\ldots,x_n]$ and $B'' = \mathbb{Q}[x_{n+1},\ldots,x_{n+m}]$. Let σ,σ',σ'' be the corresponding elementary products in each case. Then we have

$$\sigma = (1 + x_1)\cdots(1 + x_{n+m}) = \sigma'\sigma'',$$

and

$$\mathbf{F}(\sigma) = f(x_1)\cdots f(x_{n+m}) = \mathbf{F}(\sigma')\mathbf{F}(\sigma'').$$

The general result now follows easily from the algebraic independence of the σ_j's. ∎

It is easy to see that the universal multiplicative property 11.9 characterizes the sequence. The formal power series is recaptured by taking $\mathbf{F}(1 + x) = f(x)$ in the algebra $\mathbb{Q}[x]$.

The concept of a multiplicative sequence is due to F. Hirzebruch, who established their importance in the theory of characteristic classes.

Basic Construction 11.10 (multiplicative sequences of Chern classes). Let $\{F_k\}$ be a multiplicative sequence associated to the formal power series $f(x) \in \mathbb{Q}[[x]]\hat{}$. To each complex vector bundle E over a space X, we associate the **total F-class**

$$\mathbf{F}_{\mathbb{C}}(E) \equiv \mathbf{F}(c(E)) \in H^{2*}(X;\mathbb{Q})\hat{}.$$

Since $c(E \oplus E') = c(E)c(E')$, this class has the property that

$$\mathbf{F}_{\mathbb{C}}(E \oplus E') = \mathbf{F}_{\mathbb{C}}(E)\mathbf{F}_{\mathbb{C}}(E'), \tag{11.13}$$

for any two complex vector bundles E and E' over X. If we decompose $E = \ell_1 \oplus \cdots \oplus \ell_n$ according to the Splitting Principle, then

$$\mathbf{F}_{\mathbb{C}}(E) = f(x_1) \cdots f(x_n) \tag{11.14}$$

where $x_j = c_1(\ell_j)$ for each j.

Example 11.11 (the total Todd class). Associated to the formal power series

$$\mathrm{td}(x) \equiv \frac{x}{1 - e^{-x}} = 1 + \frac{1}{2}x + \frac{1}{12}x^2 + \dots$$

is the multiplicative sequence $\{\mathrm{Td}_m\}$ called the **Todd sequence.** The total Todd class is denoted by $\mathbf{Td}_{\mathbb{C}}$. Its first few terms are:

$$\mathrm{Td}_1(c_1) = \frac{1}{2}c_1$$

$$\mathrm{Td}_2(c_1,c_2) = \frac{1}{12}(c_2 + c_1^2)$$

$$\mathrm{Td}_3(c_1,c_2,c_3) = \frac{1}{24}c_2c_1.$$

If X is a compact complex manifold of dimension n, and if $E = TX$, then the number $\mathrm{Td}(X) \equiv \mathrm{Td}_n(TX)[X]$ (where $[X]$ denotes the fundamental class of X in $H_{2n}(X;\mathbb{Q})$) is called the **Todd genus** of X.

Basic Construction 11.12 (multiplicative sequences of Pontrjagin classes). Let $\{F_k\}$ be a multiplicative sequence associated to the formal power series $f(x)$. To each real vector bundle E over a space X, we associate the **total *F*-class**

$$\mathbf{F}(E) \equiv \mathbf{F}(p(E)) \in H^{4*}(X;\mathbb{Q})^{\wedge}.$$

Given two such bundles E and E' over X, we have $p(E \oplus E') = p(E)p(E')$ and so

$$\mathbf{F}(E \oplus E') = \mathbf{F}(E)\mathbf{F}(E'). \tag{11.15}$$

Assume E is oriented and of dimension $2n$, and decompose $E \otimes \mathbb{C}$ as $\ell_1 \oplus \bar{\ell}_1 \oplus \cdots \oplus \ell_n \oplus \bar{\ell}_n$ according to the Splitting Principle. Then

$$\mathbf{F}(E) = f(x_1^2) \cdots f(x_n^2) \tag{11.16}$$

where $x_k = c_1(\ell_k)$ for each k.

EXAMPLE 11.13 (the total $\hat{A}$-class). Associated to the formal power series

$$\hat{a}(x) \equiv \frac{\sqrt{x}/2}{\sinh(\sqrt{x}/2)} = 1 - \frac{1}{24}x + \frac{7}{2^7 \cdot 3^2 \cdot 5}x^2 + \ldots$$

is a multiplicative sequence $\{\hat{A}_m\}$ called the $\hat{A}$-sequence. The first few terms of the sequence are

$$\hat{A}_1(p_1) = -\frac{1}{24}p_1$$

$$\hat{A}_2(p_1,p_2) = \frac{1}{2^7 \cdot 3^2 \cdot 5}(-4p_2 + 7p_1^2)$$

$$\hat{A}_3(p_1,p_2,p_3) = -\frac{1}{2^{10} \cdot 3^3 \cdot 5 \cdot 7}(16p_3 - 44p_2p_1 + 31p_1^3).$$

Given a real bundle E, the total $\hat{A}$-class of E is the sum

$$\hat{\mathbf{A}}(E) = 1 + \hat{A}_1(p_1E) + \hat{A}_2(p_1E, p_2E) + \ldots$$

If we write $E \otimes \mathbb{C} = \ell_1 \oplus \bar{\ell}_1 \oplus \cdots \oplus \ell_n \oplus \bar{\ell}_n$ as above, then from (11.16) we see that

$$\hat{\mathbf{A}}(E) = \prod_{j=1}^{n} \frac{x_j/2}{\sinh(x_j/2)} \tag{11.17}$$

where of course $p_jE = \sigma_j(x_1^2, \ldots, x_n^2)$.

Closely related to the $\hat{A}$-sequence is the **A-sequence** $\{A_m\}$ determined by the power series $a(x) = \hat{a}(16x)$. One easily sees that $A_m = 16^m\hat{A}_m$ for each m.

The Todd class and the $\hat{A}$-class are intimately related.

Proposition 11.14. *For any oriented real vector bundle E it is true that*

$$\mathbf{Td}_{\mathbb{C}}(E \otimes \mathbb{C}) = \hat{\mathbf{A}}(E)^2.$$

Proof. We may assume dim E is even (see Remark 11.3) and we consider a formal splitting $E \otimes \mathbb{C} = \ell_1 \oplus \bar{\ell}_1 \oplus \cdots \oplus \ell_n \oplus \bar{\ell}_n$. Then by definition

$$\mathbf{Td}_{\mathbb{C}}(E \otimes \mathbb{C}) = \prod_{j=1}^{n} \frac{x_j}{1 - e^{-x_j}} \frac{(-x_j)}{1 - e^{x_j}}$$

where $x_j = c_1(\ell_j)$. Multiplying by $e^{x_j/2}e^{-x_j/2}$ in the denominator gives

$$\begin{aligned}\mathbf{Td}_{\mathbb{C}}(E \otimes \mathbb{C}) &= \prod_{j=1}^{n} \left[\frac{x_j}{e^{x_j/2} - e^{-x_j/2}}\right]^2 \\ &= \prod_{j=1}^{n} \left[\frac{x_j/2}{\sinh(x_j/2)}\right]^2 \\ &= [\hat{\mathbf{A}}(E)]^2. \quad \blacksquare\end{aligned}$$

Example 11.15 (the total L-class). Associated to the formal power series

$$\ell(x) \equiv \frac{\sqrt{x}}{\tanh\sqrt{x}} = 1 + \frac{1}{3}x - \frac{1}{45}x^2 + \ldots$$

is a multiplicative sequence $\{L_m\}$ called the **Hirzebruch L-sequence.** The first few terms of the sequence are

$$L_1(p_1) = \frac{1}{3}p_1$$

$$L_1(p_1,p_2) = \frac{1}{45}(7p_2 - p_1^2)$$

$$L_3(p_1,p_2,p_3) = \frac{1}{3^3 \cdot 5 \cdot 7}(62p_3 - 13p_1p_2 + 2p_1^3)$$

Given a real bundle E, the total L-class of E is the sum

$$\mathbf{L}(E) \equiv 1 + L_1(p_1E) + L_2(p_1E, p_2E) + \ldots .$$

If we write $E \otimes \mathbb{C} = \ell_1 \oplus \bar{\ell}_1 \oplus \cdots \oplus \ell_n \oplus \bar{\ell}_n$ according to the Splitting Principle, then from (11.28) we see that

$$\mathbf{L}(E) = \prod_{j=1}^{n} \frac{x_j}{\tanh x_j}, \tag{11.18}$$

where $p_jE = \sigma_j(x_1^2, \ldots, x_n^2)$.

Closely related to the L-sequence is the $\hat{L}$-sequence $\{L_m\}$ determined by the power series $\hat{\ell}(x) = \ell(x/4)$. One easily sees that $L_m = 4^m\hat{L}_m$. For a real oriented bundle E of dimension n, we have

$$\hat{\mathbf{L}}(E) = \prod_{j=1}^{n} \frac{x_j/2}{\tanh(x_j/2)}. \tag{11.18$'$}$$

Suppose we fix a multiplicative sequence $\{F_k\}$ as above. Then for each differentiable manifold X we define the **total F-class of X** to be

$$\mathbf{F}(X) \equiv \mathbf{F}(TX) \in H^{4*}(X;\mathbb{Q})$$

In particular, the examples above give us a **total $\hat{A}$-class** $\hat{\mathbf{A}}(X)$ and a **total L-class** $\mathbf{L}(X)$.

If we furthermore assume that X is compact, oriented and of dimension n, then we define the **F-genus of X** to be the rational number $F(X)$ obtained by evaluating $\mathbf{F}(X)$ on the fundamental homology class $[X] \in H_n(X;\mathbb{Q})$ of the manifold. In other words:

$$F(X) \equiv \mathbf{F}(X)[X] \equiv \begin{cases} F_k(p_1(X), \ldots, p_k(X))[X] & \text{if } n = 4k \\ 0 & \text{if } n \not\equiv 0 \pmod 4 \end{cases} \tag{11.19}$$

From (11.15) one easily deduces that the F-genus is multiplicative in the sense that

$$F(X \times X') = F(X)F(X') \tag{11.20}$$

for any pair of compact oriented manifolds X and X'.

Suppose now that Y is a compact oriented manifold with boundary $\partial Y = X$. Note that $TY|_X = TX \oplus$ (trivial), and so $p(Y)|_X = p(X)$. Consequently, $\mathbf{F}(Y)|_X = \mathbf{F}(X)$ and, since X is homologous to zero in Y, the F-genus of X must be zero. This proves that the F-genus of a manifold depends only on its oriented cobordism class. In fact the F-genus gives a ring homomorphism

$$F : \Omega_*^{SO} \longrightarrow \mathbb{Q} \tag{11.21}$$

from the oriented cobordism ring into $\mathbb{Q}$.

Two important examples here are the **$\hat{A}$-genus** and the **L-genus.** An important result of F. Hirzebruch says that for any compact oriented $4k$-manifold X,

$$L(X) = \text{sig}(X).$$

This can be proved by direct verification on a set of generators for the ring $\Omega_*^{SO} \otimes \mathbb{Q}$ (cf. Hirzebruch [1]). It also follows from the Atiyah-Singer Index Theorem.

Note in particular that $L(X)$ is always an integer. This is not true for $\hat{A}(X)$. The formulas above show, for example, that $\hat{A}(\mathbb{P}^2(\mathbb{C})) = -(1/8)L(\mathbb{P}^2(\mathbb{C})) = -1/8$. Nevertheless, it will follow from the Index Theorem that $\hat{A}(X)$ is an integer when X is a spin manifold. It is a fact, incidentally, that the A-genus is always an integer.

We now discuss the Chern character. Let E be a complex vector bundle of dimension n over a manifold X, and via the Splitting Principle express the total rational Chern class of E formally as

$$c(E) = 1 + c_1 + \ldots + c_n = \prod_{k=1}^{n} (1 + x_k)$$

so that $c_k = \sigma_k(x_1, \ldots, x_n)$. Consider the expression

$$\mathrm{ch}(E) = e^{x_1} + \ldots + e^{x_n} = n + \sum_{j=1}^{n} x_j + \frac{1}{2}\sum_{j=1}^{n} x_j^2 + \ldots \tag{11.22}$$

The term of degree k in this expression is just the symmetric polynomial

$$\mathrm{ch}^k E = \frac{1}{k!}\sum_{j=1}^{n} x_j^k \tag{11.23}$$

which can be rewritten as a universal polynomial expression in the elementary symmetric functions $c_1, \ldots, c_n$. In particular, $\mathrm{ch}\, E = n + \mathrm{ch}^1 E + \mathrm{ch}^2 E + \ldots$ is a well-defined element of $H^{2*}(X; \mathbb{Q})$. It is called the **Chern character** of E.

Note that if ℓ is a complex *line* bundle, then

$$\mathrm{ch}(\ell) = e^{c_1(\ell)} \tag{11.24}$$

where $c_1(\ell)$ denotes the first Chern class of ℓ.

The importance of the Chern character lies in the fact that it respects the (semi) ring structure on the set of vector bundles.

Proposition 11.16. *The Chern character has the following properties for any pair of complex vector bundles E and E' over X:*

(i) $\mathrm{ch}(E \oplus E') = \mathrm{ch}(E) + \mathrm{ch}(E')$
(ii) $\mathrm{ch}(E \otimes E') = \mathrm{ch}(E)\mathrm{ch}(E')$.

Proof. Consider formal splittings

$$c(E) = \prod_{k=1}^{n} (1 + x_k) \qquad c(E') = \prod_{j=1}^{m} (1 + x_j')$$

where as above the Chern classes are the elementary symmetric functions of the x's. Then we have corresponding splittings

$$c(E \otimes E') = \prod_{k=1}^{n} (1 + x_k) \prod_{j=1}^{m} (1 + x_j')$$

$$c(E \otimes E') = \prod_{j=1}^{m} \prod_{k=1}^{n} (1 + x_k + x_j')$$

The first is obvious. The second is a consequence of the basic fact that for complex line bundles $c_1(\ell \otimes \ell') = c_1(\ell) + c_1(\ell')$. (Compare Note 11.5

or (A.7) in App. A.) By definition we now have

$$\operatorname{ch}(E \oplus E') = \sum_{k=1}^{n} e^{x_k} + \sum_{j=1}^{m} e^{x'_j} = \operatorname{ch}(E) + \operatorname{ch}(E')$$

$$\operatorname{ch}(E \otimes E') = \sum_{j=1}^{m} \sum_{k=1}^{n} e^{x_k + x'_j} = \left(\sum_{k=1}^{n} e^{x_k}\right)\left(\sum_{j=1}^{m} e^{x'_j}\right) = \operatorname{ch}(E)\operatorname{ch}(E') \quad \blacksquare$$

Corollary 11.17. *For any compact Hausdorff space X, the Chern character descends to a ring homomorphism*

$$\operatorname{ch}: K(X) \longrightarrow H^{2*}(X; \mathbb{Q}).$$

REMARK 11.18. It is a result of Atiyah and Hirzebruch [2] that if, say, X is a finite complex, then the associated map $\operatorname{ch}: K(X) \otimes \mathbb{Q} \to H^{2*}(X; \mathbb{Q})$ is an *isomorphism*. They show, moreover, that this map extends to a ring isomorphism $\operatorname{ch}: K^*(X) \otimes \mathbb{Q} \xrightarrow{\approx} H^*(X; \mathbb{Q})$ which carries $K^1(X) \otimes \mathbb{Q}$ onto $H^{\text{odd}}(X; \mathbb{Q})$.

We now examine some basic constructions in K-theory. To make calculations we shall always assume our bundles to be a direct sum of line bundles. This is justified by the Splitting Principle 11.1 and 11.2 provided the answer is independent of the splitting. We assume throughout that X is a manifold or a finite simplicial complex.

CONSTRUCTION 11.19 (the exterior power operations). Let E be a complex vector bundle of dimension n over X and for each k, $1 \le k \le n$, consider the bundle $\Lambda^k E$. This operation on vector bundles has the property

$$\Lambda^k(E \oplus E') = \sum_{i+j=k} (\Lambda^i E) \otimes (\Lambda^j E'). \tag{11.25}$$

To extend the operation to K-theory we consider the ring $K(X)[[t]]$ of formal power series with coefficients in $K(X)$. Assigning to a vector bundle E the element

$$\lambda_t(E) = \sum_{k=0} [\Lambda^k E] t^k \tag{11.26}$$

gives a map with the property that

$$\lambda_t(E \oplus F) = \lambda_t(E)\lambda_t(F) \tag{11.27}$$

by (11.25). From the universal property of $K(X)$ the map (11.24) extends to a group homomorphism

$$\lambda_t: K(X) \longrightarrow K(X)[[t]]^{\hat{}}.$$

The k-component of this map is called the **kth power operation** on $K(X)$.

For vector bundles, $\lambda_t(E)$ is a polynomial and $\lambda_m(E)$ is a well-defined element of $K(X)$ for all $m \in \mathbb{Z}$. For example, $\lambda_t(\ell) = 1 + t[\ell]$ for any line bundle ℓ. For general elements of $K(X)$ the above statement is false. Note, for example, that $\lambda_t(-[\ell]) = (1 + t[\ell])^{-1} = \sum (-t)^m[\ell]^m$.

Fix a vector bundle E and consider a formal splitting $E = \ell_1 \oplus \cdots \oplus \ell_n$ with $x_k = c_1(\ell_k)$. From (11.27) we have that $\lambda_t(E) = \prod \lambda_t(\ell_k) = \prod (1 + t[\ell_k])$, and so

$$\mathrm{ch}(\lambda_t E) = \prod_{k=1}^{n} (1 + te^{x_k}). \tag{11.28}$$

In particular, we conclude that

$$\mathrm{ch}(\lambda_{-1} E) = \mathrm{ch}(\Lambda^{\mathrm{even}} E - \Lambda^{\mathrm{odd}} E) = \prod_{k=1}^{n} (1 - e^{x_k}). \tag{11.29}$$

CONSTRUCTION 11.20 (the Adams operations). Closely related to the power operations are a family of ring homomorphisms $\psi_k : K(X) \to K(X)$ called the **Adams operations.** For a line bundle ℓ, they are defined by setting,

$$\psi_k(\ell) \equiv \ell^k. \tag{11.30}$$

The extension is then determined by the Splitting Principle. Write $E = \ell_1 \oplus \cdots \oplus \ell_n$ and define $\psi_k(E) = \ell_1^k \oplus \cdots \oplus \ell_n^k$. We see that

$$\left.\begin{aligned} \psi_k(E \oplus E') &= \psi_k(E) \oplus \psi_k(E'), \\ \psi_k(E \otimes E') &= \psi_k(E) \otimes \psi_k(E'). \end{aligned}\right\} \tag{11.31}$$

The first is obvious. For the second, note that $\psi_k[(\sum \ell_i) \otimes (\sum \ell'_j)] = \psi_k[\sum \ell_i \otimes \ell'_j] = \sum (\ell_i \otimes \ell'_j)^k = \sum \ell_i^k \otimes (\ell'_j)^k = (\sum \ell_i^k) \otimes (\sum \ell'^k_j) = \psi_k(\sum \ell_i) \otimes \psi_k(\sum \ell'_j)$.

Note that $\mathrm{ch}\, \psi_k E = \sum e^{kx_j} = \rho_k \mathrm{ch}\, E$ where $\rho_k : H^{2*}(X; \mathbb{Q}) \to H^{2*}(X; \mathbb{Q})$ is defined by setting $\rho_k \equiv k^m$ on $H^{2m}(X; \mathbb{Q})$.

The Adams operations transform naturally under the homomorphism $f^* : K(X) \to K(Y)$ induced by a continuous map $f : Y \to X$.

It is an interesting exercise to show that $\sum_{k=1}^{\infty} (-t)^k \psi_k(E) = -t[(d/dt)\lambda_t(E)]/\lambda_t(E)$.

CONSTRUCTION 11.21 (the Clifford difference element). Let E be a real oriented riemannian vector bundle of dimension $2n$, and let $\omega_E \equiv i^n e_1 \cdots e_{2n}$ be the oriented unit volume element in the complex Clifford bundle $\mathbb{C}\ell(E) = C\ell(E) \otimes \mathbb{C}$. Since $\omega_E^2 = 1$, we have a splitting $\mathbb{C}\ell(E) = \mathbb{C}\ell^+(E) \oplus \mathbb{C}\ell^-(E)$ where $\mathbb{C}\ell^{\pm}(E) \equiv (1 \pm \omega_E)\mathbb{C}\ell(E)$. We then define the **Clifford difference element**

$$\delta(E) \equiv [\mathbb{C}\ell^+(E)] - [\mathbb{C}\ell^-(E)] \in K(X). \tag{11.32}$$

Suppose E' is another such bundle. From the fact that $\mathbb{C}\ell(E \oplus E') = \mathbb{C}\ell(E) \hat{\otimes} \mathbb{C}\ell(E')$ and $\omega_{E \oplus E'} = \omega_E \omega_{E'}$, one sees easily that

$$\delta(E \oplus E') = \delta(E)\delta(E'). \tag{11.33}$$

The proof of the Splitting Principle 11.2 shows that we may consider E to be a direct sum of 2-plane bundles $E = E_1 \oplus \cdots \oplus E_n$, and $E \otimes \mathbb{C} = \ell_1 \oplus \bar{\ell}_1 \oplus \cdots \oplus \ell_n \oplus \bar{\ell}_n$ where $E_j \otimes \mathbb{C} = \ell_j \oplus \bar{\ell}_j$.

Consider therefore the case where $\dim E = 2$. We have $E \otimes \mathbb{C} = \ell \oplus \bar{\ell}$ for some complex line bundle ℓ. In fact, if at a fixed point $x \in X$ we choose an oriented orthonormal basis (e_1, e_2) of E_x, then $\ell_x = \mathbb{C} \cdot (e_1 - ie_2)$ and $\bar{\ell}_x = \mathbb{C} \cdot (e_1 + ie_2)$. Clearly $\mathbb{C}\ell(E) = \mathbb{C} \cdot 1 \oplus \mathbb{C} \cdot \omega_E \oplus \ell \oplus \bar{\ell}$, and since $\omega_E = ie_1e_2$ we find that

$$\mathbb{C}\ell^+(E) = \mathbb{C} \cdot (1 + \omega_E) \oplus \bar{\ell} \qquad \text{and} \qquad \mathbb{C}\ell^-(E) = \mathbb{C} \cdot (1 - \omega_E) \oplus \ell.$$

Since the bundles $\mathbb{C} \cdot (1 \pm \omega_E)$ are trivial, we find that for $\dim E = 2$,

$$\delta(E) = [\bar{\ell}] - [\ell].$$

Applying the Splitting Principle and (11.33), we conclude that for $E = E_1 \oplus \cdots \oplus E_n$,

$$\delta(E) = \prod_{k=1}^{n} ([\bar{\ell}_k] - [\ell_k])$$

and so

$$\operatorname{ch}[\delta(E)] = \prod_{k=1}^{n} (e^{-x_k} - e^{x_k}) \tag{11.34}$$

where $x_k = c_1(\ell_k)$ for $k = 1, \ldots, n$. From (11.6) we know that $\chi(E) = x_1 \cdots x_n$. Consequently, one verifies that

$$\begin{aligned} \operatorname{ch}[\delta(E)] &= \chi(E) \prod_{k=1}^{n} \left(\frac{e^{-x_k} - e^{x_k}}{x_k}\right) \\ &= (-1)^n \chi(E) \prod_{k=1}^{n} \left(\frac{e^{x_k/2} - e^{-x_k/2}}{x_k}\right)(e^{x_k/2} + e^{-x_k/2}) \\ &= (-2)^n \chi(E) \prod_{k=1}^{n} \left(\frac{\sinh(x_k/2)}{x_k/2}\right)^2 \left(\frac{x_k/2}{\tanh(x_k/2)}\right) \end{aligned}$$

From (11.17) and (11.18′) we conclude the following:

Proposition 11.22. *For any oriented real vector bundle E of dimension 2n, one has the relation*

$$\boxed{\operatorname{ch}[\delta(E)] = (-2)^n \chi(E) \hat{\mathbf{L}}(E) \hat{\mathbf{A}}(E)^{-2}.}$$

CONSTRUCTION 11.23 (the spinor difference element). Let E and ω_E be as in Construction 11.21 and suppose E carries a spin structure. Let $S_{\mathbb{C}}(E)$ be the complex spinor bundle for E and consider the decomposition $S_{\mathbb{C}}(E) = S_{\mathbb{C}}^+(E) \oplus S_{\mathbb{C}}^-(E)$ where $S_{\mathbb{C}}^{\pm}(E) = (1 \pm \omega_E)S_{\mathbb{C}}(E)$. We define the **spinor difference element** to be

$$s(E) \equiv [S_{\mathbb{C}}^+(E)] - [S_{\mathbb{C}}^-(E)] \in K(X). \tag{11.35}$$

It follows easily from the material of Chapter I that if E' is another such bundle, then

$$s(E \oplus E') = s(E)s(E'). \tag{11.36}$$

As above we may now restrict attention to the case $\dim E = 2$ where $E \otimes \mathbb{C} = \ell \oplus \bar{\ell}$. Recall from (11.5) that as oriented 2-plane bundles $E \cong \ell$. The fact that E is spin is equivalent to the fact that $\chi(E) = c_1(\ell)$ is even, i.e., that ℓ has a square root $\ell^{1/2}$. The spinor bundle is given by

$$S_{\mathbb{C}}(E) = \ell^{1/2} \oplus \bar{\ell}^{1/2}$$

One verifies that $S_{\mathbb{C}}^+(E) = \bar{\ell}^{1/2}$ and $S_{\mathbb{C}}^-(E) = \ell^{1/2}$, and so

$$s(E) = [\bar{\ell}^{1/2}] - [\ell^{1/2}].$$

Passing to the general case $E = E_1 \oplus \cdots \oplus E_n$ via the Splitting Principle as above, we then have

$$s(E) = \prod_{k=1}^{n} ([\bar{\ell}_k^{1/2}] - [\ell_k^{1/2}]),$$

and so

$$\mathrm{ch}[s(E)] = \prod_{k=1}^{n} (e^{-x_k/2} - e^{x_k/2}) \tag{11.37}$$

This proves the following:

Proposition 11.24. *For any real spin vector bundle E of dimension $2n$, one has that*

$$\boxed{\mathrm{ch}[s(E)] = (-1)^n \chi(E) \hat{\mathbf{A}}(E)^{-1}}$$

§12. Thom Isomorphisms and the Chern Character Defect

We present here some material relevant to understanding the general form of the Index Theorem. It is not necessary, however, for understanding the cohomological formula for the index in the basic cases.

Let X be an oriented n-dimensional manifold which is not necessarily compact. Let $H^*_{\mathrm{cpt}}(X)$ denote the cohomology of the complex of rational singular cochains with compact support on X. (A cochain c has **compact support** if there is a compact subset $K \subset X$ such that $c(\sigma) = 0$ for any chain σ which does not meet K.) **Poincaré duality** states that there is a canonical isomorphism

$$\mathscr{D}_X : H^p_{\text{cpt}}(X) \xrightarrow{\approx} H_{n-p}(X) \tag{12.1}$$

for $p = 0, \ldots, n$ (see de Rham [1]). If Y is another such manifold of dimension m, and if $f : Y \to X$ is a continuous map, then for each p with $p \geqq m - n$ there is a linear map

$$f_! : H^p_{\text{cpt}}(Y) \longrightarrow H^{p-(m-n)}_{\text{cpt}}(X) \tag{12.2}$$

called **integration over the fibre** (or the **Gysin homomorphism**). It is defined by setting $f_!(u) = \mathscr{D}_X^{-1} f_* \mathscr{D}_Y$ where f_* is the usual map induced on homology.

An important example is provided by the following. Let E be an oriented vector bundle of fibre dimension k over X. Consider the maps

$$\pi : E \longrightarrow X \qquad \text{and} \qquad i : X \longrightarrow E$$

where π is the bundle projection and i denotes inclusion as the zero section. Since these maps are homotopy equivalences, π_* and i_* induce isomorphisms on homology with $\pi_* i_* = Id$. Consequently, the maps

$$\pi_! : H^{p+k}_{\text{cpt}}(E) \longrightarrow H^p_{\text{cpt}}(X) \qquad \text{and} \qquad i_! : H^p_{\text{cpt}}(X) \longrightarrow H^{p+k}_{\text{cpt}}(E)$$

are isomorphisms for all p. Since $\pi_! i_! = \mathscr{D}_X^{-1} \pi_* \mathscr{D}_E \mathscr{D}_E^{-1} i_* \mathscr{D}_X = \text{Id}$, we see that

$$\pi_! = (i_!)^{-1}. \tag{12.3}$$

DEFINITION 12.1. The map $i_! : H^p_{\text{cpt}}(X) \to H^{p+k}_{\text{cpt}}(E)$ is called the **Thom isomorphism** of E for compactly supported cohomology.

Note that if X is compact, then $H^p_{\text{cpt}}(X) = H^p(X)$. Furthermore, if D_E denotes the disk bundle of E, then by excision and Lefschetz duality we have

$$H^{k+p}_{\text{cpt}}(E) \cong H_{n-p}(E) \cong H_{n-p}(D_E) \cong H^{k+p}(D_E, \partial D_E). \tag{12.4}$$

We thereby recover the Thom isomorphism in its more conventional form (cf. Milnor-Stasheff [1]). We also have the following basic result:

Lemma 12.2. *If X is compact, then for all $u \in H^*(X) = H^*_{\text{cpt}}(X)$, one has*

$$i^* i_!(u) = \chi(E) \cdot u \tag{12.5}$$

where $\chi(E)$ is the Euler class of E.

Proof. We shall only outline an argument (for full details, see Milnor-Stasheff [1]). Let $Z = i_*[X] = i_* \mathscr{D}_X(1)$ be the class of the zero-section in

$H_n(E)$. The equation

$$\chi(E) = i^* i_!(1) = i^* \mathscr{D}_E^{-1} Z$$

is one of the standard characterizations of the Euler class. To wit, if $c \in H_k(X)$, then

$$(\chi(E), c) = (\mathscr{D}_E^{-1} Z, i_* c) = (1, Z \cap i_* c),$$

that is to say that χ is given by intersection *in* E with the zero section. For a general class $u \in H^p(X)$, we fix $c \in H_{k+p}(X)$ and note that

$$\begin{aligned}(\chi(E) \cdot u, c) &= (\chi(E), \mathscr{D}_X u \cap c) = (1, Z \cap i_*(\mathscr{D}_X u \cap c)) \\ &= (1, (i_* \mathscr{D}_X u) \cap i_* c) = (\mathscr{D}_E^{-1} i_* \mathscr{D}_X u, i_* c) \\ &= (i_! u, i_* c) = (i^* i_! u, c) \quad \blacksquare\end{aligned}$$

In K-theory there is a group $K_{\text{cpt}}(X)$ analogous to the group $H^*_{\text{cpt}}(X)$ discussed above. It consists of homotopy classes of triples $[E, F; \sigma]$ where E and F are complex vector bundles over X and where σ is a bundle isomorphism from E to F defined outside some compact subset of X (see I.9). If X is a compact manifold, then $K_{\text{cpt}}(X) = K(X)$. If $U \subset X$ is an open subset of any manifold, then there is a natural inclusion homomorphism $K_{\text{cpt}}(U) \to K_{\text{cpt}}(X)$.

Suppose now that $\pi: E \to X$ is a *complex* vector bundle of rank k over a manifold, and let $i: X \to E$ be the inclusion as the zero section. Then, as proved in Appendix C, Theorem C.8, there is a natural **Thom isomorphism** $i_!: K_{\text{cpt}}(X) \to K_{\text{cpt}}(E)$ of the form

$$i_!(u) = \Lambda_{-1} \cdot \pi^* u$$

where $\Lambda_{-1} = [\pi^* \Lambda_{\mathbb{C}}^{\text{even}} E, \pi^* \Lambda_{\mathbb{C}}^{\text{odd}} E; \sigma]$ and where σ is defined at each non-zero vector e in E by setting

$$\sigma_e = e \wedge - (e^*) \llcorner.$$

(The element e^* is the dual of e under some fixed hermitian metric.) If X is compact, then $\Lambda_{-1} = i_!(1)$ is a well-defined element of $K_{\text{cpt}}(E)$. When X is not compact, the product $\Lambda_{-1} \cdot \pi^* u$ can still be shown to be a well-defined element of $K_{\text{cpt}}(E)$ (cf. Karoubi [2]). Roughly speaking, $\pi^* u$ has compact support in the "X-directions" and Λ_{-1} has compact support in "fibre-directions". If we restrict the element Λ_{-1} to the zero section, we recover the element $\lambda_{-1}(E) \in K^*(X)$. This gives the following.

Lemma 12.3. *If X is compact, then for all $\xi \in K(X) = K_{\text{cpt}}(X)$, one has*

$$i^* i_!(\xi) = \lambda_{-1}(E) \cdot \xi \tag{12.6}$$

where $\lambda_{-1}(E) = [\Lambda^{\text{even}} E] - [\Lambda^{\text{odd}} E] \in K(X)$.

Assume now that X and Y are smooth manifolds and that $f: X \hookrightarrow Y$ is a smooth proper embedding. Assume furthermore that the normal bundle N to $f(X)$ is equipped with a complex structure. (Hence, $\dim Y - \dim X$ is even.) Under these circumstances we can define a natural mapping

$$f_!: K_{\text{cpt}}(X) \to K_{\text{cpt}}(Y)$$

by taking the Thom isomorphism $i_!: K_{\text{cpt}}(X) \to K_{\text{cpt}}(N)$ followed by the map $K_{\text{cpt}}(N) \to K_{\text{cpt}}(Y)$ obtained by identifying N with a regular neighborhood of X in Y.

Observe now that for any proper embedding $f: X \hookrightarrow Y$ of manifolds, the normal bundle to the associated (proper) embedding $f_*: TX \hookrightarrow TY$ has a canonical complex structure. This normal bundle is just the pull-back to TX of $N \oplus N$ where N is the normal bundle to X. The first factor is thought of as lying in "manifold-directions," the second in "fibre-directions." The complex structure is given by

$$T = \begin{pmatrix} 0 & -\text{Id} \\ \text{Id} & 0 \end{pmatrix}$$

Consequently, for any proper embedding of manifolds $f: X \hookrightarrow Y$, there is an associated map

$$f_!: K_{\text{cpt}}(TX) \longrightarrow K_{\text{cpt}}(TY) \tag{12.7}$$

which is of fundamental importance in defining the topological index of an elliptic operator.

In the last section we defined the Chern character $\text{ch}: K(X) \to H^{\text{even}}(X)$. This homomorphism has a direct extension

$$\text{ch}: K_{\text{cpt}}(X) \longrightarrow H^{\text{even}}_{\text{cpt}}(X)$$

to the case of compact supports. For any given complex vector bundle $\pi: E \to X$, we have defined Thom isomorphisms:

$$K_{\text{cpt}}(X) \xrightarrow{i_!} K_{\text{cpt}}(E) \qquad \text{and} \qquad H_{\text{cpt}}(X) \xrightarrow{i_!} H_{\text{cpt}}(E),$$

and it is natural to ask whether $i_!\text{ch} = \text{ch}\, i_!$. This is not true in general and the resulting "correction term" is of basic importance.

We assume from this point on that the manifold X is compact. Then to each complex vector bundle $\pi: E \to X$ we associate the class

$$\mathfrak{I}(E) \equiv \pi_!\text{ch}\, i_!(1) \tag{12.8}$$

Note that for any $\xi \in K(X)$ we have $\pi_!\text{ch}\, i_!\xi = \pi_!\text{ch}(i_!(1) \cdot \pi^*\xi) = \pi_![\text{ch}\, i_!(1) \cdot \text{ch}\, \pi^*\xi] = [\pi_!\text{ch}\, i_!(1)]\text{ch}\, \xi$, and so, since $\pi_! = (i_!)^{-1}$ on $H^*_{\text{cpt}}(E)$,

$$(i_!)^{-1}\text{ch}(i_!\xi) = \mathfrak{I}(E)\text{ch}\, \xi, \tag{12.9}$$

That is, $\mathfrak{I}(E)$ is just the "commutativity defect" mentioned above.

Now it is not difficult to check that $\mathfrak{T}(E)$ is natural, i.e., $f^*\mathfrak{T}(E) = \mathfrak{T}(f^*E)$ for any continuous map between manifolds. Consequently $\mathfrak{T}(E)$ is a characteristic which we shall compute. Note that $i_!\mathfrak{T}(E) = \text{ch } i_!(1) = \text{ch } \Lambda_{-1}$. Applying i^* and Propositions 12.2 and 12.3, we find that $\chi(E)\mathfrak{T}(E) = i^*i_!\mathfrak{T}(E) = i^*\text{ch } \Lambda_{-1} = \text{ch } i^*\Lambda_{-1} = \text{ch } \lambda_{-1}(E)$. This computation can be carried out for the universal bundle E over the classifying space BU_n whose cohomology ring is a polynomial ring generated by the Chern classes $c_1, \ldots, c_n$. Here we are authorized to write the equation

$$\mathfrak{T}(E) = \frac{\text{ch } \lambda_{-1}(E)}{\chi(E)} \tag{12.10}$$

If we split $c(E) = \prod (1 + x_j)$ formally as in §11, we find from equations (11.6) and (11.29) that

$$\mathfrak{T}(E) = \prod_{k=1}^{n} \frac{1 - e^{x_k}}{x_k}$$

From (11.11) and (11.9) we can rewrite this as

$$\mathfrak{T}(E) = (-1)^n \mathbf{Td}_{\mathbf{C}}(\bar{E})^{-1} \tag{12.11}$$

Assume now that E is a real oriented riemannian vector bundle of even dimension over X. Then one can define the basic element

$$\delta(E) \equiv [\pi^*\mathbb{C}\ell^+(E), \pi^*\mathbb{C}\ell^-(E); \mu] \in K_{\text{cpt}}(E) \tag{12.12}$$

where $\mu_e \equiv e\cdot$ denotes Clifford multiplication by e. It is evident that $i^*\delta(E) = \delta(E)$, the difference element considered in 11.21. Arguing as above we see that: $\chi(E)\pi_!\text{ch } \delta(E) = i^*i_!\pi_!\text{ch } \delta(E) = i^*\text{ch } \delta(E) = \text{ch } \delta(E)$. Applying the calculations of Proposition 11.22 then proves the following.

Propositon 12.4. *For any oriented real vector bundle E of dimension $2n$ on X, one has*

$$\boxed{\pi_!\text{ch } \delta(E) = (-2)^n \hat{\mathbf{L}}(E)\hat{\mathbf{A}}(E)^{-2}}$$

If we now assume that E has a spin structure, we can construct the element

$$\mathbf{s}(E) \equiv [\pi^*S_{\mathbf{C}}^+(E), \pi^*S_{\mathbf{C}}^-(E); \mu] \in K_{\text{cpt}}(E) \tag{12.13}$$

where again $\mu_e = e\cdot$ denotes Clifford multiplication by e. Clearly, $i^*\mathbf{s}(E) = \mathbf{s}(E)$ = the element discussed in Construction 11.23. Arguing as above and applying the calculation of Proposition 11.24, we find the following:

Proposition 12.5. *For any real spin vector bundle E of dimension $2n$ on X, one has*

$$\boxed{\pi_!\text{ch } \mathbf{s}(E) = (-1)^n \hat{\mathbf{A}}(E)^{-1}}$$

These last two propositions extend easily to the case of coefficients. Given any element $u \in K(X)$, one immeidately verifies that

$$\pi_! \mathrm{ch}[\boldsymbol{\delta}(E) \cdot \pi^* u] = (-2)^n \mathrm{ch}\, u \cdot \hat{\mathbf{L}}(E) \hat{\mathbf{A}}(E)^{-2} \tag{12.14}$$

$$\pi_! \mathrm{ch}[\mathbf{s}(E) \cdot \pi^* u] = (-1)^n \mathrm{ch}\, u \cdot \hat{\mathbf{A}}(E)^{-1} \tag{12.15}$$

Indeed, note that $\pi_! \mathrm{ch}(\boldsymbol{\delta}(E)\pi^* u) = \pi_![\mathrm{ch}\ \boldsymbol{\delta}(E) \mathrm{ch}\ \pi^* u] = \pi_![\mathrm{ch}\ \boldsymbol{\delta}(E)\pi^* \mathrm{ch}\ u] = [\pi_! \mathrm{ch}\ \boldsymbol{\delta}(E)] \mathrm{ch}\ u$. If $u = [E']$ is the class corresponding to a complex vector bundle E' over X, then

$$\begin{aligned} \boldsymbol{\delta}(E) \cdot \pi^* u &\cong [\pi^* \mathbb{C}\ell^+(E) \otimes E', \pi^* \mathbb{C}\ell^-(E) \otimes E'; \mu], \\ \mathbf{s}(E) \cdot \pi^* u &\cong [\pi^* S^+_{\mathbb{C}}(E) \otimes E', \pi^* S^-_{\mathbb{C}}(E) \otimes E'; \mu]. \end{aligned} \tag{12.16}$$

These elements $\boldsymbol{\delta}(E)$ and $\mathbf{s}(E)$ are fundamental. Using them, one can define **Thom isomorphisms in K-theory** for E as follows:

Proposition 12.6. *Let $\pi : E \to X$ be an oriented real vector bundle of dimension $2n$ on X. Then the map*

$$i_!^{\delta} : K(X) \otimes \mathbb{Q} \longrightarrow K_{\mathrm{cpt}}(E) \otimes \mathbb{Q} \qquad \text{given by} \quad i_!^{\delta}(u) \equiv \boldsymbol{\delta}(E) \cdot \pi^* u$$

is an (additive) isomorphism. If E is spin, then the map

$$i_!^{s} : K(X) \longrightarrow K_{\mathrm{cpt}}(E) \qquad \text{given by} \quad i_!^{s}(u) \equiv \mathbf{s}(E) \cdot \pi^* u$$

is an (additive) isomorphism.

For a proof of this proposition and a discussion of related results, the reader is referred to Appendix C.

§13. The Atiyah-Singer Index Theorem

We present here the topological formulas of Atiyah and Singer for the index of an elliptic operator on a compact manifold. We begin with the general K-theoretic formula for which we give a detailed proof. Then, using material derived above, we shall rewrite the formula in cohomological terms and work out the details in some important special cases.

Let X be a compact differentiable manifold of dimension n and consider an elliptic operator $P : \Gamma(E) \to \Gamma(F)$ where E and F are smooth complex vector bundles over X. Recall from §1 that the principal symbol $\sigma(P)$ of P defines a class

$$\boldsymbol{\sigma}(P) \equiv [\pi^* E, \pi^* F; \sigma(P)] \in K_{\mathrm{cpt}}(TX) \tag{13.1}$$

where $\pi : TX \to X$ is the tangent bundle of X. (We have identified $K_{\mathrm{cpt}}(TX)$ with $K(DX, \partial DX)$ since $DX/\partial DX$ is naturally homeomorphic to the one point compactification of TX.) Choose now a smooth embedding

$f: X \hookrightarrow \mathbb{R}^N$ into some euclidean space, and consider the induced map

$$f_!: K_{\text{cpt}}(TX) \longrightarrow K_{\text{cpt}}(T\mathbb{R}^N) \tag{13.2}$$

defined in (12.7). Follow this by the homomorphism

$$q_!: K_{\text{cpt}}(T\mathbb{R}^N) \longrightarrow K(\text{pt}) \cong \mathbb{Z} \tag{13.3}$$

where $q: T\mathbb{R}^N \to \text{pt}$ is the canonical "scrunch" map taking $T\mathbb{R}^N$ to a point. Note that $T\mathbb{R}^N = \mathbb{R}^N \oplus \mathbb{R}^N = \mathbb{C}^N$ and $q: \mathbb{C}^N \to \text{pt}$ can be considered as a vector bundle. Viewed in this way, the map $q_!$ is just the inverse of the Thom isomorphism $i_!: K(\text{pt}) \to K_{\text{cpt}}(\mathbb{C}^N)$. (It is also just the Bott periodicity map. For an alternative view, consider the embedding $T\mathbb{R}^N = \mathbb{R}^{2N} \to S^{2N} = \mathbb{R}^{2N} \cup \{\infty\}$ and the induced map $K_{\text{cpt}}(T\mathbb{R}^N) \to K(S^{2N}) \cong K(\text{pt})$.) Applying scrunch-shriek to $f_!$ gives the index.

DEFINITION 13.1. The **topological index** of P is the integer

$$\text{top-ind}(P) \equiv q_! f_! \sigma(P). \tag{13.4}$$

One must verify that this definition is independent of the choice of f. To begin consider $\tilde{f} = j \circ f$ where $j: \mathbb{R}^N \hookrightarrow \mathbb{R}^{N+N'}$ is a linear inclusion. The induced map $j_!: K_{\text{cpt}}(T\mathbb{R}^N) \to K_{\text{cpt}}(T\mathbb{R}^{N+N'})$ is just the Thom isomorphism for the bundle $\mathbb{C}^{N+N'} \to \mathbb{C}^N$, and one easily checks that $\tilde{q}_! \tilde{f}_! = q_! f_!$ where $\tilde{q}: T\mathbb{R}^{N+N'} \to \text{pt}$. If we are given two embeddings $f_0: X \hookrightarrow \mathbb{R}^{N_0}$ and $f_1: X \hookrightarrow \mathbb{R}^{N_1}$, then the embeddings $j_0 f_0: X \hookrightarrow \mathbb{R}^{N_0+N_1}$ and $j_1 \circ f_1: X \to \mathbb{R}^{N_0+N_1}$, defined as above, are isotopic. That is, $F_t = tj_1 \circ f_1 + (1-t)j_0 \circ f_0$, $0 \le t \le 1$, is a smooth family of embeddings. Applying the homotopy invariance of K_{cpt} completes the proof that (13.4) is independent of the choice of f. One of the basic results in mathematics is the following:

Theorem 13.2 (The Atiyah-Singer Index Theorem [1]). *For any elliptic operator P over a compact manifold, one has*

$$\text{ind}(P) = \text{top-ind}(P),$$

that is, the topological and analytic indices of P coincide.

Proof. For the purposes of the proof we introduce a special class of operators. Let E and F be (smooth) com lex vector bundles over a compact riemannian manifold X. An operator $P \in \Psi DO_m(E, F)$ is called **classical** if its principal symbol is homogeneous of degree m in ξ outside of some compact subset of T^*X, that is, P is classical if there is a constant c so that $\sigma_{t\xi}(P) = t^m \sigma_\xi(P)$ for all $\xi \in T^*X$ with $\|\xi\| \geqq$ c and for all $t \geqq 1$. If X is not compact, we define the classical operators to be those which have this property over every compact subdomain of X. The set of all such operators will be denoted $\Psi CO_m(E, F)$.

Given an operator $P \in \Psi CO_m(E, F)$ we can consider its **asymptotic principal symbol**

$$\hat{\sigma}_\xi(P) = \lim_{t\to\infty} \frac{\sigma_{t\xi}(P)}{t^m}$$

defined for all ξ in $\partial DX \equiv \{\xi \in T^*X : \|\xi\| = 1\}$. This gives us an exact sequence

$$0 \longrightarrow \Psi DO_{m-1}(E, F) \longrightarrow \Psi CO_m(E, F) \xrightarrow{\sigma} \Gamma(\mathrm{Hom}(\pi^*E, \pi^*F)) \longrightarrow 0 \tag{13.5}$$

where $\pi : \partial DX \to X$ is the bundle projection. The surjectivity of $\hat{\sigma}$ is seen as follows. Given a section $s \in \Gamma(\mathrm{Hom}(\pi^*E, \pi^*F))$, extend s smoothly to all of T^*X so that it is homogeneous of degree m in ξ for $\|\xi\| \geqq 1$. Given local trivializations of E and F over a coordinate chart U on X, one easily constructs an operator P in U with principal symbol s. Let $\{U_j\}$ be a finite covering of X by such charts and let $\{\chi_j^2\}$ be a partition of unity subordinate to this covering. Then the operator $P = \sum \chi_j P \chi_j \in \Psi CO_m(E, F)$ has principal symbol $\sigma(P) = s$, and the surjectivity is proved.

Our first main step in the proof will be to show that the analytic index makes sense at the symbolic level and, in fact, gives a well-defined homomorphism $\mathrm{ind} : K_{\mathrm{cpt}}(T^*X) \to \mathbb{Z}$. We begin with a technical lemma which will be useful later on. For this lemma, X is assumed to be a manifold which is not necessarily compact but is of finite topological type.

Lemma 13.3. *Let $\pi : B \to X$ be a smooth, real vector bundle over X. Then every element in $K_{\mathrm{cpt}}(B)$ can be represented by a triple of the form $(\pi^*E, \pi^*F; \sigma) \in \mathscr{L}_1(B)_{\mathrm{cpt}}$ where E and F are vector bundles on X which are trivial outside a compact set, and where $\sigma : \pi^*E \to \pi^*F$ is homogeneous of degree 0 on the fibres of B (wherever it is defined).*

Note that outside a compact subset of X, σ is defined everywhere on the fibres. At such points the homogeneity implies that σ is in fact constant on the fibres.

Proof. We know from Chapter I, §9 that any element in $K_{\mathrm{cpt}}(B)$ can be represented by a triple $(E_0, F_0; \sigma_0)$ where $\sigma_0 : E_0 \to F_0$ is a bundle equivalence defined outside a compact subset $K \subset B$. There exists a bundle $E_0^\perp$ on B so that the sum $E_0 \oplus E_0^\perp$ is trivial, and we can replace $(E_0, F_0; \sigma_0)$ with the equivalent triple $(\tilde{E}, \tilde{F}; \tilde{\sigma}) \equiv (E_0 \oplus E_0^\perp, F_0 \oplus E_0^\perp; \sigma \oplus \mathrm{Id})$. Then there exist trivializations

$$\tau_{\tilde{E}} : \tilde{E}|_{(B-K)} \xrightarrow{\approx} (B - K) \times \mathbb{C}^m \quad \text{and} \quad \tau_{\tilde{F}} : \tilde{F}|_{(B-K)} \xrightarrow{\approx} (B - K) \times \mathbb{C}^m$$

so that $\tilde{\sigma} = \tau_{\tilde{F}}^{-1} \circ \tau_{\tilde{E}}$. (Let $\tau_{\tilde{E}}$ be the assumed trivialization, and set $\tau_{\tilde{F}} = \tau_{\tilde{E}} \circ \tilde{\sigma}^{-1}$.)

Choose now a compact domain $\Omega \subset X$ so that $K \subset B|_{\Omega}$. Set $E = i^*\tilde{E}$ and $F = i^*\tilde{F}$ where $i: X \hookrightarrow B$ is the zero-section, and let τ_E and τ_F denote the restrictions of the trivializations $\tau_{\tilde{E}}$ and $\tau_{\tilde{F}}$ to E and F respectively. We claim that over B there exist bundle isomorphisms

$$f_E: \tilde{E} \xrightarrow{\approx} \pi^*E \qquad \text{and} \qquad f_F: \tilde{F} \xrightarrow{\approx} \pi^*F \tag{13.6}$$

which are compatible with the given trivializations over $B|_{(X-\Omega)}$, i.e., which have the property that

$$f_E = \tau_E^{-1} \circ \tau_{\tilde{E}} \qquad \text{and} \qquad f_F = \tau_F^{-1} \circ \tau_{\tilde{F}}.$$

at points of $B|_{(X-\Omega)}$. These isomorphisms are constructed as follows. Let $h: B \times [0,1] \to B$ be the homotopy defined by $h(b,t) = tb$, and set $\mathscr{E} = h^*\tilde{E}$ and $\mathscr{F} = h^*\tilde{F}$. Note that

$$\mathscr{E}|_{B\times\{0\}} = \pi^*E, \qquad \mathscr{F}|_{B\times\{0\}} = \pi^*F$$

$$\mathscr{E}|_{B\times\{1\}} = \tilde{E}, \qquad \mathscr{F}|_{B\times\{1\}} = \tilde{F}$$

Introduce connections on $\mathscr{E}$ and $\mathscr{F}$ which extend the canonical flat connections (compatible with the trivializations) over $B|_{(X-\Omega)}$. Parallel transport along the curves $b \times [0,1]$ gives the desired maps (13.6). The bundle map $\sigma = f_F \circ \tilde{\sigma} \circ f_E^{-1}: \pi^*E \to \pi^*F$ is an isomorphism which is defined on $B - K$ and constant on the fibres of $B - \pi^{-1}(\Omega)$. Fix $r > 0$ so that $K \subset \{b \in B|_{\Omega}: \|b\| \leqq r\}$. We now redefine σ in the set where $\|b\| \geqq r$ so that it is homogeneous of degree zero (by setting $\sigma_{t\xi} = \sigma_{\xi}$ for $\|\xi\| = r$ and $t \geqq 1$). This gives the desired triple and completes the proof of the lemma. ∎

Suppose now that X is a compact manifold and choose an element $u \in K_{\text{cpt}}(T^*X)$. Represent u by an element $(\pi^*E, \pi^*F; \sigma)$ as in Lemma 13.3. Fix an integer m. From the discussion above we know that we can choose an (elliptic) operator $P \in \Psi CO_m(E, F)$ whose asymptotic principal symbol is exactly σ, and so in particular $\boldsymbol{\sigma}(P) = u$. We now set

$$\operatorname{ind} u \equiv \operatorname{ind} P \tag{13.7}$$

and show that this definition is independent of all the choices involved. We know from §7 that ind P depends only on the homotopy class of its principal symbol. Now if $P' \in \Psi CO_m(E, F)$ satisfies $\hat{\sigma}(P') = \hat{\sigma}(P)$, then $\sigma(P')$ and $\sigma(P)$ are homotopic (rel ∞). Therefore, ind P is independent of the choice of P with a given asymptotic principal symbol. It is also independent of the homotopy class of the representative $(\pi^*E, \pi^*F; \sigma)$ of u. To see this suppose that $(\pi^*E', \pi^*F'; \sigma')$ is another such representative and that there exists an element $\mathbf{a} = (\tilde{E}, \tilde{F}; \tilde{\sigma}) \in \mathscr{L}_1(T^*X \times [0,1])_{\text{cpt}}$ whose restriction to $T^*X \times \{k\}$ is $(\pi^*E^{(k)}, \pi^*F^{(k)}; \sigma^{(k)})$ for $k = 0,1$. The argument used for Lemma 13.3 applies here to prove that $\mathbf{a}$ can be replaced with an element of the form $(\pi^*\tilde{E}, \pi^*\tilde{F}; \tilde{\sigma})$ where $\tilde{E}$ and $\tilde{F}$ are bundles on $X \times [0,1]$ and where $\tilde{\sigma}$ is homogeneous of degree zero outside a compact set. Given mth order operators P and P' associated as above to these representatives, one can easily use the element $(\pi^*\tilde{E}, \pi^*\tilde{F}; \tilde{\sigma})$ to construct a homotopy between them and thereby show that ind P = ind P' as claimed.

Suppose now that we have two distinct representatives $\mathbf{u}_k = (\pi^*E_k, \pi^*F_k; \sigma_k)$ for our class u, and that we have chosen associated zero-order operators P_k where $k = 0,1$. By the definition of the equivalence defining $L_1(T^*X)_{\text{cpt}} \cong K(T^*X)_{\text{cpt}}$ this means that there exist elementary complexes $\mathbf{e}_k = (\pi^*G_k, \pi^*G_k; \text{Id})$, $k = 0,1$, so that $\mathbf{a}_0 \oplus \mathbf{e}_0$ and $\mathbf{a}_1 \oplus \mathbf{e}_1$ are homotopic. We can choose associated operators $P'_k = P_k \oplus \text{Id}$ for $\mathbf{a}_k \oplus \mathbf{e}_k$, and one easily sees that ind P'_k = ind P_k for each k. Hence, we have ind P_0 = ind P_1, and so ind u is well defined at least if we choose operators of order $m = 0$.

The definition is also independent of the choice of the order m. To see this, suppose we are given an elliptic operator $P \in \Psi CO_m(E, F)$. Choose a metric and a unitary connection on E, and let $\nabla^*\nabla$ be the associated laplacian on E (see II.8.3). Then for any integer ℓ, we consider the composition $P_\ell = P \circ (1 + \nabla^*\nabla)^{\ell/2} \in \Psi CO_{m+\ell}(E, F)$. It is easily seen that $\hat{\sigma}(P_\ell) = \hat{\sigma}(P)$ and since $(1 + \nabla^*\nabla)^{\ell/2}$ is invertible, that ind P_ℓ = ind P. It follows that ind u is well defined using operators of any order.

We have now shown that for any compact manifold X, the analytic index (13.7) gives a well-defined homomorphism:

$$\text{ind}: K_{\text{cpt}}(T^*X) \longrightarrow \mathbb{Z} \tag{13.8}$$

Our task now is to prove that this coincides with the homomorphism top-ind defined above. This will be accomplished if we can establish the following two properties:

Property 1. *In the special case where* $X = T^*X = \text{pt}$, *the homomorphism* $\text{ind}: K(\text{pt}) \to \mathbb{Z}$ *is the identity.*

Property 2. *If* X *and* Y *are compact manifolds and* $f: X \hookrightarrow Y$ *is a smooth embedding, then*

$$\text{ind}(u) = \text{ind}(f_! u)$$

for all $u \in K_{\text{cpt}}(T^*X)$, *where* $f_!$ *is the homomorphism* (12.7).

To see that these properties suffice to prove the theorem, we first choose an embedding $f: X \hookrightarrow S^N$ and let $j: \text{pt} \hookrightarrow S^N$ denote the inclusion of a point. By Property 2 we have $\mathrm{ind}(u) = \mathrm{ind}(f_!u) = \mathrm{ind}(j_!^{-1}f_!u)$, and by Property 1 we know that $\mathrm{ind} \circ j_!^{-1} \cong j_!^{-1} = q_!$. We conclude that $\mathrm{ind}(u) = q_! f_!(u) = \text{top-ind}(u)$.

Property 1 is easily established. Each element in $K(\mathrm{pt})$ can be represented in the form $[\mathbb{C}^r] - [\mathbb{C}^s]$ where $(\mathbb{C}^r, \mathbb{C}^s) \in \mathscr{L}_1(\mathrm{pt})$ is a pair of vector spaces. An elliptic operator on this pair is just a linear map $P: \mathbb{C}^r \to \mathbb{C}^s$, and we see that $\mathrm{ind}\, P = r - s$.

Property 2 is more difficult to establish. Following Atiyah and Singer [1], we split this up into more easily established properties. The first one shows that ind is well defined over open manifolds of finite topological type.

The Excision Property 13.4. *Let $\mathcal{O}$ be an open manifold, and let*

$$f: \mathcal{O} \hookrightarrow X \qquad \text{and} \qquad f': \mathcal{O} \hookrightarrow X'$$

be two open embeddings into compact manifolds X and X'. Then $\mathrm{ind} \circ f_! = \mathrm{ind} \circ f'_!$ *on $K_{\mathrm{cpt}}(T^*\mathcal{O})$.*

Proof. Fix $u \in K_{\mathrm{cpt}}(T^*\mathcal{O})$. By Lemma 13.3 we know that u can be represented by a triple $(\pi^*E, \pi^*F; \sigma)$ where E and F are bundles over $\mathcal{O}$ which are trivial outside a compact subset of $\mathcal{O}$ and where σ is homogeneous of degree 0 outside a compact subset of $T^*\mathcal{O}$. In particular, outside a compact set $\Omega \subset \mathcal{O}$ there are trivializations

$$\tau_E: E|_{(\mathcal{O}-\Omega)} \xrightarrow{\approx} (\mathcal{O} - \Omega) \times \mathbb{C}^m \quad \text{and} \quad \tau_F: F|_{(\mathcal{O}-\Omega)} \xrightarrow{\approx} (\mathcal{O} - \Omega) \times \mathbb{C}^m \tag{13.9}$$

with respect to which $\sigma_{x,\xi} = \sigma_x = (\tau_F)_x^{-1} \circ (\tau_E)_x$ at all points $(x,\xi) \in T^*(\mathcal{O} - \Omega)$. This means that over $T^*(\mathcal{O} - \Omega)$ the morphism σ comes from a bundle map $\sigma_0: E \to F$ over the base. Moreover, with respect to the trivializations (13.9), σ_0 becomes the identity mapping, i.e., $\sigma_0(z_1, \ldots, z_m) = (z_1, \ldots, z_m)$ at all points $x \in \mathcal{O} - \Omega$. Recall that a bundle map $\sigma_0 \in \Gamma(\mathrm{Hom}(E,F))$ is just a differential operator of order zero.

We now choose a zero-order elliptic operator $P \in \Psi CO_0(E,F)$ which has symbol $\sigma(P) = \sigma$ outside a compact set in $T^*\mathcal{O}$ and which *is* the operator $\sigma_0 = \mathrm{Id}$ in $\mathcal{O} - \Omega$. Such an operator clearly exists.

Suppose now that we are given an open embedding $f: \mathcal{O} \hookrightarrow X$. Using (13.9), we extend the bundles E and F trivially over $X - f(\mathcal{O})$, and we extend the operator P to be the identity there. This defines an elliptic operator $f_!P$ on X with the property that

$$[\sigma(f_!P)] = f_![\sigma(P)] = f_!u. \tag{13.10}$$

Clearly any element in $\ker(f_!P)$ has support in Ω and hence belongs to the subspace $\ker P$ (under the natural embedding $\ker P \subset \ker f_!P$ given by extending by zero). Hence, $\dim(\ker f_! P) = \dim(\ker P)$. The same remarks apply to the adjoint $(f_! P)^*$. Assuming that X is compact, we conclude from this and from (13.10) that

$$\operatorname{ind}(f_! u) = \operatorname{ind}(f_! P) = \dim(\ker P) - \dim(\ker P^*).$$

Since the right hand side is independent of f, our assertion 13.4 is proved. ∎

The Multiplicative Property 13.5. *Let X and Y be compact manifolds. Then for all elements $u \in K_{\mathrm{cpt}}(T^*X)$ and $v \in K_{\mathrm{cpt}}(T^*Y)$ we have that*

$$\operatorname{ind}(u \cdot v) = (\operatorname{ind} u)(\operatorname{ind} v) \tag{13.11}$$

Proof. Naively the argument goes as follows. We represent u and v as above by first-order elliptic operators

$$P : \Gamma(E) \longrightarrow \Gamma(F) \qquad \text{and} \qquad Q : \Gamma(E') \longrightarrow \Gamma(F')$$

over X and Y respectively. We introduce metrics and define a "graded tensor product"

$$D : \Gamma((E \otimes E') \oplus (F \otimes F')) \longrightarrow \Gamma((F \otimes E') \oplus (E \otimes F')$$

by

$$D = \begin{pmatrix} P \otimes 1 & -1 \otimes Q^* \\ 1 \otimes Q & P^* \otimes 1 \end{pmatrix} \tag{13.12}$$

Note that $E \otimes E'$, etc., here denotes the *exterior* tensor product over $X \times Y$. The operators $P \otimes 1$, etc. are uniquely determined by requiring that for $\varphi \in \Gamma(E)$ and $\psi \in \Gamma(E')$ we have $(P \otimes 1)(\varphi(x) \otimes \psi(y)) = (P\varphi(x)) \otimes \psi(y)$. Using the fact that $P \otimes 1$ and $1 \otimes Q$ commute, one easily computes that

$$\left.\begin{aligned} D^*D &= \begin{pmatrix} P^*P \otimes 1 + 1 \otimes Q^*Q & 0 \\ 0 & PP^* \otimes 1 + 1 \otimes QQ^* \end{pmatrix} \\ DD^* &= \begin{pmatrix} PP^* \otimes 1 + 1 \otimes Q^*Q & 0 \\ 0 & P^*P \otimes 1 + 1 \otimes QQ^* \end{pmatrix} \end{aligned}\right\} \tag{13.13}$$

Note that: $D^*D\varphi = 0 \Rightarrow (D^*D\varphi, \varphi) = (D\varphi, D\varphi) = 0 \Rightarrow D\varphi = 0$. Hence, $\ker D^*D = \ker D$. Furthermore, since D^*D is diagonal, it suffices to compute $\ker D^*D$ separately on each summand, $E \otimes E'$ and $F \otimes F'$. Given $\varphi \in \Gamma(E \otimes E')$, we see that: $D^*D\varphi = 0 \Rightarrow (\widetilde{P^*P}\varphi, \varphi) + (\widetilde{Q^*Q}\varphi, \varphi) = 0 \Rightarrow \|\tilde{P}\varphi\|^2 + \|\tilde{Q}\varphi\|^2 = 0 \Rightarrow \tilde{P}\varphi = \tilde{Q}\varphi = 0$ (where $\tilde{R}$ denotes $R \otimes 1$ or $1 \otimes R$, whichever is appropriate). Note that $\ker \tilde{P} \cap \ker \tilde{Q} \cong \ker P \otimes \ker Q$.

Continuing in this fashion we deduce that

$$\ker D = \ker D^*D \cong (\ker P \otimes \ker Q) \oplus (\ker P^* \otimes \ker Q^*)$$

$$\operatorname{coker} D \cong \ker D^* = \ker DD^* \cong (\ker P^* \otimes \ker Q) \oplus (\ker P \otimes \ker Q^*)$$

are therefore in $K(\mathrm{pt})$

$$[\ker D] - [\operatorname{coker} D] = ([\ker P] - [\operatorname{coker} P])([\ker Q] - [\operatorname{coker} Q]).$$

In particular we have

$$\operatorname{ind} D = (\operatorname{ind} P)(\operatorname{ind} Q). \tag{13.14}$$

Now the principal symbol of D is exactly the (outer) tensor product of the symbols of P and Q. Therefore, naively we have established (13.11). However, there is one technical flaw in the argument. This is the fact that for $P \in \Psi CO_1(E, F)$, the operator $P \otimes 1 \in \Psi DO_1(E \otimes E', F \otimes E')$ does not in general belong to the class $\Psi CO_1(E \otimes E', F \otimes E')$ because its principal symbol is not homogeneous outside a compact set in $T^*(X \times Y)$. It is only homogeneous outside a uniform neighborhood of the "T^*Y-axes" in $T^*(X \times Y)$.

This flaw is repaired as follows. We shall construct a continuous family of operators $(P \otimes 1)_\varepsilon \in \Psi CO_1(E \otimes E', F \otimes E')$ for $\varepsilon > 0$ such that $\lim_{\varepsilon \searrow 0} (P \otimes 1)_\varepsilon = P \otimes 1$, where this limit is taken in the space of bounded linear maps from $L_1^2(E \otimes E')$ to $L_0^2(F \otimes E')$. Applying the construction to each entry in (13.12) will give us a family of elliptic operators D_ε in ΨCO^1 such that $\lim_{\varepsilon \searrow 0} D_\varepsilon = D$ as bounded (Fredholm) maps between Sobolev spaces. From the local constancy of the index (see 7.3.) we have

$$\operatorname{ind} D_\varepsilon = \operatorname{ind} D \qquad \text{for all } \varepsilon > 0.$$

On the other hand it will be evident from the construction that given any compact subset $K \subset T^*(X \times Y)$, there exists a constant $\varepsilon_K > 0$ so that $\sigma(D_\varepsilon) \equiv \sigma(D)$ on K for all $\varepsilon \leqq \varepsilon_K$. It follows by excision that $[\sigma(D_\varepsilon)] = [\sigma(D)] = u \cdot v$ for all ε sufficiently small. Hence, $\operatorname{ind}(u \cdot v) = \operatorname{ind}(D_\varepsilon) = \operatorname{ind}(D) = (\operatorname{ind} P)(\operatorname{ind} Q) = (\operatorname{ind} u)(\operatorname{ind} v)$, and the property will be established.

It remains to construct the operator $(P \otimes 1)_\varepsilon$. This is done by multiplying the symbol of $P \otimes 1$ by a function $\psi_\varepsilon(|\xi|, |\eta|)$ of the cotangent variables $(\xi, \eta) \in T^*X \times T^*Y$. This function is constructed as follows. Fix a C^∞ function $\phi: \mathbb{R}^+ \to [0, 1]$ such that $\phi(t) = 0$ for $t \leqq 1$ and $\phi(t) = 1$ for $t \geqq 2$. Then for $\varepsilon > 0$ and for $r,s \geqq 0$, set $\psi_\varepsilon(r,s) = 1 - \phi(\varepsilon\sqrt{r^2 + s^2})\phi(\varepsilon s/r)$. In multiplying the symbol of $P \otimes 1$ by $\psi_\varepsilon(|\xi|, |\eta|)$, one can use a good coordinate presentation or some symbol calculus. The choice of method is not critical. It is a straightforward and worthwhile exercise to check that the resulting family $(P \otimes 1)_\varepsilon$ has the properties claimed above. This completes the proof of 13.5. ■

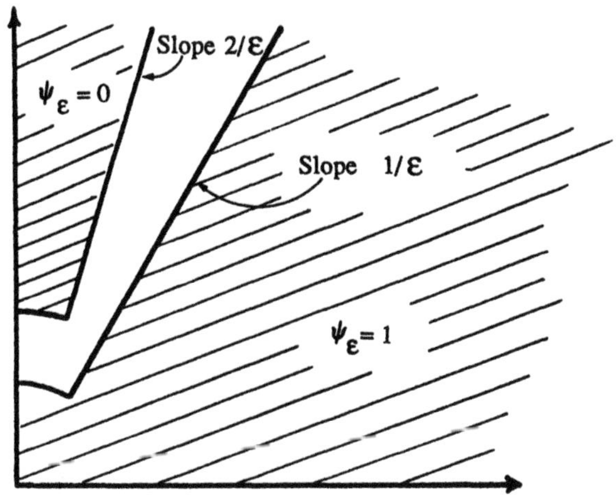

We shall actually need the multiplicative property in the more general context of twisted products, i.e., fibre bundles. However, we only need to consider the special case of sphere bundles which arise from vector bundles by adding a section at infinity. More specifically, let $\pi: \mathbf{P} \to X$ be a principal O_n-bundle over a compact manifold X and consider the associated bundles

$$V = \mathbf{P} \times_{O_n} \mathbb{R}^n \qquad Z = \mathbf{P} \times_{O_n} S^n \tag{13.15}$$

where O_n acts on $\mathbb{R}^n$ by the standard representation and acts on S^n by extending this representation to the one-point compactification on $\mathbb{R}^n$. (That is, O_n acts on S^n by trivially extending the standard representation to $\mathbb{R}^{n+1} = \mathbb{R}^n \times \mathbb{R}^1$ and then restricting to the unit sphere.) We define a product

$$K_{\text{cpt}}(T^*X) \otimes K_{O_n}(T^*S^n)_{\text{cpt}} \longrightarrow K_{\text{cpt}}(T^*Z) \tag{13.16}$$

as follows. Choosing a metric in Z we get a splitting $T^*Z = \pi^*T^*X \oplus T(Z/X)$ where $T(Z/X) = T^*Z/\pi^*T^*X$ denotes the tangent spaces along the fibres of the projection $\pi: Z \to X$. This splitting gives us a multiplication

$$K_{\text{cpt}}(T^*X) \otimes K_{\text{cpt}}T(Z/X) \longrightarrow K_{\text{cpt}}(T^*Z).$$

(Given a direct sum of vector bundles $E \oplus E'$ on X, the map on $K_{\text{cpt}}(E) \otimes K_{\text{cpt}}(E')$ is defined by first taking the outer tensor product on

$E \times E'$ over $X \times X$ and then restricting to the diagonal.) Combining this with the composition

$$K_{O_n}(T^*S^n)_{cpt} \to K_{O_n}(\mathbf{P} \times T^*S^n)_{cpt} \to K_{cpt}(\mathbf{P} \times_{O_n} T^*S^n) = K_{cpt}(T(Z/X))$$

gives the desired multiplication (13.16).

The associated bundle construction, which associates to a linear representation $\rho: O_n \to O_N$ the vector bundle $V_\rho = \mathbf{P} \times_\rho \mathbb{R}^N$, extends naturally to a homomorphism

$$\alpha_{\mathbf{P}}: R(O_n) \longrightarrow K(X) \tag{13.17}$$

The ring $K_{cpt}(T^*X)$ is naturally a $K(X)$-module. Therefore, via (13.17) it becomes an $R(O_n)$-module. We are now in a position to state the main property.

The Multiplicative Property for Sphere Bundles 13.6. *Let Z be an S^n-bundle defined as above over a compact manifold X. Then*

$$\operatorname{ind}(u \cdot v) = \operatorname{ind}(u \cdot \operatorname{ind}_{O_n} v)$$

for all $u \in K_{cpt}(T^*X)$ and $v \in K_{O_n}(T^*S^n)_{cpt}$.

Proof. The proof of this fact follows very much the argument given for 13.5. We represent u and v by first-order elliptic operators P and Q respectively. (Q is an O_n-equivariant operator on O_n-bundles.) Using local trivializations of the bundle, we cover $\mathbf{P}$ by a finite number of product neighborhoods $\{U_j \times O_n\}_{j=1}^N$. We lift the operator P back over each product and glue together with a partition of unity with respect to $\{U_j\}$ on X, to get an O_n-invariant operator $\tilde{P}$ on $\mathbf{P}$. We now consider the tensor product operator $\tilde{D}$ on $\mathbf{P} \times S^n$ defined as in (13.12) (with P replaced by $\tilde{P}$). This is an O_n-operator and can be pushed down to give an operator D on the quotient Z. Notice that "pushing down" is equivalent to restricting $\tilde{D}$ to the subspace of sections coming from the base. It is easily checked that $\sigma(D)$ represents the class $u \cdot v$ where the multiplication is that defined in (13.16) above.

It remains to compute the analytic index of D in terms of ind P and $\operatorname{ind}_{O_n} Q$. We shall work upstairs with the operator $\tilde{D}$ restricted to sections coming from the base. Using (13.13) and the arguments which follow it, we see that:

$$\begin{aligned} \ker \tilde{D} &= \ker \tilde{D}^*\tilde{D} \\ &= (\ker(\tilde{P} \otimes 1) \cap \ker(1 \otimes Q)) \oplus (\ker(\tilde{P}^* \otimes 1) \cap \ker(1 \otimes Q^*)) \end{aligned}$$

with an analogous statement for $\ker \tilde{D}^* = \operatorname{coker} \tilde{D}$. Since the operators $\tilde{P} \otimes 1$ and $1 \otimes Q$ commute, we can carry out the computation in steps, that is, we first pass to the kernel of $1 \otimes Q$ (or $1 \otimes Q^*$ whichever is relevant) and consider the operators $\tilde{P} \otimes 1$ and $\tilde{P}^* \otimes 1$ acting there. For example, on $\Gamma(E \otimes E')$ the space $\ker(1 \otimes Q)$ consists of those sections $\tilde{\varphi}$ which when restricted to each factor $\{p\} \times S^n \subset \mathbf{P} \times S^n$, lie in the space $E_{\pi(p)} \otimes \ker Q$. To say that $\tilde{\varphi}$ comes from a section on the base means that $\tilde{\varphi}$ satisfies the transformation law: $\tilde{\varphi}(pg^{-1}, gx) = \rho_g \tilde{\varphi}(p,x)$ for $g \in \mathrm{O}_n$, where ρ is the natural representation of O_n on $\ker Q$. This means precisely that $\tilde{\varphi}$ corresponds to a section φ over X of the bundle $E \otimes \mathbf{ker\ Q}$ where $\mathbf{ker\ Q}$ is the vector bundle on X associated via the principal bundle $\mathbf{P}$ to the representation ρ. Therefore, in passing to $\ker(1 \otimes Q)$ and $\ker(1 \otimes Q^*)$ the operator D descends to an operator on X on the form $P_k + P_{k^*}^*$ where

$$P_k : \Gamma(E \otimes \mathbf{ker\ Q}) \longrightarrow \Gamma(F \otimes \mathbf{ker\ Q}), \quad \text{and}$$

$$P_{k^*} : \Gamma(E \otimes \mathbf{ker\ Q^*}) \longrightarrow \Gamma(F \otimes \mathbf{ker\ Q^*}).$$

It follows that $\operatorname{ind} D = \operatorname{ind} P_k - \operatorname{ind} P_{k^*}^* = \operatorname{ind}[P \otimes (\mathbf{ker\ Q} - \mathbf{coker\ Q})] = \operatorname{ind}[u \cdot \operatorname{ind}_{\mathrm{O}_n} v]$ as claimed. ∎

From the properties we have established, matters can be easily reduced to computing some simple cases. However, it is unavoidable that one must compute the index of some operator at some point. We do this now.

Lemma 13.7. *Consider the n-sphere S^n to be an O_n-manifold under the restriction of the standard representation on $\mathbb{R}^n \oplus \mathbb{R} \supset S^n$ (i.e., by rotations about an axis). Let $i : \mathrm{pt} \hookrightarrow S^n$ denote the inclusion of one of the two fixed-points of the action. Then*

$$\operatorname{ind}_{\mathrm{O}_n}(i_!1) = 1 \in R(\mathrm{O}_n)$$

Proof. Consider the operator $D^0 : C\ell^0 \to C\ell^1$ with respect to the standard metric on S^n. Recall that D^0 is just the de Rham-Hodge operator $d + d^* : \Lambda^{\text{even}} \to \Lambda^{\text{odd}}$. This is an O_n-operator and from Hodge Theory (II.5) we see easily that $\operatorname{ind}_{\mathrm{O}_n}(D^0) = [\mathbf{H}^0] + (-1)^n[\mathbf{H}^n]$. The action of O_n on $\mathbf{H}^0 = \{\text{constant functions}\}$ is always trivial. The action on $\mathbf{H}^n = \mathbb{R}\{\text{the volume } n\text{-form}\}$ is trivial if and only if n is even. Therefore, we have

$$\operatorname{ind}_{\mathrm{O}_n} D^0 = \begin{cases} 2 & \text{if } n \text{ is even} \\ 1 - \xi & \text{if } n \text{ is odd} \end{cases}$$

where ξ represents the non-trivial 1-dimensional representation on O_n. We leave the details of the computation of the symbol class of D^0 to the reader. One finds that in $\mathrm{K}_{\mathrm{O}_n}(S^n)$

$$[\sigma(D^0)] = \begin{cases} 2i_!(1) & \text{if } n \text{ is even} \\ (1 - \xi)i_!(1) & \text{if } n \text{ is odd.} \end{cases}$$

Combined with the above, this completes the proof of the lemma. ∎

We now complete the proof of the Index Theorem. We have shown that it will suffice to establish Property 2, namely that $\operatorname{ind} = \operatorname{ind} \circ f_!$ for embeddings $f: X \hookrightarrow Y$. By the Excision Property we may replace the compact manifold Y with a tubular neighborhood of X in Y, which is diffeomorphic to the normal bundle of X in Y. Consequently it will suffice to prove that

$$\operatorname{ind} u = \operatorname{ind}(f_! u) \qquad \text{for all } u \in K_{\text{cpt}}(T^*X)$$

where V is a vector bundle over X and $f: X \hookrightarrow V$ is the inclusion of the zero-section. Again by the Excision Property we may compactify V by passing to the associated sphere bundle as in (13.15). We then apply the Multiplicative Property for Sphere Bundles 13.6, with $v = i_! 1$. Using Lemma 13.7 we find that: $\operatorname{ind}(u \cdot i_! 1) = \operatorname{ind}(u \cdot \operatorname{ind}_{O_n}(i_! 1)) = \operatorname{ind}(u)$. However, by definition $f_! u = u \cdot i_! 1$, and the proof is complete. ∎

The remainder of this section will be devoted to deriving certain cohomological formulas for the topological index. The most general one is the following:

Recall that for any manifold X, the tangent bundle TX is canonically an almost complex manifold since $T(TX) = \pi^* TX \oplus \pi^* TX \cong \pi^* TX \otimes \mathbb{C}$. This gives TX a canonical orientation as a manifold, A positively oriented basis of $T(TX)$ is of the form $(e_1, Je_1, e_2, Je_2, \ldots, e_n, Je_n)$ where $e_1, \ldots, e_n$ is a basis of $\pi^* TX$ and J carries the "horizontal" to the "vertical" factor. With this orientation we can evaluate any element $u \in H^{2n}_{\text{cpt}}(TX)$ on the fundamental class $[TX]$ of the manifold. The result is denoted by $u[TX]$.

Theorem 13.8. *Let P be an elliptic operator over a compact manifold X of dimension n. Then*

$$\boxed{\operatorname{ind} P = (-1)^n \{\operatorname{ch} \sigma(P) \cdot \hat{\mathbf{A}}(X)^2\}[TX]} \tag{13.18}$$

where $\hat{\mathbf{A}}(X)$ denotes the total $\hat{A}$-class of X pulled back to TX.

Proof. We consider first the scrunch map $q: T\mathbb{R}^N = \mathbb{R}^N \oplus \mathbb{R}^N = \mathbb{C}^N \to \text{pt}$. Consider this as a complex bundle and let $i: \text{pt} \hookrightarrow \mathbb{C}^N$ be the inclusion as the origin. Fix an element $u \in K_{\text{cpt}}(\mathbb{C}^N)$ and apply the defect formula (12.9) with $u \equiv i_! \xi$. Recalling that $q_! = (i_!)^{-1}$, that $\mathfrak{J}(\mathbb{C}^N) = 1$, and that $\operatorname{ch}: K(\text{pt}) \to H^0(\text{pt})$ is an isomorphism, we find that: $q_! \operatorname{ch} u = q_! u$. Since $q_!$ on H^*_{cpt} is just integration over the fibre, we find that

$$q_! u = \operatorname{ch} u[T\mathbb{R}^N]. \tag{13.19}$$

Consider now a real vector bundle $p: v \to X$ and let $i: X \to v$ denote the inclusion as the zero-section. Taking derivatives gives the bundle

$p: Tv \to TX$ with zero-section $i: TX \to Tv$. One sees easily that the bundle $p: Tv \to TX$ is equivalent to $\pi^*v \oplus \pi^*v = \pi^*v \otimes \mathbb{C}$. While TX is not compact, the map i is proper and equation (12.9) can be shown to hold for elements of $K_{\text{cpt}}(TX)$. Thus for any $\sigma \in K_{\text{cpt}}(TX)$ we have

$$p_! \text{ch } i_! \sigma = \mathfrak{J}(v \otimes \mathbb{C}) \text{ch } \sigma.$$

Evaluating on the fundamental class and recalling that $p_!$ is integration over the fibre give the formula:

$$(\text{ch } i_! \sigma)[Tv] = \{\mathfrak{J}(v \otimes \mathbb{C}) \text{ch } \sigma\}[TX]. \tag{13.20}$$

Consider now an embedding $f: X \hookrightarrow \mathbb{R}^N$ with normal bundle v. We identify v with an open tubular neighborhood of $f(X)$ in $\mathbb{R}^N$. Similarly we have an open embedding

$$Tv \subset T\mathbb{R}^N \tag{13.21}$$

as a tubular neighborhood of $f(TX)$. Given any $\sigma \in K_{\text{cpt}}(TX)$, the class $i_! \sigma$ has compact support in Tv and naturally extends to $T\mathbb{R}^N$ under the open inclusion (13.21). This extended class is, by definition, the element $f_! \sigma \in K_{\text{cpt}}(T\mathbb{R}^N)$ given in (13.2). In particular we have

$$\text{ch}(i_! \sigma)[Tv] = \text{ch}(f_! \sigma)[T\mathbb{R}^N]. \tag{13.22}$$

Combining (13.19)–(13.21) gives

$$\text{ind } P = \{\mathfrak{J}(v \otimes \mathbb{C}) \text{ch } \sigma(P)\}[TX].$$

It remains to identify $\mathfrak{J}(v \otimes \mathbb{C})$. For this we note that since v is the normal bundle to X, we have $TX \oplus v = (\text{trivial})$. Since $\mathfrak{J}$ is multiplicative, this means that $\mathfrak{J}(v \otimes \mathbb{C}) = \mathfrak{J}(TX \otimes \mathbb{C})^{-1}$. Applying formula (12.11) we see that

$$\mathfrak{J}(TX \otimes \mathbb{C})^{-1} = (-1)^n \mathbf{Td}_{\mathbb{C}}(TX \otimes \mathbb{C}). \tag{13.23}$$

We have used here that $TX \otimes \mathbb{C}$ is self-conjugate. For the final step we invoke Proposition 11.14. ■

Integrating over the fibre immediately gives the following.

Theorem 13.8 (the cohomological formula for the index; Atiyah-Singer [2]). *Let P be an elliptic operator on a compact oriented n-manifold X and let $\sigma = \sigma(P) \in K_{\text{cpt}}(TX)$ denote the symbol class of P. Then*

$$\boxed{\text{ind } P = (-1)^{\frac{n(n+1)}{2}} \{\pi_! \text{ch } \sigma) \cdot \hat{\mathfrak{A}}(X)^2\}[X]} \tag{13.24}$$

where $\pi: TX \to X$ is the bundle projection.

Note. The factor $(-1)^{\frac{n(n+1)}{2}}$ compensates for the difference between the orientation on TX induced by the one on X, and the canonical orientation described above.

We now consider two important special cases.

Theorem 13.9. *Let X be a compact oriented manifold of dimension $n = 2m$ and consider the signature operator $D^+ : \Gamma(\mathbb{C}\ell^+(X)) \to \Gamma(\mathbb{C}\ell^-(X))$. Then*

$$\operatorname{ind} D^+ = L(X) = \operatorname{sig}(X)$$

More generally, if E is any complex vector bundle over X, then the index of $D_E^+ : \Gamma(\mathbb{C}\ell^+(X) \otimes E) \to \Gamma(\mathbb{C}\ell^-(X) \otimes E)$ is given by

$$\operatorname{ind}(D_E^+) = \{\operatorname{ch}_2 E \cdot \mathbf{L}(X)\}[X] \tag{13.25}$$

where by definition $\operatorname{ch}_2 E = \sum_k 2^k \operatorname{ch}^k E$.

Proof. Clearly we have that $\sigma(D^+) = \delta(TX)$ where δ is defined in (12.12). By Proposition 12.4, $\pi_! \operatorname{ch} \delta(TX) = (-2)^m \hat{\mathbf{L}}(X) \hat{\mathbf{A}}(X)^{-2}$. Applying the formula (13.24) above, we find that

$$\operatorname{ind} D^+ = 2^m \hat{L}(X) = \begin{cases} L(X) & \text{if } m \text{ is even} \\ 0 & \text{if } m \text{ is odd.} \end{cases}$$

(A direct identification of ind D^+ with the signature of X was carried out in II.6.2). For the more general case we apply the formula (12.14) and (12.16). We conclude similarly that

$$\operatorname{ind}(D_E^+) = 2^m \{\operatorname{ch} E \cdot \mathbf{L}(X)\}[X].$$

Writing this out and using the fact that $\hat{L}_\ell = 2^{-2\ell} L_\ell$, we find that $2^m\{\operatorname{ch} E \cdot \hat{\mathbf{L}}(X)\}[X] = \sum \{2^m \operatorname{ch}^k E \cdot \hat{L}_\ell(X)\}[X] = \sum \{2^{m-2\ell} \operatorname{ch}^k E \cdot L_\ell(X)\}[X] = \sum \{2^k \operatorname{ch}^k E \cdot L_\ell(X)\}[X]$ since the sum is over (k,ℓ) with $2\ell + k = m$. This proves (13.25). ∎

Similar arguments give the following in the spin case:

Theorem 13.10. *Let X be a compact spin manifold of dimension $n = 2m$ and consider the Atiyah-Singer operator $\not{D}^+ : \Gamma(\$_{\mathbb{C}}^+(X)) \to \Gamma(\$_{\mathbb{C}}^-(\mathbb{C}))$. Then*

$$\operatorname{ind} \not{D}^+ = \hat{A}(X).$$

More generally, if E is any complex vector bundle over X, then the index of $\not{D}_E^+ : \Gamma(\$_{\mathbb{C}}^+(X) \otimes E) \to \Gamma(\$_{\mathbb{C}}^-(X) \otimes E)$ is given by

$$\operatorname{ind}(\not{D}_E^+) = \{\operatorname{ch} E \cdot \hat{\mathbf{A}}(X)\}[X]. \tag{13.26}$$

Proof. Note that $\boldsymbol{\sigma}(D^+) = \mathbf{s}(TX)$ where $\mathbf{s}$ is defined in (12.13). By Proposition 12.5, $\pi_! \mathrm{ch}\, \mathbf{s}(TX) = (-1)^m \hat{\mathbf{A}}(X)^{-1}$. Applying formula (13.24) above, we find that ind $\not{D}^+ = \hat{\mathbf{A}}(X)[X] = \hat{A}(X)$.

For the more general case we apply formulas (12.16) and (12.15) to conclude that $\pi_! \mathrm{ch}[\mathbf{s}(TX) \otimes E] = (-1)^m \mathrm{ch}\, E \cdot \hat{\mathbf{A}}(X)^{-1}$. Therefore $\mathrm{ind}(D_E^+) = \{\mathrm{ch}\, E \cdot \hat{\mathbf{A}}(X)\}[X]$ as claimed. ∎

REMARK 13.11. For a compact spin $2m$-manifold X, formula (13.26) is equivalent to the full Index Theorem. This is seen as follows. Any elliptic zero, $P_0 \equiv (1 + P^*P)^{-1/2}P$, which has the same index and the same symbol class $\boldsymbol{\sigma} \in K_{\mathrm{cpt}}(TX)$. By the Thom isomorphism 12.6, $\boldsymbol{\sigma}$ can be written in the form $\boldsymbol{\sigma} = \mathbf{s}(TX) \cdot \pi^* u$ for some $u \in K(X)$. (We use here that X is spin.) Writing $u = [E] - [F]$ for vector bundles E and F on X, we see that

$$\boldsymbol{\sigma} = [\pi^* \$_{\mathbb{C}}^+ \otimes E, \pi^* \$_{\mathbb{C}}^- \otimes E; \mu] - [\pi^* \$_{\mathbb{C}}^+ \otimes F, \pi^* \$_{\mathbb{C}}^- \otimes F; \mu]$$

i.e., at the level of the principal symbol, P is equivalent to the difference of two Atiyah-Singer operators with coefficients. This means essentially that $P \oplus \not{D}_F^+$ is homotopic to $\not{D}_E^+$. Therefore, ind P = ind $\not{D}_E^+$ − ind $\not{D}_F^+$ = $\{(\mathrm{ch}\, E - \mathrm{ch}\, F) \cdot \hat{\mathbf{A}}(X)\}[X] = \{\mathrm{ch}\, u \cdot \hat{\mathbf{A}}(X)\}[X]$. However, (12.15) gives $\pi_! \boldsymbol{\sigma} = \pi_!(\mathbf{s}(X)\pi^* u) = (-1)^m \mathrm{ch}\, u \cdot \hat{\mathbf{A}}(X)^{-1}$, and ind $P = \{\pi_! \boldsymbol{\sigma} \cdot \hat{\mathbf{A}}(X)^2\}[X]$ as claimed.

For non-spin manifolds, one can argue similarly by using the signature operator with coefficients.

One interesting corollary of the index formulas above is the following:

Theorem 13.12. *On an odd-dimensional compact manifold, the index of every elliptic differential operator is zero.*

Note that this result does not remain true for pseudodifferential operators.

Proof. Consider the diffeomorphism $c: TX \to TX$ given by $c(v) \equiv -v$ and note that if dim X is odd, then $c_*[TX] = -[TX]$. Let P be an elliptic differential operator of degree m with principal symbol $\sigma(P)$. Since $\sigma_{-\xi}(P) = (-1)^m \sigma_\xi(P)$, we see that $c^*\sigma(P) = (-1)^m \sigma(P)$. Since $\sigma(P)$ and $-\sigma(P)$ are regularly homotopic (by $\sigma(t,P) = e^{\pi i t}\sigma(P)$, $0 \le t \le 1$), they define the same elements in K-theory, and we conclude that $c^*\boldsymbol{\sigma}(P) = \boldsymbol{\sigma}(P)$. Applying formula (13.18) now gives

$$\begin{aligned} \mathrm{ind}\, P &= -\{\mathrm{ch}\, \boldsymbol{\sigma}(P) \cdot \hat{\mathbf{A}}(X)^2\}[TX] \\ &= -c^*\{\mathrm{ch}\, \boldsymbol{\sigma}(P) \cdot \hat{\mathbf{A}}(X)^2\} c_*[TX] \\ &= -\{\mathrm{ch}\, c^*\boldsymbol{\sigma}(P) \cdot \hat{\mathbf{A}}(X)^2\} c_*[TX] \\ &= -\{\mathrm{ch}\, \boldsymbol{\sigma}(P) \cdot \hat{\mathbf{A}}(X)^2\}(-[TX]) \\ &= -\mathrm{ind}\, P. \quad ∎ \end{aligned}$$

Many important elliptic operators on a manifold X arise in the following way. Suppose the structure group of X can be reduced to a compact, connected subgroup $G \subset \mathrm{SO}_{2m}$ (where $\dim X = 2m$), and that $P:\Gamma(E) \to \Gamma(F)$ is an elliptic operator where E and F are vector bundles associated to unitary representations, ρ_E and ρ_F respectively, of G. To apply the index formula (13.24) to P we must compute $\pi_!\mathrm{ch}\,\boldsymbol{\sigma}(P)$ where $\pi: TX \to X$ is the bundle projection.

To do this we pass to the universal case. Let $\pi:\tilde{T} \to \mathrm{BG}$ denote the universal $2m$-plane bundle associated to the inclusion $G \subset \mathrm{SO}_{2m}$, and let $\tilde{E},\tilde{F}$ be the complex bundles over BG associated to the representations ρ_E and ρ_F respectively. Let $\boldsymbol{\sigma} = [\pi^*\tilde{E}, \pi^*\tilde{F}; \sigma] \in K_{\mathrm{cpt}}(\tilde{T})$ be any elliptic symbol from $\tilde{E}$ to $\tilde{F}$. Then from (12.5) and the fact that $\pi_! = (i_!)^{-1}$, we have that

$$\chi(\tilde{T})\pi_!\mathrm{ch}\,\boldsymbol{\sigma} = i^*i_!\pi_!\mathrm{ch}\,\boldsymbol{\sigma} = i^*\mathrm{ch}\,\boldsymbol{\sigma} = \mathrm{ch}\,\tilde{E} - \mathrm{ch}\,\tilde{F}.$$

The algebra $H^*(BG;\mathbb{Q})$ always embeds in the polynomial algebra $H^*(BT;\mathbb{Q})$ where $T \subset G$ is a maximal torus. Therefore, if $\chi(\tilde{T}) \neq 0$, we can write: $\pi_!\mathrm{ch}\,\boldsymbol{\sigma} = (\mathrm{ch}\,\tilde{E} - \mathrm{ch}\,\tilde{F})/\chi(\tilde{T})$. Pulling back to X by the classifying map for TX (with its G-structure) gives the following corollary to Theorem 13.8.

Theorem 13.13. *Let X be a compact $2m$-manifold with structure group $G \subset \mathrm{SO}_{2m}$ as above. Let $P:\Gamma(E) \to \Gamma(F)$ be an elliptic operator where E and F are associated to unitary representations of G. Suppose that the image of the Euler class χ under the map $H^{2m}(\mathrm{BSO}_{2m}) \to H^{2m}(BG)$ is not zero. Then the characteristic class* $(\mathrm{ch}\,E - \mathrm{ch}\,F)/\chi(TX) \in H^*(X;\mathbb{Q})$ *is well defined, and*

$$\mathrm{ind}\,P = (-1)^m\left\{\frac{\mathrm{ch}\,E - \mathrm{ch}\,F}{\chi(TX)} \cdot \hat{\mathbf{A}}(X)^2\right\}[X].$$

EXAMPLE 13.14 (**The Riemann-Roch-Hirzebruch Formula**). Let X be a compact complex manifold with a hermitian metric and let E be a holomorphic hermitian bundle over X. The Dolbeault complex

$$\Lambda^{0,0} \otimes E \xrightarrow{\bar{\partial}} \Lambda^{0,1} \otimes E \xrightarrow{\bar{\partial}} \cdots \xrightarrow{\bar{\partial}} \Lambda^{0,m} \otimes E \tag{13.27}$$

converts to an elliptic operator

$$\Lambda^{0,\mathrm{even}} \otimes E \xrightarrow{\bar{\partial}+\bar{\partial}^*} \Lambda^{0,\mathrm{odd}} \otimes E$$

where $\bar{\partial}^*$ denotes the adjoint of $\bar{\partial}$. Theorem 13.13 can be applied with $G = \mathrm{U}_m \subset \mathrm{SO}_{2m}$ and with $P = \bar{\partial} + \bar{\partial}^*$. If we consider $T \equiv TX$ as an m-dimensional complex vector bundle, then $\Lambda^{0,*} \cong \Lambda^*_{\mathbb{C}}T$ and we see that $\mathrm{ch}\,\Lambda^{0,\mathrm{even}} - \mathrm{ch}\,\Lambda^{0,\mathrm{odd}} = \mathrm{ch}\,\lambda_{-1}(T)$. From (12.10) and (12.11) we see that $\mathrm{ch}\,\lambda_{-1}(T)/\chi(T) = (-1)^m\mathbf{Td}_{\mathbb{C}}(\bar{T})^{-1}$. From Proposition 11.14 we have $\hat{\mathbf{A}}(X)^2 = \mathbf{Td}_{\mathbb{C}}([T]_{\mathbb{R}} \otimes \mathbb{C}) = \mathbf{Td}_{\mathbb{C}}(\bar{T} \oplus T) = \mathbf{Td}_{\mathbb{C}}(\bar{T})\mathbf{Td}_{\mathbb{C}}(T)$. Plugging into

13.13 immediately gives the following:

$$\operatorname{ind}(\bar\partial + \bar\partial^*) = \{\operatorname{ch} E \cdot \mathbf{Td}_{\mathbb{C}}(X)\}[X]$$

where $\mathbf{Td}_{\mathbb{C}}(X) = \mathbf{Td}_{\mathbb{C}}(T)$ is the total Todd class of X.

Observe that $\ker P = \ker P^*P = \ker(\bar\partial\bar\partial^* + \bar\partial^*\bar\partial) = \{\varphi \in \Lambda^{0,\text{even}} \otimes E : \bar\partial\varphi = \bar\partial^*\varphi = 0\}$, and similarly $\operatorname{coker} P = \ker P^* = \{\varphi \in \Lambda^{0,\text{odd}} \otimes E : \bar\partial\varphi = \bar\partial^*\varphi = 0\}$. Applying the Hodge Decomposition Theorem (II.5.6) gives the following:

Theorem 13.15. *Let $H^*(X; E)$ denote the* kth *cohomology group of the Dolbeault complex* (13.27) *over the compact complex manifold X. Then*

$$\sum (-1)^k \dim H^k(X; E) = \{\operatorname{ch} E \cdot \mathbf{Td}_{\mathbb{C}}(X)\}[X].$$

EXAMPLE 13.16. Let X be a compact oriented riemannian manifold of dimension $2m$, and consider the operator $D^0 : \Gamma(C\ell^0 X) \to \Gamma(C\ell^1 X)$ given in II.6.1. We leave as an exercise to the reader the verification that $\operatorname{ind} D^0 = \chi(X)$.

§14. Fixed-Point Formulas for Elliptic Operators

The proof of the Index Theorem outlined above carries over, almost without change, to the cases of G-operators, $C\ell_k$-linear operators, families of operators, etc. The main point is always to find the right K-theoretic setting in which to work.

In this section we consider the case of G-operators and the associated G-index Theorem. We assume throughout that X is a compact **G-manifold**, i.e., a manifold equipped with a given smooth action $\mu : X \times G \to X$ of a compact Lie group G. By a **G-bundle** on X we shall mean a complex vector bundle $E \to X$ with a G-action which carries fibres to fibres linearly and projects to μ. The Grothendieck group of equivalence classes of such G-bundles (cf. I.9) is called the **equivariant K-theory** of X and is denoted $K_G(X)$.

Equivariant K-theory has the same properties as ordinary K-theory if one restricts to the category of G-spaces and G-equivariant maps. Hence, given a G-operator P on X, one can pass through the same constructions as above (using a G-equivariant embedding $X \hookrightarrow \mathbb{R}^N$ and the Thom isomorphism) to define the **topological G-index** $\text{top-ind}_G(P)$ in $K_G(\text{pt}) \cong R(G)$.

Theorem 14.1 (Atiyah and Singer [1]). *For any elliptic G-operator P on a compact G-manifold, one has that*

$$\operatorname{ind}_G(P) = \text{top-ind}_G(P)$$

The proof of this "G-index Theorem" follows precisely the arguments outlined above for the basic case where $G = \{e\}$.

In the case that G acts trivially on X, there is a cohomological formula for the index which is deduced in analogy with the basic case. An important fact is that when G acts trivially on X, there is a natural isomorphism

$$K_G(X) \xrightarrow{\approx} K(X) \otimes R(G) \tag{14.1}$$

determined as follows. Every finite-dimensional representation V of G can be written in the form $\bigoplus_i \mathrm{Hom}_G(V_i, V) \otimes V_i$ where the direct sum ranges over the set $\{V_i\}$ of equivalence classes of irreducible representations of G. Similarly, if G acts trivially on X, then any G-bundle E can be written as

$$E = \bigoplus_i \mathrm{Hom}_G(E_i, E) \otimes E_i \tag{14.2}$$

where E_i denotes the trivial bundle $E_i = X \times V_i$. This association induces the isomorphism (14.1) (see Segal [1]). Composing this isomorphism with $\mathrm{ch} \otimes \mathrm{Id}$ gives a homomorphism

$$\mathrm{ch}_G : K_G(X) \longrightarrow H^*(X; \mathbb{Q}) \otimes R(G).$$

For non-compact spaces, such as TX, this extends to K-theory and cohomology with compact supports.

For each $g \in G$ there is a homomorphism $\chi_g : R(G) \to \mathbb{C}$ determined by setting $\chi_g(\rho) \equiv \mathrm{trace}(\rho(g))$ for each finite dimensional representation $\rho : G \to \mathrm{Hom}(V, V)$. Composing with ch_G gives a homomorphism

$$\mathrm{ch}_g : K_G(X) \longrightarrow H^*(X; \mathbb{C}) \tag{14.3}$$

which, for an element $u = \sum u_i \otimes r_i \in K(X) \otimes R(G)$ is written $\mathrm{ch}_g u = \sum (\mathrm{ch}\, u_i)\chi_g(r_i)$. The isomorphism (14.1) together with the arguments for the basic case given in §13 show the following:

Proposition 14.2. *Let X be a compact n-manifold on which G acts trivially, and let P be an elliptic G-operator on X with symbol class $\sigma = \sigma(P) \in K_{G,\mathrm{cpt}}(TX)$. Then one has that*

$$\mathrm{ind}_G(P) = (-1)^n \{\mathrm{ch}_G \sigma \cdot \hat{\mathbf{A}}(X)^2\}[TX]. \tag{14.4}$$

In particular, for each $g \in G$,

$$\mathrm{ind}_g(P) = (-1)^n \{\mathrm{ch}_g \sigma \cdot \hat{\mathbf{A}}(X)^2\}[TX]. \tag{14.5}$$

In the case that G acts non-trivially on X, the formula (14.5) can be replaced by one which involves the data of the action in a neighborhood of the fixed-point set. The result is a grand generalization of the classical Lefschetz Fixed-Point Formula for maps of finite order. The key to the computation is a certain "localization" theorem of Atiyah and Segal [1].

An excellent summary of the arguments involved and a large collection of illuminating examples are given in the book of Shanahan [1]. This exposition, together with the highly readable original literature, are recommended to the reader interested in full details. We shall present the result here in concise form.

To begin, we recall that for each $g \in G$, the **fixed-point set of g** is defined to be the set

$$F_g \equiv \{x \in X : gx = x\}.$$

Since G is compact, we know by averaging that G acts as isometries for some riemannian metric on X. An elementary argument using the exponential map then shows that for each $g \in G$, the set F_g is a smooth closed submanifold of X (see Kobayashi [1]). In general, however, F_g is not connected and the dimensions of the components can vary.

We wish to derive a cohomological formula for $\text{ind}_g(P)$, where P is an elliptic G-operator. For this purpose, we may replace G by the closure of the cyclic subgroup generated by g. This is a compact abelian Lie group for which g is a "topological generator." With this assumption we see that $F_g = \{x \in X : g'x = x \text{ for all } g' \in G\} \equiv F_G$, the fixed-point set for the entire group. This is a trivial G-space; however, the normal bundle N of F_g is a *non-trivial* real G-bundle. The normal bundle to the induced embedding

$$i : TF_g \hookrightarrow TX$$

is the complex G-bundle $\pi^*N \otimes \mathbb{C}$ (where $\pi : TF_g \to F_g$ is the bundle projection). The element $\lambda_{-1}(N_{\mathbb{C}}) \equiv \lambda_{-1}(\pi^*N \otimes \mathbb{C})$ is the Thom class for this bundle and is used, as in (12.7), to define a homomorphism

$$i_! : K_{G,\text{cpt}}(TF_g) \longrightarrow K_{G,\text{cpt}}(TX) \tag{14.6}$$

The fundamental result of Atiyah and Segal [1] is that after localizing at g, i.e., after introducing formal inverses for all elements $r \in R(G)$ with $\chi_g(r) \neq 0$, the map (14.6) becomes an isomorphism with inverse given by

$$i_!^{-1} = \frac{i^*}{\lambda_{-1}(N_{\mathbb{C}})}.$$

This leads to the following.

Theorem 14.3 (The Atiyah-Segal-Singer Fixed-Point Formula). *Let X be a compact G-manifold, where G is a compact Lie group, and let P be an elliptic G-operator on X with symbol class $\sigma = \sigma(P) \in K_{G,\text{cpt}}(TX)$. For each element $g \in G$, the "Lefschetz number" $\text{ind}_g(P) \equiv \text{trace}(g|_{\ker P}) - \text{trace}(g|_{\text{coker } P})$ is given by the formula*

$$\text{ind}_g(P) = (-1)^d \left\{ \frac{\text{ch}_g(i^*\sigma)}{\text{ch}_g(\lambda_{-1}(N_{\mathbb{C}}))} \cdot \hat{\mathfrak{A}}(F_g)^2 \right\} [TF_g] \tag{14.7}$$

where F_g denotes the fixed-point set of g, where $\lambda_{-1}(N_{\mathbb{C}})$ denotes the Thom class of the complexified normal bundle of F_g, and where d denotes the dimension of F_g (an integer-valued function which varies from component to component).

Note that G is not assumed to be connected; in particular, Theorem 14.3 applies when G is a finite group. Some of the most important applications of the result come from this case. This theorem also gives rise to a number of intriguing relations between topology and elementary number theory (see Hirzebruch-Zagier [1]).

The most basic example of a G-operator comes from the de Rham complex. Let G act by isometries on a compact riemannian manifold X, and consider the Dirac operator $D^0: \mathbb{C}\ell^0(X) \to \mathbb{C}\ell^1(X)$. (Recall that this is exactly the operator $d + \delta: \Lambda_{\mathbb{C}}^{\text{even}}(X) \to \Lambda_{\mathbb{C}}^{\text{odd}}(X)$). This operator is G-equivariant and for each $g \in G$, the number $\text{ind}_g(D^0)$ is just the classical Lefschetz number of g, i.e., $\text{ind}_g(D^0) = L(g) \equiv \text{trace}(g|_{H^{\text{even}}}) - \text{trace}(g|_{H^{\text{odd}}})$. The symbol class of D^0 is just the class $\sigma(D^0) = \lambda_{-1}(\pi^* TX \otimes \mathbb{C})$. Restricting to TF_g we have $\pi^* TX \otimes \mathbb{C} = (\pi^* TF_g \otimes \mathbb{C}) \oplus (\pi^* N \otimes \mathbb{C}) \stackrel{\text{def}}{=} T_{\mathbb{C}} \oplus N_{\mathbb{C}}$, and so $i^*\sigma(D^0) = \lambda_{-1}(T_{\mathbb{C}} \oplus N_{\mathbb{C}}) = \lambda_{-1}(T_{\mathbb{C}})\lambda_{-1}(N_{\mathbb{C}})$. It follows that $\text{ch}_g(i^*\sigma(D^0)) = \text{ch}_g(\lambda_{-1}(T_{\mathbb{C}}))\text{ch}_g(\lambda_{-1}(N_{\mathbb{C}}))$, and formula (14.7) becomes

$$L(g) = \chi(F_g).$$

In particular, if g has isolated fixed points, then $L(g) = \text{card}(F_g)$. This result uses strongly the fact that g is contained in a compact group of diffeomorphisms. The general Lefschetz formula applies to any diffeomorphism f with isolated non-degenerate fixed points, i.e., points where $\det(I - df) \neq 0$. In this case points are added with the weight factor $\text{sign}[\det(I - df)]$, so that the Lefschetz number becomes equal to the algebraic sum of the fixed points.

One might ask whether the fixed-point formula (14.7) can be extended to cover geometric automorphisms of an elliptic operator which do not lie in a compact group. Such a formula was given by Atiyah and Bott [3,4] for the case of automorphisms with isolated non-degenerate fixed points.

To give the reader some feeling for the fixed-point formula, we shall work out the details for the signature operator $D^+: \Gamma(\mathbb{C}\ell^+(X)) \to \Gamma(\mathbb{C}\ell^-(X))$ and the Atiyah-Singer operator $\not{D}^+: \Gamma(\$^+(X)) \to \Gamma(\$^-(X))$ (where X is spin).

Let us suppose that X is a compact oriented riemannian manifold and that $g: X \to X$ is an orientation preserving isometry. Let N denote the normal bundle of the fixed-point set F_g, and note that the differential dg of g gives a bundle isometry

$$dg: N \longrightarrow N. \tag{14.8}$$

Fix $x \in F_g$ and note that since g is an isometry, we have $g(\exp_x v) = \exp_x(dg \cdot v)$ for all $v \in N_x$. It follows that $dg \cdot v \neq v$ for each $v \neq 0$, since otherwise $\exp_x(tv)$ would lie in F_g for all $t \in \mathbb{R}$. Since g lies in a compact abelian Lie group, we know from elementary representation theory that there is an orthogonal decomposition $N_x = N_x(\pi) \oplus \bigoplus_{0<\theta<\pi} N_x(\theta)$, where $dg_x|_{N_x(\pi)} = -\mathrm{Id}$, and where the space $N_x(\theta)$ splits into 2-dimensional subspaces in which dg_x rotates every vector by θ. This decomposition is "constant" on each component of F_g. This follows from (14.2) or by simply observing that parallel translation along any curve joining x to y in F_g gives a g-equivariant isometry of $N_x(\theta)$ with $N_y(\theta)$. (This follows without difficulty from the fact that g is an isometry.) Consequently, we conclude that the bundle N admits a decomposition

$$N = N(\pi) \oplus \bigoplus_{0<\theta<\pi} N(\theta) \tag{14.9}$$

where dg acts on $N(\pi)$ by multiplication by -1, and where for each θ, $0 < \theta < \pi$, $N(\theta)$ is a complex bundle in which dg acts by multiplication by $e^{i\theta}$.

REMARK 14.4. Note that the bundle $N(\pi)$ may not be orientable but it carries the same orientation class as F_g, i.e., for every loop $\gamma \subset F_g$, the orientation of F_g changes along γ if and only if the orientation of $N(\pi)$ also changes along γ. (This is because X is orientable and each $N(\theta)$, $0 < \theta < \pi$, is complex and hence also orientable.)

For any real vector bundle E, we introduce the characteristic class

$$\mathbf{L}_\pi(E) \equiv \chi(E)\mathbf{L}^{-1}(E) \tag{14.10}$$

where $\mathbf{L}(E)$ is the total L-class of Hirzebruch and where $\chi(E)$ is the Euler class of E. If E is not orientable, then $\chi(E)$ lives in cohomology with twisted coefficients.

For any complex vector bundle E, and any θ, $0 < \theta < \pi$, we define $\mathbf{L}_\theta(E)$ to be the total class associated to the multiplicative sequence of Chern classes with formal power series $\coth(x + \frac{1}{2}i\theta)$. Therefore, if $E = \ell_1 \oplus \ldots \oplus \ell_k$ is a formal splitting into complex line bundles with $c_1(\ell_j) = x_j$, then

$$\mathbf{L}_\theta(E) = \prod_{j=1}^{k} \coth(x_j + \tfrac{1}{2}i\theta) \tag{14.11}$$

With these definitions we can state the following.

We recall now that if X is oriented and of even-dimension, then there is a canonical splitting $C\ell(X) = C\ell^+(X) \oplus C\ell^-(X)$ and the Dirac construction gives the "signature operator" $D^+ : \Gamma(C\ell^+(X)) \to \Gamma(C\ell^-(X))$. This operator is preserved by the group G of isometries of X. For each $g \in G$ we denote

$$\mathrm{sig}(X, g) \equiv \mathrm{ind}_g(D^+)$$

to emphasize that this invariant depends only on X and g. If $\dim X = 4k$ and $g = 1$, then $\operatorname{sig}(X, g) = \operatorname{sig}(X)$.

Theorem 14.5. (The G-signature Theorem). *Let $g: X \to X$ be an orientation-preserving isometry of a compact oriented 2m-dimensional manifold X. Then*

$$\operatorname{sig}(X, g) = \left\{ \prod_{0 < \theta \leq \pi} \mathbf{L}_\theta(N(\theta)) \cdot \mathbf{L}(F_g) \right\} [F_g] \tag{14.12}$$

where $N = \oplus N(\theta)$ is the decomposition of the normal bundle to F_g given in (14.9).

Note. The cohomology class in (14.12) factors into a product of an ordinary cohomology class with $\chi(N(\pi))$. As observed in Remark 14.4, $N(\pi)$ and F_g have the same orientation type. Hence, the pairing in (14.12) is well defined.

Proof. We begin with a remark on dimensions. Since g is orientation preserving, we see that $\dim_{\mathbb{R}}(N(\pi))$ is even. Hence we may write

$$\dim F_g = 2d, \qquad \dim N(\pi) = 2r, \qquad \dim_{\mathbb{R}} N(\theta) = 2s(\theta)$$

where d,r and $s(\theta)$ are integers which vary from component to component on F_g.

Recall that $\sigma(D^+) = \delta(TX)$, and therefore by the multiplicativity (11.33) of δ we have $i^*\delta(TX) = \delta(TF_g \oplus N) = \delta(TF_g)\delta(N)$. By pushing forward to F_g and using Proposition 12.4, the Fixed-point Formula (14.7) becomes

$$\operatorname{ind}_g(D^+) = 2^d \left\{ \frac{\operatorname{ch}_g(\delta(N))}{\operatorname{ch}_g(\lambda_{-1}(N \otimes \mathbb{C}))} \cdot \hat{\mathbf{L}}(F_g) \right\} [F_g]. \tag{14.13}$$

It remains to evaluate the term $\operatorname{ch}_g(\delta(N)\lambda_{-1}(N \otimes \mathbb{C})^{-1})$. Since everything is multiplicative, we can consider the factors $N(\theta)$ separately. Via the Splitting Principle it suffices to compute in the case where ν is an oriented real 2-plane bundle on which g acts by rotation by θ. We write $\nu \otimes \mathbb{C} = \ell \oplus \bar{\ell}$ where ℓ is a complex line bundle, and we set $x = c_1(\ell) = \chi(\nu)$. Then we have

$$\delta(\nu) = \bar{\ell} - \ell \qquad \text{and} \qquad \lambda_{-1}(\nu \otimes \mathbb{C}) = (1 - \ell)(1 - \bar{\ell}).$$

From the fact that $\operatorname{ch}_g(\ell) = \operatorname{ch}(\ell)e^{i\theta} = e^{x + i\theta}$, we find that

$$\begin{aligned} \operatorname{ch}_g(\delta(\nu))\lambda_{-1}(\nu \otimes \mathbb{C})^{-1} &= \frac{e^{-x-i\theta} - e^{x+i\theta}}{(1 - e^{x+i\theta})(1 - e^{-x-i\theta})} \\ &= \coth \tfrac{1}{2}(x + i\theta). \end{aligned}$$

This leads us to introduce the multiplicative sequence of Chern classes $\hat{\mathbf{L}}_\theta$ associated to the formal power series $\coth \frac{1}{2}(x + i\theta)$. The above computation shows that

$$\mathrm{ch}_g(\delta(N(\theta))\lambda_{-1}(N(\theta) \otimes \mathbb{C})^{-1}) = \hat{\mathbf{L}}_\theta(N(\theta)).$$

This computation is equally valid when $\theta = \pi$ provided that $N(\pi)$ is orientable. In fact, under this assumption we take a formal splitting $N(\pi) = \nu_1 \oplus \cdots \oplus \nu_r$ into oriented 2-plane bundles, and with $x_j = \chi(\nu_j)$ we find that

$$\begin{aligned}\mathrm{ch}_g(\delta(N(\pi))\lambda_{-1}(N(\pi) \otimes \mathbb{C})^{-1}) &= \prod_{j=1}^{r} \coth \tfrac{1}{2}(x_j + i\pi) \\ &= \prod_{j=1}^{r} \tanh(x_j/2) \\ &= \prod_{j=1}^{r} \frac{x_j}{2} \prod_{j=1}^{r} \left[\frac{x_j/2}{\tanh(x_j/2)}\right]^{-1} \\ &= \hat{\chi}(N(\pi))\hat{\mathbf{L}}(N(\pi))^{-1}\end{aligned}$$

where $\hat{\chi}(N(\pi)) \equiv 2^{-r}\chi(N(\pi))$. Assembling these calculations, we find that

$$\mathrm{sig}(X, g) = 2^d \left\{ \prod_{0 < \theta \leqq \pi} \hat{\mathbf{L}}_\theta(N(\theta)) \cdot \hat{\mathbf{L}}(F_g) \right\} [F_g]$$

where we have set $\hat{\mathbf{L}}_\pi = \hat{\chi} \cdot \hat{\mathbf{L}}^{-1}$. Multiplying the appropriate homogeneous component of this cohomology class by 2^d allows us to remove the hats from the $\hat{\mathbf{L}}$'s and gives us the desired formula (14.12). ∎

REMARK 14.6. This formula undergoes only minor modification if one takes coefficients in a complex G-bundle E. Specifically, if $D_E^+ : \Gamma(\mathbb{C}\ell^+ \otimes E) \to \Gamma(\mathbb{C}\ell^- \otimes E)$ is the twisted signature operator, then setting $\mathrm{sig}(X,E,g) \equiv \mathrm{ind}_g(D_E^+)$, we have

$$\mathrm{sig}(X,E,g) = \{\mathrm{ch}\, E \cdot \mathscr{L}\}[F_g] \tag{14.14}$$

where $\mathscr{L}$ is the cohomological expression appearing in formula (14.12).

The applications of Theorem 14.5 are numerous and varied (see Atiyah-Singer [2], Shanahan [1], and Hirzebruch-Zagier [1] for example). We mention here just a few corollaries. Note to begin, that the expression $\mathbf{L}_\pi = \chi \cdot \mathbf{L}^{-1}$ involves the Euler class. Consequently we have the following.

Corollary 14.7. *Let X, D^+, g, etc., be as in Theorem 14.5. If* $\dim N(\pi) > \dim F_g$, *then* $\mathrm{sig}(X, g) = 0$.

Note that when g is an involution, (i.e, $g^2 = \mathrm{Id}$), we have $N = N(\pi)$. In this case the corollary states that $\dim F_g < \frac{1}{2}\dim X \Rightarrow \mathrm{sig}(X, g) = 0$. This statement has a pretty generalization.

Corollary 14.8 (Atiyah-Singer [2]). *Let X be a compact oriented 4m-mainfold, and let $g: X \to X$ be an orientation preserving involution with fixed-point set F_g. Let $F_g \cdot F_g$ denote the closed oriented manifold obtained by intersecting F_g with (a generic displacement of) itself. Then we have*

$$\operatorname{sig}(X, g) = \operatorname{sig}(F_g \cdot F_g).$$

The right hand side of the formula is independent of the transversal displacement of F_g used to define $F_g \cdot F_g$. It depends, of course, only on the oriented cobordism class of $F_g \cdot F_g$.

Proof. Since $g^2 = \mathrm{Id}$, we have $N = N(\pi)$. One can verify easily that $\{\mathbf{L}(F_g)\mathbf{L}^{-1}(N)\chi(N)\}[F_g] = \{\mathbf{L}(F_g)\mathbf{L}^{-1}(N)\}[F_g \cdot F_g] = \{\mathbf{L}(F_g \cdot F_g)\}[F_g \cdot F_g] = \operatorname{sig}(F_g \cdot F_g)$. (We use here that $N|_{F_g \cdot F_g}$ is the normal bundle to $F_g \cdot F_g$ in F_g.) Applying (14.12) completes the proof. ∎

The proof of the following corollary is left as an exercise for the reader.

Corollary 14.9 (Atiyah-Singer [2]). *Let X be a compact connected oriented 4-manifold, and suppose that $g: X \to X$ is a diffeomorphism of odd order (necessarily orientation preserving). Suppose F_g consists of oriented embedded surfaces $S_1, \ldots, S_n$ and that for each k, dg rotates the oriented normal bundle to S_k through an angle θ_k. Then*

$$\operatorname{sig}(g, X) = \prod_{k=1}^{m} \sin^{-2}(\tfrac{1}{2}\theta_k) S_k \cdot S_k$$

where $S_k \cdot S_k$ denotes the homological self-intersection number of S_k.

Our second basic set of examples will come from considering the Atiyah-Singer operator $D\!\!\!/^{+}: \Gamma(\$_{\mathbb{C}}^{+}) \to \Gamma(\$_{\mathbb{C}}^{-})$ acting on spinors over a compact even-dimensional spin manifold X. We shall assume that G is a compact Lie group acting by orientation preserving isometries of X. Note that there is an induced action of G on the bundle $P_{\mathrm{SO}}(X)$ of oriented orthonormal tangent frames. Recall that the spin structure is a 2-fold covering $P_{\mathrm{spin}}(X) \to P_{\mathrm{SO}}(X)$ which is non-trivial on the fibres. It corresponds to an element $u \in H^1(P_{\mathrm{SO}}(X); \mathbb{Z}_2)$.

DEFINITION 14.10. The action of G **preserves the spin structure** of X if it lifts to an action on the bundle $P_{\mathrm{Spin}}(X)$. An individual isometry $g: X \to X$ is said to preserve the spin structure of X if the closed subgroup $G \subset \mathrm{Isom}(X)$ generated by g preserves this structure.

Note that if g preserves the spin structure, then $g^*u = u$, where $u \in H^1(P_{\mathrm{SO}}(X); \mathbb{Z}_2)$ is the element corresponding to the structure.

If G is connected, then it follows from elementary covering space theory that either G or a 2-fold covering group of G preserves the spin structure. In particular, if $\pi_1 G = 0$, then G preserves every spin structure on X.

Whenever G preserves the spin structure, it acts on the bundle of spinors and commutes with the Atiyah-Singer operator $\not{D}^+$ above. For each $g \in G$ one defines

$$\operatorname{Spin}(X, g) \equiv \operatorname{ind}_g(\not{D}^+),$$

and the Fixed-Point Theorem gives us a cohomological formula for this invariant. Its expression involves the sequences of Chern classes $\hat{\mathbf{A}}_\theta$ associated to the formal power series $[2 \sinh \frac{1}{2}(x + \theta)]^{-1}$ for $0 < \theta < \pi$. If E is a complex vector bundle with formal splitting $E = \ell_1 \oplus \ldots \oplus \ell_k$ into line bundles with $c_1(\ell_j) = x_j$, then

$$\hat{\mathbf{A}}_\theta(E) = 2^{-k} \prod_{j=1}^{k} \frac{1}{\sinh \frac{1}{2}(x_j + i\theta)} = \prod_{j=1}^{k} \frac{e^{\frac{1}{2}(x_j + i\theta)}}{e^{x_j + i\theta} - 1}$$

In the special case where $\theta = \pi$ we define a characteristic class $\hat{\mathbf{A}}_\pi(E)$ for any oriented real $2k$-dimensional bundle E as follows. Let $E = E_1 \oplus \ldots \oplus E_k$ be a formal splitting into oriented 2-plane bundles, and set $x_j = \chi(E_j)$. Then

$$\hat{\mathbf{A}}_\pi(E) \equiv 2^{-k} \prod_{j=1}^{k} \frac{1}{\sinh \frac{1}{2}(x_j + i\pi)} = (2i)^{-k} \prod_{j=1}^{k} \frac{1}{\cosh(x_j/2)}$$

Theorem 14.11 (The G-spin Theorem). *Let $g : X \to X$ be a spin structure preserving isometry of a compact even-dimensional spin manifold X. Then*

$$\operatorname{Spin}(X, g) = (-1)^\sigma \left\{ \prod_{0 < \theta \leq \pi} \hat{\mathbf{A}}_\theta(N(\theta)) \cdot \hat{\mathbf{A}}(F_g) \right\} [F_g]$$

where $N = \oplus N(\theta)$ is the decomposition of the normal bundle to F_g given in (14.9), and where $\sigma : F_g \to \{0,1\}$ is a locally constant function which depends on the action of g on the spin structure.

REMARK 14.12. In general the calculation of the sign function σ is a complicated and subtle affair. We refer the interested reader to Atiyah-Bott [3],[4] and Atiyah-Hirzebruch [3] for details.

Proof. We proceed much as in the proof of the G-signature Theorem. Recall that $\sigma(D^+) = \mathfrak{s}(TX)$ and therefore by the multiplicativity (11.36) of $\mathfrak{s}$ we have $i^*\sigma(D^+) = \mathfrak{s}(TF_g \oplus N) = \mathfrak{s}(TF_g)\mathfrak{s}(N)$. Pushing forward to F_g and using Proposition 12.5 converts the Fixed-Point Formula to

$$\operatorname{ind}_g(D^+) = \left\{ \frac{\operatorname{ch}_g(\mathfrak{s}(N))}{\operatorname{ch}_g(\lambda_{-1}(N \otimes \mathbb{C}))} \cdot \hat{\mathbf{A}}(F_g) \right\} [F_g].$$

Computing as before, we consider the case of an oriented 2-plane bundle ν with $\nu \otimes \mathbb{C} = \ell \oplus \bar{\ell}$. Here we have that $\mathfrak{s}(\nu) = \bar{\ell}^{1/2} - \ell^{1/2}$ and

$\lambda_{-1}(\nu \otimes \mathbb{C}) = (1 - \ell)(1 - \bar{\ell})$, and therefore that $\mathrm{ch}_g(\mathbf{s}(\nu)\lambda_{-1}(\nu \otimes \mathbb{C})^{-1}) = \pm[e^{\frac{1}{2}(x+i\theta)} - e^{-\frac{1}{2}(x+i\theta)}]^{-1}$, if g acts on ν by rotation by θ. It follows directly that

$$\mathrm{ch}_g(\mathbf{s}(N(\theta))\lambda_{-1}(N(\theta) \otimes \mathbb{C})^{-1}) = \pm\hat{\mathbf{A}}_\theta(N(\theta))$$

for each θ, $0 < \theta \leq \pi$. This completes the proof. ■

The G-Spin Theorem can be enhanced by taking coefficients in a G-bundle E. The formula changes exactly as in the G-signature case (cf. Remark 14.6).

§15. The Index Theorem for Families

Consider now a family of elliptic operators P on a compact manifold X parameterized by a compact Hausdorff space A. In §8 it was shown that this family has a well-defined analytic index $\mathrm{ind}\, P \in K(A)$. On the other hand the constructions of §13 can be generalized to define a topological index in $K(A)$. The main result of this section asserts that these two indices coincide.

The topological index of the family P is defined as follows. Let $\pi : \mathscr{X} \to A$ denote the underlying family of manifolds (recall that this is a bundle with fibre X and structure group $\mathrm{Diff}(X)$), and let $T\mathscr{X} \to A$ denote the associated family of tangent bundles (so that $(T\mathscr{X})_a = T(\mathscr{X}_a)$ where $\mathscr{X}_a \equiv \pi^{-1}(a)$). It is not difficult to see that for sufficiently large m we can find a map $f : \mathscr{X} \to A \times \mathbb{R}^m$ which restricts to a smooth embedding $f_a : \mathscr{X}_a \hookrightarrow \{a\} \times \mathbb{R}^m$ for each $a \in A$. This induces a map $T\mathscr{X} \hookrightarrow A \times T\mathbb{R}^m$ which for each $a \in A$, restricts to an embedding $T\mathscr{X}_a \hookrightarrow \{a\} \times T\mathbb{R}^m$ with normal bundle $N_a \oplus N_a \cong N_a \otimes \mathbb{C}$. Here N_a denotes the normal bundle of $f_a(\mathscr{X}_a) \subset \mathbb{R}^m$ pulled back to $T\mathscr{X}_a$. We now define a map

$$f_! : K_{\mathrm{cpt}}(T\mathscr{X}) \longrightarrow K_{\mathrm{cpt}}(A \times \mathbb{C}^m)$$

by taking the composition

$$K_{\mathrm{cpt}}(T\mathscr{X}) \longrightarrow K_{\mathrm{cpt}}(N \otimes \mathbb{C}) \longrightarrow K_{\mathrm{cpt}}(A \times \mathbb{C}^m)$$

where the first map is the Thom isomorphism, and where the second map is induced via an embedding $N \otimes \mathbb{C} \hookrightarrow A \times T\mathbb{R}^m \cong A \times \mathbb{C}^m$, which is constructed fibrewise by identifying $N_a \otimes \mathbb{C}$ with a tubular neighborhood of $f_a(\mathscr{X}_a)$ in $\{a\} \times T\mathbb{R}^m$ as before.

The projection $q : A \times \mathbb{C}^m \to A$ induces a Thom isomorphism (or **Bott periodicity** map)

$$q_! : K_{\mathrm{cpt}}(A \times \mathbb{C}^m) \xrightarrow{\approx} K(A).$$

The composition $q_! \circ f_! : K_{\text{cpt}}(T\mathscr{X}) \to K(A)$ is independent of the choice of the embedding f.

DEFINITION 15.1. Let P be a family of elliptic operators on a compact manifold X parameterized by a compact Hausdorff space A as in §8. Let $\sigma(P) \in K_{\text{cpt}}(T\mathscr{X})$ be the class determined by the principal symbol of the family. Then the **topological index** of P is the element

$$\text{top-ind}(P) \equiv q_! f_!(\sigma(P)) \in K(A).$$

Straightforward generalization of the arguments given in §13 proves the following:

Theorem 15.2 (The Atiyah-Singer Index Theorem for Families [3]). *For any P as above one has that*

$$\text{ind}(P) = \text{top-ind}(P).$$

Applying the Chern character and arguing as in §13 establishes the following cohomological form of this result.

Theorem 15.3 (Atiyah-Singer [3]). *If P is as above, then*

$$\text{ch}(\text{ind } P) = (-1)^{\frac{n(n+1)}{2}} \pi_!\{\text{ch } \sigma(P) \cdot \hat{\mathbf{A}}(T\mathscr{X})^2\}$$

where $n = \dim X$ *and where* $\pi : T\mathscr{X} \to A$ *is the natural projection.*

Suppose now that $\pi : \mathscr{X} \to A$ is a **family of riemannian manifolds**, i.e., a family of manifolds as above together with a family of riemannian metrics introduced continuously in the fibres.

Corollary 15.4. *Let* $\pi : \mathscr{X} \to A$ *be a family of compact oriented riemannian manifolds of dimension 2m, and let* $\mathscr{D}^+ : \Gamma(\mathbb{C}\ell^+(T\mathscr{X})) \to \Gamma(\mathbb{C}\ell^-(T\mathscr{X}))$ *be the associated family of signature operators. Then*

$$\text{ind}(\mathscr{D}^+) = 2^m \pi_!\{\hat{\mathbf{L}}(T\mathscr{X})\}.$$

More generally, if we take coefficients in a family of hermitian vector bundles $\mathscr{E}$ *with connection, then the associated operator* $\mathscr{D}^+_{\mathscr{E}} : \Gamma(\mathbb{C}\ell^+(T\mathscr{X}) \otimes \mathscr{E}) \to \Gamma(\mathbb{C}\ell^-(T\mathscr{X}) \otimes \mathscr{E})$ *has index given by*

$$\text{ind}(\mathscr{D}^+_{\mathscr{E}}) = 2^m \pi_!\{\text{ch } \mathscr{E} \cdot \hat{\mathbf{L}}(T\mathscr{X})].$$

Proof. The symbol of $\mathscr{D}^+$ is $\sigma(\mathscr{D}^+) = \delta(T\mathscr{X})$. Using 12.4 to push forward over the projection $\pi_0 : T\mathscr{X} \to \mathscr{X}$, we find that $(\pi_0)_!\{\text{ch } \sigma(\mathscr{D}^+) \cdot \hat{\mathbf{A}}(T\mathscr{X})^2\} = (-2)^m \hat{\mathbf{L}}(T\mathscr{X})\hat{\mathbf{A}}(T\mathscr{X})^{-2}\hat{\mathbf{A}}(T\mathscr{X})^2 = (-2)^m \hat{\mathbf{L}}(T\mathscr{X})$. Projecting on to A then gives the first formula. The second formula follows similarly from the fact that $\sigma(\mathscr{D}^+_{\mathscr{E}}) = \delta(T\mathscr{X}) \cdot \pi_0^*(\mathscr{E})$. ∎

It is an interesting exercise to apply this formula to the Lusztig family (8.3).

Carrying through the computations above with $\boldsymbol{\delta}$ replaced by $\mathbf{s}$ gives the following.

Corollary 15.5. *Let $\pi: \mathscr{X} \to A$ be a family of compact oriented spin manifolds of dimension $2m$, and let $\mathscr{D}_{\mathscr{E}}^{+}: \Gamma(\mathscr{S}^{+} \otimes \mathscr{E}) \to \Gamma(\mathscr{S}^{-} \otimes \mathscr{E})$ be the family of Atiyah-Singer operators twisted by a family of coefficient bundles $\mathscr{E}$. Then*

$$\operatorname{ind}(\mathscr{D}_{\mathscr{E}}^{+}) = \pi_!\{\operatorname{ch} \mathscr{E} \cdot \hat{\mathbf{A}}(T\mathscr{X})\}.$$

All of these results go through for families of G-operators. One replaces K by K_G in this case.

§16. Families of Real Operators and the $C\ell_k$-Index Theorem

In discussing the Index Theorem thus far we have considered only complex differential operators. Given a real operator, say P, defined using real vector bundles, one can always complexify and apply the index theorem in the form above. Since $\dim_{\mathbb{R}}(\ker P) = \dim_{\mathbb{C}}(\ker P \otimes \mathbb{C})$ (and similarly for P^*), we see that the ordinary real and complex indices of p coincide. Thus, in the basic case no information is lost under complexification.

This is not true, however, if one passes to the index theorem for families. The index of a family of real operators takes its value in the group $KO(A)$, and the complexification map $KO(A) \to K(A)$ is not always injective. Note for example that $\widetilde{KO}(S^n) \cong \mathbb{Z}_2$ for $n \equiv 1 (\bmod 8)$, but $\tilde{K}(S^n) = \{0\}$ in these dimensions.

For this reason Atiyah and Singer established a separate index theorem for families of real operators. It is a more subtle and profound result than one might naively expect. The constructions and arguments outlined above for complex families go through essentially unchanged in the real case, provided one employs the appropriate "K-theory." It is here that matters become interesting, for the appropriate theory is not KO-theory, the straightforward theory of real bundles. It is the more general KR-theory which is defined on any space with involution and reduces to KO-theory when the involution is trivial. Recall (cf. I. 10) that if X is a space with involution $f: X \to X$, then $KR(X)$ is the Grothendieck group of pairs (E, f_E) where E is a complex vector bundle over X and where $f_E: E \to E$ is an involution which covers f and is $\mathbb{C}$-antilinear on the fibres.

The introduction of this theory is motivated by the simple fact that the principal symbol of a real operator is in general *not* real. Consider, for example, the operator $\partial/\partial\theta: C^\infty(S^1) \to C^\infty(S^1)$ whose symbol is $\sigma_\xi = i\xi$.

Recall also that for any real Dirac operator D, one has that $\sigma_\xi(D) = i\xi\cdot$ (cf. Example 1.5). The appearance here of the complex number i cannot be ignored. It is essential in the calculus of pseudo-differential operators and must be retained if the proofs discussed above are to carry over.

Suppose now that $P:\Gamma(E) \to \Gamma(F)$ is a real differential operator between real bundles E and F over a compact manifold X. In local terms, we have that $P = \sum A^\alpha(x)\partial^{|\alpha|}/\partial x^\alpha$ plus lower order terms, where the A^α are real matrix-valued functions. Consequently for any tangent vector ξ, we have $\sigma_\xi(P) = \sum A^\alpha(x)(i\xi)^\alpha$, from which it follows that

$$\sigma_{-\xi}(P) = \overline{\sigma_\xi(P)}. \tag{16.1}$$

This is the key to defining the symbol class of a real operator.

Definition 16.1. Given a compact manifold X, consider the tangent bundle $\pi: TX \to X$ to be equipped with the canonical involution $f: TX \to TX$ defined by $f(\xi) = -\xi$. Given any real bundle $E \to X$, consider $\pi^*(E \otimes \mathbb{C})$ to be equipped with the antilinear involution defined by complex conjugation. Then for any real elliptic operator $P:\Gamma(E) \to \Gamma(F)$ the **Real symbol class** of P is defined to be the element

$$[\pi^*(E \otimes \mathbb{C}), \pi^*(F \otimes \mathbb{C}); \sigma(P)] \in KR_{\mathrm{cpt}}(TX). \tag{16.2}$$

Here $\pi^*(E \otimes \mathbb{C})$ and $\pi^*(F \otimes \mathbb{C})$ are Real bundles on the Real space TX, and (16.1) says that $\sigma(P)$ is an isomorphism of real bundles outside the zero section of TX. Hence, (16.2) naturally defines an element in $KR_{\mathrm{cpt}}(TX$.)

In general if one wants to remember that a given bundle is the complexification of a real bundle, one carries along the associated complex conjugation map. The new point here is that when the bundle is pulled back over TX we think of this conjugation as covering the involution $\xi \mapsto -\xi$. With this little refinement everything works as one hopes.

To define the topological index of a real elliptic operator P we first choose an embedding $f: X \hookrightarrow \mathbb{R}^m$. The associated embedding $TX \hookrightarrow T\mathbb{R}^m$ is compatible with the involutions i.e., is a mapping of Real spaces. If N is the normal bundle to X in $\mathbb{R}^m$, then $\pi^*N \oplus \pi^*N \cong \pi^*N \otimes \mathbb{C}$ is the normal bundle to TX in $T\mathbb{R}^m$. We consider this to be a Real vector bundle on TX as in Definition 16.1 above. For such bundles, the **Thom isomorphism** holds for KR-theory. (It is defined exactly as it was for K-theory in §12. We need only note that the de Rham element Δ_{-1} of a Real bundle is itself Real. See Atiyah [2].) We can now define a map

$$f_!: KR_{\mathrm{cpt}}(TX) \longrightarrow KR_{\mathrm{cpt}}(T\mathbb{R}^m)$$

as before by composing the Thom isomorphism with the map induced by the inclusion of the normal bundle as a tubular neighborhood of TX in $T\mathbb{R}^m$. This inclusion can be easily chosen to be compatible with the involutions.

If $\mathscr{X} \to A$ is a family of smooth manifolds over the compact space A, this construction extends, by using local triviality of the family, to give a map

$$f_! : KR_{\mathrm{cpt}}(T\mathscr{X}) \longrightarrow KR_{\mathrm{cpt}}(A \times T\mathbb{R}^m).$$

We now identify $T\mathbb{R}^m \cong \mathbb{R}^m \oplus \mathbb{R}^m \cong \mathbb{C}^m$ by associating (x,ξ) to $x + i\xi$. Clearly the involution on $T\mathbb{R}^m$ becomes complex conjugation on $\mathbb{C}^m$. The fundamental (1,1)-Periodicity Theorem (I,10.3), says that there is a natural isomorphism

$$q_! : KR_{\mathrm{cpt}}(A \times \mathbb{C}^m) \longrightarrow KR(A)$$

for any compact (Real) space A. As before, the composition $q_! \circ f_!$ can be shown to be independent of the choices involved in the construction.

DEFINITION 16.2. Let P be a family of real elliptic differential operators on a compact manifold X parameterized by a compact Hausdorff space A. Let $\sigma(P) \in KR_{\mathrm{cpt}}(T\mathscr{X})$ be the symbol class of the family (where, as before, $\mathscr{X} \to A$ is the underlying family of manifolds). The **topological index of the family P** is defined to be the element

$$\text{top-ind}(P) = q_! f_! \sigma(P) \in KR(A) \cong KO(A).$$

(Note that A carries the trivial involution.)

With this definition, the arguments for the Index Theorem discussed above go through easily to prove the following.

Theorem 16.3 (Atiyah and Singer [3], [4]). *Let P be a family of real elliptic operators on a compact manifold parameterized by a compact Hausdorff space A. Let* $\mathrm{ind}(P) \in KO(A)$ *be the analytic index of the family defined as in* 8.5 *by replacing complex objects with real ones. Then*

$$\mathrm{ind}(P) = \text{top-ind}(P).$$

It should be remarked that the index theorem for families is a useful tool, much more powerful than the standard index theorem. A number of applications are given in the next chapter.

One important consequence of this result is the derivation of a topological formula for the Clifford index discussed in §10. No such formula appears in the current literature even though the derivation was known to Atiyah and Singer. We shall present the details here.

Assume from this point on that $E = E^0 \oplus E^1$ is a real, $\mathbb{Z}_2$-graded $C\ell_k$-bundle over a compact riemannian manifold X. Assume E carries a bundle metric for which Clifford multiplication by unit vectors in $\mathbb{R}^k$ is orthogonal. Let $P : \Gamma(E) \to \Gamma(E)$ be an elliptic self-adjoint operator and assume that P is $C\ell_k$-linear and $\mathbb{Z}_2$-graded. In 10.4 we defined an analytic index

$\text{ind}_k P \in KO^{-k}(\text{pt})$ for P in terms of the $C\ell_k$-module ker P. We shall now give a topological formula for this index.

To do this we construct the following family $\mathscr{P}$ of elliptic operators parameterized by $\mathbb{R}^k$. We assume that P has degree zero by replacing P with $(1 + P^*P)^{-1/2}P$. We recall from 10.2 that with respect to the splitting $E = E^0 \oplus E^1$ we can write P as

$$P = \begin{pmatrix} 0 & P^1 \\ P^0 & 0 \end{pmatrix}$$

where $P^1 = (P^0)^*$. The product family $\mathscr{P}$ on $\mathbb{R}^k \times X$ is now constructed by assigning to each $v \in \mathbb{R}^k$ the operator

$$\mathscr{P}_v^0 : \Gamma(E^0) \to \Gamma(E^1) \tag{16.3}$$

defined by the restriction to E^0 of the operator

$$\mathscr{P}_v \equiv v + P \tag{16.4}$$

where "v" denotes Clifford multiplication by v. Note that there is a "conjugate family" of operators $\bar{\mathscr{P}}_v$ defined by $\bar{\mathscr{P}}_v \equiv v - P$. Since P commutes with Clifford multiplication, we have that

$$\bar{\mathscr{P}}_v \mathscr{P}_v = \mathscr{P}_v \bar{\mathscr{P}}_v = -\{\|v\|^2 + P^2\}. \tag{16.5}$$

In particular, $\mathscr{P}_v^0$ is *invertible* for all $v \neq 0$. Since the invertible operators on Hilbert space form a contractible set, we could easily pass to a family parameterized by S^k. However, the calculations will be more transparent if we treat $\mathscr{P}^0$ as a family with "compact support" in $\mathbb{R}^k$, whose index lies in $KO_{\text{cpt}}(\mathbb{R}^k) \cong KO^{-k}(\text{pt})$. The main result is the following:

Theorem 16.4. *Let P be an elliptic self-adjoint graded $C\ell_k$-operator on a compact manifold. Then*

$$\text{ind}_k(P) = \text{top-ind}(\mathscr{P}^0)$$

where $\mathscr{P}$ is the family over $\mathbb{R}^k$ defined by (16.4).

Proof. Set $K^0 \equiv \ker P^0 \subset \Gamma(E^0)$ and $K^1 \equiv \ker P^1 \cong \text{coker}\, P^0 \subset \Gamma(E^1)$. By Theorem 5.5 there are L^2-orthogonal direct sum decompositions

$$\Gamma(E^0) = V^0 \oplus K^0 \qquad \text{and} \qquad \Gamma(E^1) = V^1 \oplus K^1 \tag{16.6}$$

where $P^0 : V^0 \xrightarrow{\approx} V^1$ is an isomorphism. Since P commutes which Clifford multiplication, we have that $v \cdot K^0 = K^1$ for all $v \neq 0$ in $\mathbb{R}^k$. Furthermore, since multiplication by $v/\|v\|$ is L^2-orthogonal, we have $v \cdot V^0 = V^1$ for all such v. It follows that the family $\mathscr{P}_v^0 = v + P^0 : \Gamma(E^0) \to \Gamma(E^1)$ decomposes, with respect to (16.6), as a direct sum of two operators. By (16.5) the first summand

$$V^0 \xrightarrow[\approx]{\mathscr{P}_v^0} V^1$$

is an isomorphism for all $v \in \mathbb{R}^k$, and thus can be ignored for the purposes of computing the index. The second summand is just

$$K^0 \xrightarrow{\mathscr{P}_v^0 = v} K^1$$

This operator is independent of variables on X. Its index equals the index of $\mathscr{P}^0$ and is given by the element

$$[K^0, K^1; v] \in KO_{\text{cpt}}(\mathbb{R}^k) \cong KO^{-k}(\text{pt}).$$

Under the Atiyah-Bott-Shapiro isomorphism $KO^{-k}(\text{pt}) \xrightarrow{\approx} \hat{\mathfrak{M}}_k / i^* \hat{\mathfrak{M}}_{k+1}$, this corresponds exactly to the element represented by the $\mathbb{Z}_2$-graded $C\ell_k$-module $\ker P = K^0 \oplus K^1$, i.e., it corresponds exactly to $\text{ind}_k P$. ■

Theorem 16.4 can be applied to give topological formulas for the index of any of the graded $C\ell_k$ Dirac operators discussed in Chapter II, §7. In particular this includes the Atiyah-Singer operator whose index is a basic invariant which we shall now discuss in detail.

Let X be a compact spin manifold of dimension n, and recall that X carries a canonical graded $C\ell_n$ Dirac bundle

$$\mathfrak{S}(X) \equiv P_{\text{Spin}}(X) \times_{\ell} C\ell_n,$$

whose associated Dirac operator $\mathfrak{D}$ is called the **Atiyah-Singer operator** (see equation (II.7.1) forward). This operator has an index $\text{ind}_n(\mathfrak{D}) \in KO^{-n}(\text{pt})$ which coincides, by 16.4, with the index of the family $\mathscr{D}$ on $\mathbb{R}^n \times X$ defined by setting

$$\mathscr{D}_{v,x} = v + \mathfrak{D}_x.$$

To compute the topological index of this family we must understand its symbol class $\sigma(\mathscr{D}) \in KR_{\text{cpt}}(\mathbb{R}^n \times TX)$. For this we note first that $\pi: \mathbb{R}^n \times TX \to X$ is a Real bundle over X whose fibre at $x \in X$ is $\mathbb{R}^n \times T_xX$ with involution $(v,\xi) \mapsto (v, -\xi)$. The fibre of the bundle $\mathfrak{S}(X)$ at x is the Clifford algebra $C\ell(T_xX) \cong C\ell_n$. Vectors $\xi \in T_xX$ act by left Clifford multiplication as usual, and vectors $v \in \mathbb{R}^n$ act by right multiplication. The principal symbol of $\mathscr{D}$ is the map $\sigma(\mathscr{D}): \Gamma(\pi^*\mathfrak{S}_{\mathbb{C}}^0) \to \Gamma(\pi^*\mathfrak{S}_{\mathbb{C}}^1)$ defined by

$$\sigma_{\xi,v}(\mathscr{D}) = v + i\xi \quad (= R_v + iL_\xi),$$

where $\pi^*\mathfrak{S}_{\mathbb{C}}^k \equiv \pi^*\mathfrak{S}^k \otimes \mathbb{C}$ is treated as a Real bundle on $\mathbb{R}^n \times TX$ with involution given by conjugation. The symbol class. $[\pi^*\mathfrak{S}_{\mathbb{C}}^0, \pi^*\mathfrak{S}_{\mathbb{C}}^1; \sigma(\mathscr{D})]$ when restricted to any fibre $\mathbb{R}^n \times T_xX \cong \mathbb{R}^n \times \mathbb{R}^n \cong \mathbb{C}^n$, becomes exactly the element,

$$[C\ell_n^0, C\ell_n^1; v + i\xi] \in KR_{\text{cpt}}(\mathbb{C}^n)$$

which, as we have seen in I.10.12 is the generator of the group $KR_{\text{cpt}}(\mathbb{C}^n) \cong \mathbb{Z}$. (It corresponds exactly to the deRham element Λ_{-1} under the usual isomorphism $C\ell^* \simeq \Lambda^*$.) Consequently the symbol class

$$\sigma(\not{\mathfrak{D}}) \equiv [\pi^*\not{\mathfrak{S}}_{\mathbb{C}}^0, \pi^*\not{\mathfrak{S}}_{\mathbb{C}}^1; \sigma(\not{\mathfrak{D}})] \in KR_{\text{cpt}}(\mathbb{R}^n \times TX)$$

is a Thom class or "orientation class" for the bundle $\pi: \mathbb{R}^n \times TX \to X$ in KR-theory. Under the Thom isomorphism $i_!: KR(X) \to KR_{\text{cpt}}(\mathbb{R}^n \times TX)$ we have

$$\sigma(\not{\mathfrak{D}}) = i_!(1) \tag{16.7}$$

Choose now a smooth embedding $f: X \hookrightarrow \mathbb{R}^{n+8\ell}$ and let N denote its normal bundle. This induces a smooth embedding $f: TX \hookrightarrow T\mathbb{R}^{n+8\ell}$ with normal bundle $\pi^*N \oplus \pi^*N \cong \pi^*N \otimes \mathbb{C}$, and we get the following commutative diagram of bundle maps:

$$\begin{array}{ccc} \mathbb{R}^n \times TX & \xleftarrow{\tilde{p}} & \mathbb{R}^n \times (\pi^*N \otimes \mathbb{C}) \\ {\scriptstyle \pi}\downarrow & & \downarrow{\scriptstyle \pi} \\ X & \xleftarrow[p]{} & N \end{array}$$

The vertical arrows are Real bundles. Taking zero-sections and embedding the normal bundles as tubular neighborhoods give the following commutative diagram of embeddings:

$$\begin{array}{ccccc} \mathbb{R}^n \times TX & \overset{\tilde{j}}{\subset} & \mathbb{R}^n \times (\pi^*N \otimes \mathbb{C}) & \overset{\tilde{k}}{\subset} & \mathbb{R}^n \times \mathbb{C}^{n+8\ell} \\ \cup & & \cup & & \cup \\ X & \underset{j}{\subset} & N & \underset{k}{\subset} & \mathbb{R}^{n+8\ell} \end{array} \tag{16.8}$$

We now recall from II.1.15 that N has a canonically induced spin structure. Since $\dim_{\mathbb{R}} N = 8\ell$ there is a Thom isomorphism $j_!: KO(X) \to KO_{\text{cpt}}(N)$ defined by setting

$$j_!(u) = (p^*u) \cdot \mathfrak{s}(N)$$

where $\mathfrak{s}(N) \equiv [S^+(N), S^-(N); \mu]$ is defined as in (12.13) using the real spinor bundle of N (see Appendix C). We could view this as an isomorphism in KR-theory where the spaces X and N carry the trivial involution. There is a completely analogous Thom isomorphism $\tilde{j}_!: KR_{\text{cpt}}(\mathbb{R}^n \times TX) \to KR_{\text{cpt}}(\mathbb{R}^n \times (\pi^*N \otimes \mathbb{C}))$ defined using the spin structure on π^*N and the Real bundle $\pi^*S_{\mathbb{C}}(N) = \pi^*S(N) \otimes \mathbb{C}$. One easily checks that

$$i_! j_! = \tilde{j}_! i_! \qquad \text{and} \qquad i_! k_! = \tilde{k}_! i_!$$

where $k_!$ and $\tilde{k}_!$ denote the natural maps induced by the open inclusions

k and $\tilde{k}$. Consequently, (16.8) leads to a commutative diagram

$$\begin{array}{ccccc} KR_{\text{cpt}}(\mathbb{R}^n \times TX) & \xrightarrow{\tilde{f}_!} & KR_{\text{cpt}}(\mathbb{R}^n \times \mathbb{C}^{n+8\ell}) & & \\ \uparrow{\scriptstyle i_!} & & \uparrow{\scriptstyle i_!}\,\wr\wr & \searrow^{\tilde{q}_!}_{\approx} & \\ & & & & KR^{-n}(pt) \\ & & & \nearrow^{\approx}_{q_!} & \\ KR(X) & \xrightarrow{f_!} & KR_{\text{cpt}}(\mathbb{R}^{n+8\ell}) & & \end{array} \qquad (16.9)$$

where $f_! \equiv k_! j_!$, $\tilde{f}_! = \tilde{k}_! \tilde{j}_!$, and where $q_!$ and $\tilde{q}_!$ denote the Bott periodicity isomorphisms.

The topological index of the family $\not{\mathscr{D}}$ is defined to be the element $\tilde{q}_! \tilde{f}_! \sigma(\not{\mathscr{D}})$. Applying (16.7) and (16.9) we conclude that

$$\text{top-ind}(\not{\mathscr{D}}) = \tilde{q}_! \tilde{f}_! i_!(1) = q_! f_!(1). \qquad (16.10)$$

DEFINITION 16.5. For a compact spin manifold X of dimension n, the **Atiyah-Milnor-Singer invariant** is the element

$$\hat{\mathscr{A}}(X) \equiv q_! f_!(1) \in KO^{-n}(\text{pt})$$

where $f: X \hookrightarrow \mathbb{R}^{n+8\ell}$ is a smooth embedding for some ℓ, and where $q_!$ and $f_!$ are defined as above.

This definition is easily seen to agree with the one given in II.3.22. In particular, we see that $\hat{\mathscr{A}}(X)$ depends only on the spin cobordism class of X. Combining (16.10) with Theorem 16.4 gives the following main result:

Theorem 16.6. *Suppose X is a compact spin manifold of dimension n, and let $\not{\mathfrak{D}}: \Gamma(\not{\mathfrak{S}}) \to \Gamma(\not{\mathfrak{S}})$ be the canonical $C\ell_n$-linear Atiyah-Singer operator. Then*

$$\text{ind}_n(\not{\mathfrak{D}}) = \hat{\mathscr{A}}(X).$$

From Theorem II.7.11 and the spin-cobordism invariance of $\hat{\mathscr{A}}(X)$, we see that $\hat{\mathscr{A}}: \Omega_*^{\text{spin}} \to KO^{-*}(\text{pt})$ is a graded **ring** homomorphism.

Suppose now that E is any real vector bundle on X. Furnish E with any orthogonal connection and introduce the tensor product connection on $\not{\mathfrak{S}} \otimes E$. This is again a graded $C\ell_n$-Dirac bundle and has an associated Dirac operator $\not{\mathfrak{D}}_E$. Let $\not{\mathscr{D}}_E = v + \not{\mathfrak{D}}_E$ be its associated family. One easily checks that $\sigma(\not{\mathscr{D}}_E) = i_!([E])$. Passing through the arguments above one finds that $\text{top-ind}(\not{\mathscr{D}}_E) = q_! f_!([E])$. This proves the following result:

DEFINITION 16.7. Let X be a compact spin manifold of dimension n. For each real vector bundle E over X, the **associated Atiyah-Milnor-Singer invariant** is the element

$$\hat{\mathscr{A}}(X;E) \equiv q_! f_!([E]) \in KO^{-n}(\text{pt})$$

where $q_! f_!: KO(X) \to KO^{-n}(\text{pt})$ is defined as above.

Theorem 16.8. *Suppose X is a compact spin manifold of dimension n, and let E be a real vector bundle over X. Let $\not{\mathfrak{D}}_E: \Gamma(\not{\mathfrak{S}} \otimes E) \to \Gamma(\not{\mathfrak{S}} \otimes E)$ be the*

$C\ell_n$-linear Atiyah-Singer operator with coefficients in E (defined using any orthogonal connection on E.) Then

$$\operatorname{ind}_n(\mathfrak{D}_E) = \hat{\mathscr{A}}(X;E).$$

§17. Remarks on Heat and Supersymmetry

Since the basic work of Gilkey and Patodi it has been known that index formulas for classical elliptic operators (such as the Atiyah-Singer operator with coefficients in a bundle) can be derived from the asymptotics of the heat kernel. As explained at the end of §6, it is necessary to make detailed computations of the density ρ_n occurring in the asymptotic expansion trace $K_t(x,x) \sim \sum \rho_k(x)t^{(k-n)/2}$ as $t \searrow 0$. Expositions of this can be found in Gilkey [2] and Atiyah-Bott-Patodi [1].

In 1979 W. Allard showed us a simple and elementary calculation of these densities obtained by working with certain canonical operators on the underlying principal bundle and using the natural filtration of the Clifford algebra. (His resulting proof of the classical Index Theorem is completely elementary.)

In 1982, E. Witten found a different approach to these formulas through considerations of symplectic geometry and supersymmetry. There is a fixed-point formula for S^1-actions on (finite dimensional) symplectic manifolds due to Duistermaat and Heckman [1]. Witten considered the analogue of this formula for the canonical S^1-action on the free loop space of a manifold. Using this together with some ideas from supersymmetry he outlined a proof of the index theorem for the Atiyah-Singer operator (see Atiyah [13]). His ideas have engendered a series of interesting papers on the subject, notably by L. Alvarez-Gaumé, E. Getzler, N. Berline and M. Vergne, and J. M. Bismut.

It should be pointed out however that none of these methods applies to prove the index theorem for families or the $C\ell_k$-index Theorem (in their strong forms). These theorems in general involve torsion elements in K-theory which are not detectable by cohomological means. Moreover, it is known that the $\mathbb{Z}_2$-invariants appearing in the $C\ell_k$-index Theorem are not computable from local densities, for this would imply a certain multiplicativity under coverings which examples show not to be true.

CHAPTER IV

Applications in Geometry and Topology

In this chapter we shall use the results of our previous discussion to derive a series of consequences in differential topology and geometry. Some of the theorems we shall derive have other, completely different proofs, and some (to date) can only be proved by these means. Nevertheless we hope to make clear that spin geometry and the Index Theorem give a unified approach to a wide range of geometric problems.

The applications presented here fall into several categories: integrality and divisibility theorems for characteristic numbers, immersions of manifolds and the vector field problem, group actions on manifolds with positive scalar curvature, Kähler geometry, pure spinors and basic twistor geometry, the theory of calibrations, and the study of riemannian metrics with reduced holonomy. A few of these results are already mentioned in previous chapters in order to add spice to the exposition there. On the other hand there are applications given in previous chapters which do not appear here. Notable among them are the results of §§7 and 8 in Chapter I, the new curvature identities 5.15–16 of Chapter II and their application to the Homology Sphere Theorem (II.7.6) of Gallot and Meyer.

The first part of this chapter is concerned with applications of the Index Theorem to differential topology. Recall that the ordinary index of an elliptic is an integer. However, the Atiyah-Singer formula computes this index in terms of topological invariants which are in general just rational numbers. The fact that these rational numbers are always integers under certain hypotheses, plays a significant role in topology. Its importance stems from the underlying elliptic operators. Two general problems to which such integrality theorems often apply are: finding the smallest codimension for immersing a manifold into euclidean space, and finding the largest number of pointwise linearly independent vector fields on a manifold. We shall show how to prove results in these areas by directly constructing the relevant operators via Clifford multiplication.

Using more sophisticated forms of the Index Theorem, one can prove interesting results about smooth compact group actions on spin manifolds. For example, applying the G-Index Theorem to the Atiyah-Singer operator shows that on spin manifolds admitting an S^1-action the $\hat{A}$-

genus is zero. Applying the $C\ell_k$-Index Theorem to the $C\ell_k$-linear Atiyah-Singer operator and invoking some basic riemannian geometry yields a more refined result about S^3-actions. All of this is done in §3.

Sections 4, 5 and 6 are devoted to the study of riemannian manifolds of positive scalar curvature. For compact simply-connected spin manifolds, the KO-index of the Atiyah-Singer operator gives a nearly complete set of invariants for deciding whether or not there exists a metric of positive scalar curvature. This is discussed in §4. In §5 the analogous question is discussed for manifolds with non-trivial fundamental group. Here there is an interesting interaction between spin geometry and the fundamental group which is mediated by an appropriately twisted Atiyah-Singer operator. The ideas developed here carry over to give results about the existence of complete metrics with positive scalar curvature on non-compact spin manifolds. For this it will be necessary to develop a "relative version" of the Index Theorem over open manifolds. This is discussed in §6.

The techniques of spin geometry can be applied to say something about the topology of the space of all riemannian metrics with positive scalar curvature on a given manifold. This is done in §7.

On an even-dimensional riemannian manifold there is a concept, due to É. Cartan, of pure spinors. These are related directly to almost complex structures on the manifold and to the Penrose twistor construction. This together with theorems on integrability are examined in §9.

A Kähler manifold is a riemannian manifold with a parallel almost complex structure. (Such a structure is always integrable, i.e., the underlying manifold is always complex.) When X is a Kähler manifold, the Clifford bundle $\mathbb{C}\ell(X)$ has a very pretty decomposition. There are differential operators $\mathscr{D}$ and $\bar{\mathscr{D}}$ such that $\mathscr{D}^2 = \bar{\mathscr{D}}^2 = 0$ and $\mathscr{D} + \bar{\mathscr{D}} = D$ (the Dirac operator on $\mathbb{C}\ell(X)$). There are also 0-order operators $\mathscr{L}$ and $\bar{\mathscr{L}}$ such that $\langle \mathscr{L}, \bar{\mathscr{L}}, [\mathscr{L}, \bar{\mathscr{L}}] \rangle \cong \mathfrak{sl}_2(\mathbb{C})$ and $\mathscr{L} + \bar{\mathscr{L}} = L$ (cf. (II.5.20)). From these one can define certain Clifford cohomology groups and establish within them all the classical identities of Kähler geometry (see §8). One can also decompose holomorphic Dirac bundles and define spinor cohomology groups. A wide variety of vanishing theorems are then proved using Clifford formalism in §11. These are applied to prove the existence of compact manifolds with Sp_m-holonomy.

There is an intimate relationship between riemannian manifolds with reduced holonomy, calibrations, and spin geometry. This is presented in detail in §10 with special emphasis on the exceptional cases of G_2 and Spin_7 manifolds.

In the last section we shall examine the role played by spinors in a proof of the Positive Mass Conjecture in general relativity.

§1. Integrality Theorems

The point of this section is to use the Index Theorem to prove the integrality and divisibility of certain rational characteristic numbers. These results were known before the Index Theorem and, in fact, they provided part of the motivation and insight for its proof. We begin with the following classical result of Atiyah and Hirzebruch [1].

Theorem 1.1. *Let X be a compact spin manifold of dimension $4k$. Then $\hat{A}(X)$ is an integer. Furthermore, if* $\dim(X) \equiv 4 \pmod 8$, *then $\hat{A}(X)$ is an even integer.*

Proof. The first statement follows from the fact that on a spin manifold, the $\hat{A}$-genus is the index of the Atiyah-Singer operator. The second statement follows similarly from Theorem 6.16 of Chapter II but can be deduced directly as follows. Notice from the classification of Clifford algebras (Theorem I.4.3), that in dimensions 4 (mod 8), the complex spinor representations are actually quaternionic. The connection in the spinor bundle is always such that multiplication by quaternion scalars is parallel. Hence, the kernel and cokernel of the Dirac operator are quaternion vector spaces, and their complex dimension is even. ∎

For manifolds which are not spin, the $\hat{A}$-genus is very often not an integer.

Recall that for a 4-manifold the signature is always 8 times the $\hat{A}$-genus (see II.6.17). Thus, Theorem 1.1 generalizes the following result of Rochlin:

Corollary 1.2. *The signature of a compact smooth spin 4-manifold is a multiple of 16.*

It was pointed out in Chapter II (see 2.12ff) that on a spin 4-manifold X, the intersection form is even. That is, $x \cup x \equiv 0 \pmod 2$ for all $x \in H^2(X; \mathbb{Z})$. This implies easily that the signature of X is a multiple of 8 (see Milnor-Husemoller [1], p. 242). The importance of smoothness in Rochlin's Theorem was recently underlined by Mike Freedman's profound results [1] which proved, among other things, the existence of a topological spin 4-manifold of index 8.

Another integrality result comes from our discussion of Spin^c-manifolds.

Theorem 1.3. *Let X be a compact orientable manifold and suppose $c \in H^2(X; \mathbb{Z})$ is a class such that $c \equiv w_2(X)$* (mod 2). *(Hence, X is a Spin^c-manifold.) Then the rational number*

$$\{e^{\frac{1}{2}c}\hat{\mathbf{A}}(X)\}[X]$$

is an integer.

Proof. This number is the index of an elliptic operator (see App. D). ■

Another immediate consequence is the following classical theorem of Bott [1], [2]:

Theorem 1.4. *Let E be a complex vector bundle over the $2n$-sphere S^{2n}. Then the top Chern class of E is divisible by $(n-1)!$, that is,*

$$\frac{1}{(n-1)!}c_n(E)[S^{2n}] \in \mathbb{Z}.$$

Proof. Consider the twisted signature operator

$$D^+:(C\ell^+ \otimes E) \longrightarrow \Gamma(C\ell^- \otimes E)$$

over S^{2n}. Then the index of this operator is

$$\begin{aligned}\operatorname{ind}(D^+) &= \{\operatorname{ch} E \cdot \mathbf{L}(S^{2n})\}[S^{2n}] \\ &= (\operatorname{ch} \mathrm{E})[S^{2n}] \\ &= \operatorname{ch}_n(E)[S^{2n}] \\ &= \frac{1}{(n-1)!}c_n(E)[S^{2n}],\end{aligned}$$

since all the Pontrjagin classes of S^{2n} are zero and since $H^{2k}(S^{2n}) = 0$ for $k \neq 0,n$. Of course, the index is an integer. ■

This result is useful for many things, among them the study of immersions of manifolds into $\mathbb{R}^n$. Here, however, there is a direct approach using spin geometry.

§2. Immersions of Manifolds and the Vector Field Problem

In this section we shall present a unified approach to the following two problems. Let X^n be a compact manifold of dimension n and without boundary.

I. Find the smallest integer q such that there exists an immersion $X^n \looparrowright \mathbb{R}^{n+q}$.
II. Find the largest integer q such that there exist q pointwise linearly independent vector fields on X

In the first case we always have that $q \leqq n-1$ by the classical work of Whitney. Furthermore, using Stiefel-Whitney classes and the product formula $w(T)w(N) = 1$, one can sometimes find very good lower bounds for q (see App. B, for a discussion of the product formula). Nevertheless, for better results it is sometimes useful to use real characteristic classes or KO-theory. This is the approach we shall take.

The techniques presented here generalize fundamental ideas of Atiyah [9]. They appeared in Lawson-Michelsohn [1]. A second approach, also using spin geometry, can be found in Mayer [1].

We shall assume throughout that X is a compact oriented manifold of dimension n. For simplicity we let T denote the tangent bundle of X.

We begin with the first problem. Suppose that X admits an immersion $X \looparrowright \mathbb{R}^{n+q}$. Denote by N the normal bundle to this immersion, and note that

$$T \oplus N = \tau \tag{2.1}$$

where τ denotes the trivial $(n+q)$-plane bundle. Let $\langle \cdot,\cdot \rangle$ and $\bar{\nabla}$ be the metric and connection on τ induced from euclidean space. Using the decomposition (2.1) we introduce on τ a new riemannian connection ∇, called the **projected connection,** as follows. For a section $\varphi = (V,\nu) \in \Gamma(T \oplus N)$, we set

$$\nabla \varphi \equiv ((\bar{\nabla} V)^T, (\bar{\nabla} \nu)^N) \tag{2.2}$$

where $(\ )^T, (\ )^N$ denote pointwise orthogonal projection onto T and N respectively. The connection thus induced on T is the canonical riemannian one for the induced metric. The difference $\bar{\nabla} - \nabla$ on τ is the **second fundamental form** of the immersion.

We now consider the bundle $C\ell(\tau)$ as a bundle of left-modules over $C\ell(T)$ under the obvious inclusion $C\ell(T) \subset C\ell(\tau)$. This makes $C\ell(\tau)$, with the connection induced from ∇ on τ, into a Dirac bundle over X (cf. II.4.8). If $n = 4k$, then we can split the bundle $C\ell(\tau)$ via the parallel volume form $\omega = e_1 \cdots e_n$, and get the elliptic operator

$$D^+ : \Gamma(C\ell^+(\tau)) \longrightarrow \Gamma(C\ell^-(\tau)). \tag{2.3}$$

We shall prove below (Lemma 2.5) that

$$\operatorname{ind}(D^+) = 2^q A(X) \tag{2.4}$$

where $A(X) = 2^n \hat{A}(X)$ is the A-genus of X. It is a characteristic number which is always an integer.

Notice now that while $C\ell(\tau)$ is not in general trivial as a bundle of left $C\ell(\tau)$-modules, it is trivial as an abstract vector bundle. In fact, we can find $n + q$ pointwise orthonormal sections $\varepsilon_1, \ldots, \varepsilon_{n+q}$ of τ. These sections generate a finite multiplicative subgroup $G \equiv \{1\} \cup \{\varepsilon_{i_1} \cdots \varepsilon_{i_p}\}_{i_1 < \cdots < i_p}$ in the algebra of sections $\Gamma(C\ell(\tau))$. Evidently, the group algebra $\mathbb{R} \cdot G$ is just the Clifford algebra $C\ell_{n+q}$.

We now consider the group G acting on $C\ell(\tau)$ by right multiplication $R_g(\varphi) = \varphi \cdot g$. This action does not quite commute with the Dirac operator, so we take the average

$$\tilde{D} \equiv \frac{1}{|G|} \sum_{g \in G} R_g \circ D \circ R_g^{-1} \tag{2.5}$$

and note that

$$\tilde{D} \circ R_g = R_g \circ \tilde{D} \qquad \text{for all } g \in G. \tag{2.6}$$

One can easily see that each commutator $D \circ R_g - R_g \circ D$ is a differential operator of order zero. Hence, the averaged operator $\tilde{D}$ has the *same principal symbol* as D.

Of course right multiplication commutes with left multiplication, so the bundles

$$C\ell^{\pm}(\tau) \equiv (1 \pm \omega)C\ell(\tau)$$

are clearly G-invariant. It follows that $\tilde{D}$ is also of the form

$$\tilde{D} = \begin{pmatrix} 0 & \tilde{D}^- \\ \tilde{D}^+ & 0 \end{pmatrix}$$

with respect to the splitting $C\ell(\tau) = C\ell^+(\tau) \oplus C\ell^-(\tau)$, and the operator $\tilde{D}^+$ has the same symbol as D^+. From the topological invariance of the index we conclude that

$$\operatorname{ind}(\tilde{D}^+) = \operatorname{ind}(D^+)$$

The main step now is to observe that since $\tilde{D}^+$ commutes with G, the kernel and cokernel of $\tilde{D}^+$ are $\mathbb{R}G \cong C\ell_{n+q}$-modules, and so the index of D^+ is divisible by a large power of 2.

There is one further observation which sharpens the result. Recall that $C\ell(\tau)$ carries a $\mathbb{Z}_2$-grading $C\ell(\tau) = C\ell^0(\tau) \oplus C\ell^1(\tau)$ determined as the ± 1-eigenbundles of the parallel automorphism α (which extends $v \mapsto -v$ on τ). This grading clearly carries over to $C\ell^+(\tau)$ and $C\ell^-(\tau)$, and makes each of these a bundle of $\mathbb{Z}_2$-graded modules under **right** Clifford multiplication. Furthermore, since $\alpha(D\varphi) = -D(\alpha\varphi)$, for any $\varphi \in \Gamma(C\ell(\tau))$, we see that $\ker D^+$ and $\ker D^-$ are each $\mathbb{Z}_2$-graded under right multiplication by $\mathbb{R}G = C\ell_{n+q}$. This proves the following:

Theorem 2.1. *Let X be a compact oriented manifold of dimension $n = 4k$. If there exists an immersion $X \looparrowright \mathbb{R}^{n+q}$, then*

$$2^{q-1}\hat{A}(X) \equiv 0 (\operatorname{mod} a_{n+q})$$

where $2a_{n+q}$ is the dimension of an irreducible real $\mathbb{Z}_2$-graded $C\ell_{n+q}$-module.

See the table below for values of a_{n+q}. Note that by Proposition I.5.20, a_{n+q} is the dimension of an irreducible, real *ungraded* module over $C\ell_{n+q-1}$.

As a quick illustration we note that $\hat{A}(\mathbb{P}^2(\mathbb{C})) = 2$ and that $2^q \equiv 0 \pmod{a_{4+q}}$ only for $q \geq 3$. Hence, $\mathbb{P}^2(\mathbb{C})$ immerses only in codimension ≥ 3, the range guaranteed by the Whitney Immersion Theorem. Moreover, since the $\hat{A}$-genus is an oriented cobordism invariant, the conclusion is much more general. In dimension four, $\hat{A}(X) = 2\operatorname{sig}(X)$ and we have the following:

Corollary 2.2. *If a compact oriented 4-manifold X can be immersed into* $\mathbb{R}^6$, *then the signature of X is even.*

These results can be somewhat strengthened by two tricks. The first trick allows us to deal with embeddings. Note that $C\ell(\tau) = C\ell(T \oplus N) = C\ell(T) \hat{\otimes} C\ell(N)$ where $\hat{\otimes}$ denotes the graded tensor product of Clifford bundles. Clearly we have $C\ell^{\pm}(\tau) = [(1 \pm \omega)C\ell(T)] \hat{\otimes} C\ell(N) = C\ell^{\pm}(T) \otimes C\ell(N)$.

Suppose now that the dimension of the normal bundle $q \equiv 0 \pmod 4$, and let $\omega_N = e_1 \cdots e_q$ be the normal volume element. Then ω_N is parallel in the projected connection and commutes with all tangent vectors. Hence, ω_N commutes with the Dirac operator D. We can define bundles

$$C\ell^{\pm\pm}(\tau) = C\ell^{\pm}(T) \hat{\otimes} C\ell^{\pm}(N) = (1 \pm \omega)C\ell(T) \hat{\otimes} (1 \pm \omega_N)C\ell(N) \tag{2.7}$$

and we get Dirac operators

$$D^{+\cdot}: \Gamma(C\ell^{+\cdot}(\tau)) \longrightarrow \Gamma(C\ell^{-\cdot}(\tau)) \tag{2.8}$$

where "$\cdot$" can be either $+$ or $-$. Evidently, $D^{+} = D^{++} \oplus D^{+-}$, and so

$$\operatorname{ind}(D^{+}) = \operatorname{ind}(D^{++}) + \operatorname{ind}(D^{+-}). \tag{2.9}$$

Furthermore, we shall prove below the following fact:

Lemma 2.3. *If the Euler class $\chi(N)$ of the normal bundle is zero, then*

$$\operatorname{ind}(D^{++}) = \operatorname{ind}(D^{+-}). \tag{2.10}$$

Carrying out the averaging process above, we may replace all D's by $\tilde{D}$'s. (Note that each of $C\ell^{\pm\pm}(\tau)$ is invariant under right-multiplication by $C\ell(\tau)$, and in particular by G.) It is an elementary fact that for an *embedded* submanifold, $\chi(N) = 0$ (see Milnor-Stasheff [1]). Consequently, we have the following:

Theorem 2.4. *Let X be as in Theorem 2.1 and suppose* $q \equiv 0 \pmod 4$. *If there exists an embedding* $X \hookrightarrow \mathbb{R}^{n+q}$, *then*

$$2^{q-2}\hat{A}(X) \equiv 0 \pmod{a_{n+q}}.$$

The same conclusion holds for any immersion $X \looparrowright R^{n+q}$ *for which the Euler class of the normal bundle is zero.*

Our second trick is to simply take coefficients in a vector bundle over X. This is technically quite easy and has the effect of introducing all of the rational K-theory of the space into the problem. We get conditions which are not simply in terms of characteristic numbers, but essentially involve the entire cohomology ring. We proceed as follows.

Let E be a vector bundle over X. Here E may be real, complex or quaternionic, and we assume it is equipped with an appropriate inner product and connection. Then we take $\tau = T \oplus N$ as above and consider the bundle

$$C\ell(\tau) \otimes E$$

with the tensor product connection. This remains a bundle of right and left modules over X. It is easy to check that *all of the constructions above can be carried out with* $C\ell(\tau)$ *replaced by* $C\ell(\tau) \otimes E$.

We need only compute the index of D^+. We shall state the result here, and give the proof below.

Lemma 2.5. *The index of* $D^+ : \Gamma(C\ell^+(\tau) \otimes E) \to \Gamma(C\ell^-(\tau) \otimes E)$ *is given by the formula*

$$\operatorname{ind}(D^+) = 2^q\{\operatorname{ch}_2 E \cdot \mathbf{\hat{A}}(X)\}[X] \tag{2.10}$$

where $\mathbf{\hat{A}}(X)$ *is the total $\hat{A}$-class of* X, *and where* $\operatorname{ch}_2 E$ *is defined as follows. Let* $\operatorname{ch} E = 1 + \operatorname{ch}^1 E + \operatorname{ch}^2 E + \ldots \in H^{2*}(X; \mathbb{Q})$, *denote the Chern character of E. Then for any* $t \in \mathbb{R}$,

$$ch_t E \equiv \sum_{k \geqq 0} (\operatorname{ch}^k E) t^k \tag{2.11}$$

In the case that $\chi(N) = 0$,

$$\operatorname{ind}(D^{++}) = \tfrac{1}{2} \operatorname{ind}(D^+) \tag{2.12}$$

where $\operatorname{ind}(D^+)$ *is given by formula* (2.10).

We note that when E is real, we set $\operatorname{ch} E = \operatorname{ch}(E \otimes \mathbb{C})$, and when E is quaternionic, we set $\operatorname{ch} E = \operatorname{ch}([E]_{\mathbb{C}})$ where $[\cdot]_{\mathbb{C}}$ denotes restriction of scalars from $\mathbb{H}$ to $\mathbb{C}$.

To state our results we shall need the following table of numbers which is easily computed from the data given in Chapter I, §5. Let $2a_k$, $2b_k$ and c_k denote the dimension over K of an irreducible $\mathbb{Z}_2$-graded K-module for $C\ell_k$ where $K = \mathbb{R}, \mathbb{C}$ and $\mathbb{H}$ respectively. Note that by Proposition I.5.20, the numbers a_k, b_k and $\frac{1}{2}c_k$ correspond to the dimensions of irreducible *ungraded* modules for $C\ell_{k-1}$. For all values of k we have that

$$a_{k+8} = 16a_k, \qquad b_{k+8} = 16b_k, \qquad c_{k+8} = 16c_k.$$

The values for $k \leqq 8$ are given in the following table.

n	1	2	3	4	5	6	7	8
a_n	1	2	4	4	8	8	8	8
b_n	1	1	2	2	4	4	8	8
c_n	2	2	2	2	4	8	16	16

The method given above proves the following general result:

Theorem 2.6. *Let X be a compact orientable manifold of dimension n, and suppose there exists an immersion $X \looparrowright \mathbb{R}^{n+q}$. Then for any complex bundle E over X,*

$$2^{q-1}\{\mathrm{ch}_2 E \cdot \mathbf{A}(X)\}[X] \equiv 0 \pmod{b_{n+q}}. \tag{2.13}$$

Furthermore, if q is even and the normal Euler class $\chi(N) = 0$, then

$$2^{q-2}\{\mathrm{ch}_2 E \cdot \mathbf{A}(X)\} \equiv 0 \pmod{b_{n+q}}. \tag{2.14}$$

When $n \equiv 0 \pmod 4$, there are the following refinements: If $E = E_R \otimes \mathbb{C}$ for a real bundle E_R, or if $E = [E_H]_{\mathbb{C}}$ for a quaternionic bundle E_H, then (2.13) holds with b_{n+q} replaced by a_{n+q} and c_{n+q} respectively. If, furthermore, $q \equiv 0 \pmod 4$ and $\chi(N) = 0$, the analogous improvements can be made in (2.14).

Proof. Again the argument goes as follows. Let $\tau = T \oplus N$ and consider the Dirac bundle $C\ell(\tau) \otimes E$. Average the Dirac operator over the group in $\Gamma(C\ell(\tau))$ generated by the global orthonormal sections $\varepsilon_1, \ldots, \varepsilon_{n+q}$ of τ. Then the kernel and cokernel of the operator are $\mathbb{Z}_2$-graded modules over the algebra $C\ell_{n+q}$, generated by $\varepsilon_1, \ldots, \varepsilon_{n+q}$. (The $\mathbb{Z}_2$-grading comes from the even-odd grading of $C\ell(\tau)$. In the complex case there are volume forms in every even dimension: $\omega_{\mathbb{C}} = i^m e_1 \cdots e_{2m}$). Applying the formula from Lemma 2.5 gives the first part of the theorem. Splitting the operator with the normal volume form and applying (2.12) gives the second part of the theorem. ■

We point out that the improvements in this method obtained by introducing coefficients are substantial. For example, they give results in every even dimension.

The congruences (2.13) and (2.14) give a relatively simple machine for calculating lower bounds on embedding and immersing dimensions.

For the complex and quaternionic projective spaces. $\mathbb{P}^n(\mathbb{C})$ and $\mathbb{P}^n(\mathbb{H})$, these methods give the best results of this kind for every n (see Lawson-Michelsohn [1] for details). Similar non-immersion theorems were found by K. H. Mayer [1]. There were earlier, slightly weaker results of this type due to Atiyah and Hirzebruch [1], whose methods did not involve elliptic operators.

Proof of Lemma 2.5. We note that the operator in question is just the twisted signature operator (cf. II.6). More precisely, since $C\ell^{\pm}(T \oplus N) = C\ell^{\pm}(T) \otimes C\ell(N)$, we see that we can write D^+ as

$$D^+ : \Gamma(C\ell^+(T) \otimes \tilde{E}) \longrightarrow \Gamma(C\ell^-(T) \otimes \tilde{E})$$

where $\tilde{E} \equiv E \otimes C\ell(N)$. It follows that

$$\operatorname{ind}(D^+) = \{\operatorname{ch}_2\tilde{E} \cdot \mathbf{L}(X)\}[X], \tag{2.15}$$

from the basic formula for the signature complex (see III.13.9). Now $\operatorname{ch}_2\tilde{E} = \operatorname{ch}_2(E \otimes C\ell(N)) = \operatorname{ch}_2(E)\operatorname{ch}_2(C\ell(N))$, and since $T \oplus N = \tau = \mathbb{R}^{n+q}$, we see that $C\ell(T)C\ell(N) = \mathbf{Cl}_{n+q} = \mathbf{R}^{2^{n+q}}$ (trivial bundles). It follows that

$$\operatorname{ch}(C\ell(N)) = 2^{n+q}\operatorname{ch}(C\ell(T))^{-1}. \tag{2.16}$$

To compute this last expression, we apply the Splitting Principle and express the total Chern class of $T \otimes \mathbb{C}$ formally as

$$c(T \otimes \mathbb{C}) = \prod_{k=1}^{n/2} (1 + x_k)(1 - x_k)$$

so that the Pontrjagin classes are

$$p_k(T) = \sigma_k(x_1^2, \ldots, x_{n/2}^2)$$

where σ_k denotes the kth elementary symmetric function. In this language, we have

$$\mathbf{L}(X) = \prod_{k=1}^{n/2} \frac{x_k}{\tanh(x_k)}.$$

Recall that $C\ell(T) \otimes \mathbb{C} \cong \Lambda^*_{\mathbb{C}}(T \otimes \mathbb{C})$ and that therefore

$$\begin{aligned} \operatorname{ch}(C\ell(T)) &= \prod_{k=1}^{n/2} (1 + e^{-x_k})(1 + e^{x_k}) \\ &= 2^n \prod_{k=1}^{n/2} \cosh^2(x_k/2). \end{aligned}$$

Consequently,

$$\operatorname{ch}(C\ell(N)) = 2^q \prod_{k=1}^{n/2} \cosh^{-2}(x_k/2),$$

and for any $t \in \mathbb{R}$,

$$\operatorname{ch}_t(C\ell(N)) = 2^q \prod_{k=1}^{n/2} \cosh^{-2}(x_k t/2).$$

It follows that

$$\begin{aligned} \operatorname{ch}_2(C\ell(N))\mathbf{L}(X) &= 2^q \prod_{k=1}^{n/2} \frac{x_k}{\sinh(x_k)\cosh(x_k)} \\ &= 2^q \prod_{k=1}^{n/2} \frac{2x_k}{\sinh(2x_k)} \\ &\equiv 2^q \mathbf{\hat{A}}(X), \end{aligned}$$

since the A-class is the multiplicative series given by the formal power series $2x/\sinh(2x)$. Formula (2.10) now follows immediately from (2.15).

For the second half of the lemma we make the following observation. Let E be a complex vector bundle with $\dim_{\mathbb{C}} E = 2m$. Write $c(E \otimes_{\mathbb{R}} \mathbb{C}) = \prod (1 - x_k^2)$ so that $p_k(E) = \sigma_k(x_1^2, \ldots, x_m^2)$ as above. Then we have

$$\begin{aligned} \mathrm{ch}(\mathrm{C}\ell^+ E) - \mathrm{ch}(\mathrm{C}\ell^- E) &= \prod_{k=1}^{2m} (e^{x_k} - e^{-x_k}) \\ &= 2^m x_1 \cdots x_m \prod_{k=1}^{m} \left\{ \frac{e^{x_k} - e^{-x_k}}{2x_k} \right\} \\ &= 2^m \chi(E) \alpha(E) \end{aligned}$$

where α is a multiplicative sequence. It follows that

$$\begin{aligned} \mathrm{ind}(D^{++}) - \mathrm{ind}(D^{+-}) &= \{(\mathrm{ch}_2 \mathrm{C}\ell^+ N - \mathrm{ch}_2 \mathrm{C}\ell^- N) \cdot \mathrm{ch}_2 E \cdot \mathbf{L}(X)\}[X] \\ &= \{\chi(N) \cdot C\}[X] \end{aligned}$$

for some class $C \in H^{\mathrm{even}}(X;\mathbb{Q})$. Hence, if $\chi(N) = 0$, we have $\mathrm{ind}(D^{++}) = \mathrm{ind}(D^{+-}) = \frac{1}{2}\mathrm{ind}(D^+)$ (cf. (2.9)). This completes the proof. ∎

Using the techniques developed above, we shall now analyze the second problem mentioned at the outset of this section. Suppose our manifold X carries q pointwise linearly independent vector fields $e_1, \ldots, e_q$. We may assume $e_1, \ldots, e_q$ to be pointwise orthonormal with respect to a riemannian metric on X. Let us suppose for the moment that X is of dimension $n = 4k$ (and is orientable). Then we have the signature operator $D^+ : \Gamma(\mathrm{C}\ell^+(X) \to \Gamma(\mathrm{C}\ell^- X)$. As above, we may average D over the finite group $G \subset \Gamma(\mathrm{C}\ell\, X)$ generated by $e_1, \ldots, e_q$. That is, we set

$$\tilde{D} \equiv \frac{1}{|G|} \sum_{g \in G} R_g \circ D \circ R_g^{-1}$$

where R_g denotes right Clifford multiplication by g. This multiplication clearly preserves the sub-bundles $\mathrm{C}\ell^{\pm}(T) = (1 \pm \omega)\mathrm{C}\ell(T)$, and we achieve an operator $\tilde{D}^+ : \Gamma(\mathrm{C}\ell^+(T) \to \Gamma(\mathrm{C}\ell^-(T))$ which is G-equivariant and has the same symbol as D^+. As above we find that the kernel and cokernel of $\tilde{D}^+$ are $\mathbb{Z}_2$-graded modules over the Clifford algebra $\mathrm{C}\ell_q \cong \mathbb{R}G$. Since $\mathrm{ind}(\tilde{D}^+) = \mathrm{ind}(D^+) = \mathrm{sig}(X)$, we have the following:

Theorem 2.7 (Atiyah [9] Frank [1]). *Let X be a compact oriented manifold of dimension $4k$. If X carries q pointwise linearly independent vector fields, then*

$$\mathrm{sig}(X) \equiv 0 \pmod{2a_q}$$

If, furthermore, $q \equiv 0 \pmod 4$, then

$$\mathrm{sig}(X) \equiv 0 \pmod{4a_q}.$$

As an example, note that if X carries 3 such vector fields, then $\text{sig}(X) \equiv 0 \pmod 8$. If it carries 5, then $\text{sig}(X) \equiv 0 \pmod{16}$. In general, if X carries $2m$ such vector fields, then $\text{sig}(X) \equiv 0 \pmod{2^m}$.

Proof of Theorem 2.7. The first statement follows from the argument above. For the second statement, note that there is an orthogonal splitting $T(X) = T_0 \oplus T_1$ where $T_1 \equiv \text{span}(e_1, \ldots, e_q)$. Let w be the oriented volume element of T_0. Since $\dim T_0 \equiv 0 \pmod 4$, we have that $w^2 = 1$ and w commutes with $e_1, \ldots, e_q$. By averaging $\mathscr{D}$ as above, over the group $\mathbb{Z}_2$ generated by R_w, we may assume $\tilde{D}$ commutes with R_w.

Since $R_w^2 = 1$, we see that R_w has ± 1 eigenbundles on each of $C\ell^+$ and $C\ell^-$. They are given by $C\ell^{+\pm} = C\ell^+(1 \pm w)$ and $C\ell^{-\pm} = C\ell^-(1 \pm w)$. Each bundle $C\ell^{\pm\pm}$ is G-invariant, and we obtain two G-equivariant elliptic operators

$$\tilde{D}^{++}: \Gamma(C\ell^{++}) \longrightarrow \Gamma(C\ell^{-+}) \qquad \text{and} \qquad \tilde{D}^{+-}: \Gamma(C\ell^{+-}) \longrightarrow \Gamma(C\ell^{--}),$$

with the property that

$$\tilde{D}^+ = \tilde{D}^{++} \oplus \tilde{D}^{+-}.$$

We know, therefore, that $\text{ind}(\tilde{D}^+) = \text{ind}(\tilde{D}^{++}) + \text{ind}(\tilde{D}^{+-})$. A straightforward computation, in the spirit of that given for the second part of Lemma 2.5, shows that $\text{ind}(\tilde{D}^{++}) - \text{ind}(\tilde{D}^{+-}) = 2^{q/2}\chi(T_0)[X] = 0$. Thus,

$$\text{ind}(\tilde{D}^{++}) = \text{ind}(\tilde{D}^{+-}) = \tfrac{1}{2}\,\text{ind}(\tilde{D}^+)$$

and since the kernel and cokernel of $\tilde{D}^{++}$ are $\mathbb{Z}_2$-graded $C\ell_q$-modules, the result follows as before. ∎

When Atiyah first presented these arguments (in [9]), he mentioned that they could not achieve the additional power of 2 (when $q \equiv 0(4)$) obtainable by other means. However, the last argument given above succeeds in getting the final 2 and gives the best known results of this type.

Theorem 2.7 is very pleasant because $|\text{sig}(X)|$ is a homotopy invariant. It is also an oriented cobordism invariant. However, for any given manifold which carries a q-frame field, there is much more information available. For this we simply need to take coefficients in a vector bundle E. Applying the arguments given above directly to $C\ell(X) \otimes E$ proves the following result (cf. Lawson-Michelsohn [1]).

Theorem 2.8. *Let X be a compact oriented manifold of even dimension, and suppose that X carries q pointwise linearly independent vector fields. Then for every complex vector bundle E over X,*

$$\{\text{ch}_2 E \cdot \mathbf{L}(X)\}[X] \equiv 0 \pmod{2b_q} \tag{2.16}$$

If, furthermore, q is even and $\text{ch}^{q/2}E = 0$, then

$$\{\text{ch}_2 E \cdot \mathbf{L}(X)\}[X] \equiv 0 \pmod{4b_q} \tag{2.17}$$

When $\dim(X) = 4k$, *there are the following refinements. If* $E = E_R \otimes \mathbb{C}$ *for a real bundle* E_R, *or if* $E = [E_H]_{\mathbb{C}}$ *for a quaternion bundle* E_H, *then* (2.16) *holds with* b_q *replaced by* a_q *or* c_q *respectively. If, furthermore,* $q \equiv 0 \pmod 4$ *and* $\text{ch}^{q/2}E = 0$, *the analogous improvement can be made in* (2.17).

Proof. The only point which needs attention here is the proof that $\text{ind}(\tilde{D}^{++}) = \text{ind}(\tilde{D}^{+-})$ when $\text{ch}^{q/2}E = 0$. This follows from the equation

$$\text{ind}(\tilde{D}^{++}) - \text{ind}(\tilde{D}^{+-}) = 2^{q/2}\{\text{ch}\, E \cdot \chi(T_0)\}[X],$$

which can be verified by straightforward calculation. ■

It is an interesting observation of Atiyah [9] that all of the above carries over without change to prove the following generalization of Theorems 2.7 and 2.8.

Theorem 2.9. *Let X be a compact oriented manifold whose tangent bundle can be decomposed into oriented subbundles*

$$T(X) = T_0 \oplus T_1 \oplus T_2 \oplus \cdots \oplus T_q$$

where

$$\dim_{\mathbb{R}}(T_k) \equiv 1 \pmod 4 \qquad \text{for } k = 1, \dots, q.$$

Then all the conclusions of Theorems 2.7 and 2.8 hold for X.

Proof. Let e_k denote the oriented volume element for the plane field T_k, $k = 1, \dots, q$. Since these planes are mutually orthogonal and each has dimension one (mod 4), the elements $e_1, \dots, e_q \in \Gamma(C\ell(X))$ generate a finite group G with $\mathbb{R} \cdot G = C\ell_q$. The arguments now proceed exactly as before. ■

As a final application of these methods we consider the refined Clifford index for $C\ell_k$-Dirac bundles introduced in §7 of Chapter II. Let X be a compact oriented manifold of dimension $4k + 1$. Recall that the **Kervaire semi-characteristic** of X is then defined to be the residue class

$$\begin{aligned}\sigma(X) &\equiv \sum_{j=1}^{2k} b_{2j}(X) \pmod 2\\ &\equiv \dim H^{\text{even}}(X) \pmod 2\\ &\equiv \tfrac{1}{2}\chi(X) \pmod 2.\end{aligned}$$

Theorem 2.10 (Atiyah [9]). *Let X be a compact oriented* $(4k + 1)$*-manifold. If X carries an oriented tangent p-plane field for some $p \equiv 2$ or $3 \pmod 4$, then $\sigma(X) = 0$.*

Proof. By assumption there exists a splitting $T(X) = T_0 \oplus T_1$ where $\dim(T_0) \equiv 2 \pmod 4$, and where T_0 and T_1 are oriented. This decomposition can be assumed orthogonal for some riemannian metric on X. Let ω_0 be the oriented volume form for T_0, and note that $\omega_0^2 \equiv -1$.

We consider now the Euler characteristic operator $D^0 : \Gamma(C\ell^0(X)) \to \Gamma(C\ell^1(X))$. Since $\dim(T_0)$ is even, each $C\ell^k(X)$ is invariant under right multiplication by ω_0. Averaging the Dirac operator over the group $\mathbb{Z}_4$ generated by R_{ω_0} makes $\ker D^0$ an R_{ω_0}-invariant space, i.e., $\ker D^0$ becomes a $C\ell_1 = \mathbb{C}$-module. In particular, $\ker(D^0) \cong \mathbf{H}^{\text{even}} \cong \bigoplus_k H^{2k}(X;\mathbb{R})$ has even dimension (over $\mathbb{R}$), and so $\sigma(X) = 0$. ■

§3. Group Actions on Manifolds

This section is concerned with smooth group actions. Recall that a Lie group G is said to **act effectively** on a manifold X if there is a differentiable map $\Phi : G \times X \to X$ so that the corresponding map $\varphi(g) \equiv \Phi(g, \cdot)$ gives a continuous and injective homomorphism $\varphi : G \hookrightarrow \text{Diff}(X)$. Here $\text{Diff}(X)$ denotes the group of C^∞-diffeomorphisms of X in the standard C^∞ topology.

There are two basic groups we shall consider here, namely,

$$S^1 \equiv \{z \in \mathbb{C} : |z| = 1\} \tag{3.1}$$

$$S^3 \equiv \{q \in \mathbb{H} : |q| = 1\} \tag{3.2}$$

where multiplication is induced from multiplication in the field. There are isomorphisms $S^1 \cong SO_2 \cong U_1$ and $S^3 \cong \text{Spin}_3 \cong SU_2 \cong Sp_1$. These two groups are basic for the following reason. We say that a Lie group G has an **S^k subgroup** (for $k = 1$ or 3) if G contains a Lie subgroup isomorphic to a finite quotient of S^k. (The only possibilities are S^1 when $k = 1$ and S^3 or $S^3/\mathbb{Z}_2$ when $k = 3$.)

Proposition 3.1. *Let G be a compact connected Lie group (of dimension >0). Then G contains an S^1-subgroup. In fact the S^1-subgroups are dense in G.*

Moreover, if G is non-abelian, then it contains an S^3-subgroup.

Proof. These statements follow directly from the elementary theory of compact Lie groups (see Adams [1]). To prove the first, one notes that every element of G is contained in a maximal torus, where S^1-subgroups are dense. To prove the last statement one takes the subgroup corresponding to the subalgebra spanned by a root space and a root vector. ■

DEFINITION 3.2. We say that a manifold X **admits an S^k-action** ($k = 1$ or 3) if there exists an effective action of a finite quotient of S^k on X.

Two main results concerning connected group actions on spin manifolds are the following:

Theorem 3.3 (Atiyah and Hirzebruch [3]). *Let X be a compact spin manifold which admits an S^1-action. Then $\hat{A}(X) = 0$.*

Theorem 3.4 (Lawson and Yau [1]). *Let X be a compact spin manifold which admits an S^3-action. Then $\hat{\mathscr{A}}(X) = 0$.*

Recall that $\hat{\mathscr{A}}$ is the generalization of $\hat{A}$ to KO-theory given in (II.3.22). In dimensions $4k$, the two invariants are essentially the same. However, in dimensions 1 and 2 (mod 8), the invariant $\hat{\mathscr{A}}$ takes values in $\mathbb{Z}_2$ and can be non-trivial while, of course, $\hat{A} = 0$.

REMARK 3.5. The stronger hypothesis of Theorem 3.4 is required for its stronger conclusion. We can see this as follows. Let X^8 be a compact spin 8-manifold with $\hat{A}(X^8) = 1$ (see the paragraph before II.7.12). Equip S^1 with the interesting spin structure (cf. II.7.8), and take the product $Y = X \times S^1$ with the product spin structure. Then *Y admits an S^1-action, but $\hat{\mathscr{A}}(Y) \neq 0$.*

The two theorems above could be combined into the following statement by using Proposition 3.1.

Theorem 3.6 *Let X be a compact spin n-manifold with $\hat{\mathscr{A}}(X) \neq 0$. If $n \equiv 1$ or 2 (mod 8), then the only compact, connected Lie transformation groups of X are tori. If $n \equiv 0$ (mod 4), then the only such group is the trivial one.*

As we have noted before (cf. II.2.18), the spin manifolds with non-zero $\hat{\mathscr{A}}$ constitute half of the exotic spheres in dimensions 1 and 2 (mod 8). This gives the following:

Corollary 3.6. *Let Σ^n be an exotic n-sphere which does not bound a spin manifold ($n \equiv 1$ or 2 (mod 8)). Then the only compact connected Lie symmetry groups of Σ^n are tori of dimension $\leqq [\frac{1}{2}(n+1)]$.*

Proof. The kernel of the surjective homomorphism $\hat{\mathscr{A}} : \Theta_n \to \mathbb{Z}_2$, for $n \equiv 1,2$ (mod 8), is the subgroup of homotopy spheres which do not bound spin manifolds. This follows from the work of Milnor [7] and Adams [3]. If n is even, any toral transformation group T^k must have a fixed point set, and the induced linear action on the normal spaces must be effective. Therefore, $2k \leqq n$. If n is odd and the fixed-point set is non-empty, a similar argument applies. For the general case, see Ku [1]. ∎

Using other techniques, people have investigated the non-existence of torus actions on exotic spheres. For example, using results of R. Schultz

[1] it can be shown that there are three exotic 10-spheres whose largest connected transformation group is S^1 or $S^1 \times S^1$. These spheres are indeed very unsymmetric!

It is interesting to note that this asymmetry is contagious. The $\hat{\mathscr{A}}$-invariant is additive with respect to the connected sum operation #. Thus, if X is a spin n-manifold with $\hat{\mathscr{A}}(X) = 0$ and if Σ is an exotic n-sphere as above ($n \equiv 1$ or $2 \pmod 8$), then

$$\hat{\mathscr{A}}(X \# \Sigma) = \hat{\mathscr{A}}(\Sigma) \neq 0$$

and $X \# \Sigma$ only admits symmetry groups of toral type. This is particularly striking when $X = G/H$ is homogeneous. The manifold $(G/H) \# \Sigma$ is homeomorphic to G/H, but has almost no symmetries. A good example is the complex projective space $X = \mathbb{P}^{4k+1}(\mathbb{C})$.

Another interesting phenomenon takes place under coverings. Fix an exotic sphere Σ^n as above, and let d be its order in the group Θ_n of homotopy spheres. Consider the homogeneous space $X = (S^3/\mathbb{Z}_d) \times S^{n-3}$. The manifold $Y = X \# \Sigma^n$ is quite inhomogeneous, however, its universal (d-fold) covering space is

$$\tilde{Y} = (S^3 \times S^{n-3}) \# \underbrace{(\Sigma^n \# \cdots \# \Sigma^n)}_{d \text{ times}} = S^3 \times S^{n-3},$$

a symmetric space. This instability under finite coverings is not true of the $\hat{A}$-genus.

Proof of Theorem 3.3. The argument is based on the Atiyah-Segal-Singer Fixed-Point Theorem or, more specifically, the G-Spin Theorem (III.14.11). It proceeds as follows. Let X be an even-dimensional compact spin manifold with an S^1-action. We may assume by averaging that the action preserves the riemannian metric. By passing, if necessary, to a 2-fold covering group, we may assume that the S^1-action preserves the spin structure, i.e., that it lifts to an action on the bundle $P_{\mathrm{Spin}}(X)$. This produces an S^1-action on the canonical complex spinor bundle $\$_{\mathbb{C}}$, which projects to the given one on X. This action commutes with the Atiyah-Singer operator $\not{D}$, i.e., $g \circ \not{D} \circ g^{-1} = \not{D}$ for all $g \in S^1$. The volume form is S^1-invariant, and so the splitting $\$_{\mathbb{C}} = \$_{\mathbb{C}}^+ \oplus \$_{\mathbb{C}}^-$ is preserved. Thus the operator $\not{D}^+$: $\Gamma(\$_{\mathbb{C}}^+) \to \Gamma(\$_{\mathbb{C}}^-)$ becomes an S^1-operator, and we have defined an S^1-index:

$$\mathrm{ind}_{S^1}(\not{D}^+) = [\ker \not{D}^+] - [\ker \not{D}^-] \in R(S^1) \tag{3.3}$$

where $R(S^1)$ denotes the representation ring of S^1. There is a natural isomorphism of $R(S^1)$ with the Laurent polynomial ring

$$\mathrm{tr}: R(S^1) \longrightarrow \mathbb{Z}[t, t^{-1}] \tag{3.4}$$

obtained by taking the trace of the representation. Here $t = e^{i\theta}$ corresponds to the complex-valued function, or "character," on S^1 given by the natural inclusion $S^1 \subset \mathbb{C}$. Under the isomorphism (3.4) the index defined in (3.3) becomes a Laurent polynomial which we denote by

$$i_{\not{D}^+}(t) = \sum_{k=-N}^{N} n_k t^k.$$

Note that for any $g \in S^1$, we have

$$i_{\not{D}^+}(g) = \operatorname{tr}(g|_{\ker \not{D}^+}) - \operatorname{tr}(g|_{\ker \not{D}^-}). \tag{3.5}$$

Now the G-Spin Theorem (III.14.11) gives a formula for $i_{\not{D}^+}(g)$ of the form

$$i_{\not{D}^+}(g) = \mathscr{E}_g[F_g]$$

where F_g is the fixed-point set of g. For almost all $g \in S^1$, in fact, for all non-torsion elements, the subgroup generated by g is dense. In this case $F_g = F =$ the fixed-point set of S^1, and the expression $\mathscr{E}_g$ varies only with "normal rotation angles" as follows. The normal bundle to F has an S^1-invariant decomposition $N = \bigoplus_{m \in \mathbb{Z}} N_m$ where each N_m carries the structure of a complex bundle and where S^1 acts on N_m by scalar multiplication by g^m. For any given k-dimensional complex vector bundle E we define the characteristic function

$$\begin{aligned}\mathbf{V}(E, t) &= \prod_{j=1}^{k} \left(t^{\frac{1}{2}} e^{\frac{1}{2}x_j} - t^{-\frac{1}{2}} e^{-\frac{1}{2}x_j}\right)^{-1} \\ &= \prod_{j=1}^{k} \frac{t^{\frac{1}{2}} e^{\frac{1}{2}x_j}}{t e^{x_j} - 1} \\ &= \prod_{j=1}^{k} \frac{t^{-\frac{1}{2}} e^{-\frac{1}{2}x_j}}{1 - t^{-1} e^{-x_j}}\end{aligned}$$

where t is taken to be an indeterminate and where, as usual, the x_j's are formal roots of the total Chern class of E. Because of the square roots, the expression (3.6) is defined only up to sign.

Now for $g \in S^1 \subset \mathbb{C}$ a non-torsion element as above, the G-Spin Theorem states that

$$i_{\not{D}^+}(g) = (-1)^{\sigma} \left\{ \prod_m \mathbf{V}(N_m, g^m) \cdot \hat{\mathbf{A}}(F) \right\} [F]$$

where the signs on each component of F are determined by the action of S^1 on the spin structure. It follows that we have an equality of rational functions

$$i_{\not{D}^+}(t) = (-1)^{\sigma} \left\{ \prod_m \mathbf{V}(N_m, t^m) \cdot \hat{\mathbf{A}}(F) \right\} [F]$$

From (3.6) it is evident that the expression of the right is zero for $t = 0$ and $t = \infty$. Hence, the Laurent polynomial $i_{D^+}(t)$ must vanish identically. In particular, from (3.5) we have that $i_{D^+}(1) = \dim(\ker \not{D}^+) - \dim(\ker \not{D}^-) = \hat{A}(X) = 0$. This completes the proof. ∎

The above argument actually proves the following stronger result.

Theorem 3.7 (Atiyah-Hirzebruch [3]). *Let X be a compact spin manifold of even dimension. For every non-trivial S^1-action which preserves the spin structure on X, the index*

$$\operatorname{ind}_{S^1}(\not{D}^+) = 0 \qquad \text{in } R(S^1).$$

This result on S^1-actions has an interesting refinement observed by P. Landweber, R. Stong and S. Weinberger. To state it we must differentiate between actions of even and odd type. An S^1-action on a spin manifold X is said to be of **even type** if it lifts to an action on $P_{\text{Spin}}(X)$. Otherwise it is said to be of **odd type.** (In the latter case the action of the 2-fold covering group lifts to $P_{\text{Spin}}(X)$.) These notions can be reformulated by considering a lifting of the element $-1 \in S^1 \subset \mathbb{C}$ to $P_{\text{Spin}}(X)$. The action is of even type if and only if -1 has order 2 on $P_{\text{Spin}}(X)$. The action is of odd type if and only if -1 has order 4 on $P_{\text{Spin}}(X)$.

Suppose that the element -1 acts freely on X. Then one sees easily that the S^1-action is even if and only if the quotient $X/\mathbb{Z}_2$ is a spin manifold.

If the fixed-point set $X^{\mathbb{Z}_2}$ of -1 is not empty, then

$$\operatorname{codim}(X^{\mathbb{Z}_2}) \equiv \begin{cases} 0 \pmod 4 & \text{if the action is even,} \\ 2 \pmod 4 & \text{if the action is odd.} \end{cases} \tag{3.7}$$

We see this as follows. Recall that -1 acts on its normal bundle N by scalar multiplication in the fibres. Since the action of -1 is orientation-preserving (it is connected to the identity in Diff(X)), the rank of N is even. Furthermore, there is a $\mathbb{Z}_2$-equivariant diffeomorphism of N with a tubular neighborhood of $X^{\mathbb{Z}_2}$ in X. From this we see immediately that the action on X is even if and only if the action on the normal sphere bundle to $X^{\mathbb{Z}_2}$ (with the induced spin structure) is even. Here the $\mathbb{Z}_2$-action is free and by the paragraph above the question comes down to whether the quotient of the normal sphere is a spin manifold. By Remark II.2.4 we know that real projective $(n-1)$-space is spin if and only if $n \equiv 0 \pmod 4$. This proves (3.7). ∎

Theorem 3.8 (Landweber and Stong [1]). *Let X be a compact spin manifold of dimension $4k$ with an S^1-action. If the action is of odd type, then*

$$\operatorname{sig}(X) = 0$$

Proof. A result of Hattori and Taniguchi [1] states that for any S^1-action on an oriented manifold,

$$\text{sig}(X) = \text{sig}(X^{S^1})$$

where X^{S^1} is the fixed-point set of the action with an appropriately chosen orientation. Consider now the inclusions: $X^{S^1} \subset X^{\mathbb{Z}_2} \subset X$, and note that $(X^{\mathbb{Z}_2})^{S^1} = X^{S^1}$. By a result of Edmonds [1], the manifold $X^{\mathbb{Z}_2}$ is orientable. Therefore, applying the argument again shows that $\text{sig}(X) = \text{sig}(X^{\mathbb{Z}_2})$. However, by (3.7) we have that $\dim(X^{\mathbb{Z}_2}) \equiv 2 \pmod 4$, and so $\text{sig}(X^{\mathbb{Z}_2}) = 0$. ■

This result is analogous to the fact that on manifolds which admit orientation reversing diffeomorphisms, all the Pontrjagin numbers are zero. For S^1-actions on spin manifolds which are free outside the fixed-point set, there are additional characteristic classes which vanish (see Landweber and Stong [1]). This leads into the fascinating area of elliptic genera (cf. Chudnovsky and Chudnovsky [1]).

Proof of Theorem 3.4. Recall that any spin manifold X which carries a metric of positive scalar curvature has $\hat{\mathscr{A}}(X) = 0$ (see II.8.12). Thus Theorem 3.4 is a corollary of the following more general result. ■

Theorem 3.9 (Lawson and Yau [1]). *Any compact spin manifold which admits a non-trivial S^3-action, carries a riemannian metric of positive scalar curvature.*

When the action is free, the proof of this theorem is rather easy. In this case, the manifold is the total space of a smooth principal S^3-bundle $X \xrightarrow{\pi} M = X/S^3$. Introduce an invariant splitting $TX = \mathscr{V} \oplus \mathscr{H}$ where $\mathscr{V}$ denotes the field of 3-planes tangent to the orbits. The bundle $\mathscr{V}$ has a natural trivialization $\mathscr{V} \cong X \times \mathbb{R}^3$ given as usual by choosing an orthonormal basis e_1, e_2, e_3 of $T_1(S^3)$ and defining sections E_j of $\mathscr{V}$ by $E_j(x) = d/dt\,(\exp(te_j)\cdot x)_{t=0}$. We fix a metric on M and for each $\varepsilon > 0$ we construct a metric g_ε on X as follows. We declare $\mathscr{V}$ and $\mathscr{H}$ to be everywhere orthogonal; we lift the given metric on M to $\mathscr{H}$ via π; and we set $\langle E_i, E_j \rangle \equiv \varepsilon^2\delta_{ij}$. This uniquely determines the metric g_ε. In this metric the orbits of S^3 are all totally geodesic and isometric to the euclidean sphere of radius ε. Applying the fundamental equations of O'Neill [1] for a riemannian submersion, one easily sees that the scalar curvature κ_ε of the metric g_ε is of the form $\kappa_\varepsilon(x) = 6/\varepsilon^2 + a(x) + \varepsilon^2 b(x)$ for continuous functions a and b on X. Taking ε sufficiently small completes the proof for this case.

The general case, where the action of S^3 is not free, is more difficult and is handled as follows. Consider the diagonal action of S^3 on $X \times S^3$. This action is free with quotient X. (Of course, this orbit space map

$X \times S^3 \to X$ is different from the "product" projection $\mathrm{pr}: X \times S^3 \to X$.) We now fix an S^3-invariant metric of X, and using this metric we construct, for each $\varepsilon > 0$, a metric g_ε on $X \times S^3$ in the manner above. The metric g_ε is, in fact, $S^3 \times S^3$-invariant and thus makes the product projection $\mathrm{pr}: X \times S^3 \to X$ into a riemannian submersion. Detailed calculations show that for ε sufficiently small, this submersed metric has positive scalar curvature. The estimates in this calculation are most delicate at the fixed point set.

Theorem 3.9 actually provides a wealth of non-existence theorems for S^3-actions. We clearly have the following:

Corollary 3.10. *Any compact manifold which cannot carry a riemannian metric of positive scalar curvature has no S^3-actions.*

In the next section we shall produce large families of manifolds which do not carry positive scalar curvature. One consequence of this will be a proof of the fact (due originally to Browder and Hsiang [1]) that *if a spin manifold X admits an S^3-action, then all the "higher $\hat{A}$-genera" of X are zero.*

§4. Compact Manifolds of Positive Scalar Curvature

One of the most important applications of the Atiyah-Singer operator in riemannian geometry is in the study of manifolds of positive scalar curvature. Recall that the scalar curvature of a riemannian manifold X is a function $\kappa: X \to \mathbb{R}$ defined at each point x by averaging all the sectional curvatures at x. Specifically we define

$$\kappa(x) = \sum_{i,j=1}^{n} \langle R_{e_i,e_j}(e_j), e_i \rangle \tag{4.1}$$

where $\{e_1, \ldots, e_n\}$ is an orthonormal basis of $T_x X$ and R is the Riemann curvature tensor. Note that κ can also be written as

$$\kappa(x) = \mathrm{trace}(\mathrm{Ric}_x) \tag{4.2}$$

where Ric_x denotes the Ricci transformation at x. Thus, the condition of **positive scalar curvature,** i.e., $\kappa > 0$, is the weakest of the various "positive curvature" assumptions.

There has already been some discussion of manifolds with $\kappa > 0$ given in §8 of Chapter II. We recall the basic results proved there. There is a graded ring homomorphism

$$\hat{\mathscr{A}}: \Omega_*^{\mathrm{Spin}} \longrightarrow KO^{-*}(\mathrm{pt}) \tag{4.3}$$

which in dimensions $4k$ is a multiple of the $\hat{A}$-genus. This homomorphism is defined in (3.22) of Chapter II. It coincides with the KO-index of the

Atiyah-Singer operator, and can be computed in terms of the space of harmonic spinors (see Theorem II.7.10). Applying the Bochner-type formula (II.8.18) gives the following result of Atiyah, Hitchin, Lichnerowicz and Singer.

Theorem 4.1. *Let X be a compact spin manifold with $\kappa > 0$, then $\hat{\mathscr{A}}(X) = 0$.*

This theorem implies, in particular, that a spin $4k$-manifold with non-vanishing $\hat{A}$-genus cannot carry a metric of positive scalar curvature. There are many examples of such manifolds. For example, let $V^{2k}(d)$ denote a non-singular complex hypersurface of degree d in $\mathbb{P}^{2k+1}(\mathbb{C})$. Then $V^{2k}(d)$ is spin if and only if d is even. (See II.2.7.) However, we have that

$$\hat{A}(V^{2k}(d)) = \frac{1}{2^{2k}(2k+1)!} \prod_{j=-k}^{k} (d - 2j), \tag{4.4}$$

so that for all even $d > 2k$, and all $k \geqq 1$, the $4k$-dimensional spin manifold $V^{2k}(d)$ cannot carry $\kappa > 0$, and therefore cannot carry metrics of positive Ricci or positive sectional curvature.

It is a consequence of the Calabi Conjecture, proved by S. T. Yau [2], that each of the hypersurfaces $V^{2k}(d)$ for $d \leqq 2k + 1$ carries a metric of positive Ricci curvature. This shows that the spin condition in Theorem 4.1 is necessary. It also shows, together with Theorem 4.1, that the polynomial $p(d) \equiv \hat{A}(V^{2k}(d))$ must have zeros for $d = 2,4,\ldots,2k$. It is interesting that with this information and some general considerations, it is rather easy to compute formula (4.4) for $p(d)$.

An important phenomenon occurs in the borderline case $V^{2k}(2k+2)$ where Yau proves that there exist metrics of zero Ricci curvature. In this case we see that $\hat{A} = 2$. It follows from Corollary II.8.10, that in any Ricci flat metric, $V^{2k}(2k+2)$ carries at least two parallel spinor fields. Such fields automatically reduce the holonomy group of the metric. We shall pursue these considerations further in §10. We now return to the discussion of Theorem 4.1.

Recall that the invariant $\hat{\mathscr{A}}$ is non-zero also in dimensions 1 and 2 (mod 8). Here it takes values in $\mathbb{Z}_2$. Among the spin manifolds on which it is non-zero, are half of the exotic spheres in dimensions 1 and 2 (mod 8). (A general discussion of positive curvature metrics on exotic spheres is given in II.8.) This fact has the following consequence:

Corollary 4.2. *In dimensions* 1 *and* 2 (mod 8), *every compact spin manifold is homeomorphic to a manifold which cannot carry positive scalar curvature.*

Proof. Let X be a spin k-manifold with $k = 1$ or 2 (mod 8), and let Σ be an exotic k-sphere with $\hat{\mathscr{A}}(\Sigma) \neq 0$. Since $\hat{\mathscr{A}}(X \# \Sigma) = \hat{\mathscr{A}}(X) + \hat{\mathscr{A}}(\Sigma)$, we may apply Theorem 4.1 to either X or $X \# \Sigma$. ∎

To keep these non-existence theorems in perspective it is useful to examine manifolds which are known to carry positive scalar curvature. The classical examples of spaces with curvature $\geqq 0$ are the homogeneous spaces with normal metrics given as follows. Let G be a compact Lie group with Lie algebra $\mathfrak{g}$, and let $(\cdot,\cdot)$ be any Ad_G-invariant inner product of $\mathfrak{g}$. Then for any closed subgroup $H \subset G$, there is a G-invariant metric naturally introduced on the coset space $X = G/H$. The sectional curvatures of this metric are always $\geqq 0$, and apart from the case of the flat torus, the scalar curvature is always >0. (For details, see Kobayashi-Nomizu [2] or Cheeger-Ebin [1].)

These homogeneous spaces represent special cases of the general phenomenon of Theorem 3.8. Any compact manifold which admits a compact connected non-abelian Lie transformation group, carries a metric with $\kappa > 0$.

There is a general surgery procedure for constructing metrics with $\kappa > 0$. Recall (cf. Milnor [8]) that a **surgery of codimension q** on an n-manifold X is a modification of X of the following type. Let $\Sigma^{n-q} \subset X$ be a smoothly embedded $(n-q)$-sphere with a **trivialized tubular neighborhood.** By this we mean a neighborhood U of Σ^{n-q} together with a diffeomorphism $f: U \to S^{n-q} \times D^q$ such that $f(\Sigma^{n-q}) = S^{n-q} \times \{0\}$. The surgery operation now consists of removing the neighborhood $U \cong S^{n-q} \times D^q$ and replacing it with the product $D^{n-q+1} \times S^{q-1}$ by gluing in the obvious, canonical way along the boundary $S^{n-q} \times S^{q-1}$.

Proposition 4.3 (Gromov-Lawson [2] and Schoen-Yau [2]). *Let X be a manifold which carries a metric of positive scalar curvature. Then any manifold obtained from X by surgery in codimension $\geqq 3$, also carries a metric of positive scalar curvature.*

The method of proof in Gromov-Lawson [2] is an elementary construction, the proof in Schoen-Yau [2] uses techniques of partial differential equations.

Combining this result with techniques of the h-cobordism Theorem (cf. Milnor [8]) and facts concerning the cobordism ring Ω_*^{SO}, it is possible to prove the following result:

Theorem 4.4 (Gromov-Lawson [2]). *Let X be a compact simply-connected manifold of dimension $\geqq 5$.*

(A) *If X is not spin, then X carries a metric with $\kappa > 0$.*
(B) *If X is spin, and is spin cobordant to a manifold with $\kappa > 0$, then X carries a metric with $\kappa > 0$.*

Remark. Contrasting part (A) of this theorem with Corollary 4.2 makes it evident that the question of positive scalar curvature is intimately related to spin geometry. This relationship will soon be examined in detail.

Proof of Theorem 4.4. Suppose X is spin and spin cobordant to a manifold X_0 which carries $\kappa > 0$. By a sequence of surgeries (on embedded circles) we can make X_0 simply-connected while preserving the spin corbordism class. By Proposition 4.3, the new X_0 will also carry $\kappa > 0$.

Our cobordism assumption means that there exists a compact spin manifold W with $\partial W \approx X \amalg X_0$ (as spin manifolds). As above, we can kill $\pi_1(W)$ by surgery. Moreover, since $\dim W = n \geqq 6$ and W is spin, every 2-sphere embedded in W has a trivial normal bundle (cf. II.2.11), and we can kill $\pi_2(W)$ by surgery. It follows that the groups $H_{n-k}(W,X_0) \cong H^k(W,X)$ are zero for $k = 1,2$, and torsion-free for $k = 3$. By the basic theory of S. Smale (cf. Milnor [8]), there exists a smooth function $f: W \to [0,1]$ with the following properties:

(i) $f|_{X_0} \equiv 0$ and $f|_X \equiv 1$,
(ii) all critical points of f are non-degenerate and lie in the interior of W,
(iii) all critical points of f have index $\leqq n - 3$.

It now follows from the fundamental theorem of Morse Theory (cf. Milnor [6]) that X can be obtained from X_0 by surgeries in codimension $\geqq 3$. We then apply Proposition 4.3.

When X is not spin, the argument is similar. From detailed knowledge of the oriented cobordism ring we know that there exists a compact oriented manifold W with $\partial W = X \amalg X_0$, where X_0 carries $\kappa > 0$. We may assume that $\pi_1(X_0) = 0$ as before. We now proceed to kill $\pi_1(W)$ by surgery. We then have $H_2(W) \cong \pi_2(W)$, and the second Stiefel-Whitney class gives a homomorphism

$$w_2 : \pi_2(W) \longrightarrow \mathbb{Z}_2. \tag{4.5}$$

Since $\dim W \geqq 6$, the group $\pi_2(W)$ is generated by embedded 2-spheres, and w_2 exactly measures the non-triviality of the normal bundle to these embeddings. Thus, we can kill the subgroup $\ker(w_2) \subset \pi_2(W)$. The homomorphism (4.5) then becomes injective. Observe now that w_2 restricts to be the second Stiefel-Whitney class of X, which is non-zero since X is not spin. It follows that (4.5) is an isomorphism and that the map $H_2(X_0) \cong \pi_2(X_0) \to \pi_2(W) \cong H_2(W)$ is surjective. In particular, $H_k(W,X_0) = 0$ for $k = 1,2$. Consequently $H^k(W,X) = 0$ for $k = 1,2$, and $H^3(W,X)$ is torsion free. The argument now proceeds as before. ■

Let X be a compact manifold of dimension $\geqq 5$. Theorem 4.4 states that there is an obstruction to the existence of a positive scalar curvature metric on X only if X is spin, and, furthermore, that obstruction depends only on the spin cobordism class of X. This leads us to consider the set $\mathfrak{P}_n$ of spin cobordism classes $\alpha \in \Omega_n^{\text{Spin}}$ with the property that some manifold in α carries positive scalar curvature. We claim that $\mathfrak{P}_* \equiv \oplus \mathfrak{P}_n$ *is an ideal in*

Ω_*^{Spin}. Closure under addition (which corresponds to disjoint union) is clear. Suppose now that $\alpha \in \mathfrak{P}_n$ and choose any class $\beta \in \Omega_m^{\mathrm{Spin}}$. Represent α by a spin manifold X with a metric g_X with $\kappa_X > 0$. Represent β by a spin manifold Y with some metric g_Y. Then for all $\varepsilon > 0$, the product metric $g = \varepsilon g_X + g_Y$ on $X \times Y$ has scalar curvature $\kappa = \varepsilon^{-1}\kappa_X + \kappa_Y$ which is >0 for all sufficiently small ε. This shows that $\beta \cdot \alpha \in \mathfrak{P}_{n+m}$, and $\mathfrak{P}_*$ is an ideal as claimed.

Consequently, setting $\mathfrak{ko}_* \equiv \Omega_*^{\mathrm{Spin}}/\mathfrak{P}_*$, we get a graded ring homomorphism

$$\hat{\mathfrak{a}} : \Omega_*^{\mathrm{Spin}} \longrightarrow \mathfrak{ko}_*. \tag{4.6}$$

Note that in each dimension n, $\mathfrak{ko}_n$ is a finitely generated group. Theorem 4.4 states the following.

Corollary 4.5. *The classes $\hat{\mathfrak{a}}_*$ constitute a complete set of invariants for determining whether or not a simply-connected manifold (of dimension $\geqq 5$) can carry a metric of positive scalar curvature.*

The fundamental Theorem 4.1 implies that there is a factoring of graded ring homomorphisms

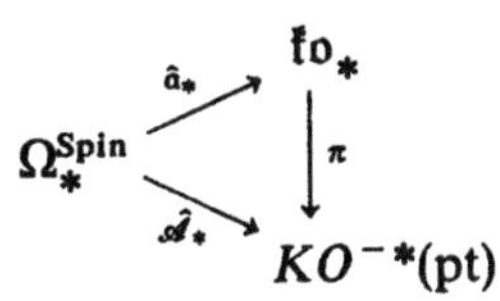

and it seems likely that π is an isomorphism, i.e., that $\hat{\mathfrak{a}} \cong \hat{\mathscr{A}}$. (We know $\pi \otimes \mathbb{Q}$ to be an isomorphism; see Gromov-Lawson [2].) If this were so, then the $\hat{\mathscr{A}}$-class would be exactly the obstruction to the existence of positive scalar curvature metrics in the simply-connected case.

Since $\Omega_n^{\mathrm{Spin}} = 0$ for $n = 5$, 6 and 7, Theorem 4.4 has the following consequence.

Corollary 4.6. *Every compact simply-connected manifold of dimension* 5, 6, *or* 7 *carries a metric of positive scalar curvature.*

We shall next take up the question of positive scalar curvature on non-simply-connected manifolds. Here the situation can be quite subtle, as the following example, due to L. Bérard-Bergery [2], illustrates. Let Σ be an exotic 9-sphere such that $\hat{\mathscr{A}}(\Sigma) \neq 0$, and consider the spin manifold $X \equiv (\mathbb{P}^7(\mathbb{R}) \times S^2) \# \Sigma$, with $\pi_1 X \cong \mathbb{Z}_2$. *This manifold does not carry positive scalar curvature* (since $\hat{\mathscr{A}}(X) = \hat{\mathscr{A}}(\Sigma) \neq 0$), *however its two-fold universal covering manifold does.* To see this, one uses the fact that $\Omega_9^{\mathrm{Spin}} \cong \mathbb{Z}_2 \oplus \mathbb{Z}_2$ (cf. Stong [1]) which shows that the universal covering $\tilde{X} \equiv (S^7 \times S^2) \# \Sigma \# \Sigma \cong S^7 \times S^2$.

§5. Positive Scalar Curvature and the Fundamental Group

We now take up the case of manifolds with non-trivial fundamental group. To set the stage we note that the product of any manifold with S^2 carries a metric with $\kappa > 0$. Consequently, in dimensions $\geqq 6$, the presence of positive scalar curvature places no restriction on the isomorphism class of the fundamental group. (This is also true in dimensions 4 and 5; cf. R. Carr [1].) Nevertheless, if one takes into account the interaction of the fundamental group with the geometry of the manifold, then tremendous restrictions arise.

The interaction we are seeking lies in the geometry of the covering spaces. In particular, we shall be interested in manifolds with covering spaces which are "large in all directions, or put another way, whose universal covering spaces contain big "cubes." To give an idea of what this should mean, we examine a notion due to Gromov. Let I^n denote the standard metric cube in $\mathbb{R}^n$. We shall say that a riemannian n-manifold X contains a **cube of size** L if there exists a continuous map $f: I^n \to X$ such that $\operatorname{dist}(f(\Sigma), f(\Sigma')) \geqq L$ for each pair of opposite faces Σ, Σ' of the cube I^n. We can then say that a riemannian manifold X has **big covering spaces,** if for any $L > 0$, there is a covering space which (in the lifted metric) contains a cube of size L. For a compact manifold this concept is independent of the metric chosen (since any two metrics are uniformly commensurate). A good example of a manifold with big covering spaces is a torus T^n.

This idea can be usefully generalized. However, for the purposes here it is better to "dualize" the concept by mapping out of the manifold rather than into it.

Definition 5.1. A C^1-map $f: X \to Y$ between riemannian manifolds is **ε-contracting** (for a given $\varepsilon > 0$) if $\|f_* v\| \leqq \varepsilon \|v\|$ for all tangent vectors v to X. This means that for any piecewise smooth curve γ in X, one has $\operatorname{length}(f(\gamma)) \leqq \varepsilon \operatorname{length}(\gamma)$.

Such ε-contracting maps are not hard to find (consider the constant maps). Suppose, however, that X and Y are compact oriented manifolds of the same dimension. Then the existence of an ε-contracting map $f: X \to Y$ of **non-zero degree** means that in a strong sense X is "bigger than Y" on an order of at least $1/\varepsilon$. Now every oriented n-manifold admits maps of degree 1 onto the n-sphere. Thus, we let $S^n(1)$ denote the standard riemannian n-sphere of constant curvature 1, and we replace the above notion of big covering spaces with the following (cf. Gromov-Lawson [1],[3]):

Definition 5.2. A compact riemannian n-manifold is said to be **enlargeable** if for every $\varepsilon > 0$ there exists an orientable riemannian covering space which admits an ε-contracting map onto $S^n(1)$ which is constant at infinity and of non-zero degree.

If for each $\varepsilon > 0$, there is a finite covering space with these properties, we call the manifold **compactly enlargeable.**

Note. A map is **constant at infinity** if it is constant outside a compact set. The **degree** of such a map $f: X \to S^n$ is defined as

$$\deg(f) = \frac{\int_X f^*\omega}{\int_{S^n} \omega} \tag{5.1}$$

where ω is an n-form on S^n with non-zero integral. The degree can also be defined as usual in terms of regular values of f.

The square flat torus $T^n = \mathbb{R}^n/\mathbb{Z}^n$ is certainly enlargeable since the universal covering space has the required mappings for all $\varepsilon > 0$. This torus is, in fact, compactly enlargeable. We see this as follows. For each $k \in \mathbb{Z}^+$, the lattice $(k \cdot \mathbb{Z})^n \subset \mathbb{Z}^n$ gives a k^n-fold covering torus $\tilde{T}^n \equiv \mathbb{R}^n/(k \cdot \mathbb{Z})^n$, which admits the (π/k)-contracting map to $S^n(1)$ of degree 1 pictured below.

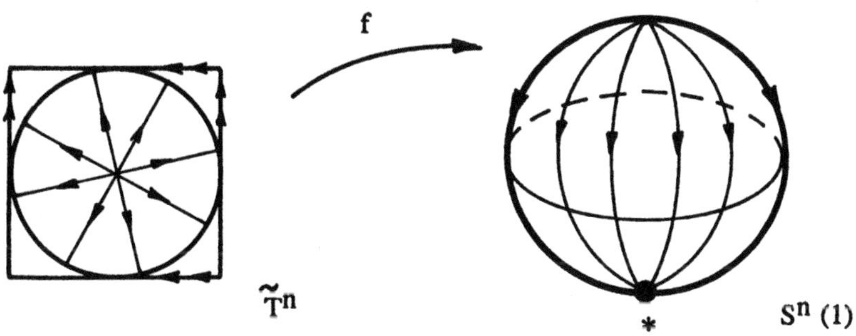

Theorem 5.3. *The following statements hold in the category of compact manifolds:*

(A) *Enlargeability is independent of the riemannian metric.*
(B) *Enlargeability depends only on the homotopy-type of the manifold.*
(C) *The product of enlargeable manifolds is enlargeable.*
(D) *The connected sum of any manifold with an enlargeable manifold is again enlargeable.*
(E) *Any manifold which admits a map of non-zero degree onto an enlargeable manifold is itself enlargeable.*

Proof. It is evident that (E) $\Rightarrow$ (B) $\Rightarrow$ (A) and that (E) $\Rightarrow$ (D). To prove (E) we consider two compact oriented riemannian n-manifolds X and Y, and a map $F: X \to Y$ of non-zero degree. By compactness there exists a $c > 0$ so that $\|dF\| \leq c$ on X (i.e., F is c-contracting). Given $\varepsilon > 0$, there is a riemannian covering space $p:\tilde{Y} \to Y$ which admits a (ε/c)-contracting map $f:\tilde{Y} \to S^n(1)$ which is constant outside a compact set $\tilde{K} \subset \tilde{Y}$ and of non-zero degree. Taking the fibre product of p and F gives a covering space $p':\tilde{X} \to X$ and a proper mapping $\tilde{F}:\tilde{X} \to \tilde{Y}$ so that the diagram

$$\begin{array}{ccc} \tilde{X} & \xrightarrow{\tilde{F}} & \tilde{Y} \\ {\scriptstyle p'}\downarrow & & \downarrow{\scriptstyle p} \\ X & \xrightarrow{F} & Y \end{array}$$

commutes. Since $\tilde{F}$ is a lifting of F, we have $\|\nabla\tilde{F}\| \leqq c$ on $\tilde{X}$. Hence, the composition $f \circ \tilde{F}:\tilde{X} \to S^n(1)$ is ε-contracting. Since $\tilde{F}$ is proper, we see that $f \circ \tilde{F}$ is constant outside the compact set $\tilde{F}^{-1}(\tilde{K})$. It is easy to see that: $\deg(f \circ \tilde{F}) = \deg(f)\deg(F) \neq 0$. Hence, X is enlargeable as claimed.

To prove (C), we fix a degree-1 map $\phi:S^n(1) \times S^m(1) \to S^{n+m}(1)$ and let $c = \sup\|d\phi\|$. This map is chosen to be constant on the set $(S^n(1) \times \{*\}) \cup (\{*\} \times S^m(1))$, where each "$*$" denotes a distinguished point in the sphere. Suppose now that we are given (ε/c)-contracting maps, $f:X^n \to S^n(1)$ and $g:X^m \to S^m(1)$, which are constant ($= *$) at infinity and of non-zero degree. Then the map $\phi \circ (f \times g): X^n \times Y^m \longrightarrow S^{n+m}(1)$ is ε-contracting, constant at infinity and of non-zero degree. From here the argument is straightforward. ∎

Theorem 5.3 says that the category of enlargeable manifolds is closed under products, connected sums (with anything), changes of differentiable structure, etc. The category is also quite rich. It contains, as basic building blocks, many families of important $K(\pi,1)$-manifolds.

Theorem 5.4 (Gromov-Lawson [1]). *The following manifolds are enlargeable:*

(A) *Any compact "solvmanifold", i.e., a compact manifold diffeomorphic to G/Γ where G is a solvable Lie group and Γ is a discrete subgroup.*
(B) *Any compact manifold which admits a metric of non-positive sectional curvature.*
(C) *Any "sufficiently large" 3-manifold (X is sufficiently large if there is a compact surface of positive genus $\Sigma \subset X$ with $\pi_1\Sigma \to \pi_1 X$ injective.)*

Proof. We begin with (B). It is easy to see that a manifold X of non-positive curvature is enlargeable. By the Cartan-Hadamard Theorem (cf. Milnor [6]), the exponential map at any point of the universal covering $\tilde{X}$ is a diffeomorphism whose inverse $\exp_x^{-1}:\tilde{X} \to T_x\tilde{X}$ is distance decreasing onto $T_x\tilde{X} \cong \mathbb{R}^n$ with its *euclidean* metric. Composing with a family of ε-contracting maps $f_\varepsilon: \mathbb{R}^n \to S^n(1)$ as above completes the proof of (B).

To prove (A) we invoke the basic fact that every compact solvmanifold X admits a fibration $\pi: X \to T^k$ over a torus of positive dimension with a (compact) solvmanifold X_0 as fibre (see Raghunathan [1] for example). We introduce the standard flat metric on $T^k \equiv \mathbb{R}^k/\mathbb{Z}^k$, and we fix a metric g on X. By replacing g with tg for $t > 0$ sufficiently large, we can assume that π is distance-decreasing.

Consider now the covering $\tilde{X} \to X$ induced by lifting π over the covering $\mathbb{R}^k \to T^k$. The induced fibration $\pi: \tilde{X} \to \mathbb{R}^k$ is again distance decreasing. Furthermore, there exists a diffeomorphism

$$\varphi: \tilde{X} \longrightarrow \mathbb{R}^k \times X_0$$

such that $\pi = \mathrm{pr}_1 \circ \varphi$ where $\mathrm{pr}_1: \mathbb{R}^k \times X_0 \to \mathbb{R}^k$ is the projection map. To see this when $k = 1$, lift a non-vanishing vector field from T^1 to $\tilde{X}$ and integrate to get the coordinate transverse to the fibres. When $k > 1$, apply induction.

Suppose that $\varepsilon > 0$ is given. Consider the ball $B = \{x \in \mathbb{R}^k : \|x\| \leq \pi/\varepsilon\}$ and fix a map $f: \mathbb{R}^k \to S^k(1)$ which is ε-contracting, constant on $\mathbb{R}^k - B$ and of degree 1 (as pictured in the figure above). Choose a metric g_0 on X_0 so that the composition

$$\tilde{X} \xrightarrow{\varphi} \mathbb{R} \times X_0 \xrightarrow{\mathrm{pr}_2} X_0$$

is distance decreasing on the compact subset $\pi^{-1}(B) \subset \tilde{X}$. (A sufficiently small multiple of any metric will do.) We may assume that X_0 is enlargeable by induction on dimension. Hence there exists a covering $\pi_0: \tilde{X}_0 \to X_0$ and an ε-contracting map $g: \tilde{X}_0 \to S^{n-k}(1)$ which is constant at infinity and of non-zero degree. Let $q: \tilde{\tilde{X}} \to \tilde{X}$ be the covering induced by lifting φ over $\mathrm{Id} \times \pi_0$, so there is a commutative diagram (below).

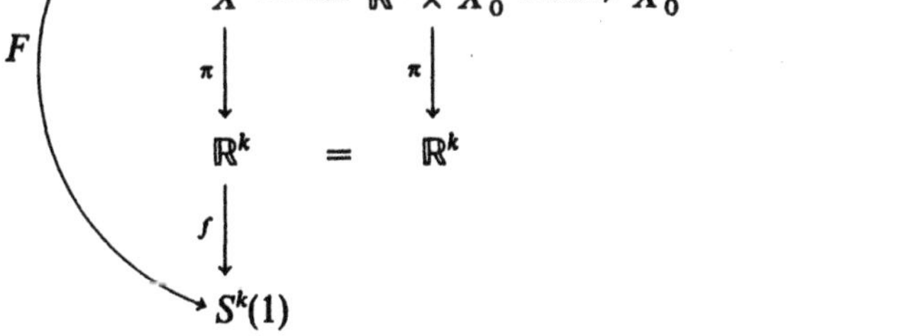

We now take the "smash product"

$$\tilde{\tilde{X}} \xrightarrow{F \wedge G} S^n(1)$$

exactly as we did in the proof of Theorem 5.3(C). This map is 4ε-contracting, constant at infinity and of non-zero degree. This completes the proof of part (A). For part (C), the reader is referred to Gromov-Lawson [1]. ∎

Note that the tori $T^n = S^1 \times \cdots \times S^1$, $n \geqq 1$, play a special role in this theory. For example, they belong to each of the three classes of manifolds in Theorem 5.4. Furthermore, there is the following "stability" phenomenon:

If X is an enlargeable manifold, so is $X \times T^n$ for any $n \geqq 1$. (5.2)

If a manifold X carries a metric with $\kappa > 0$, so does $X \times T^n$ for any $n \geqq 1$. (5.3)

There is an (adversary) relationship between enlargeability and positive scalar curvature, and the mediating agent is the Atiyah-Singer operator. Here, however, the operator has coefficients in a vector bundle. The first main result is the following:

Theorem 5.5 (Gromov-Lawson [1], [3]). *An enlargeable spin manifold X cannot carry a metric of positive scalar curvature. In fact any metric with $\kappa \geqq 0$ on X must be flat.*

Note. Here the spin assumption is not vital. It suffices to know that an appropriate covering space is spin.

Note. Results of this type were first proved in dimensions $\leqq 7$ by R. Schoen and S. T. Yau [1], [2] using minimal surface techniques.

Theorems 5.4 and 5.5 together produce the following nice "exclusion theorem."

Corollary 5.6. *A compact manifold which carries a metric with sectional curvature* $\leqq 0$ (or <0), *cannot carry a metric with scalar curvature* >0 ($\geqq 0$ *respectively*).

Proof of 5.5. For clarity's sake we shall present here a proof for the case of compactly enlargeable manifolds. The more general result requires some tools from analysis (the Relative Index Theorem) and will be proved with more general results in the next section.

Let X be a compactly enlargeable n-manifold, and suppose X carries a metric with $\kappa \geqq \kappa_0$ for a constant $\kappa_0 > 0$. From (5.2) and (5.3) we may assume that X has even dimension $2n$. (If not, replace X by $X \times S^1$.)

Choose a complex vector bundle E_0 over the sphere $S^{2n}(1)$ with the property that the top Chern class $c_n(E_0) \neq 0$. This is certainly possible, since on S^{2n} the Chern character

$$\text{ch } E = \dim E + \frac{1}{(n-1)!} c_n(E)$$

gives an isomorphism $\text{ch}: K(S^{2n}) \to H^*(S^{2n}; \mathbb{Z})$ (cf. Atiyah-Hirzebruch [2]). We now fix a unitary connection ∇^{E_0} on E_0 and we let R^{E_0} denote the curvature 2-form.

Let $\varepsilon > 0$ be given and choose a finite orientable covering $\tilde{X} \to X$ which admits an ε-contracting map $f: \tilde{X} \to S^{2n}(1)$ of non-zero degree. Using f, we pull back the bundle E_0, with its connection, to $\tilde{X}$. This gives us a bundle $E \equiv f^*E_0$ with connection $\nabla^E \equiv f^*\nabla^{E_0}$. We then consider the complex spinor bundle $\$_{\mathbb{C}}$ of $\tilde{X}$ with its canonical riemannian connection, and consider the Atiyah-Singer operator $\not{D}$ on the tensor product $\$_{\mathbb{C}} \otimes E$ (as in II.5.10). We know from Theorem II.8.17 that there is a "Bochner formula"

$$\not{D}_E^2 = \nabla^*\nabla + \tfrac{1}{4}\kappa + \Re^E \tag{5.4}$$

where $\Re^E(\sigma \otimes \varepsilon) \equiv \sum_{j<k} (e_j e_k \sigma) \otimes (R^E_{e_j,e_k}\varepsilon)$ depends universally and linearly on the components of the curvature tensor R^E of E. The operator $\Re^E$ is a symmetric bundle endomorphism of $\$_{\mathbb{C}} \otimes E$, and if its pointwise norm everywhere satisfies the inequality

$$\|\Re^E\| < \tfrac{1}{4}\kappa_0, \tag{5.5}$$

then we will have $\not{D}_E^2 > 0$, and the index of the operator

$$\not{D}_E^+ : \Gamma(\$_{\mathbb{C}}^+ \otimes E) \longrightarrow \Gamma(\$_{\mathbb{C}}^- \otimes E), \tag{5.6}$$

will be zero. To achieve this estimate we first note that there is a constant k_n depending only on dimension, such that

$$\|\Re^E\| \leqq k_n \|R^E\|. \tag{5.7}$$

We then observe that the curvature of E is the pull-back of the curvature of E_0, and therefore

$$\|R^E\| \leqq \varepsilon^2 \|R^{E_0}\|. \tag{5.8}$$

To prove (5.8), we fix a point $x \in \tilde{X}$ and let $\{\psi_1, \ldots, \psi_{n(n-1)/2}\}$ be an orthonormal basis of $\Lambda^2 T_x\tilde{X}$ which diagonalizes the symmetric bilinear form $\beta(\psi,\psi') \equiv \langle f_*\psi, f_*\psi'\rangle$. This means that $\langle f_*\psi_j, f_*\psi_k\rangle = \lambda_j^2\delta_{jk}$ where, since f is ε-contracting, we have $|\lambda_j| \leqq \varepsilon^2$ for all j. It follows that there exists an orthonormal basis $\{\tilde{\psi}_1, \ldots, \tilde{\psi}_{n(n-1)/2}\}$ of $\Lambda^2 T_{f(x)}S^{2n}$ such that $f_*\psi_j = \lambda_j\tilde{\psi}_j$, for all j. We then compute that $\|R^E\|_x^2 = \sum \|R^E_{\psi_j}\|^2 = \sum \|R^{E_0}_{f_*\psi_j}\|^2 = \sum \lambda_j^2 \|R^{E_0}_{\tilde{\psi}_j}\|^2 \leqq \varepsilon^4 \|R^{E_0}\|^2$.

Combining (5.7) and (5.8) shows that the positivity condition (5.5) is satisfied for all $\varepsilon < \sqrt{\kappa_0/4k_n}$. The index of the operator (5.6) must then vanish. However, this index is given by

$$\begin{aligned}
\operatorname{index}(\not{D}_E^+) &= \{\operatorname{ch} E \cdot \hat{\mathbf{A}}(\tilde{X})\}[\tilde{X}] \qquad (5.9)\\
&= \left\{\left(d + \frac{1}{(n-1)!} c_n(E)\right) \cdot \hat{\mathbf{A}}(\tilde{X})\right\}[\tilde{X}]\\
&= d\hat{A}(\tilde{X}) + \frac{1}{(n-1)!} c_n(E)[\tilde{X}]\\
&= \frac{1}{(n-1)!} c_n(f^*E_0)[\tilde{X}]\\
&= \frac{1}{(n-1)!} f^*(c_n(E_0))[\tilde{X}]\\
&= \frac{1}{(n-1)!} \deg(f) c_n(E_0)[S^{2n}]\\
&\neq 0.
\end{aligned}$$

This contradiction completes the proof of the first statement of the theorem.

To prove the second statement, we call upon the following result which appears in Kazdan-Warner [1]:

Theorem 5.7 (J. P. Bourguignon). *Let X be a compact manifold which carries no metric with $\kappa > 0$. Then any metric with $\kappa \geqq 0$ on X is Ricci flat.*

To complete the argument we now apply the following corollary of the Splitting Theorem of Cheeger and Gromoll [2].

Proposition 5.8. *A Ricci flat enlargeable manifold is flat.*

Proof. Let X be a (compact) enlargeable n-manifold with a Ricci flat metric. The Splitting Theorem implies that the universal cover of X splits as a riemannian product $\mathbb{R}^m \times Y$ of a euclidean space with a compact simply-connected (Ricci-flat) manifold Y. We claim that $\dim(Y) = 0$. If not, set $\delta \equiv \text{diameter}(Y) > 0$, and choose a covering $\tilde{X}$ of X which admits a $(\pi/4\delta)$-contracting map $f: \tilde{X} \to S^n(1)$ which is constant at infinity and of non-zero degree. This covering is a quotient $\tilde{X} = (\mathbb{R}^m \times Y)/\Gamma$ by a subgroup Γ of the deck group. The deck group acts freely and properly discontinuously by isometries which preserve the factors. By passing to a subgroup of finite index we may assume that Γ acts freely and properly discontinuously on $\mathbb{R}^m$. Then $\tilde{X}$ is a fibre bundle $p: \tilde{X} \to \mathbb{R}^m/\Gamma$ over the quotient manifold with fibre Y.

Since f is $(\pi/4\delta)$-contracting, we see that each fibre is mapped into a ball of radius $\pi/4$ in $S^n(1)$. Hence, there is a continuous map $f_0: \mathbb{R}^m/\Gamma \to S^n(1)$

so that

$$\text{dist}(f_0(p(x)), f(x)) \leqq \pi/2$$

for all $x \in X$. (Choose a section of the disk bundle whose fibre at $y \in \mathbb{R}^m/\Gamma$ is the convex hull of $f(Y_p) = f(p^{-1}(y))$.) This map can be chosen to agree with f at infinity. By pushing along the unique geodesic arc joining $f(x)$ to $f_0(p(x))$, we construct a homotopy from f to $f_0 \circ p$ which is constant at infinity. Since $\dim(Y) > 0$, it follows that $\deg(f) = \deg(f_0 \circ p) = 0$ contrary to assumption. Therefore, we conclude that $\dim(Y) = 0$, and both the proposition and the theorem are proved. ■

The argument given for Theorem 5.5 actually proves much more than is claimed. To use the full force of the argument we only need to readjust our definitions. In doing this we shall unify these results with the previous ones concerning the $\hat{A}$-genus.

We begin by generalizing the notion of the degree of a map. Let $f: X \to Y$ be a smooth map between compact oriented manifolds, and suppose $\dim X - \dim Y = 4k \geqq 0$. The inverse image $f^{-1}(y)$ of a regular value $y \in Y$ is a compact 4k-manifold whose oriented cobordism class is independent of y. To see this, join two regular values y and y' by a smoothly embedded are γ, and wiggle f, away from $f^{-1}(y) \cup f^{-1}(y')$, so that it becomes transverse to γ. The inverse image of γ provides a cobordism between $f^{-1}(y)$ and $f^{-1}(y')$. It now follows that the $\hat{A}$-genus $\hat{A}(f^{-1}(y))$ is independent of y. We call this number the **$\hat{A}$-degree of f.**

The $\hat{A}$-degree can be computed by a formula of type (5.1). Let ω be a volume form on Y with non-zero integral, and let $\hat{A}_k(X)$ be a de Rham representative (a closed $4k$-form) of the kth-component of the total $\hat{A}$-class of X. Then

$$\hat{A}\text{-deg}(f) = \frac{\int_X f^*\omega \wedge \hat{A}_k(X)}{\int_Y \omega} \tag{5.10}$$

Notice that the $\hat{A}$-degree makes sense when X is not compact, provided that f is constant at infinity.

We now generalize the notion of enlargeability by replacing the word "degree" in 5.2 by the word "$\hat{A}$-degree".

DEFINITION 5.9. A compact riemannian n-manifold is said to be **$\hat{A}$-enlargeable** if given any $\varepsilon > 0$, there exists a riemannian covering space which admits an ε-contracting map onto $S^m(1)$ (for some $m \equiv n \pmod 4$) which is constant at infinity and of non-zero $\hat{A}$-degree. The basic facts are as before.

Theorem 5.10. *The statements (A), (C) and (D) of Theorem 5.3 remain true with the word "enlargeability" replaced by the word "$\hat{A}$-enlargeability."*

Proof. The proof of part (*C*) of Theorem 5.3 goes through after making the obvious linguistic changes. Parts (*A*) and (*D*) are easy. ■

Of course, an enlargeable manifold is $\hat{A}$-enlargeable. Moreover, an orientable manifold of non-zero $\hat{A}$-genus is also $\hat{A}$-enlargeable since the constant map to S^0 has non-zero $\hat{A}$-degree. More generally we have the following:

Corollary 5.11. *A compact manifold which admits a mapping of non-zero $\hat{A}$-degree onto an enlargeable manifold is $\hat{A}$-enlargeable. In particular, the product of an enlargeable manifold and a manifold of non-zero $\hat{A}$-genus is $\hat{A}$-enlargeable.*

We now have a unification of many of the previous results.

Theorem 5.12 (Gromov-Lawson [1], [3])). *An $\hat{A}$-enlargeable spin manifold cannot carry a metric with $\kappa > 0$. In fact, any metric with $\kappa \geqq 0$ on such a manifold is Ricci flat.*

Proof. The proof of Theorem 5.5 goes through with no essential changes. The main point is that the computation (5.9) now becomes

$$\begin{aligned} \operatorname{ind}(\not{D}_E^+) &= \left\{\left(d + \frac{1}{(m-1)!}c_m(E)\right)\cdot \hat{\mathbf{A}}(\tilde{X})\right\}[\tilde{X}] \qquad (5.11)\\ &= d\hat{A}(\tilde{X}) + \frac{1}{(m-1)!}\{c_m(E)\cdot \hat{A}_k(\tilde{X})\}[X] \\ &= \frac{1}{(m-1)!}\{f^*c_m(E_0)\cdot \hat{A}_k(\tilde{X})\}[\tilde{X}] \\ &= \frac{1}{(m-1)!}(\hat{A}\text{-deg}\, f)c_m(E_0)[S^{2m}] \\ &\neq 0. \quad ■ \end{aligned}$$

This theorem implies, in particular, that a spin manifold of the form $X \times Y$ where X is enlargeable and $\hat{A}(Y) \neq 0$, cannot carry positive scalar curvature. In a sense, it interpolates between the two previous results which concerned enlargeability and $\hat{A}$ separately.

A good interpretation of Theorem 5.12 can be made in terms of the higher $\hat{A}$-genera. These are defined for any compact differentiable manifold X as follows. Fix a $K(\pi,1)$-space K and a map $f: X \to K$ (corresponding to a homomorphism $\pi_1 X \to \pi_1 K$). For each cohomology class $u \in H^*(K;\mathbb{Z}) \cong H^*(\pi;\mathbb{Z})$, **the higher $\hat{A}$-genus associated to u** is the number

$$\hat{A}_{f,u}(X) \equiv \{f^*u \cdot \hat{\mathbf{A}}(X)\}[X]$$

where, as usual, $\hat{\mathbf{A}}(X)$ is the total $\hat{A}$-class of X. We could, of course, pass to the "universal" space $K_X = K(\pi_1(X),1)$ and the map $f_X : X \to K_X$ through which every $f : X \to K(\pi,1)$ can be factored. These "universal" higher $\hat{A}$-genera are simply indexed by the cohomology classes of the group $\pi_1(X)$.

EXAMPLE 5.13. When the space $K = K(\pi,1)$ is a compact oriented m-manifold and when $u \in H^m(K; \mathbb{Z})$ is the fundamental cohomology class, then it is evident that:

$$\hat{A}_{f,u}(X) = (\hat{A}\text{-deg})(f). \tag{5.12}$$

Observe now that many fundamental examples of enlargeable manifolds (such as solvmanifolds, manifolds of non-positive curvature, and sufficiently large prime 3-manifolds) are manifolds of $K(\pi,1)$-type. Theorem 5.12 states that for a spin manifold X of positive scalar curvature, the higher $\hat{A}$-genera, coming as in (5.12) from maps to an enlargeable manifold, must be zero.

It is at the moment unknown whether every compact $K(\pi,1)$-manifold is enlargeable; however, any counterexample to this statement is likely to be quite complicated. Assuming it is so, one might be led to conjecture the following:

Conjecture 5.14. *For any compact spin manifold with positive scalar curvature, all the higher $\hat{A}$-genera must vanish.*

There is much evidence for this conjecture if one restricts to "geometric" $K(\pi,1)$-spaces. At very least, the conjecture serves as a rough map of the wilderness.

There is, in fact, a general theory which allows one to transform this conjecture to a more general conjecture in the K-theory of C^*-algebras. Again it involves the Atiyah-Singer operator.

To each discrete group Γ there is a canonically associated C^*-algebra, denoted $C^*(\Gamma)$ and called the **C^*-algebra of Γ.** Over any manifold with fundamental group Γ we can construct the flat $C^*(\Gamma)$-bundle $\mathscr{E} \to X$ corresponding to the representation of Γ on $C^*(\Gamma)$ by right multiplication. This is just the associated bundle

$$\mathscr{E} \equiv \tilde{X} \times_\Gamma C^*(\Gamma) \tag{5.13}$$

where $\tilde{X}$ is the universal covering space of X (a principal Γ-bundle). Note that $\mathscr{E}$ is a bundle of left $C^*(\Gamma)$-modules. Each fibre is free of rank one.

At this point, our discussion becomes rather general. A guide to the formidible technical details required to make it rigorous is found in J. Rosenberg [1]. Suppose the manifold X is spin and consider the twisted Atiyah-Singer operator

$$\not{D}^+ : \Gamma(\$_{\mathbb{C}}^+ \otimes \mathscr{E}) \longrightarrow \Gamma(\$_{\mathbb{C}}^- \otimes \mathscr{E}) \tag{5.14}$$

where $\mathscr{E}$ carries its canonical flat connection. The kernel and cokernel of $\not{D}^+$ are (after possibly a compact perturbation of $\not{D}^+$) finitely generated projective $C^*(\Gamma)$-modules. The formal difference gives an element in the algebraic K-group, $K_0(C^*(\Gamma))$. This element depends only on the homotopy class of the operator $\not{D}^+$ and is called the **analytic index of $\not{D}^+$.** Mishchenko and Fomenko [1] have given a formula for this index of the Atiyah-Singer type:

$$\operatorname{ind}(\not{D}^+) = \{\operatorname{ch}\mathscr{E}\cdot\hat{\mathbf{A}}(X)\}[X] \in K_0(C^*(\Gamma)) \otimes \mathbb{Q} \tag{5.15}$$

The claim is that for "reasonable" fundamental groups Γ, this index carries all the higher $\hat{A}$-genera of the manifold X. To prove this it is necessary to study a certain universal map

$$e_\Gamma : K_*(K(\Gamma,1)) \longrightarrow K_*(C^*(\Gamma)), \tag{5.16}$$

defined by using the analogous bundle $\mathscr{E}$ over $K(\Gamma,1)$. (Here K_* denotes K-homology theory extended to infinite CW-complexes.) Our assertion is true whenever this map is injective after tensoring with $\mathbb{Q}$.

Observe now that since the bundle $\mathscr{E} \to X$ is flat, the curvature term $\mathfrak{R}^{\mathscr{E}}$ in the Bochner formula (5.4) is zero. This implies (by Rosenberg [1]) that if X carries positive scalar curvature, then $\operatorname{ind}(\not{D}^+) = 0$. Consequently, Conjecture 5.14 is true whenever the map $e_\Gamma \otimes \mathbb{Q}$ is injective.

The injectivity of e_Γ has been established by Kasparov [1] for any Γ which arises as a discrete subgroup of a connected Lie group.

Readers may have noticed the direct parallel between the higher $\hat{A}$-genera and the Novikov higher signatures, or "higher L-genera." (Such classes can be formulated for any multiplicative sequence.) S. P. Novikov has conjectured that these higher signatures are, like the signature itself, homotopy invariants of a manifold. A proof of this conjecture for certain fundamental groups has been given by G. Lusztig [1] using a family of Dirac operators of the form $D_t^+ : (C\ell^+(X) \otimes E_t) \to \Gamma(C\ell^-(X) \otimes E_t)$ where E_t is a family of flat bundles over X induced from $K(\Gamma,1)$. Much more generally, one can consider the operator

$$D^+ : \Gamma(C\ell^+(X) \otimes \mathscr{E}) \to \Gamma(C\ell^-(X) \otimes \mathscr{E})$$

in analogy with (5.14). The assertion here is that the index

$$\operatorname{ind}(D^+) = \{\operatorname{ch}\mathscr{E}\cdot\mathbf{L}(X)\}[X] \in K_*(C^*(\Gamma))$$

carries all the higher signatures of X. This again will follow from the injectivity of the map $e_\Gamma \otimes \mathbb{Q}$. For this reason, the conjecture concerning e_Γ is called the **Strong Novikov Conjecture.**

We shall end this section with some remarks on a possible classification of manifolds with $\pi_1 = \Gamma$ which admit positive scalar curvature. The homomorphism $\hat{\mathscr{A}} : \Omega_*^{\mathrm{Spin}} \to KO_*(\mathrm{pt})$ induces a transformation of generalized homology theories. Thus, setting $h_*(\Gamma) \equiv h_*(K(\Gamma,1))$, we get a transformation

$$\hat{\mathscr{A}}:\Omega_*^{\mathrm{Spin}}(\Gamma) \longrightarrow KO_*(\Gamma). \tag{5.17}$$

Given any compact spin n-manifold X with $\pi_1 X \cong \Gamma$, there is a canonical mapping $X \to K(\Gamma,1)$, inducing an isomorphism on π_1. This map determines an element $[X] \in \Omega_n^{\mathrm{Spin}}(\Gamma)$.

Again as a guide to the forest, one could conjecture that as in the case where $\Gamma = \{e\}$, the classes $\hat{\mathscr{A}}([X])$ constitute a complete set of invariants for the existence of positive scalar curvature on X. There is some evidence for this. The analogue of Theorem 4.3 has been proved by Rosenberg [2] and Miyazaki [1]. They show that the question indeed only depends on the fundamental class $[X] \in \Omega_n^{\mathrm{Spin}}(\pi_1 X)$ in the spin case. In the non-spin case, it depends only on the class $[X] \in \Omega_n^{\mathrm{SO}}(\pi_1 X)$, and often, for example when $\pi_1 X \cong \mathbb{Z}$ or $\mathbb{Z}_p$, one can show that such X always carry $\kappa > 0$.

Unhappily, the conjecture that $\hat{\mathscr{A}} = 0$ in the presence of positive scalar curvature fails for torsion groups Γ. Rosenberg [2] shows that every spin 5-manifold with $\pi_1 \cong \mathbb{Z}_3$ carries $\kappa > 0$, but that $\mathrm{Im}\, \hat{\mathscr{A}} \neq 0$ in this case. Nevertheless, for torsion-free groups Γ there is reasonable evidence for the following:

CONJECTURE. For torsion free groups Γ, the classes $\hat{\mathscr{A}}([X])$ (from (3.17)) constitute a complete set of obstructions to the existence of positive scalar curvature on compact spin manifolds X with $\pi_1 X \cong \Gamma$.

§6. Complete Manifolds of Positive Scalar Curvature

We now consider the question of the existence and structure of complete metrics of positive scalar curvature on non-compact manifolds. We shall touch only a few results. The reader is referred to Gromov-Lawson [3] for a much more extensive discussion.

The first important concept here is the following:

DEFINITION 6.1 A **bundle with compact support** on a manifold X is a vector bundle $E \to X$ which is trivialized at infinity together with a connection which is compatible with that trivialization.

This means that outside a compact subset of X, there is a given isomorphism of E with the product bundle. This isomorphism identifies the given connection with the canonical flat one on the product. The connection will always be assumed to be orthogonal or unitary (depending on whether E is real or complex).

Suppose now that X is a complete riemannian spin manifold of even dimension $n = 2k$. Let $\mathbb{S}_{\mathbb{C}}$ denote the spinor bundle of X with its canonical riemannian connection, and let $E \to X$ be any bundle with compact support. Then the Atiyah-Singer operator $\not{D}_E$ on L^2-sections of $\$_{\mathbb{C}} \otimes E$ is essentially self-adjoint and satisfies $\ker \not{D}_E = \ker \not{D}_E^2$ (see Theorem II.5.7). This operator satisfies the pointwise formula

$$\not{D}_E^2 = \nabla^*\nabla + \tfrac{1}{4}\kappa + \mathfrak{R}^E \tag{6.1}$$

derived in II.8.17. Since E is flat at infinity, this reduces to the equation $\not{D}_E^2 = \nabla^*\nabla + \frac{1}{4}\kappa$ outside a compact subset of X. Using this fact and some standard arguments from analysis, we prove the following. We say that a function κ is **uniformly positive** on a set A if $\kappa \geqq \kappa_0$, for some constant $\kappa_0 > 0$, on A.

Proposition 6.2. *Suppose X has uniformly positive scalar curvature outside of a compact set. Then the Dirac operator $\not{D}_E$ on $L^2(\$_{\mathbb{C}} \otimes E)$ has a finite-dimensional kernel and a bounded Green's function.*

Corollary 6.3. *The operator*

$$\not{D}_E^+ : L^2(\$_{\mathbb{C}}^+ \otimes E) \longrightarrow L^2(\$_{\mathbb{C}}^- \otimes E) \tag{6.2}$$

and its adjoint $\not{D}_E^-$ have finite-dimensional kernels. Hence the index

$$\operatorname{ind}(\not{D}_E^+) = \dim(\ker \not{D}_E^+) - \dim(\ker \not{D}_E^-) \tag{6.3}$$

is a well-defined integer.

Proof of 6.2. Let $K \subset X$ be a compact set outside of which E is trivialized and $\frac{1}{4}\kappa \geqq \kappa_0$ for some $\kappa_0 > 0$. Let $c > 0$ be a constant such that $c\ \mathrm{Id} \geqq -(\frac{1}{4}\kappa + \mathfrak{R}^E)$ over K. Then for any element $\varphi \in \ker(\not{D}_E)$, formula (6.1) implies that $\nabla^*\nabla\varphi = -\frac{1}{4}\kappa\varphi - \mathfrak{R}^E(\varphi)$. Integrating by parts, and estimating gives the inequality

$$\int_X \|\nabla\varphi\|^2 + \kappa_0 \int_{X-K} \|\varphi\|^2 \leqq c\int_K \|\varphi\|^2. \tag{6.4}$$

Let $\|\cdot\|_A$ denote the L^2-norm over a set $A \subset X$. If we assume $\|\varphi\|_X^2 \,(= \|\varphi\|_K^2 + \|\varphi\|_{X-K}^2) = 1$, then (6.4) implies that

$$\frac{1}{\kappa_0 + c}\|\nabla\varphi\|_X^2 + \frac{\kappa_0}{\kappa_0 + c} \leqq \|\varphi\|_K^2. \tag{6.5}$$

Consider now the **uniform** C^1-norm $\|\varphi\|_{C^1,K}$ over K. By Theorem III.5.4 we know that there exists a constant $C > 0$ such that

$$\|\varphi\|_{C^1,K} \leqq C\|\varphi\|_X \tag{6.6}$$

for all $\varphi \in \ker(D_E)$. Let us fix a number $\varepsilon > 0$ and consider and ε-dense subset of points $\{x_m\}_{m=1}^d$ in K. If $\dim(\ker D_E) \geqq d$, then there exists an element $\varphi \in \ker D_E$ with $\|\varphi\|_X = 1$, such that $\varphi(x_j) = 0$ for $j = 1, \ldots, d$. It then follows from (6.6) that the pointwise norm of φ is uniformly less than $C\varepsilon$ on K, i.e. $\|\varphi\|_{C^0,K} \leqq C\varepsilon$. For ε sufficiently small, this violates (6.5), and we conclude that $\dim(\ker D_E)$ is finite.

The invertibility of the Green's function will not be used directly here. We refer to the reader to Gromov-Lawson [3] for the proof of this fact. ■

Proof of 6.3. With respect to the bundle splitting $\$_{\mathbb{C}} \otimes E = (\$_{\mathbb{C}}^+ \otimes E) \oplus (\$_{\mathbb{C}}^- \otimes E)$ the operator $\not{D}_E$ can be written (pointwise) as

$$\not{D}_E = \begin{pmatrix} 0 & \not{D}_E^- \\ \not{D}_E^+ & 0 \end{pmatrix}$$

where $\not{D}_E^\pm : \Gamma(\$_{\mathbb{C}}^\pm \otimes E) \to \Gamma(\$_{\mathbb{C}}^\mp \otimes E)$. Passing to L^2-sections, we get an orthogonal decomposition $L^2(\$_{\mathbb{C}} \otimes E) = L^2(\$_{\mathbb{C}}^+ \otimes E) \oplus L^2(\$_{\mathbb{C}}^- \otimes E)$, and $\not{D}_E$ continues to interchange factors as above. Consequently

$$\ker \not{D}_E = \ker \not{D}_E^+ \oplus \ker \not{D}_E^-$$

and the finite dimensionality of $\ker \not{D}_E^\pm$ is clear.

Since $\not{D}_E$ is self-adjoint, the operators $\not{D}_E^\pm$ are adjoints of one another, and so $\ker \not{D}_E^- \cong \operatorname{coker} \not{D}_E^+$. ■

Of course we will need a version of the vanishing theorems from the compact case.

Proposition 6.4. *Let X and E be as above, and suppose there is a constant $c > 0$ so that*

$$\tfrac{1}{4}\kappa + \Re^E \geqq c\,\mathrm{Id}. \tag{6.7}$$

on X. Then $\ker \not{D}_E = 0$, *and in particular,* $\operatorname{ind}(\not{D}_E^+) = 0$.

Proof. As in the proof of II.5.7, the completeness of X allows us to integrate by parts. Thus formula (6.1) implies that

$$\|\not{D}_E\varphi\|_X^2 = \|\nabla\varphi\|_X^2 + \int_X \langle(\tfrac{1}{4}\kappa + \Re^E)(\varphi), \varphi\rangle \geqq c\|\varphi\|_X^2$$

for all $\varphi \in L^2(\$ \otimes E)$, and it is evident that $\ker \not{D}_E = 0$. ■

In order to apply the arguments of the previous section, we need something to play the role of the Atiyah-Singer Index Theorem. This will be done by a Relative Index Theorem. We will present here only the special case that we need. More general versions of this result have been proved by J. Cheeger and by H. Donnelly [3].

Theorem 6.5 (Gromov-Lawson [3]). *Let X be a complete even-dimensional spin manifold whose scalar curvature is uniformly positive at infinity. Let E_0 and E_1 be two bundles with compact support on X, and assume that*

$\dim E_0 = \dim E_1$. *Then the difference of the indices of the Dirac operators* $D\!\!\!/^+_{E_0}$ *and* $D\!\!\!/^+_{E_1}$ *is a topological invariant, given by the formula*

$$\operatorname{ind} D\!\!\!/^+_{E_1} - \operatorname{ind} D\!\!\!/^+_{E_0} = \{(\operatorname{ch} E_1 - \operatorname{ch} E_0)\cdot \hat{\mathbf{A}}(X)\}[X]. \tag{6.8}$$

where, via the trivializations at infinity, the class $\operatorname{ch} E_1 - \operatorname{ch} E_0$ *has compact support on* X.

Via Chern-Weil Theory (see Kobayashi-Nomizu [1].). the classes $\operatorname{ch} E_0$, $\operatorname{ch} E_1$, and $\hat{\mathbf{A}}(X)$ can be represented canonically in terms of the curvature forms of the given connections. Formula (6.8) can then be rewritten as

$$\operatorname{ind} D\!\!\!/^+_{E_1} - \operatorname{ind} D\!\!\!/^+_{E_0} = \int_X (\operatorname{ch} E_1 - \operatorname{ch} E_0) \wedge \hat{\mathbf{A}}(X). \tag{6.8}$$

(Recall that E_0 and E_1 are flat outside a compact set.)

Another method to compute the relative index is as follows. Chop off the manifold X outside the support of E_0 and E_1 to obtain a compact manifold $\hat{X}$ with boundary. Let Y be the double of $\hat{X}$ and let E be the bundle on Y given by E_0 on one piece and E_1 on the other piece with E_0 and E_1 glued together at the "seam" $\partial\hat{X}$ by the trivialization. Then the right hand side of (6.8) becomes $\{\operatorname{ch} E \cdot \hat{\mathbf{A}}(Y)\}[Y]$.

An important special case of Theorem 6.5 is where $E_1 = E$ is some bundle with compact support on X and where $E_0 = X \times \mathbb{C}^k$ is the trivialized bundle of dimension $k = \dim E$. Note that $D\!\!\!/_{E_0} = D\!\!\!/ \oplus \cdots \oplus D\!\!\!/$ (k-times) where $D\!\!\!/$ is the usual Atiyah-Singer operator. We then define the **reduced Chern character** of E to be

$$\widehat{\operatorname{ch}}\, E = \operatorname{ch} E - \operatorname{ch} \mathbb{C}^k = \operatorname{ch}^1 E + \operatorname{ch}^2 E + \ldots \tag{6.9}$$

The formula (6.8) then becomes

$$\operatorname{ind}(D\!\!\!/^+_E) - k \operatorname{ind}(D\!\!\!/^+) = \{\widehat{\operatorname{ch}}\, E \cdot \hat{\mathbf{A}}(X)\}[X]. \tag{6.10}$$

For the proof of Theorem 6.5 the reader is referred to Gromov-Lawson [3].

Completion of the proof of Theorems 5.5 and 5.12. Armed with Theorem 6.5 we are able to extend the arguments of the last chapter from the "compactly enlargeable" to the "enlargeable" case. Let X be a compact riemannian manifold with $\kappa \geqq \kappa_0$ for a constant $\kappa_0 > 0$. Let $\varepsilon > 0$ be given and suppose $\tilde{X}$ is a spin covering space which admits an ε-contracting map $f: \tilde{X} \to S^{2m}(1)$ of non-zero $\hat{A}$-degree. Fix a bundle E_0 with connection over $S^{2m}(1)$ with $c_m(E_0) \neq 0$ and set $E = f^*E_0$ with its induced connection. Note that E is a bundle with compact support on X, and so we can construct operators $D\!\!\!/$ and $D\!\!\!/_E$ as above.

Since $\kappa \geqq \kappa_0 > 0$, we have $\text{ind}(\not{D}^+) = 0$. However, for ε sufficiently small, the inequality (6.7) will be satisfied, and we conclude that $\text{ind}(\not{D}_E^+) = 0$. Consequently, by (6. 10) we have

$$\begin{aligned} 0 = \{\widehat{\text{ch}}\, E \cdot \hat{\mathbf{A}}(\tilde{X})\}[\tilde{X}] &= \{c_m(E) \cdot \hat{\mathbf{A}}(\tilde{X})\}[\tilde{X}] \\ &= \{f^*c_m(E_0) \cdot \hat{\mathbf{A}}(\tilde{X})\}[\tilde{X}] \\ &= (\hat{A}\text{-deg})(f)c_m(E_0)[S^{2m}] \end{aligned}$$

contrary to assumption. ∎

Theorem 6.5 allows us to prove a number of results about complete metrics on non-compact manifolds. We begin by focusing on some important facts. The first fact is that in all the arguments of §5 we only used the fact that the mappings f were uniformly contracting on the curvature **2-form** of a connection. This leads to the following notion:

DEFINITION 6.6. A smooth map $f: X \to Y$ between riemannian manifolds is said to be **ε-contracting on 2-forms** or **(ε,Λ²)- contracting** if

$$\|f_*\xi\| \leqq \varepsilon\|\xi\|$$

for all elements $\xi \in \Lambda^2 TX$, or equivalently if

$$\sup_X \|f^*\psi\| \leqq \varepsilon \sup_Y \|\psi\|$$

for any differential 2-form ψ on Y.

The strength of this is that to be (ε,Λ^2)-contracting the map needs only to be contracting in $(n-1)$ of the directions at each point, where $n = \dim(X)$. For example, the linear map $L: \mathbb{R}^n \to \mathbb{R}^n$ with matrix

$$\begin{pmatrix} 1 & & & & 0 \\ & \varepsilon & & & \\ & & \varepsilon & & \\ & 0 & & \ddots & \\ & & & & \varepsilon \end{pmatrix}$$

is (ε,Λ^2)-contracting but not ε-contracting.

All of the discussion of §5 remains valid if the hypothesis "ε-contracting" is replaced by "(ε,Λ^2)-contracting". The real power of this observation will be seen in the non-compact case.

DEFINITION 6.7. An orientable riemannian manifold X is called **(ε,Λ²)-hyperspherical** if there exists an (ε,Λ^2)-contracting map $f: X \to S^m(1)$ which is constant at infinity and of non-zero $\hat{A}$-degree.

The basic argument given above proves the following result.

Theorem 6.8 (Gromov-Lawson [3]). *There is a constant $\varepsilon_n > 0$ so that any spin n-manifold carrying a complete metric with $\kappa \geqq 1$, cannot be $(\varepsilon_n, \Lambda^2)$-hyperspherical.*

The number ε_n can be explicitly estimated for each n.

We now introduce a topological property of manifolds.

DEFINITION 6.9. A manifold X is called **weakly enlargeable** if for each riemannian metric on X and each $\varepsilon > 0$, there exists an orientable covering space of X which is (Λ^2, ε)-hyperspherical in the lifted metric.

Any (compact) enlargeable manifold is weakly enlargeable. However, there are many non-compact examples .

Proposition 6.10. *If X is enlargeable, then $X \times \mathbb{R}$ is weakly enlargeable.*

Proof. Fix a metric g on $X \times \mathbb{R}$ and consider the composition $X \times \mathbb{R} \to \mathbb{R} \to S^1(1)$ where the first map is projection and where the second map is constant ($= *$) outside the interval $(-1,1)$, and of degree 1. Call this composition μ. Since X is compact, μ is δ-contracting for some $\delta \geqq 1$.

Consider also the projection $X \times [-1,1] \to X$, and choose a metric g_0 on X so that this map is 1-contracting from the given metric g on $X \times [-1,1]$.

Fix now an $\varepsilon > 0$ and choose a riemannian covering $(\tilde{X}, \tilde{g}_0)$ of (X, g_0) which admits an (ε/δ)-contracting map $f: \tilde{X} \to S^m(1)$ which is constant ($= *'$) at infinity and of non-zero $\hat{A}$-degree. Extend the map f trivially to $\tilde{X} \times \mathbb{R}$ and note that for the lift $\tilde{g}$ of the original metric g, this map is (ε/δ)-contracting over $\tilde{X} \times [-1,1]$.

Fix now a 1-contracting map $\sigma: S^m(1) \times S^1(1) \to S^{m+1}(1)$ which is constant on the set $(S^m(1) \times \{*\}) \cup (\{*'\} \times S^1(1))$. Then the composition

$$\tilde{X} \times \mathbb{R} \xrightarrow{f \times \mu} S^m(1) \times S^1(1) \xrightarrow{\sigma} S^{m+1}(1) \tag{6.11}$$

is (ε, Λ^2)-contracting with respect to the metric $\tilde{g}$, constant at infinity and of the same $\hat{A}$-degree as f. ∎

We now have, for example, that any manifold of the form $X^n \times \mathbb{R}$, where X^n is compact and carries non-positive curvature, is weakly enlargeable. A basic case is that of $T^n \times \mathbb{R}$.

The property of weak enlargeability is contagious.

Proposition 6.11. *Let U be an open submanifold of X such that the map $\pi_1 U \to \pi_1 X$ is injective. If U is weakly enlargeable, so is X.*

Proof. This is an immediate consequence of the definition of weak enlargeability. ■

Corollary 6.12. *Any manifold X which contains a (compact) enlargeable hypersurface X_0 with $\pi_1 X_0 \to \pi_1 X$ injective is weakly enlargeable.*

Proof. Let U be a tubular neighborhood of X_0 and apply 6.10, 6.11. ■

As an example, let $X = T^n - A$ where A is any closed subset such that $A \cap T^{n-1} = \emptyset$ for some linear hypertorus $T^{n-1} \subset T^n$.

This example gives substance to the following:

Theorem 6.13 (Gromov-Lawson [3]). *A weakly enlargeable manifold cannot carry a complete metric of positive scalar curvature.*

REMARK. J. Kazdan [1] has proved a version of Theorem 5.7 for noncompact complete manifolds. This strengthens Theorem 6.13 by adding the statement: *Any complete metric with $\kappa \geqq 0$ on a weakly enlargeable manifold must be Ricci flat.*

Proof. Note that if the scalar curvature were *uniformly* positive, the result would follow easily from Theorem 6.8. To achieve the uniform positivity of κ we shall multiply by a large 2-sphere. The estimates then become more delicate.

Let X be weakly enlargeable and suppose X carries a complete metric g with $\kappa > 0$. For any given $\varepsilon > 0$, we can find a covering $\tilde{X}$ of X which admits an (ε,Λ^2)-contracting map $f:\tilde{X} \to S^m(1)$, constant at infinity and of non-zero $\hat{A}$-degree. In fact, we may assume this map to be (ε,Λ^2)-contracting *with respect to the (not necessarily complete) metric $g' \equiv \kappa g$.* Hence, the map f is pointwise $\varepsilon\kappa(x)$ contracting on 2-forms with respect to the (lift of the) metric g.

We now take the riemannian product $\tilde{X} \times S^2(r)$ and consider the composition

$$\underbrace{\tilde{X} \times S^2(r) \xrightarrow{f \times (\frac{1}{r})} S^m(1) \times S^2(1) \xrightarrow{\sigma} S^{m+2}(1)}_{F} \qquad (6.12)$$

where σ is a fixed "smashing" map as above (cf. (6.11)). Call this composition F. We want to derive a pointwise estimate for the contraction of F on 2-forms. Since f has compact support on $\tilde{X}$, it is c-contracting on tangent vectors for some $c > 0$. The dilation map on $S^2(r)$ is of course $(1/r)$-contracting on tangent vectors and $(1/r^2)$-contracting on 2-forms. We may assume σ to be essentially 1-contracting. It follows that at any point

(x,y), the map F is $e(x)$-contracting on 2-forms, where $e(x) \equiv \max\{\varepsilon\kappa(x), c/r, 1/r^2\}$. The map F is supported in a compact set K. Set $\kappa_0 = \inf\{\kappa(x): (x,y) \in K \text{ some } y\} > 0$, and choose $r > 1$ so that $\max\{c/r, 1/r^2\} \leqq \varepsilon\kappa_0$. Then the map F will be $\varepsilon\kappa(x)$-contracting on 2-forms at every point (x,y).

We now proceed as before to pull back a fixed bundle with connection from $S^{m+2}(1)$, via the map F. The curvature term (6.7) will be of the form

$$\frac{1}{4}\kappa(x) + \frac{2}{r^2} + \mathfrak{R}^E \tag{6.13}$$

where $E = F^*(E_0)$ is the induced bundle. There is a constant γ depending only on dimensions so that

$$\|\mathfrak{R}^E\| \leqq \gamma\|R^E\|$$

pointwise on $\tilde{X} \times S^2(r)$. However, since F is pointwise $\varepsilon\kappa(x)$-contracting on 2-forms, we see that

$$\|\mathfrak{R}^E\| \leqq \gamma\varepsilon\kappa(x)\|R^{E_0}\|_\infty$$

pointwise, where $\|R^{E_0}\|_\infty$ denotes the C^0-norm over $S^{m+2}(1)$. We may assume that ε satisfies $\varepsilon < \frac{1}{4}\gamma\|R^{E_0}\|_\infty$ since these constants are fixed at the beginning. The curvature term (6.13) then satisfies the inequality

$$\frac{1}{4}\kappa(x) + \frac{2}{r^2} + \mathfrak{R}^E \geqq \frac{2}{r^2},$$

on all of $\tilde{X} \times S^2(r)$. Hence, Proposition 6.4 applies to the twisted Atiyah-Singer operator $\not{D}_E^+$, and we conclude that $\text{index}(\not{D}_E^+) = 0$. Of course, $\text{index}(\not{D}^+) = 0$ for the standard Atiyah-Singer operator. We now apply the Relative Index Theorem as before to complete the proof. To do this, it is necessary that the bundle E_0 over $S^{m+2}(1)$ have a non-trivial Chern character, and therefore that m be even. If m is odd, however, we may simply replace $S^2(r)$ in the construction above, with $S^3(r)$, and everything goes through.

We now complete the argument. Let $k = 2$ or 3, so that $m + k = 2N$, for some integer N. Then since $\text{ind}(\not{D}_E^+) = \text{ind}(\not{D}^+) = 0$, Theorem 6.5 together with formula (6.10) implies that

$$\begin{aligned}
0 &= \{\widehat{\text{ch}}\, E \cdot \hat{\mathbf{A}}(\tilde{X} \times S^k)\}[\tilde{X} \times S^k] \\
&= \{c_N(E) \cdot \hat{\mathbf{A}}(\tilde{X} \times S^k)\}[\tilde{X} \times S^k] \\
&= \{F^*c_N(E_0) \cdot \hat{\mathbf{A}}(\tilde{X} \times S^k)\}[\tilde{X} \times S^k] \\
&= (\hat{A}\text{-deg})(F)c_N(E_0)[S^{2N}] \\
&= (\hat{A}\text{-deg})(f)c_N(E_0)[S^{2N}]
\end{aligned}$$

contrary to assumption. ■

EXAMPLES. As a consequence of Theorem 6.13 we know that there is a large collection of manifolds which cannot carry complete metrics with $\kappa > 0$. This includes $X \times \mathbb{R}$ for any enlargeable manifold X. It also includes manifolds of the form $T^n - C$ where C is any closed subset which misses some geodesic subtorus T^{n-1} of codimension one. (For example, let C be finite.) One can also replace (T^n, T^{n-1}) here by any compact pair (X^n, X^{n-1}) where X^n carries a metric of non-positive sectional curvature in which X^{n-1} is totally geodesic. The property of not admitting complete metrics with $\kappa > 0$ remains after taking products of these manifolds and then taking connected sums with countable families of arbitrary spin manifolds. It persists also after "boundary" connected sums with any open spin manifold.

The "exclusion theorem" 5.6 of the last chapter has a nice generalization to the open case.

Corollary 6.14. *A manifold which carries a complete hyperbolic metric of finite volume cannot carry a complete metric with positive scalar curvature.*

Proof. A hyperbolic manifold X of finite volume has ends of the form $N \times \mathbb{R}$ where N has a finite covering by a nilmanifold and $\pi_1 N \to \pi_1 X$ is injective. Since $N \times \mathbb{R}$ is weakly enlargeable, so is X by Proposition 6.11, and Theorem 6.14 applies. ■

By introducing one further trick we are able to "localize" the results above. To do this we recall the notion of a warped product. Let X_1 and X_2 be riemannian manifolds with metrics g_1 and g_2 respectively, and let $f: X_1 \to \mathbb{R}^+$ be a smooth function. The **warped product of X_1 and X_2 with warping function** f is the cartesian product manifold $X = X_1 \times X_2$ with riemannian metric

$$g \equiv g_1 + f^2 g_2 \tag{6.14}$$

It is an elementary exercise to compute that for such a warped-product metric, the scalar curvature is

$$\kappa = \kappa_1 + \frac{1}{f^2}\left\{\kappa_2 - 2nf\nabla^2 f - n(n-1)\|\nabla f\|^2\right\} \tag{6.15}$$

where $\varkappa_j$ is the scalar curature of X_j and $n = \dim(X_2)$. If we let $X_2 = S^n(1)$, the euclidean n-sphere of radius 1, then on $X = X_1 \times S^n(1)$ we have

$$\kappa = \kappa_1 + \frac{n(n-1)}{f^2}\left(1 - \|\Delta f\|^2\right) - 2n\frac{\nabla^2 f}{f}. \tag{6.16}$$

By choosing f small and constant over a compact set, we can make κ positive there. On a compact manifold X_0 we can then slowly change f to suit our purposes. This is the basic idea in proving the next result.

Theorem 6.15 (Gromov-Lawson [3]). *Let X be a non-compact, connected spin n-manifold containing a compact, connected, orientable hypersurface $X_0 \subset X$. Suppose that X_0 separates X into two components. Suppose further that there is a map $F: X \to Y$, onto an enlargeable $(n-1)$-manifold which, when restricted to X_0, has non-zero degree. Then X carries no complete metric with $|\mathrm{Ric}| \leqq$ constant and with κ uniformly positive outside a compact subset.*

Proof. Suppose that X carries a complete metric with $\kappa \geqq 2$ outside a compact set. Let X_+ and X_- denote the two connected components of $X - X_0$. Since X is non-compact, at least one of these components, say X_+, is unbounded in the given metric. Define a function $\rho: X \to \mathbb{R}$ by setting

$$\rho(x) = \begin{cases} \operatorname{dist}(x, X_0) & \text{for } x \in X_+ \\ -\operatorname{dist}(x, X_0) & \text{for } x \in X_-. \end{cases}$$

By the assumption on Ric there is a smoothing $\tilde{\rho}$ of ρ with $\|\nabla \tilde{\rho}\| \leqq 2$ and with $|\nabla^2 \tilde{\rho}| \leqq C$ for some constant $C \geqq 2$. Fix a number ρ_0 so that $\kappa \geqq 2$ outside the compact set $K \equiv \{x \in : |\tilde{\rho}(x)| \leqq \rho_0\}$. Let ε_n be the number given by Theorem 6.8, and choose $R > 0$ so that $1/R < \varepsilon_n$. Let $\kappa_{\inf} = \inf\{\kappa(x): x \in K\}$, and choose r, $0 < r < 1$, so that $1/r^2 > 1 + |\kappa_{\inf}|$. Choose now a C^∞ function $\phi: \mathbb{R} \to \mathbb{R}$ so that

$$\begin{cases} \phi(t) = r & \text{for } |t| \leqq \rho_0 \\ \phi(t) = R & \text{for } |t| > \rho_0 + 2R/\varepsilon \\ 0 \leqq |\phi'| \leqq \varepsilon & \\ |\phi''| \leqq \varepsilon & \end{cases} \tag{6.17}$$

where ε is the constant

$$\varepsilon \equiv r/3C.$$

We take the warped product of X with $S^2(1)$ using the function f on X given by

$$f(x) \equiv \phi(\tilde{\rho}(x)).$$

By (6.16) the scalar curvature $\hat{\kappa}$ of this metric satisfies: $\hat{\kappa} \geqq |\kappa_{\inf}| + 2/r^2 > 1$ over $K \times S^2$, and $\hat{\kappa} > 2 + 6\varepsilon^2 C^2 r^{-2} > 1$ over $(X - K) \times S^2$. Hence, $\hat{\kappa} > 1$ on all of $X \times S^2$.

We shall now find a covering of $X \times S^2$ which is $(\varepsilon_n, \Lambda^2)$-hyperspherical, in contradiction to Theorem 6.8. To begin we fix a number $b > \rho_0 + 2R/\varepsilon$ and we consider the compact set $\Omega \equiv \{x \in X : b \leqq \tilde{\rho}(x) \leqq b + 4\pi\}$. Fix a metric on Y. Let $\delta \equiv \sup\{\|F_*\|_x : x \in \Omega\}$ and choose a riemannian covering $\pi: \tilde{Y} \to Y$ which admits an (ε_n/δ)-contracting map $g: \tilde{Y} \to S^{n-1}(1)$ which is constant at infinity and of non-zero degree. Taking the fibre product of F and π gives a commutative diagram

$$\begin{array}{ccc} \tilde{X} & \xrightarrow{\tilde{F}} & \tilde{Y} \\ {\scriptstyle \tilde{\pi}}\downarrow & & \downarrow{\scriptstyle \pi} \\ X & \xrightarrow{F} & Y \end{array}$$

where $\tilde{\pi}$ is a covering map and $\tilde{F}$ is proper. Let $\tilde{g} = g \circ \tilde{F}$ and let $\tilde{h} = p \circ \tilde{\rho} \circ \tilde{\pi} : \tilde{X} \to S^1(1)$ where $p : \mathbb{R} \to S^1(1)$ is the degree-1 map given by collapsing everything outside the interval $(b, b + 4\pi)$ to a point. (This map is $(\frac{1}{2})$-contracting). Let $\tilde{G}$ denote the composition

$$\tilde{X} \xrightarrow{\tilde{g} \times \tilde{h}} S^{n-1}(1) \times S^1(1) \xrightarrow{\sigma} S^n(1), \qquad \tilde{G} = \sigma \circ (\tilde{g} \times \tilde{h})$$

where σ is a "smashing" map which collapses the axes $S^{n-1}(1) \times \{*\} \cup \{*\} \times S^1(1)$ to a point. The map $\tilde{G}$ is a constant outside a compact set contained in $\tilde{\Omega} = \tilde{\pi}^{-1}(\Omega)$. Since $\tilde{h}$ is 1-contracting and $\tilde{g}$ is ε_n-contracting, it follows that $\tilde{G}$ is, at every point, $\sqrt{(1 + \varepsilon_n^2)}$-contracting and ε_n-contracting in the hyperplane $\ker(\tilde{h}_*)$.

We now lift the warped metric to $\tilde{X} \times S^2$ and let $\tilde{H}$ denote the composition

$$\tilde{X} \times S^2 \xrightarrow{\tilde{G} \times \mathrm{Id}} S^n(1) \times S^2(1) \xrightarrow{\sigma'} S^{n+2}(1), \qquad \tilde{H} = \sigma' \circ (\tilde{G} \times \mathrm{Id})$$

where σ' is again a "smashing" map. Since the warping function $\tilde{f} = f \circ \tilde{\pi} = \text{constant} = R$ on the support of $\tilde{G}$, and since $1/R < \varepsilon_n$, we see that $\tilde{H}$ is again, like $\tilde{G}$, 1-contracting and "ε_n-contracting in a hyperplane" at each point. In particular, *the map $\tilde{H}$ is ε_n-contracting on 2-forms.*

To complete the proof we must show that $\tilde{H}$ is of non-zero degree. The main point here is the following. Let t be any regular value of $\tilde{\rho}$ and consider the compact manifold $X_t \equiv \tilde{\rho}^{-1}(t)$. Clearly X_t is homologous to $X_0 \equiv \tilde{\rho}^{-1}(0)$, and so by the hypothesis on F, its restriction gives a map $F : X_t \to Y$ of non-zero degree. The rest is straightforward. Set $\tilde{X}_t = \tilde{\pi}^{-1}(X_t)$ and note that the lift $\tilde{F}$ is proper on $\tilde{X}_t$ and has the same degree as F. Now the degree of $\tilde{g} \times \tilde{h}$ equals the degree of $\tilde{g}$ restricted to a regular level set of $\tilde{h}$, and $\deg(\tilde{g}) = \deg(g)\deg(\tilde{F}) \neq 0$. Hence, $\deg(\tilde{G}) = \deg(\tilde{H}) \neq 0$. ∎

There are many interesting applications of Theorem 6.15. For example, let X be an open n-manifold and $X_0 \subset X$ a compact hypersurface.

Suppose X_0 disconnects X and let $\mathscr{E}$ be an unbounded component of $X - X_0$ with $\partial\mathscr{E} = X_0$ (see diagram below).

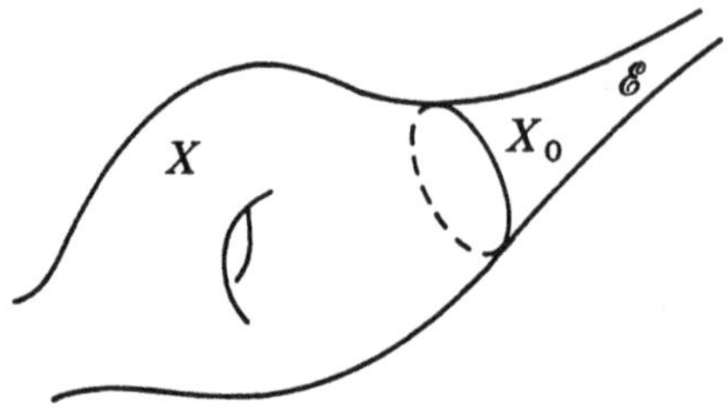

DEFINITION 6.16. We say that $\mathscr{E}$ is a **bad end** if there exists a map $F: \mathscr{E} \to Y$ onto an enlargeable manifold Y so that $F|_{X_0}$ is of non-zero degree.

Theorem 6.17 (Gromov-Lawson [3]). *Suppose X is a spin manifold with a bad end $\mathscr{E}$. Then there is no complete metric on X which has $|\mathrm{Ric}|$ bounded and κ uniformly positive on $\mathscr{E}$.*

Note. Actually, only the end $\mathscr{E}$ needs to be spin.

Proof. Let X be given a complete metric. Consider the double $\mathbb{D}(\mathscr{E}) = \mathscr{E} \cup_{X_0} \mathscr{E}$ with a metric which agrees, outside a neighborhood of the "seam" $\partial\mathscr{E} = X_0$, with the given one. This manifold satisfies all the hypotheses of Theorem 6.15, and therefore cannot have $|\mathrm{Ric}|$ bounded and κ uniformly positive outside the compact neighborhood of X. ∎

Theorem 6.18. *A compact 3-manifold of $K(\pi, 1)$-type cannot carry a metric with positive scalar curvature. In fact, no compact 3-manifold which can be written as a connected sum with a $K(\pi, 1)$-manifold, can carry positive scalar curvature.*

Proof. Let X be a compact $K(\pi, 1)$ 3-manifold. We may assume that X is orientable and therefore spin. Choose an embedded curve $\gamma \subset X$ which is not homotopic to zero, and consider the covering space $\tilde{X} \to X$ corresponding to the cyclic subgroup of $\pi_1 X$ generated by $[\gamma]$. There is a lifting of γ to an embedded curve $\tilde{\gamma} \subset \tilde{X}$ which generates $\pi_1\tilde{X}$. Note that $\tilde{X}$ is a $K(\pi_1\tilde{X}, 1)$-manifold. In particular, $\pi_1\tilde{X} \cong \mathbb{Z}$ and the inclusion $\tilde{\gamma} \subset \tilde{X}$ is a homotopy equivalence. (It is not possible that $\pi_1\tilde{X} = \mathbb{Z}_m$ since $H^{2k}(K(\mathbb{Z}_m, 1); \mathbb{Z}_m) \cong \mathbb{Z}_m$ for all k.)

Suppose now that X carries a metric g with $\kappa > 0$. Then g lifts to a complete metric $\tilde{g}$ on $\tilde{X}$ of uniformly positive scalar curvature with $|\mathrm{Ric}|$ bounded.

Choose a tubular neighborhood $\tilde{U}$ of $\tilde{\gamma}$, and set $\mathscr{E} \equiv \tilde{X} - \tilde{U}$. We claim that $\mathscr{E}$ is a bad end. To prove this, we first show that the inclusion $\partial\mathscr{E} \hookrightarrow \mathscr{E}$ induces an isomorphism on H_1. This is an easy consequence of the Mayer-Vietoris sequence for $\tilde{X} = \tilde{U} \cup \mathscr{E}$ which gives

$$0 = H_2(\tilde{X}) \longrightarrow H_1(\partial\mathscr{E}) \longrightarrow H_1\mathscr{E} \oplus H_1\tilde{U} \longrightarrow H_1\tilde{X} \longrightarrow 0.$$

(Recall that the inclusions $\tilde{\gamma} \subset \tilde{U} \subset \tilde{X}$ are homotopy equivalences.) By general theory the surjective homomorphism $\pi_1\mathscr{E} \to H_1(\mathscr{E}) \cong \mathbb{Z} \oplus \mathbb{Z}$ is represented by a continuous map $F:\mathscr{E} \to S^1 \times S^1 = K(\mathbb{Z} \oplus \mathbb{Z}, 1)$. Restricted to $\partial\mathscr{E} \cong S^1 \times S^1$ this map induces an isomorphism on π_1, and therefore has non-zero degree. Thus, $\mathscr{E}$ is a bad end, and the first statement of Theorem 6.18 is proved.

For the second statement consider a compact oriented 3-manifold $Z = X \# Y$ where X is as above. Let $s: X \# Y \to X$ be a map given by collapsing the Y-summand to a point $* \in X$. Choose $\gamma \subset X - \{*\}$ and pass to the covering $\pi: \tilde{X} \to X$ with the lifted curve $\tilde{\gamma} \subset \tilde{X}$ as above. Taking the fibre product of s and π gives a commutative diagram

$$\begin{array}{ccc} \widetilde{X \# Y} & \xrightarrow{\tilde{s}} & \tilde{X} \\ {\scriptstyle\tilde{\pi}}\downarrow & & \downarrow{\scriptstyle\pi} \\ X \# Y & \xrightarrow{s} & X \end{array}$$

where $\tilde{\pi}$ is a covering map and $\tilde{s}$ is proper. The map $\tilde{s}$ simply consists of collapsing a family of "Y-summands" to the discrete family of points $\pi^{-1}(*)$. The neighborhood $\tilde{U}$ of $\tilde{\gamma}$ can be chosen to lie in $\tilde{X} - \pi^{-1}(*)$ where $\tilde{s}$ is a diffeomorphism. Thus we can lift $\tilde{U}$ back to $\widetilde{X \# Y}$ and consider the end $\mathscr{E}' \equiv \widetilde{X \# Y} - \tilde{s}^{-1}(\tilde{U})$. By restriction we have a proper degree-one map $\tilde{s}: \mathscr{E}' \to \mathscr{E}$ onto the end $\mathscr{E}$ constructed above so that $\tilde{s}: \partial\mathscr{E}' \xrightarrow{\approx} \partial\mathscr{E}$ is a diffeomorphism. Composing with the map $F: \mathscr{E} \to S^1 \times S^1$ above shows that the end $\mathscr{E}'$ is a bad one, and Theorem 6.17 applies as before. ∎

Recall that every compact orientable 3-manifold has a "prime decomposition"

$$X = \Sigma_1 \# \ldots \# \Sigma_m \# (S^1 \times S^2) \# \ldots \# (S^1 \times S^2) \# K_1 \# \ldots \# K_n$$

where the manifolds Σ_i have finite fundamental groups (in SO_4) and where each K_j is a $K(\pi,1)$-manifold (see Milnor [4] and Hempel [1]). The theorem above says that *if X carries positive scalar curvature, then there are no $K(\pi,1)$-factors in the prime decomposition of X.* There are alternative proofs of this fact using minimal surfaces (see Gromov-Lawson [3], Schoen-Yau [1]). If standard conjectures in the theory of 3-manifolds were true, this result combined with Proposition 4.3 would settle the question of which 3-manifolds carry $\kappa > 0$.

Arguing as in the proof of Theorem 5.5, we see that a compact 3-manifold X with $\kappa \geqq 0$, which does not carry $\kappa > 0$, must be flat. There are six such manifolds up to diffeomorphism. (See Calabi [1].)

Results for non-compact 3-manifolds M can be obtained by these methods. We say that M **contains an incompressible surface** if there is an embedding $\Sigma \hookrightarrow M$ of a compact surface of genus > 0, such that the induced homomorphism $\pi_1(\Sigma) \to \pi_1(M)$ is injective. We say that M **carries a small circle** if there is an embedding $S^1 \hookrightarrow M$ whose class is of infinite order in $H_1(M; \mathbb{Z})$ and such that the class of the small normal circle is of infinite order in $H_1(M - S^1; \mathbb{Z})$. Thus, $S^1 \times \mathbb{R}^2$ carries a small circle, but $S^1 \times S^2$ does not. From 6.12 and 6.13 we see that *any 3-manifold which carries an incompressible surface cannot carry a complete metric with* $\kappa > 0$ (a result due originally to Schoen and Yau [6]). Furthermore, *any 3-manifold which carries a small circle cannot carry a complete metric with* $\kappa \geqq 1$. In particular, the manifold $S^1 \times \mathbb{R}^2$ carries no such metric (see Gromov-Lawson [3]).

§7. The Topology of the Space of Positive Scalar Curvature Metrics

Fix a compact m-dimensional manifold X and let $\mathcal{M}(X)$ denote the space of all riemannian metrics on X. Note that $\mathcal{M}(X)$ is a convex cone in the linear space $\Gamma(T^*X \otimes T^*X)$ and is therefore contractible. It is acted on naturally by the group $\mathrm{Diff}(X)$ of diffeomorphisms of X.

We consider here the subspace $\mathcal{P}(X) \subset \mathcal{M}(X)$ of those metrics which have positive scalar curvature. This subspace is stable under $\mathrm{Diff}(X)$, and as we have seen, could possibly be empty. The first results on the topology of $\mathcal{P}(X)$ were due to Hitchin [1] who defined for spin manifolds X a homomorphism

$$h: \pi_{n-1}(\mathcal{P}(X)) \longrightarrow KO^{-n-m}(\mathrm{pt}) \tag{7.1}$$

for each $n \geqq 1$ as follows. Suppose $\alpha \in \pi_{n-1}(\mathcal{P}(X))$ is represented by a map $f: S^{n-1} \to \mathcal{P}(X)$. Since $\mathcal{M}(X)$ is contractible, we can extend this to a map $f: D^n \to \mathcal{M}(X)$ where $S^{n-1} = \partial D^n$. (This extension process realizes the isomorphism $\pi_{n-1}(\mathcal{P}(X)) \xrightarrow{\approx} \pi_n(\mathcal{M}(X), \mathcal{P}(X))$.) Fix a spin structure on X and associate to each metric $f(y)$ the canonical $C\ell_m$-linear Atiyah-Singer operator $\mathfrak{D}_y^+$ for that metric. This gives a family of elliptic operators over D^n which at each point of ∂D^n are invertible (since the scalar curvature is positive). The invertible operators form a contractible space, so we can deform f to be constant on S^{n-1}. Taking ind_m of the resulting family gives an element $h(f) \in KO^{-m}(D^n, S^{n-1}) \cong KO^{-m}(S^n) \cong KO^{-m-n}(\mathrm{pt})$ which depends only on the homotopy class of f.

Hitchin is able to show that his homomorphism (7.1) is often nontrivial by the following means. Fix a metric $\gamma \in \mathcal{P}(X)$ without isometries and consider the embedding $i_\gamma: \mathrm{Diff}(X) \hookrightarrow \mathcal{P}(X)$ given by $i_\gamma(g) = g^*(\gamma)$.

Composing with h leads to a homomorphism $\pi_{n-1}(\text{Diff}(X)) \to KO^{-n-m}(\text{pt})$ whose value on an element $u \in \pi_{n-1}(\text{Diff}(X))$ is $\hat{\mathscr{A}}(Z_u)$ where $Z_u \to S^n$ is the fibre bundle with fibre X constructed by clutching together two copies of $D^n \times X$ along $S^{n-1} \times X$ via u. Using deep results from topology, Hitchin is able to construct sufficiently many non-trivial examples to prove the following:

Theorem 7.1 (Hitchin [1]). *Let X be a compact spin manifold of dimension n with $\mathscr{P}(X) \neq \emptyset$, then*

$$\pi_0(\mathscr{P}(X)) \neq 0 \qquad \text{if } n \equiv 0 \text{ or } 1 \pmod 8$$

$$\pi_1(\mathscr{P}(X)) \neq 0 \qquad \text{if } n \equiv 0 \text{ or } -1 \pmod 8$$

Note that these classes do not survive to the quotient $\mathscr{P}(X)/\text{Diff}(X)$.

A slightly different approach to the study of the topology of $\mathscr{P}(X)$ was given in Gromov-Lawson [3]. The cleanest statements require the Relative Index Theorem 6.5. However, the basic idea is based on the following elementary construction. Fix $\gamma \in \mathscr{P}(X)$ and suppose Y is a compact manifold with $\partial Y = X$. We say that γ **extends to** Y if there exists a metric $\tilde{\gamma} \in \mathscr{P}(Y)$ which coincides with the product metric $\gamma + dt^2$ on a collar neighborhood $U \approx X \times (0,1]$ of the boundary. This extendability depends only on the homotopy class of γ in $\mathscr{P}(X)$. To see this, consider a smooth family of metrics $\gamma_t \in \mathscr{P}(X)$, $t \in \mathbb{R}$, which is constant for $t \leqq 0$ and for $t \geqq 1$. It is easily checked that the metric $\gamma_{t/c} + dt^2$ on $X \times \mathbb{R}$ has $\kappa > 0$ for all sufficiently large constants c. Adding the collar $X \times [0, c]$ (with this metric) to Y shows that γ_0 extends to Y if and only if γ_1 does.

Proposition 7.2. *Suppose that X is a spin manifold. Let $\{\gamma_\alpha\}_{\alpha \in A} \subset \mathscr{P}(X)$ be a family of metrics, and suppose that there exists an associated family $\{Y_\alpha\}_{\alpha \in A}$ of compact spin manifolds with $\partial Y_\alpha = X$ such that γ_α extends over Y_α for each $\alpha \in A$. If*

$$\hat{A}(Y_\alpha \cup_X (-Y_\beta)) \neq 0 \qquad \text{for all } \alpha \neq \beta,$$

then γ_α is not homotopic to γ_β in $\mathscr{P}(X)$ for all $\alpha \neq \beta$.

Proof. If γ_α is homotopic to γ_β in $\mathscr{P}(X)$, then γ_α extends over both Y_α and Y_β. The extended metrics fit together to give a metric of positive scalar curvature on $Y_\alpha \cup (-Y_\beta)$, and therefore $\hat{A}(Y_\alpha \cup (-Y_\beta)) = 0$. ∎

The "extension pairs" $(\gamma_\alpha, Y_\alpha)$ discussed above are not difficult to construct. There are two good sources.

Lemma 7.3. *Let $\pi: V \to X$ be a riemannian vector bundle of (fibre) dimension k over a compact manifold X, and let $D(V) \equiv \{v \in V : \|v\| = 1\}$. If $k > 2$, then there exists a metric $\gamma \in \mathscr{P}(\partial D(V))$ which extends over $D(V)$.*

Proof. Choose an O_k-invariant metric γ_0 on $\mathbb{R}^k$ which has $\kappa \geqq 1$ and is isometric to the standard product metric on $S^{k-1} \times [1,\infty)$ outside the unit disk. Choose an orthogonal connection on V and let $\mathscr{H}$ denote the corresponding field of horizontal planes on V. Choose any metric γ_1 on X and lift this metric to $\mathscr{H}$ via π. Introduce the metric γ_0 on the fibres of V. Using the formulas of O'Neill [1], the sum $\gamma_1 + t^2\gamma_0$ is seen to have positive scalar curvature for all $t > 0$ sufficiently small. ∎

More elaborate arguments prove the following in any manifold Y:

Theorem 7.4 (Carr [1]). *The boundary of a regular neighborhood U of any smoothly embedded finite subcomplex of codimension > 2 in Y, admits a metric of positive scalar curvature which extends over U.*

EXAMPLE 7.5. Let $V \to S^4$ be a real vector bundle of dimension four with Euler number χ and Pontrjagin number p_1. It is elementary that if $\chi = \pm 1$, the manifold $\Sigma_V = \partial D(V)$ is a homotopy sphere. Milnor showed that $\Sigma_V = S^7$, the standard 7-sphere, if and only if $p_1^2 \equiv 4 \pmod{896}$. Let V_k be the bundle with $\chi = 1$ and $p_1 = 4 + 896k$. Calculations in Milnor [7] show that

$$\hat{A}(V_k \cup D^8) = k \qquad \text{for each } k.$$

From Lemma 7.3, for each k we can construct a metric γ_k on S^7 which extends over V_k. By Proposition 7.2 these metrics are mutually non-homotopic in $\mathscr{P}(S^7)$ and so we have that

$$\pi_0(\mathscr{P}(S^7)) \text{ is infinite.}$$

Since $\pi_0(\text{Diff}\, S^7)$ is finite, we also have that

$$\pi_0(\mathscr{P}(S^7)/\text{Diff}(S^7)) \text{ is infinite.}$$

The analogous comments apply to each of the non-standard Milnor spheres Σ_V.

EXAMPLE 7.6. Fix an integer $n > 1$ and let Y denote the compact $4n$-manifold with boundary constructed by plumbing together eight copies of the tangent disk bundle of S^{2n} according to the Dynkin diagram for E_8. Let Y_k denote the boundary connected sum of k-copies of Y. It is elementary to see that $\Sigma = \partial Y$ is a homotopy sphere. By the finiteness of the group of homotopy spheres, there is an integer $m = m(n)$ such that $\partial Y_{km} = \Sigma \# \cdots \# \Sigma$ (km-times) $\cong S^{4n-1}$ for all k. Using Theorem 7.4, we can construct for each k a metric $\gamma_k \in \mathscr{P}(S^{4n-1})$ which extends over Y_{km}. On the other hand, as shown in Carr [1],

$$\hat{A}(Y_{km} \cup (-Y_{k'm})) \neq 0 \qquad \text{if } k \neq k'. \tag{7.2}$$

In fact, since $Y_{km} \cup (-Y_{k'm})$ is a $(2n-1)$-connected $4n$-manifold, its $\hat{A}$-genus and its L-genus are both certain universal (non-zero) multiples of p_n, the nth Pontrjagin number. It suffices therefore to show that the signature $\neq 0$. Since $\mathrm{sig}(Y) = 1$, we see that $\mathrm{sig}(Y_{km} \cup (-Y_{k'm})) = (k - k')m$. This establishes (7.2) and from Proposition 7.2 we conclude that

$$\pi_0(\mathscr{P}(S^{4n-1})) \text{ and } \pi_0(\mathscr{P}(S^{4n-1})/\mathrm{Diff}(S^{4n-1})) \text{ are infinite}$$

for all $n > 1$. From this we can deduce the following:

Theorem 7.7. *Let X be any compact spin manifold of dimension $4n - 1 \geqq 7$ with $\mathscr{P}(X) \neq \emptyset$. Then $\pi_0(\mathscr{P}(X))$ is infinite.*

Proof. Let $Z_k = (X \times [0,1]) \natural Y_{km}$ where $\natural$ denotes boundary connected sum. Note that $\partial Z_k = X \amalg (X \# S^{4n-1}) \approx X \amalg X$. Fix a metric $\gamma \in \mathscr{P}(X)$ and take connected sum with the metric γ_k constructed above on the component "$X \# S^{4n-1}$" (cf. 4.3). This metric extends over Z_k. (Essentially this extension is constructed by taking the connected sum of the product metric on $X \times [0,1]$ with that on Y_{km} and modifying it; see Carr [1] for example.) Since $Z_k \cup (-Z_{k'}) = (X \times S^1) \# (Y_{km} \cup (-Y_{k'm}))$, we have $\hat{A}(Z_k \cup (-Z_{k'})) = \hat{A}(Y_{km} \cup (-Y_{k'm})) \neq 0$ and 4.3 applies. ■

The arguments given above can be nicely encapsuled by using the Relative Index Theorem. Let X be a compact spin manifold of dimension $4n - 1$. For each pair of metrics $g_0, g_1 \in \mathscr{P}(X)$ we define a relative index $i(g_0, g_1) \in \mathbb{Z}$ as follows. Introduce on $X \times \mathbb{R}$ a metric g which equals the product metric $g_0 + dt^2$ on $X \times (-\infty, 0]$ and $g_1 + dt^2$ on $X \times [1, \infty)$. Set

$$i(g_0, g_1) \equiv \mathrm{ind}(\not{D}^+)$$

where $\not{D}^+ : \Gamma(\$^+) \to \Gamma(\$^-)$ is the Atiyah-Singer operator on $X \times \mathbb{R}$. By Theorem 6.5 this index is independent of the choice of g in $X \times [0,1]$. If g_0 is homotopic to g_1 in $\mathscr{P}(X)$, then we can choose g to have $\kappa > 0$ (as above) and so $i(g_0, g_1) = 0$. Thus, $i(g_0, g_1)$ measures components in $\mathscr{P}(X)$. It is shown in Gromov-Lawson [3] that

$$i(g_0, g_1) + i(g_1, g_2) + i(g_2, g_0) = 0$$

for all $g_0, g_1, g_2 \in \mathscr{P}(X)$.

Suppose further that $X = \partial Y$. Then given $g \in \mathscr{P}(X)$ we introduce a complete metric on $Y - \partial Y$ which is the product $g + dt^2$ on the boundary collar $\partial Y \times [0, \infty) = X \times [0, \infty)$. Using the Atiyah-Singer operator for this metric we define an invariant:

$$i(g, Y) \equiv \mathrm{ind}(\not{D}^+)$$

which by 6.5 is independent of the extension and which has the property that

$$i(g,Y) = 0 \qquad \text{if } g \text{ extends over } Y.$$

By the Relative Index Theorem we have that

$$i(g,Y) - i(g,Y') = \hat{A}(Y \cup (-Y')).$$

Using Browder-Novikov Theory and proceeding as above, one can detect non-trivial elements in $\pi_j(\mathscr{P}(X))$ for higher values of j.

§8. Clifford Multiplication and Kähler Manifolds

Until now we have paid little attention to complex manifolds. This is not because the topic is irrelevant to spin geometry. In fact combining complex and spin structures yields the richer study of Spin^c manifolds, which are discussed in Appendix D. In this chapter we restrict attention to the fundamental case of Kähler manifolds.

If Clifford multiplication enters naturally into riemannian geometry and leads to basic identities, then in Kählerian geometry it should enter in an even more interesting way. This is indeed true. For Kähler manifolds there is a rich algebraic formalism which relates Clifford multiplication and the complex structure. We shall sketch here the principal results. We refer the reader to Michelsohn [1] for complete details.

Let X be a $2n$-dimensional manifold equipped with an almost complex structure, that is, equipped with a bundle automorphism $J: TX \to TX$ such that $J^2 = -\mathrm{Id}$.

DEFINITION 9.1. A riemannian metric $\langle \cdot, \cdot \rangle$ on X is said to be **Kählerian** if J is pointwise orthogonal, i.e., $\langle JV, JW \rangle = \langle V, W \rangle$ for all $V, W \in T_xX$ at all points x, and if

$$\nabla J = 0$$

where ∇ denotes the canonical riemannian connection.

Note. It is a basic fact that if X admits a Kähler metric, then the almost complex structure is integrable, i.e., it comes from a system of holomorphically related coordinate charts on X.

Suppose now that X carries such a metric and let $D: \Gamma(\mathbb{C}\ell(X)) \to \Gamma(\mathbb{C}\ell(X))$ denote the associated Dirac operator on the complexified Clifford bundle $\mathbb{C}\ell(X) \equiv C\ell(TX) \otimes_{\mathbb{R}} \mathbb{C}$ of X (considered as a real manifold). The first interesting fact is that there is a natural decomposition

$$D = \mathscr{D} + \bar{\mathscr{D}}$$

where $\mathscr{D}$ and $\bar{\mathscr{D}}$ are first-order operators which are formal adjoints of one another ($\mathscr{D}^* = \bar{\mathscr{D}}$) and satisfy

$$\mathscr{D}^2 = 0, \qquad \bar{\mathscr{D}}^2 = 0. \tag{8.1}$$

Similarly, the zero-order operator L defined in Chapter II can be expanded into a pair of operators $\mathscr{L}$ and $\bar{\mathscr{L}}$ which are formal adjoints of one another and which canonically generate an $\mathfrak{sl}_2(\mathbb{C})$-subalgebra. In particular, if we set $\mathscr{H} = [\mathscr{L}, \bar{\mathscr{L}}]$, then the three endormorphisms of $\mathbb{C}\ell(X)$ satisfy the identities

$$[\mathscr{H}, \mathscr{L}] = 2\mathscr{L}, \qquad [\mathscr{H}, \bar{\mathscr{L}}] = -2\bar{\mathscr{L}}, \qquad [\mathscr{L}, \bar{\mathscr{L}}] = \mathscr{H}. \tag{8.2}$$

These operators are defined by setting

$$\mathscr{D}\varphi = \sum_j \varepsilon_j \nabla_{\bar{\varepsilon}_j}\varphi \qquad \bar{\mathscr{D}}\varphi = \sum_j \bar{\varepsilon}_j \nabla_{\varepsilon_j}\varphi$$

$$\mathscr{L}\varphi = -\sum_j \varepsilon_j \varphi \bar{\varepsilon}_j \qquad \bar{\mathscr{L}}\varphi = -\sum_j \bar{\varepsilon}_j \varphi \varepsilon_j$$

where

$$\varepsilon_j = \tfrac{1}{2}(e_j - iJe_j) \qquad \text{and} \qquad \bar{\varepsilon}_j = \tfrac{1}{2}(e_j + iJe_j)$$

for any local orthonormal frame field of the form $e_1, Je_1, \ldots, e_n, Je_n$. These complex vector fields satisfy the "supersymmetry" relations

$$\varepsilon_i\varepsilon_j + \varepsilon_j\varepsilon_i = \bar{\varepsilon}_i\bar{\varepsilon}_j + \bar{\varepsilon}_j\bar{\varepsilon}_i = 0$$

$$\varepsilon_i\bar{\varepsilon}_j + \bar{\varepsilon}_j\varepsilon_i = -\delta_{ij}$$

The endomorphism $J: TX \to TX$ extends to $\mathbb{C}\ell(X)$ as a $\mathbb{C}$-linear map which is a derivation with respect to Clifford multiplication. If we set $\mathscr{J} \equiv -iJ$, then the commuting endomorphisms $\mathscr{J}$ and $\mathscr{H}$ are each diagonalizable with integral eigenvalues, and we can define the subbundles

$$\mathbb{C}\ell^{p,q}(X) \equiv \{\varphi \in \mathbb{C}\ell(X) : \mathscr{J}(\varphi) = p\varphi \text{ and } \mathscr{H}(\varphi) = q\varphi\}$$

The representation theory for $\mathfrak{sl}_2(\mathbb{C})$ is used to show that these bundles are non-zero exactly for pairs $(p,q) \in \mathbb{Z} \times \mathbb{Z}$ with $|p + q| \leqq n$ and $p + q \equiv n \pmod 2$. This gives us the bigraded "Clifford diamond" below.

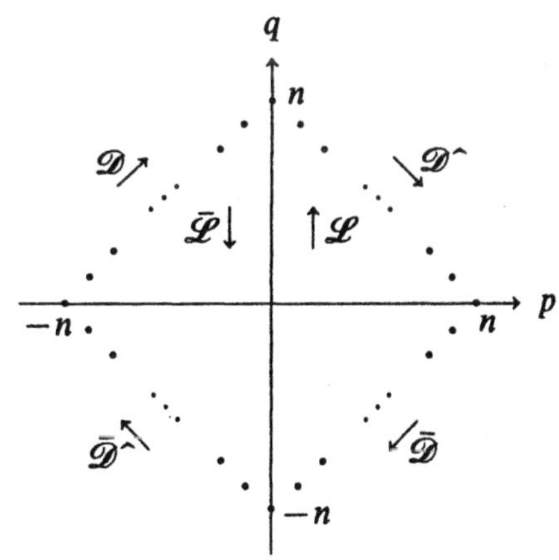

with "raising" and "lowering" operators:

$$\mathscr{L}:\mathbb{C}\ell^{p,q}(X)\longrightarrow \mathbb{C}\ell^{p,q+2}(X),\qquad \bar{\mathscr{L}}:\mathbb{C}\ell^{p,q}(X)\longrightarrow \mathbb{C}\ell^{p,q-2}(X),$$

and with the differential operators $\mathscr{D}$ and $\bar{\mathscr{D}}$ acting diagonally

$$\mathscr{D}:\Gamma(\mathbb{C}\ell^{p,q})\longrightarrow \Gamma(\mathbb{C}\ell^{p+1,q+1}),\qquad \bar{\mathscr{D}}:\Gamma(\mathbb{C}\ell^{p,q})\longrightarrow \Gamma(\mathbb{C}\ell^{p-1,q-1}).$$

The bigrading is nicely related to multiplication. One has that

$$\mathbb{C}\ell^{p,*}\cdot\mathbb{C}\ell^{p',*}\subset\mathbb{C}\ell^{p+p',*}$$

for all p,p', and furthermore that

$$\begin{cases}\mathbb{C}\ell^{p,q}\cdot\mathbb{C}\ell^{p',q'}=\{0\} & \text{if } q-q'\neq p+p', \text{ and}\\ \mathbb{C}\ell^{p,q}\cdot\mathbb{C}\ell^{p',q-(p+p')}\subset\mathbb{C}\ell^{p+p',q-p'}\end{cases}$$

for all p,p',q,q'. These facts follow easily from the identities

$$\mathscr{J}\varphi=\omega\varphi-\varphi\omega;\qquad \mathscr{H}\varphi=\omega\varphi+\varphi\omega$$

where by definition $2i\omega=\sum e_j\cdot Je_j$ for any orthonormal tangent frame $\{e_1,Je_1,\dots,e_n,Je_n\}$ as above.

We now examine the differential operators. It is easily seen that when restricted to any diagonal line in the Clifford diamond, $\mathscr{D}$ gives an **elliptic** complex

$$\cdots\xrightarrow{\mathscr{D}}\Gamma(\mathbb{C}\ell^{p-1,q-1})\xrightarrow{\mathscr{D}}\Gamma(\mathbb{C}\ell^{p,q})\xrightarrow{\mathscr{D}}\Gamma(\mathbb{C}\ell^{p+1,q+1})\xrightarrow{\mathscr{D}}\cdots$$

DEFINITION 8.2. The quotient $\mathscr{H}^{*,*}(X)\equiv\ker\mathscr{D}/\operatorname{Im}\mathscr{D}$ is called the **Clifford cohomology** of X.

The general Hodge Decomposition Theorem gives us the following. We define a $\mathscr{D}$-laplacian Δ by

$$\Delta\equiv\mathscr{D}\bar{\mathscr{D}}+\bar{\mathscr{D}}\mathscr{D},$$

and consider the associated **harmonic spaces**

$$\mathbf{H}^{p,q}(X)\equiv(\ker\Delta)\cap\Gamma(\mathbb{C}\ell^{p,q}(X)).$$

Theorem 8.3. *If X is compact, then $\mathscr{H}^{*,*}(X)$ is finite dimensional and there are natural isomorphisms*

$$\mathscr{H}^{p,q}(X)\cong\mathbf{H}^{p,q}(X)$$

for all p,q.

REMARK 8.4. If W is a holomorphic vector bundle over X, then the algebraic formalism above carries over easily to $\mathbb{C}\ell(X)\otimes_{\mathbb{C}}W$ and leads to Clifford cohomology groups $\mathscr{H}^{p,q}(X;W)$ with coefficients in W. If X is compact, Hodge Theory applies to give finite dimensional harmonic spaces $\mathbf{H}^{p,q}(X;W)$ isomorphic to $\mathscr{H}^{p,q}(X;W)$ for each p,q.

We assume from this point forward that X is compact. Our main point now is that the Clifford formalism of Chapter I leads easily to interesting structure in Clifford cohomology. The simplest operation is that of complex conjugation $c:\mathbb{C}\ell(X) \to \mathbb{C}\ell(X)$. One easily checks that $c \circ \mathscr{J} \circ c = -\mathscr{J}$ and that $c \circ \mathscr{H} \circ c = -\mathscr{H}$. This leads to the following "Serre Duality" Theorem:

Theorem 8.5 (Michelsohn [1]). *Complex conjugation in $\mathbb{C}\ell(X)$ induces isomorphisms*

$$\mathscr{H}^{p,q}(X) \cong \mathscr{H}^{-p,-q}(X)$$

for all p,q. More generally, there are isomorphisms

$$\mathscr{H}^{p,q}(X;W) \cong \mathscr{H}^{-p,-q}(X;W^*)$$

for any holomorphic vector bundle W over X.

Let $\bigstar:\mathbb{C}\ell(X) \to \mathbb{C}\ell(X)$ denote the $\mathbb{C}$-linear antiautomorphism given by the transpose. Recall that for generators $v_1,\dots,v_p \in T_xX$, we have $\bigstar(v_1 \cdots v_p) = v_p \cdots v_1$.

Theorem 8.6 (Michelsohn [1]). *The transpose antiautomorphism induces isomorphisms*

$$\bigstar:\mathbf{H}^{p,q}(X) \longrightarrow \mathbf{H}^{p,-q}(X)$$

for all p,q.

Clifford multiplication leads to the following interesting duality. Let $(\cdot,\cdot)$ denote the usual L^2-inner product on sections.

Theorem 8.7 (Michelsohn [1]). *The $\mathbb{C}$-bilinear pairing β defined on* $\mathbf{H}^{p,q}(X) \times \mathbf{H}^{p,-q}(X)$ by

$$\beta(\varphi,\psi) = (\varphi \cdot \psi, 1)$$

is non-degenerate.

In all of the above we have considered $\mathbb{C}\ell(X)$ as a left module over itself. Considering $\mathbb{C}\ell(X)$ as a right module produces a different Dirac operator $D^\wedge$ (as in II.5) and a corresponding decomposition $D^\wedge = \mathscr{D}^\wedge + \bar{\mathscr{D}}^\wedge$. These are related to $\mathscr{D}$ and $\bar{\mathscr{D}}$ by the antiautomorphism $\bigstar$:

$$\mathscr{D}^\wedge = \bigstar \circ \mathscr{D} \circ \bigstar \qquad \text{and} \qquad \bar{\mathscr{D}}^\wedge = \bigstar \circ \bar{\mathscr{D}} \circ \bigstar.$$

This leads to cohomology groups $\mathscr{H}^{*,*}(X;W)^\wedge$, and there are isomorphisms

$$\mathscr{H}^{p,q}(X;W) \cong \mathscr{H}^{p,-q}(X;W)^\wedge$$

for any holomorphic vector bundle W. The operators $\mathscr{D}$ and $\bar{\mathscr{D}}$ enter naturally when considering the $\mathfrak{sl}_2(\mathbb{C})$-structures

Proposition 8.8. *Let W be a holomorphic vector bundle over X. Then the following relations hold on $\mathbb{C}\ell(X) \otimes W$:*

$$\mathscr{D}\mathscr{L} + \mathscr{L}\mathscr{D} = 0 \qquad \bar{\mathscr{D}}\bar{\mathscr{L}} + \bar{\mathscr{L}}\bar{\mathscr{D}} = 0$$

$$\mathscr{D}\bar{\mathscr{L}} + \bar{\mathscr{L}}\mathscr{D} = \mathscr{D}^{\wedge} \qquad \bar{\mathscr{D}}\mathscr{L} + \mathscr{L}\bar{\mathscr{D}} = \bar{\mathscr{D}}^{\wedge}$$

$$[\mathscr{H},\mathscr{D}] = \mathscr{D} \qquad [\mathscr{H},\bar{\mathscr{D}}] = \bar{\mathscr{D}}$$

$$[\mathscr{H},\Delta] = 4\Delta$$

Theorem 8.9 (Michelsohn [1]). *The cohomology groups $\mathscr{H}^{p,q}(X;W)$ admit an intrinsic filtration*

$$\{0\} \subseteq \mathscr{F}_1^{p,q} \subseteq \mathscr{F}_2^{p,q} \subseteq \ldots \subseteq \mathscr{H}^{p,q}(X;W)$$

where

$$\mathscr{F}_k^{p,q} \equiv \{[\varphi] \in \mathscr{H}^{p,q}(X;W) : \mathscr{L}^k\varphi = 0\}.$$

Furthermore, the representation of the Lie algebra $\langle \mathscr{H},\mathscr{L},\bar{\mathscr{L}} \rangle$ preserves the subspaces

$$J^{p,q}(X;W) \equiv \ker(\Delta) \cap \ker(\Delta^{\wedge}) \cap \Gamma\mathbb{C}\ell^{p,q}(X;W) = \mathbf{H}^{p,q}(X;W) \cap \mathbf{H}^{p,q}(X;W)^{\wedge}$$

In the basic case when W is trivial it is shown that $\Delta = \Delta^{\wedge}$, and so we have the following:

Theorem 8.10. *The Lie algebra $\langle \mathscr{H},\mathscr{L},\bar{\mathscr{L}} \rangle$ acting on $\Gamma\mathbb{C}\ell(X)$ preserves the subspace* $\ker(\Delta) = \mathbf{H}^{*,*}(X)$. *Hence, there is a canonical $\mathfrak{sl}_2(\mathbb{C})$-structure on the Clifford cohomology of X.*

The induced operators on cohomology are of the form

$$\mathscr{H}\big|_{\mathbf{H}^{p,q}(X)} = q$$

$$\mathscr{L} : \mathbf{H}^{p,q}(X) \longrightarrow \mathbf{H}^{p,q+2}(X) \qquad \bar{\mathscr{L}} : \mathbf{H}^{p,q}(X) \longrightarrow \mathbf{H}^{p,q-2}(X).$$

DEFINITION 8.11. A class $\varphi \in \mathbf{H}^{p,q}(X)$ is called **primitive** if $\mathscr{L}\varphi = 0$.

Theorem 8.12 (Michelsohn [1]). *For all $q > 0$ and all p, the qth power $\bar{\mathscr{L}}^q : \mathbf{H}^{p,q}(X) \to \mathbf{H}^{p,-q}(X)$ is an isomorphism. Every $\varphi \in \mathbf{H}^{p,q}(X)$ can be written uniquely as*

$$\varphi = \sum_{k \geq 0} \bar{\mathscr{L}}^k \varphi_k$$

where $\varphi_k \in \mathbf{H}^{p,q+2k}(X)$ is primitive.

One might imagine that there is a direct relationship between Clifford cohomology and the standard Dolbeault cohomology of X. This is indeed true. There is an explicit element in the Hodge automorphism group which relates the two. This is computed in Michelsohn [1]. It shows in particular that

$$\mathbf{H}^{r-s,\, n-r-s}(X;W) \cong \mathbf{H}^{s,r}_{\mathrm{Dol}}(X;W) \cong H^r(X;\Omega^s(W)),$$

and so the Clifford cohomology is independent of the Kähler metric chosen on X. The role played by Clifford multiplication in $\mathbf{H}^{*,*}(X)$ does however depend on the metric.

§9. Pure Spinors, Complex Structures, and Twistors

Thus far we have said relatively little about the role of spinors in local riemannian geometry although there remains much to be discovered in this area. There are, however, several places where spinors enter naturally. For example there is a notion due to É. Cartan of spinors of "pure" type which are related to almost complex structures. These spinors are also related to the calibrations introduced in Harvey-Lawson [3]. The interesting fact is that the simplest spinors give rise, under "squaring," to the most complicated differential forms.

Whenever there is a parallel spinor on a manifold, there is a reduction of the holonomy group. Via Bochner's method, spinors play a central role in constructing and understanding manifolds with reduced holonomy. This is particularly true of the exceptional cases of G_2 and Spin_7 holonomy.

This section and the next one are devoted to a discussion of the topics just mentioned. We begin with the notion of a pure spinor. Fix $\mathbb{R}^n$ with its standard inner product $\langle\cdot,\cdot\rangle$ and extend this metric $\mathbb{C}$-linearly to $\mathbb{C}^n = \mathbb{R}^n \otimes_{\mathbb{R}} \mathbb{C}$. Let $\mathbb{C}\ell_n = C\ell_n \otimes_{\mathbb{R}} \mathbb{C}$ be the associated Clifford algebra, and let $\$_{\mathbb{C}}$ be a fundamental $\mathbb{C}\ell_n$-module, i.e., an irreducible complex spinor space. For each spinor $\sigma \in \$_{\mathbb{C}}$, we consider the $\mathbb{C}$-linear map

$$j_\sigma : \mathbb{C}^n \longrightarrow \$_{\mathbb{C}} \qquad \text{given by } j_\sigma(v) \equiv v \cdot \sigma. \tag{9.1}$$

Generically, this map is injective. However, there are interesting spinors for which $\dim(\ker j_\sigma) > 0$, and it is these that we shall now examine. We begin with the following definition.

DEFINITION 9.1. A complex subspace $V \subset \mathbb{C}^n$ is said to be **isotropic** (with respect to the bilinear form $\langle\cdot,\cdot\rangle$) if $\langle v,w\rangle = 0$ for all $v,w \in V$.

We define a hermitian inner product $(\cdot,\cdot)$ on $\mathbb{C}^n$ by setting $(v,w) = \langle v,\bar{w}\rangle$. Clearly, if $V \subset \mathbb{C}^n$ is an isotropic subspace, then $V \perp \bar{V}$ in this hermitian inner product. In particular, therefore, we have

$$2 \dim_{\mathbb{C}} V \leqq n. \tag{9.2}$$

Lemma 9.2. *For any non-zero spinor σ, the subspace* $\ker j_\sigma$ *is isotropic.*

Proof. If $v \cdot \sigma = w \cdot \sigma = 0$, then $(v \cdot w + w \cdot v) \cdot \sigma = -2\langle v,w\rangle\sigma = 0$. ∎

DEFINITION 9.3. A spinor σ is **pure** if $\ker j_\sigma$ is a maximal isotropic subspace, i.e., if $\dim(\ker j_\sigma) = [n/2]$.

Denote by $P\$$ the subset of pure spinors in $\$_{\mathbb{C}}$, and denote by $\mathscr{I}_n$ the set of maximal isotropic subspaces of $\mathbb{C}^n$. Both $P\$$ and $\mathscr{I}_n$ are naturally acted upon by the group Pin_n, and the assignment $\sigma \mapsto \ker j_\sigma$ gives a Pin_n-equivariant map

$$K: P\$ \longrightarrow \mathscr{I}_n. \tag{9.3}$$

To see that K is equivariant note that for $v \in \mathbb{C}^n$ and $\sigma \in S_{\mathbb{C}}$, we have $g(v)\cdot g \cdot \sigma = g \cdot v \cdot g^{-1} \cdot g \cdot \sigma = g \cdot v \cdot \sigma$ for all $g \in \mathrm{Pin}_n \subset C\ell_n^\times$. Hence, $\ker j_{g\sigma} = g(\ker j_\sigma)$ as claimed.

REMARK 9.4. Note that in fact we can define a Pin_n-invariant "filtration" of the spinor spaces

$$\$_0 \subset \$_1 \subset \$_2 \subset \ldots \subset \$_{[n/2]} = \$_{\mathbb{C}} \tag{9.4}$$

where $\$_k = \{\sigma : \dim(\ker j_\sigma) \geqq [n/2] - k\}$ for each k. The subset $P\$ = \$_0 - \{0\}$ consists exactly of the pure spinors. The subsets $\$_k$ are *not* linear subspaces. As we shall see, there are orthonormal bases of $\$_{\mathbb{C}}$ which consist entirely of pure spinors.

At this point our discussion divides as usual into the two cases, n even and n odd. We shall discuss in detail only the even case. (The reader can easily carry over the arguments and constructions to the odd case.) From this point on we shall assume that $n = 2m$ is an even integer, and furthermore that $\mathbb{R}^{2m}$ is oriented.

DEFINITION 9.5. An **orthogonal almost complex structure** on $\mathbb{R}^{2m}$ is an orthogonal transformation $J: \mathbb{R}^{2m} \to \mathbb{R}^{2m}$ which satisfies $J^2 = -\mathrm{Id}$. For any such J, an associated **unitary basis** of $\mathbb{R}^{2m}$ is an ordered orthonormal basis of the form $\{e_1, Je_1, \ldots, e_m, Je_m\}$. Any two unitary bases for a given J determine the same orientation. This is called the **canonical orientation** associated to J.

Let $\mathscr{C}_m$ denote the set of all orthogonal almost complex structures on $\mathbb{R}^{2m}$. It is easily seen that $\mathscr{C}_m$ is a homogeneous space for the group O_{2m}. It falls into two connected components $\mathscr{C}_m^+$ and $\mathscr{C}_m^-$ where $\mathscr{C}_m^+ \cong \mathrm{SO}_{2m}/\mathrm{U}_m$ consists of those almost complex structures whose canonical orientation is **positive** (i.e., agrees with the given one on $\mathbb{R}^{2m}$).

Associated to any $J \in \mathscr{C}_m$ there is a decomposition

$$\mathbb{C}^{2m} = V(J) \oplus \overline{V(J)}, \qquad \text{where}$$

$$V(J) \equiv \{v \in \mathbb{C}^{2m} : Jv = -iv\} = \{v_0 + iJv_0 : v_0 \in \mathbb{R}^{2m}\}. \tag{9.5}$$

For any $v_0 \in \mathbb{R}^{2m}$ we have that $\langle v_0 + iJv_0, v_0 + iJv_0\rangle = \langle v_0, v_0\rangle - \langle Jv_0, Jv_0\rangle + 2i\langle v_0, Jv_0\rangle = 0$, and so the space $V(J)$ is isotropic. Conversely, given any m-dimensional isotropic subspace $V \subset \mathbb{C}^{2m}$, there is a unique $J \in \mathscr{C}_m$ so that $V = V(J)$. To see this note that if one writes $\mathbb{C}^{2m} = R^{2m} \oplus iR^{2m}$, then $V(J)$ is just the graph of J. Hence, there is an O_{2m}-equivariant bijection

$$\mathscr{C}_m \xrightarrow[\approx]{V} \mathscr{I}_{2m} \tag{9.6}$$

which associates to J the isotropic subspace $V(J)$. Let $\mathscr{I}_{2m}^+$ denote the component corresponding to $\mathscr{C}_m^+$.

Recall now that with respect to the complex volume element $\omega_{\mathbb{C}} = i^m e_1 \cdots e_{2m}$ we have a decomposition $\$_{\mathbb{C}} = \$_{\mathbb{C}}^+ \oplus \$_{\mathbb{C}}^-$ into $+1$ and -1 eigenspaces respectively.

Lemma 9.6. *If $\sigma \in \$_{\mathbb{C}}$ is pure, then either $\sigma \in \$_{\mathbb{C}}^+$ or $\sigma \in \$_{\mathbb{C}}^-$.*

Proof. Let $v_j = (1/\sqrt{2})(e_j + if_j)$, $1 \leq j \leq m$, be a hermitian orthonormal basis of $V = \ker j_\sigma$. Since $v_i \perp v_j$ for $i \neq j$ and $v_i \perp \bar{v}_j$ for all i,j, we see that $e_1, f_1, \ldots, e_m, f_m$ is an orthonormal basis of $\mathbb{R}^{2m}$. Note that $(e_j + if_j)\sigma = 0 \Rightarrow ie_jf_j\sigma = \sigma$ for each j. Hence, $i^m e_1 f_1 \cdots e_m f_m \sigma = \pm\omega_{\mathbb{C}}\sigma = \sigma$, and $\sigma \in \$_{\mathbb{C}}^+$ or $\sigma \in \$_{\mathbb{C}}^-$ as claimed. ∎

This lemma gives a decomposition $P\$ = P\$^+ \amalg P\$^-$ of the pure spinor space into positive and negative types. Let $\mathbb{P}(P\$^+)$ denote the projectivization of the positive pure spinor space, i.e., $\mathbb{P}(P\$^+) = P\$^+/\sim$ where we say that $\sigma \sim \sigma'$ if $\sigma = t\sigma'$ for some $t \in \mathbb{C}$. Each of the spaces $\mathbb{P}(P\$^\pm)$, $\mathscr{C}_m^\pm$ and $\mathscr{I}_{2m}^\pm$ are acted upon by Spin_{2m}, in fact by SO_{2m}.

Proposition 9.7. *The maps $\sigma \mapsto K(\sigma)$ and $J \mapsto V(J)$ induce SO_{2m}-equivariant diffeomorphisms*

$$\mathbb{P}(P\$^+) \xrightarrow[\approx]{K} \mathscr{I}_{2m}^+ \xrightarrow[\approx]{V} \mathscr{C}_m^+ \qquad \textit{and} \qquad \mathbb{P}(P\$^-) \xrightarrow[\approx]{K} \mathscr{I}_{2m}^- \xrightarrow[\approx]{V} \mathscr{C}_m^-.$$

Proof. The second line follows immediately from the first by a change of orientation. The map V has already been shown to be an equivariant bijection between homogeneous spaces, and is therefore a diffeomorphism. It remains only to show that the equivariant map K is a bijection. We construct K^{-1} as follows. Fix $V \in \mathscr{I}_{2m}^+$ and let $J \in \mathscr{C}_m^+$ be the associated complex structure. Choose a unitary basis $\{e_1, Je_1, \ldots, e_m, Je_m\}$ of $\mathbb{R}^{2m}$ and set

$$\varepsilon_j = \tfrac{1}{2}(e_j - iJe_j) \qquad \bar{\varepsilon}_j = \tfrac{1}{2}(e_j + iJe_j)$$

as in §8. Define

$$\omega_j = -\varepsilon_j\bar{\varepsilon}_j \qquad \text{and} \qquad \bar{\omega}_j = -\bar{\varepsilon}_j\varepsilon_j \tag{9.7}$$

and note that by (8.3):

All of the elements ω_j, $\bar{\omega}_k$ for $1 \leqq j,k \leqq n$, commute in $\mathbb{C}\ell_{2m}$. (9.8)

Furthermore, by (8.3) these elements satisfy the identities

$$\omega_j + \bar{\omega}_j = 1 \tag{9.9}$$

$$\omega_j^2 = \omega_j \qquad \text{and} \qquad \bar{\omega}_j^2 = \bar{\omega}_j \tag{9.10}$$

$$\omega_j\bar{\omega}_j = \bar{\omega}_j\omega_j = 0 \tag{9.11}$$

$$e_j\omega_j = \bar{\omega}_j e_j \tag{9.12}$$

$$(\omega_j\varphi, \psi) = (\bar{\varepsilon}_j\varphi, \bar{\varepsilon}_j\psi) = (\varphi, \omega_j\psi) \qquad \text{for all } \varphi,\psi \in \$_{\mathbb{C}} \tag{9.13}$$

for each j, where (v,w) denotes any Spin_{2m}-invariant hermitian inner product in $\$_{\mathbb{C}}$.

These identities easily imply the following. Fix any j and let W be a linear subspace invariant under multiplication by e_j and Je_j. Then there is a hermitian orthogonal direct sum decomposition

$$W = W_j \oplus W'_j$$

where

$$W_j = \bar{\omega}_j \cdot W = \ker(\mu_{\bar{\varepsilon}_j}|_W) \qquad \text{and} \qquad W'_j = \omega_j \cdot W = \ker(\mu_{\varepsilon_j}|_W)$$

and where $\mu_{\bar{\varepsilon}_j} : W \to W$ is defined by $\mu_{\varepsilon_j}(w) = \varepsilon_j \cdot w$. By (9.12) we see that μ_{e_j} maps W_j to W'_j isomorphically and so

$$\dim W_j = \dim W'_j.$$

We first apply this process with $W = \$_{\mathbb{C}}$ and $j = 1$ to obtain a decomposition $\$_{\mathbb{C}} = \$_1 \oplus \$'_1$ where $\$_1 = \ker \mu_{\bar{\varepsilon}_1}$ is invariant under $e_2, Je_2, \ldots, e_m, Je_m$. We next set $W = \$_1$ and $j = 2$ to obtain a decomposition $\$_1 = \$_2 \oplus \$'_2$ where $\$_2 = \ker(\mu_{\bar{\varepsilon}_2}) \cap \ker(\mu_{\bar{\varepsilon}_1})$ and $\dim_{\mathbb{C}}\$_2 = 2^{m-2}$. Proceeding inductively we eventually construct

$$\$_m = \ker(\mu_{\bar{\varepsilon}_1}) \cap \ldots \cap \ker(\mu_{\bar{\varepsilon}_m}) \qquad \text{with } \dim_{\mathbb{C}}\$_m = 1. \tag{9.14}$$

The complex volume form $\omega_{\mathbb{C}} = i^m e_1 Je_1 \cdots e_m Je_m$ has the value $+1$ on $\$_m$ because: $\bar{\varepsilon}_j\sigma = 0 \Rightarrow -ie_jJe_j\sigma = \sigma$. Therefore, $\$_m \subset \$_{\mathbb{C}}^+$

We clearly have that $V(J) = \ker j_\sigma$ for $\sigma \in \$_m$. Hence, $\$_m$ is independent of the choice of unitary basis and the map $V \mapsto [\$_m]$ gives the desired map K^{-1}. ■

Globalizing this result to manifolds gives the following:

Proposition 9.8. *Let X be an oriented riemannian manifold of dimension $2m$. Then the orthogonal almost complex structures on X, whose canonical orientation is positive, are in natural one-to-one correspondence with cross-sections of the projectivized bundle $\mathbb{P}(P\$^+)$ of positive pure spinors on X. In particular, X is Kähler if and only if there is a parallel cross section of $\mathbb{P}(P\$^+)$.*

Note. $\mathbb{P}(P\$^+)$ is an SO_{2m}-bundle and is globally defined whether or not X is a spin manifold.

DEFINITION 9.9. The bundle $\tau(X) \equiv \mathbb{P}(P\$^+)$ is called the **twistor space** of X.

The total space of $\tau(X)$ carries a canonical almost complex structure defined as follows. Note to begin that the fibres of the projection $\pi : \tau(X) \to X$ are naturally homogeneous complex manifolds ($\cong SO_{2m}/U_m$). The riemannian connection of X determines a field of horizontal planes $\mathscr{H}$ for π, that is, it determines a canonical decomposition

$$T(\tau(X)) = \mathscr{V} \oplus \mathscr{H}$$

where $\mathscr{V}$ is the field of tangent planes to the fibres. As noted, $\mathscr{V}$ has an almost complex structure integrable on the fibres. The bundle $\mathscr{H}$ has a "tautological" almost complex structure defined, via the identification $\pi_* : \mathscr{H}_J \to T_*X$, to be the structure J itself. (Here we consider $\tau(X)$ as the bundle of positively oriented almost complex structures on the tangent spaces of X.)

In spinor terms the almost complex structure on $\tau(X)$ can be written as follows. Note that $\mathscr{H} = \pi^* TX$ and so $\$^+_{\mathbb{C}}(\mathscr{H}) = \pi^* \$^+_{\mathbb{C}}(X)$. Now whenever, a vector bundle is pulled over its own projectivization, a line bundle splits off tautologically. Let $\lambda_{\mathscr{H}}$ denote the tautological line bundle in $\pi^* \$^+_{\mathbb{C}}(X)$ and let $\lambda_{\mathscr{V}}$ denote the line bundle in $\$^+_{\mathbb{C}}(\mathscr{V})$ corresponding to the complex structure on the fibres. Then the line bundle $\lambda = \lambda_{\mathscr{V}} \otimes \lambda_{\mathscr{H}} \subset \$^+_{\mathbb{C}}(\mathscr{V}) \otimes \$^+_{\mathbb{C}}(\mathscr{H}) \subset \$^+_{\mathbb{C}}(\mathscr{V} \oplus \mathscr{H}) = \$^+_{\mathbb{C}}(\tau(X))$ gives the projective pure spinor field defining the almost complex structure on $\tau(X)$.

We now consider the question of integrability. Let J be an almost complex structure defined over an oriented riemannian $2m$-manifold X, and let $V \subset TX \otimes \mathbb{C}$ be the associated field of totally isotropic m-planes. The structure J is said to be **integrable** if it comes from an honest complex structure on X, i.e., if there is a holomorphically related system of local complex coordinate charts so that in each coordinate system $z_1, \ldots, z_m$, the field V is given by $V = \text{span}_{\mathbb{C}} \{\partial/\partial \bar{z}_1, \ldots, \partial/\partial \bar{z}_m\}$.

The fundamental theorem of Newlander and Nirenberg states that J is integrable if and only if

$$[\Gamma(V), \Gamma(V)] \subseteq \Gamma(V) \tag{9.15}$$

i.e., if and only if for any pair of complex vector fields v,w with values in V, the Lie bracket $[v,w]$ again has values in V.

Suppose now that σ is a (locally defined) pure spinor field whose projective class defines J. Then since $V \equiv \ker j_\sigma$ we see that locally

$$\Gamma(V) = \{v \in \Gamma(TX \otimes \mathbb{C}) : v \cdot \sigma = 0\}.$$

Proposition 9.10. *The almost complex structure defined by a pure spinor field σ is integrable if and only if σ satisfies the equation*

$$v \cdot \nabla_w \sigma - w \cdot \nabla_v \sigma = 0. \tag{9.16}$$

for all $v,w \in \Gamma(V)$ where $V = \ker j_\sigma$.

Proof. Fix $v,w \in \Gamma(V)$ and differentiate the condition $w \cdot \sigma = 0$ with respect to v. This gives the equation

$$0 = \nabla_v(w \cdot \sigma) = (\nabla_v w) \cdot \sigma + w \cdot \nabla_v \sigma.$$

Interchanging v and w and then subtracting shows that

$$0 = (\nabla_v w - \nabla_w v) \cdot \sigma + w \cdot \nabla_v \sigma - v \cdot \nabla_w \sigma.$$

Since $\nabla_v w - \nabla_w v = [v,w]$, we conclude that $v \cdot \nabla_w \sigma = w \cdot \nabla_v \sigma \Leftrightarrow [v,w] \cdot \sigma = 0 \Leftrightarrow [v,w] \in \Gamma(V)$. ■

Note that the equations (9.16) depend only on the projective class of σ, since for any smooth function f, we have $(v \cdot \nabla_w - w \cdot \nabla_v)(f\sigma) = f(v \cdot \nabla_w - w \cdot \nabla_v)\sigma$. The equations (9.16) also depend only on the *conformal class* of the underlying riemannian metric.

Proposition 9.10 has an interesting reformulation in terms of the twistor space $\tau(X)$. As mentioned above, $\tau(X)$ carries a canonical almost complex structure. Now a C^1-map between almost complex manifolds $f : (X,J_X) \to (Y,J_Y)$ will be called **holomorphic** if its differential f_* is everywhere J-linear, i.e., if $f_* \circ J_X = J_Y \circ f_*$.

Theorem 9.11. (Michelsohn [3]) *Let X be an oriented (even-dimensional) riemannian manifold with an almost complex structure determined by a projective spinor field $s \in \Gamma(\tau(X))$. Then this almost complex structure is integrable if and only if s is holomorphic.*

Note. More succinctly one could say that *cross-sections of $\tau(X)$ induce almost complex structures, and holomorphic cross-sections induce integrable ones.* However, the condition that a cross-section s be holomorphic is *not* linear since the complex structure on X depends itself on s. It is rather of the form: $\bar{\partial}^s s = 0$, where $\bar{\partial}^s$ denotes the Cauchy-Riemann operator associated to s. In this form the equation is reminiscent of many basic equations in geometry, such as the minimal surface equation $\Delta^f f = 0$, the Yang-Mills equation $\Delta^\nabla R^\nabla = 0$ (cf, Lawson [1], [2]), and the condition for balanced metrics $(d^*)^\omega \omega = 0$ (cf. Michelsohn [2]).

Proof of Theorem 9.11. It suffices to work locally, so we may choose a positive spinor field $\sigma \in \Gamma(P\$^+)$ with $[\sigma] = s$. Let $V = \ker j_\sigma$ be the field of (0,1)-subspaces as above. The condition that $s = [\sigma]$ be holomorphic is equivalent to the condition that

$$\nabla_v \sigma = \lambda_v \sigma \qquad \text{for all } v \in \Gamma(V) \tag{9.17}$$

where λ_v is a function depending on v and where ∇ is the riemannian connection on $P\$^+$. We must show that conditions (9.16) and (9.17) are equivalent. Let $\{\bar{\varepsilon}_1, \dots, \bar{\varepsilon}_n\}$ be a local hermitian frame field for V, and recall from (9.14) that the complex line field s is given by $s = \cap_i \ker(\mu_{\bar{\varepsilon}_i})$. Now equation (9.16) and the fact that $\bar{\varepsilon}_j^2 = 0$ show that $-\bar{\varepsilon}_i\bar{\varepsilon}_j\nabla_{\bar{\varepsilon}_j}\sigma = \bar{\varepsilon}_j\bar{\varepsilon}_i\nabla_{\bar{\varepsilon}_j}\sigma = \bar{\varepsilon}_j\bar{\varepsilon}_j\nabla_{\bar{\varepsilon}_i}\sigma = 0$ for all i,j. Therefore we have $\bar{\varepsilon}_j\nabla_{\bar{\varepsilon}_j}\sigma \in \cap \ker(\mu_{\bar{\varepsilon}_i}) = s \subset \$^+$, and since $\bar{\varepsilon}_j\nabla_{\bar{\varepsilon}_j}\sigma \in \$^-$, we conclude that

$$\bar{\varepsilon}_j\nabla_{\bar{\varepsilon}_j}\sigma = 0 \qquad \text{for all } j. \tag{9.18}$$

Condition (9.16) says that the $C^\infty(X)$-bilinear form on $\Gamma(V)$ given by $\beta(v,w) = w\nabla_v\sigma$ is symmetric. Hence, (9.18) $\Rightarrow \beta = 0$, and we conclude that $\nabla_v\sigma \in \bigcap \ker(\mu_{\bar{\varepsilon}_i}) = [\sigma]$ for all $v \in \Gamma(V)$. Therefore, (9.16) $\Rightarrow$ (9.18) $\Rightarrow$ (9.17). On the other hand if (9.17) holds, then $w\nabla_v\sigma = 0$ for all $v,w \in \Gamma(V)$ and so (9.16) holds. ■

REMARK. Theorem 9.11 could be modified by changing the almost complex structure on $\tau(X)$ to be the one induced by s itself, i.e., by lifting J_s uniformly to all horizontal spaces above each point (instead of twisting along the fibre). At the point $s \in \tau(X)$ these two complex structures agree.

In low dimensions these constructions are particularly interesting for the following reason:

REMARK 9.12. *In dimensions* $2m \leq 6$ *every non-zero positive* (*or negative*) *spinor is pure*, i.e., $P\$^\pm = \$_{\mathbb{C}}^\pm - \{0\}$. This is simply because the group Spin_{2m} acts transitively on the unit sphere in $\$_{\mathbb{C}}^\pm$ in these dimensions. (Transitivity also holds for $2m = 8$ if one restricts to the **real** spinors $\$^\pm$, but it does not hold for the complex ones.) As a consequence we have the following fact:

Proposition 9.13. *On an oriented manifold of dimension four or six, every nowhere-vanishing spinor field uniquely determines an almost complex structure* (*via* (*9.3*)).

Let X be an oriented riemannian 4-manifold. A 2-form φ on X is said to be **self-dual** (of **anti-self-dual**) if $*\varphi = \varphi$ ($*\varphi = -\varphi$ respectively). Since the Hodge $*$-operator satisfies $(*)^2 = 1$, there is an orthogonal decomposition

$$\Lambda^2(X) = \Lambda_+ \oplus \Lambda_-$$

where $\Lambda_\pm = \{\varphi \in \Lambda^2(X) : *\varphi = \pm\varphi\}$. Now the twistor space $\tau(X) = \mathbb{P}(P\$^+) = \mathbb{P}(\$_{\mathbb{C}}^+)$ can be identified with the bundle of unit self-dual 2-forms by associating to a complex structure J its Kähler form $\omega_J \in \Lambda_+$ (given by $\sqrt{2}\omega_J(U,W) = \langle JU,W\rangle$). It is a 2-sphere bundle over X with a canonical almost complex structure.

The following theorem has played an important role in understanding the structure of the self-dual Yang-Mills equations over riemannian 4-manifolds. The seminal ideas for these applications were due to R. Penrose and R. Ward (see Ward and Wells [1]). A riemannian 4-manifold is called **self-dual** if its Weyl conformal tensor W is self-dual, i.e., if $*W = W$ where W is considered as a 2-form with values in $\mathrm{Hom}(TX,TX)$.

Theorem 9.14 (Atiyah-Hitchin-Singer [1]). *The canonical almost complex structure on the twistor space of a riemannian 4-manifold X is integrable if and only if X is self-dual.*

We refer the reader to the original paper for details. The arguments involve the so-called "twistor equation" which is roughly the complement of the Dirac equation.

One class of manifolds whose twistor space always has an integrable complex structure is the set of manifolds of constant sectional curvature. For the case of the sphere S^{2m}, one has $\tau(S^{2m}) \cong \mathscr{I}_{2m+1} \cong \mathrm{SO}_{2m+1}/\mathrm{U}_{2m} \cong \mathrm{SO}_{2m+2}/\mathrm{U}_{m+1}$. The map $\pi:\mathscr{I}_{2m+1} \to S^{2m}$ associates to an isotropic m-plane $V \subset \mathbb{C}^{2m+1}$ the unit vector $e \in \mathbb{R}^{2m+1}$ such that $\mathbb{C}^{2m+1} = V \oplus \bar{V} \oplus \mathbb{C}e$. (This determines e up to sign. Since $\mathbb{R}^{2m+1} = \mathrm{Re}\{V \oplus \bar{V}\} \oplus \mathbb{R}e$, this sign is determined by fixing an orientation on $\mathbb{R}^{2m+1}$.) In two beautiful papers, E. Calabi [2], [3] used $\tau(S^{2m})$ to classify the harmonic maps from S^2 to S^n. He first showed that a harmonic map $\varphi: S^2 \to S^n$ must lie in a geodesic subsphere of even dimension. He then showed that any harmonic map $\varphi: S^2 \to S^{2m}$ lifts to a *holomorphic* map $\Phi: S^2 \to \mathscr{I}_{2m+1}$ which is horizontal for the projection π.

We now return to the question of almost complex structures. Suppose that $\sigma \in \Gamma(\$_{\mathbb{C}}^+)$ is a field of pure spinors over a $2m$-manifold X. Then as we have seen, the projective class $[\sigma] \in \Gamma(\mathbb{P}(P\$_{\mathbb{C}}^+))$ determines an almost complex structure on X. However, the field itself carries more information. It determines a trivialization of the canonical bundle $\Lambda_{\mathbb{C}}^m V$ associated to the structure. To see this we need the following lemma. Let $\$_{\mathbb{C}}$ be the irreducible complex representation for $\mathbb{C}\ell_{2m}$. For any $\sigma \in \$_{\mathbb{C}}$ we define the

isotropy group of σ by

$$G_\sigma \equiv \{g \in \mathrm{Spin}_{2m} : g\sigma = \sigma\}.$$

Lemma 9.15. *For any pure spinor* $\sigma \in P\$^+$, *one has that*

$$G_\sigma = \mathrm{SU}_m$$

where $SU_m \subset \mathrm{Spin}_{2m}$ *is conjugate to the lifting of the standard embedding* $\mathrm{SU}_m \subset \mathrm{SO}_{2m}$.

Proof. Set $V = \ker j_\sigma \subset \mathbb{C}^{2m}$ as above, and note that if $g\sigma = \sigma$, then $\tilde{g}(V) = V$, where by definition $\tilde{g}(v) = g \cdot v \cdot g^{-1} = \mathrm{Ad}_g(v)$. Now $V = \{x + iJx : x \in \mathbb{R}^{2m}\}$ where J is the complex structure determined by σ. The action of Spin_{2m} on $\mathbb{C}^{2m} = \mathbb{R}^{2m} \oplus i\mathbb{R}^{2m}$ given by Ad, preserves real and imaginary parts. Therefore, if $g \in G_\sigma$, then $\tilde{g}(x + iJx) = \tilde{g}(x) + i\tilde{g}(Jx) \in V$ for all x, and so

$$\tilde{g}(Jx) = J(\tilde{g}(x)) \qquad \text{for all } x \in \mathbb{R}^{2m}.$$

Hence, the image of G_σ under the covering map $\mathrm{Ad}: \mathrm{Spin}_{2m} \to \mathrm{SO}_{2m}$ is contained in the subgroup $\mathrm{U}_m = \{\gamma \in \mathrm{SO}_{2m} : \gamma J = J\gamma\}$.

We now show that $\mathrm{Ad}(G_\sigma) \subseteq \mathrm{SU}_m$. Given $g \in G_\sigma$, there exists a unitary basis $\{e_1, Je_1, \ldots, e_m, Je_m\}$ for $\mathbb{R}^{2m}$ which "diagonalizes" $\tilde{g}$, i.e., so that $\tilde{g}(e_k + iJe_k) = e^{i2\theta_k}(e_k + iJe_k)$ for real numbers $0 \leqq \theta_k < \pi$ and $1 \leqq k \leqq m$. This means that

$$g = \pm \prod_k (\cos\theta_k - \sin\theta_k e_k Je_k).$$

Now, $(e_k + iJe_k)\sigma = 0 \Rightarrow e_k\sigma = -iJe_k\sigma \Rightarrow e_kJe_k\sigma = -i\sigma$ for each k. Hence,

$$g\sigma = \pm \prod_k (\cos\theta_k + i\sin\theta_k)\sigma = \pm e^{i\Sigma\theta_k}\sigma.$$

Since $g\sigma = \sigma$, we have $2\sum\theta_k \equiv 0 \pmod{2\pi}$ and so $\det_{\mathbb{C}}(\tilde{g}) = e^{i2\Sigma\theta_k} = 1$ as claimed.

The above argument shows that for any $g \in \mathrm{Ad}^{-1}(\mathrm{SU}_m)$ we have that $g\sigma = \pm\sigma$. Since SU_m is simply-connected, $\mathrm{Ad}^{-1}(\mathrm{SU}_m)$ has two connected components. The identity component fixes σ, the other component contains -1 and hence cannot fix σ. ∎

Alternatively we could consider the Lie algebra

$$\mathfrak{g}_\sigma = \{\varphi \in \Lambda^2\mathbb{R}^{2m} : \varphi\sigma = 0\}$$

of G_σ. For any $\varphi \in \mathfrak{g}_\sigma$, there is a unitary basis of $\mathbb{R}^{2m}$ so that $\varphi = \sum \lambda_k e_k Je_k$. Since $e_kJe_k\sigma = -i\sigma$, we have that

$$\varphi\sigma = -i\sum\lambda_k\sigma$$

and so $\sum\lambda_k = 0$. This condition characterizes $\mathfrak{su}_m$ in $\mathfrak{u}_m$.

It is incidentally not true that: $G_\sigma \cong SU_m \Rightarrow \sigma$ is pure. For example, in dimension eight we could take $\sigma = \sigma_0 + \sigma_0'$ where σ_0 and σ_0' are pure and determine almost complex structures J_0 and $-J_0$ respectively.

For any $2m$-dimensional spin manifold X, Lemma 9.15 implies

Proposition 9.16. *Each globally defined pure spinor field σ on X determines a unique reduction of the structure group of X to* SU_m.

Proposition 9.17. *X admits a pure spinor field which is parallel if and only if X is Kähler and Ricci flat.*

Proof. If σ is parallel, then the holonomy group of X is contained in $Ad(G_\sigma) = SU_m$. This means that the almost complex structure is parallel and the canonical bundle is flat.

Conversely, if X is Kähler and Ricci-flat, then $\$_{\mathbb{C}} = \Lambda^*_{\mathbb{C}}(X)$ and both $\Lambda^0_{\mathbb{C}}$ and $\Lambda^m_{\mathbb{C}}$ provide parallel pure spinors. ∎

Corollary 9.18. *Suppose* $\dim(X) = 4$ or 6. *Then X admits a non-zero parallel spinor field if and only if X is Kähler and Ricci-flat.*

Proof. If $\sigma \in \Gamma(\$_{\mathbb{C}})$ is parallel, then each component $\sigma^\pm \in \Gamma(\$^\pm_{\mathbb{C}})$ is parallel, and as noted in 9.12 each component is pure. ∎

Using II.8.10, we now conclude the following:

Theorem 9.19. *Let X be a compact spin 4-manifold with $\hat{A}(X) \neq 0$. Then any riemannian metric on X with scalar curvature $\kappa \geqq 0$ is Kähler and Ricci-flat. In fact, any such metric admits a parallel quaternion structure, i.e., parallel orthogonal complex structures I and J with $IJ = -JI$.*

Proof. If $\kappa \geqq 0$, then by II.8.8 and II.8.10 and the Index Theorem, we see that X carries a parallel spinor field σ. Each component σ^+ and σ^- is also parallel. Assume that $\sigma^+ \neq 0$. Since $\$^+_{\mathbb{C}}$ is an $\mathbb{H}$-line bundle (and the $\mathbb{H}$-structure is preserved by the connection), we see that σ^+, $i\sigma^+$, $j\sigma^+$ and $k\sigma^+$ give a complete parallelization of $\$^+_{\mathbb{C}}$, that is, if one positive spinor is parallel, then all positive spinors are parallel. We conclude that X carries a family $\cong \mathbb{P}(\$^+_x) \cong S^2$ of parallel complex structures, each of which makes X into a Kähler manifold. ∎

It should be pointed out that by Yau's solution of the Calabi Conjecture [2], [3], we know that Ricci-flat Kähler manifolds exist in all dimensions.

Remark 9.20. A theorem similar to 9.19 was asserted in Hitchin [1] for all dimensions. It was based on an incorrect calculation that $G_\sigma \cong SU_m$ for all σ. Indeed, one finds counterexamples already in dimension eight. This will be discussed in the next section.

We conclude this section with a remark about what happens in odd dimensions. Let $\$_{\mathbb{C}}$ be one of the two irreducible complex spinor spaces for $\mathbb{C}\ell_{2m+1}$ and let $\sigma \in \$_{\mathbb{C}}$ be a pure spinor. The totally isotropic subspace $V = \ker j_\sigma \subset \mathbb{C}^{2m+1}$ has dimension m and we get a hermitian orthogonal decomposition

$$\mathbb{C}^{2m+1} = V \oplus \bar{V} \oplus \mathbb{C}e$$

where $e \in \mathbb{R}^{2m+1}$. This gives an orthogonal decomposition

$$\mathbb{R}^{2m+1} = V_{\mathbb{R}} \oplus \mathbb{R}e$$

where $V_{\mathbb{R}} \otimes \mathbb{C} = V \oplus \bar{V}$ (i.e., $V_{\mathbb{R}}$ has a complex structure). Assuming $\|e\| = 1$, we see that $e \in V_{\mathbb{R}}^{\perp}$ is determined up to sign. A choice for e corresponds to a choice of orientation on $\mathbb{R}^{2m+1}$. In summary then, each pure spinor $\sigma \in \$_{\mathbb{C}}$ determines a pair (e,J) where $e \in \mathbb{R}^{2m+1}$ is a unit vector and J is an orthogonal almost complex structure on $e^{\perp}$. A global pure spinor field σ on a spin $(2m+1)$-manifold X determines, as above, a reduction of the structure group to SU_m. If $\nabla\sigma = 0$, then X splits locally as a riemannian product $X \cong X_0 \times \mathbb{R}$ where X_0 is Kähler and Ricci-flat.

In dimensions three and five every non-zero spinor is pure.

§10. Reduced Holonomy and Calibrations

The holonomy group of a connected complete riemannian n-manifold X is defined as follows. Fix a point $x \in X$ and to each piecewise smooth loop γ based at x, let $h_\gamma : T_xX \to T_xX$ be the orthogonal transformation given by parallel translation around γ. These transformations form a subgroup $\mathscr{H}_x \subset \mathrm{O}(T_xX) \cong \mathrm{O}_n$ whose conjugacy class as a subgroup of O_n is independent of the choice of base point x. This conjugacy class $\mathscr{H}(X)$ is called the **holonomy group** of X. Its identity component $\mathscr{H}(X)^0$ is called the **local holonomy group** of X (see Kobayashi-Nomizu [1].) If $X = Y^k \times Z^{n-k}$ is a riemannian product, then one easily sees that $\mathscr{H}(X) = \mathscr{H}(Y) \times \mathscr{H}(Z) \subset \mathrm{O}_k \times \mathrm{O}_{n-k} \subset \mathrm{O}_n$, and if $\pi_1(X) = 0$, then by a theorem of de Rham the converse is true. It is therefore sensible to consider manifolds which are **irreducible**. This means the universal covering is not a riemannian product. In 1955 Marcel Berger [1] classified the possible holonomy groups for irreducible riemannian manifolds. There has been subsequent work refining Berger's list (see Bryant [1], [2] for a brief history). We now know that if X is irreducible and not locally symmetric, then $\mathscr{H}(X)^0$ must be one of the following:

SO_n (the generic case)

U_m ($n = 2m$; Kähler)	$\mathrm{Sp}_1 \cdot \mathrm{Sp}_m$ ($n = 4m$; Quaternionic Kähler)
SU_m (Kähler and $\mathrm{Ric} \equiv 0$)	Sp_m (Quaternionic Kähler and $\mathrm{Ric} \equiv 0$)

or must be one of the two exceptional cases:

$$G_2\ (n=7) \qquad \mathrm{Spin}_7\ (n=8)$$

where by definition $\mathrm{Sp}_1 \cdot \mathrm{Sp}_m = \mathrm{Sp}_1 \times \mathrm{Sp}_m/\mathbb{Z}_2$ and $\mathbb{Z}_2$ is generated by $(-1,-1)$. Metrics of each of the above types are known to exist.

We have already seen that:

$\mathscr{H}(X) \subset \mathrm{U}_m \iff$ there exists a parallel projective pure spinor field on X

$\mathscr{H}(X) \subset \mathrm{SU}_m \iff$ there exists a parallel pure spinor field on X.

In this and the next section we shall show that parallel spinors are also related to the existence of Sp_m, G_2, and Spin_7 holonomy metrics. These results are due to the authors and Reese Harvey (see Michelsohn [3]). The fundamental elementary point is the following:

Proposition 10.1. *Let X be an n-dimensional (riemannian) spin manifold on which there exists a globally parallel spinor field σ. Then at any point x the holonomy group satisfies*

$$\mathscr{H}_x \subset G_{\sigma_x} \tag{10.1}$$

where $G_{\sigma_x} = \{g \in \mathrm{Spin}(T_xX) : g\sigma_x = \sigma_x\}$. In other words,

$$\mathscr{H}(X) \subset G_\sigma$$

where G_σ denotes the conjugacy class of G_{σ_x} in Spin_n *(which is defined independently of x because σ is parallel).*

Conversely, if (10.1) is satisfied for a spinor *σ_x at some point x, then σ_x extends to a globally parallel field σ on X.*

Proof. If $\nabla\sigma = 0$, then the holonomy transformation $h_\gamma \in \mathrm{Spin}(T_xX)$ generated by parallel translation around any loop γ must preserve σ. Hence, $\mathscr{H}_x \subset G_{\sigma_x}$. Conversely, if $\mathscr{H}_x \subset G_{\sigma_x}$, then parallel translating σ_x from x to y is path-independent and defines a globally parallel field. ■

We know that on any spin manifold with $\hat{A} \neq 0$, any metric with zero scalar curvature has parallel spinor fields. Therefore in light of Proposition 10.1, it would be interesting to understand the structure of the isotropy groups $G_\sigma \subset \mathrm{Spin}_n$. For $n \leqq 6$ we have already seen in §9 that $G_\sigma \subset \mathrm{SU}_{[n/2]}$, so the first interesting case occurs when $n = 7$. Recall that the irreducible real representation $\$$ of Spin_7 has dimension 8. Furthermore, Spin_7 acts transitively on the unit sphere giving a diffeomorphism $S^7 \cong \mathrm{Spin}_7/G_2$ (see I.8.2 forward). Hence, all non-zero spinors are essentially equivalent, and we have the following:

Proposition 10.2. *Let $\$$ be the irreducible real spinor space for* Spin_7. *Then for any non-zero $\sigma \in \$$,*

$$G_\sigma \cong G_2$$

Corollary 10.3. *A 7-dimensional spin manifold has holonomy* $\subset G_2$ *if and only if it carries a non-trivial parallel spinor field.*

The group Spin_8 has three irreducible 8-dimensional real representations

$$\mathrm{Ad}:\mathrm{Spin}_8 \longrightarrow \mathrm{SO}(\mathbb{R}^8) \qquad \begin{matrix} \Delta^+:\mathrm{Spin}_8 \longrightarrow \mathrm{SO}(\$^+) \\ \Delta^-:\mathrm{Spin}_8 \longrightarrow \mathrm{SO}(\$^-) \end{matrix}$$

related by the triality automorphism (see I.8). Each is transitive on the unit sphere and therefore determines a unique conjugacy class of subgroup. That is, in each representation the isotropy subgroups of any two non-zero elements are conjugate.

Proposition 10.4. *There are three distinct conjugacy classes of subgroups*

$$\mathrm{Spin}_7 \subset \mathrm{Spin}_8 \begin{matrix} \supset \mathrm{Spin}_7^+ \\ \supset \mathrm{Spin}_7^- \end{matrix}$$

defined as the isotropy groups $\mathrm{Spin}_7 = G_v$, $\mathrm{Spin}_7^\pm = G_{\sigma^\pm}$ *for non-zero elements* $v \in \mathbb{R}^8$ *and* $\sigma^\pm \in \$^\pm$. *These three conjugacy classes are cyclically permuted by the triality outer automorphism of* Spin_8. *Furthermore, the conjugacy classes* $\mathrm{Spin}_7^\pm$ *are conjugate in the enlarged group* Pin_8. *In fact,*

$$\mathrm{Spin}_7^+ = e \cdot \mathrm{Spin}_7^- \cdot e^{-1} \tag{10.2}$$

for any $e \in S^7 \subset \mathrm{Pin}_8$ *(where* $S^7 \subset \mathbb{R}^8 \subset C\ell_8$ *are the standard embeddings).*

The covering homomorphism $\mathrm{Ad}:\mathrm{Spin}_8 \to \mathrm{SO}_8$ *gives embeddings*

$$\mathrm{Ad}:\mathrm{Spin}_7^\pm \hookrightarrow \mathrm{SO}_8 \tag{10.3}$$

which are not conjugate in SO_8 *but which are conjugate in* O_8.

Proof. Let $Z \approx \mathbb{Z}_2 \oplus \mathbb{Z}_2$ denote the center of Spin_8. Then the non-conjugacy of Spin_7, Spin_7^+ and Spin_7^- follows from the fact that they intersect Z in three distinct subgroups (cf. I.8). Since the representations are permuted by triality, so are these subgroups. Given $e \in S^7 \subset \mathbb{R}^8$, and $\sigma \in \$^+$, we have that $e\sigma \in \$^-$ and clearly that $G_{e\sigma} = eG_\sigma e^{-1}$. This proves (10.2).

To see that (10.3) is an embedding, note that $(\ker \mathrm{Ad}) \cap \mathrm{Spin}_7^\pm = \{1\}$. To see that $\mathrm{Ad}(\mathrm{Spin}_7^+)$ and $\mathrm{Ad}(\mathrm{Spin}_7^-)$ are not conjugate in SO_8, note that: $g\mathrm{Ad}(\mathrm{Spin}_7^+)g^{-1} = \mathrm{Ad}(\mathrm{Spin}_7^-)$ for some $g \in \mathrm{SO}_8$ would imply that $\gamma\,\mathrm{Spin}_7^+\gamma^{-1} = \mathrm{Spin}_7^-$ for $\gamma \in \mathrm{Spin}_8$ with $\mathrm{Ad}(\gamma) = g$. On the other hand (10.2) shows that $\mathrm{Ad}(\mathrm{Spin}_7^\pm)$ *are* conjugate by elements $\mathrm{Ad}(e)$ (= reflection in the hyperplane $e^\perp$) in O_8. ∎

The embeddings (10.3) give a single conjugacy class of the subgroup $\mathrm{Spin}_7 \subset \mathrm{O}_8$ and it is exactly this representation which appears in Berger's list. Consequently we have

Corollary 10.5. *An 8-dimensional spin manifold has holonomy* $\subset \mathrm{Spin}_7$ *if and only if it carries a non-trivial parallel spinor field.*

Note that if σ is a parallel real spinor field on an 8-manifold, then σ decomposes $\sigma = \sigma^+ + \sigma^-$ where both σ^+ and σ^- are parallel. If both σ^+ and σ^- are non-trivial, we get a further reduction of the holonomy group to $\mathrm{Spin}_7^+ \cap \mathrm{Spin}_7^- \cong G_2$. (To see this, note simply that under triality we have $\mathrm{Spin}_7^+ \cap \mathrm{Spin}_7^- \cong \mathrm{Spin}_7^+ \cap \mathrm{Ad}^{-1}(\mathrm{SO}_7) \cong G_\sigma$ for σ as in Proposition 10.2.) Recall that the complex spinor spaces $\$_{\mathbb{C}}^{\pm}$ for Spin_8 are just the complexifications of the real ones. Write $\$_{\mathbb{C}}^{+} = \$^+ \oplus i\$^+$ and consider an element $\sigma_{\mathbb{C}} = \sigma_0 + i\sigma_1 \in \$_{\mathbb{C}}^{+}$. Then for $g \in \mathrm{Spin}_8$, $g\sigma_{\mathbb{C}} = \sigma_{\mathbb{C}}$ if and only if $g\sigma_0 = \sigma_0$ and $g\sigma_1 = \sigma_1$, and so

$$G_{\sigma_{\mathbb{C}}} = G_{\sigma_0} \cap G_{\sigma_1}$$

If $\sigma_0, \sigma_1 \in \$^+$ are linearly dependent (over $\mathbb{R}$), then $G_{\sigma_{\mathbb{C}}} \cong \mathrm{Spin}_7$. If they are linearly independent, then $G_{\sigma_{\mathbb{C}}} \cong \mathrm{Spin}_6 \cong \mathrm{SU}_4$. In this second case $G_{\sigma_{\mathbb{C}}}$ remains unchanged if we replace $\sigma_{\mathbb{C}}$ by $\hat{\sigma}_{\mathbb{C}} = \hat{\sigma}_0 + i\hat{\sigma}_1$ where $\{\hat{\sigma}_0, \hat{\sigma}_1\}$ is any orthonormal basis of $\mathrm{span}_{\mathbb{R}}(\sigma_0, \sigma_1)$. Up to real scalars this orthonormality is equivalent to the fact that $\langle \hat{\sigma}_{\mathbb{C}}, \hat{\sigma}_{\mathbb{C}} \rangle = 0$. One easily checks that in this dimension, the pure spinors are exactly those which are isotropic with respect to the natural real structure on $\$_{\mathbb{C}}^{+}$.

REMARK. It should be pointed out that R. Bryant [1], [2] has found a number of beautiful examples of manifolds with G_2 and Spin_7-holonomy.

We shall now show that the propositions above give a simple necessary and sufficient condition for the existence of a topological G_2 or Spin_7-structure. By this we mean the following. Let X be a differentiable n-manifold and fix a Lie group $G \subset \mathrm{GL}_n(\mathbb{R})$. Then by a **topological G-structure** on X we mean a topological reduction of the structure group of TX from $\mathrm{GL}_n(\mathbb{R})$ to G. This is defined to be a G-equivariant embedding $P_G \subset P_{\mathrm{GL}}(X)$ of a principal G-bundle into the frame bundle of X. Reducing the structure group to $\mathrm{O}_n \subset \mathrm{GL}_n(\mathbb{R})$ is equivalent to choosing a riemannian metric on X. If $G \subset \mathrm{O}_n$ is a closed subgroup, then reducing the structure group from O_n to G is tautologically equivalent to finding a cross-section of the bundle

$$P_{\mathrm{O}}(X)/G \longrightarrow X$$

whose fibres are the homogeneous spaces O_n/G.

Theorem 10.6. *Let X be a differentiable 7-manifold. Then X carries a topological G_2-structure if and only if it is a spin manifold.*

Proof. If the structure group of TX can be reduced to G_2, then because $\pi_1 G_2 = 0$, X is a spin manifold (see Remark I.1.10). Conversely, suppose

X is spin and let $\$$ be the irreducible real spinor bundle of X. Then since fibre-dim($\$$) $= 8 > \dim(X)$, there exists a nowhere vanishing cross-section σ of $\$$ (cf. Milnor-Stasheff [1]). By Proposition 10.2 we can identify the unit sphere bundle in $\$$ with $P_{\text{Spin}}(X)/G_2$. Therefore the normalized section $\sigma/\|\sigma\|$ gives us a G_2-reduction. ∎

Note that if X is a spin 7-manifold, then the topological G_2-structures on X are in one-to-one correspondence with global spinor fields $\sigma \in \Gamma(\$)$ such that $\|\sigma\| \equiv 1$. The story for Spin_7-reductions is quite similar.

Theorem 10.7. *Let X be a differentiable 8-manifold. Then X carries a topological* Spin_7*-structure if and only if X is spin and either $\chi(\$^+) = 0$ or $\chi(\$^-) = 0$, where $\$^\pm$ denote the irreducible real spinor bundles on X. Equivalently, X carries a* Spin_7*-structure if and only if $w_1(X) = w_2(X) = 0$ and for an appropriate choice of orientation on X we have that*

$$p_1(X)^2 - 4p_2(X) + 8\chi(X) = 0. \tag{10.4}$$

Note. Reversing the orientation of X leaves the Pontrjagin classes unchanged but reverses the sign of $\chi(X)$ in $H^8(X)$. Thus, condition (10.4) could be replaced by the condition that for any orientation on X, one of the two equations

$$p_1(X)^2 - 4p_2(X) \pm 8\chi(X) = 0$$

holds in $H^8(X)$.

Note that for manifolds of the type $X^8 = S^1 \times Y^7$, equation (10.4) is automatically satisfied. Furthermore, X^8 is spin iff Y^7 is spin, and Theorem 10.6 applies. There are other special cases of some interest.

Corollary 10.8. *Let X be a complex manifold of dimension 4. Then X carries a topological* Spin_7^+*-structure if and only if*

$$c_1[c_1^3 - 4c_1c_2 + 8c_3] = 0$$

Corollary 10.9. *Let M and N be compact spin 4-manifolds. Then the product $X = M \times N$ carries a topological* Spin_7*-structure if and only if*

$$9\text{sig}(M)\text{sig}(N) = 4\chi(M)\chi(N)$$

In particular, $M \times M$ has such a structure if and only if

$$3\text{sig}(M) = \pm 2\chi(M).$$

Proof of Theorem 10.7. If X is spin and $\chi(\$^+) = 0$, then there is a nowhere vanishing section $\sigma \in \Gamma(\$^+)$ which, via Proposition 10.4, reduces the structure group of Spin_7^+. In fact, by 10.4 we have the identification

$$P_{\text{Spin}_8}(X)/\text{Spin}_7^+ \cong S(\$^+) \tag{10.5}$$

where $S(\$^+)$ denotes the unit sphere bundle in $\$^+$. Hence, the Spin_7^+-reductions are in one-to-one correspondence with spinor fields $\sigma \in \Gamma(\$^+)$ such that $\|\sigma\| \equiv 1$. If $\chi(\$^-) = 0$, the analogous discussion holds.

Suppose conversely that the structure group of TX has a Spin_7-reduction, i.e., suppose there is a Spin_7-equivariant embedding $P_{\mathrm{Spin}_7} \hookrightarrow P_{\mathrm{O}_8}(X)$ of a principal Spin_7-bundle into the orthogonal frame bundle of X (for some riemannian metric). Since Spin_7 is simply-connected, this determines a spin structure: $P_{\mathrm{Spin}_8}(X) \to P_{\mathrm{SO}_8}(X) \subset P_{\mathrm{O}_8}(X)$, in which P_{Spin_7} lifts to a Spin_7-equivariant embedding $P_{\mathrm{Spin}_7} \subset P_{\mathrm{Spin}_8}(X)$, which in turn corresponds by (10.5) to a cross-section of $S(\$^+)$. Hence, $\chi(\$^+) = 0$.

Note. The reader may find the lack of ambiguity in this last paragraph somewhat unsettling. If so, let us examine the argument more closely. In saying that $P_{\mathrm{Spin}_7} \subset P_{\mathrm{O}_8}(X)$ is "Spin_7-equivariant" we must specify an embedding $\mathrm{Spin}_7 \subset \mathrm{O}_8$. This choice is important only up to conjugation in O_8. (To see this, fix $g \in \mathrm{O}_8$ and consider the map $g: P_{\mathrm{O}_8}(X) \to P_{\mathrm{O}_8}(X)$ which converts the given O_8-action to the conjugated "$g\,\mathrm{O}_8 g^{-1}$"-action.) Now the two subgroups $\mathrm{Ad}(\mathrm{Spin}_7^\pm)$ are conjugate in O_8 but *not* in SO_8. We choose an embedding whose class in SO_8 is $\mathrm{Ad}(\mathrm{Spin}_7^+)$. Since Spin_7 is connected, the embedding $P_{\mathrm{Spin}_7} \hookrightarrow P_{\mathrm{O}_8}(X)$ clearly fixes an orientation on X, i.e., it selects a sub-bundle $P_{\mathrm{SO}_8}(X) \subset P_{\mathrm{O}_8}(X)$. With this choice of orientation, P_{Spin_7} lifts to $P_{\mathrm{Spin}_7^+} \subset P_{\mathrm{Spin}_8}(X)$. Suppose on the other hand that we had chosen an embedding $\mathrm{Spin}_7 \subset \mathrm{SO}_8$ of $\mathrm{Ad}(\mathrm{Spin}_7^-)$-type, which differed from the first choice by an element $g \in \mathrm{O}_8$ (with $\det g = -1$). Our given reduction $f: P_{\mathrm{Spin}_7} \hookrightarrow P_{\mathrm{O}_8}(X)$ must now be conjugated to $g \circ f \circ g^{-1}$: $P_{\mathrm{Spin}_7} \hookrightarrow P_{\mathrm{O}_8}(X)$. This new reduction chooses the ***opposite*** orientation on X and lifts to a Spin_7^--subbundle of $P_{\mathrm{Spin}_8}(X)$.

The proof that $\chi(\$^\pm) = p_1(X)^2 - 4p_2(X) \pm 8\chi(X)$ is a good exercise in the algebra of characteristic classes, which we leave to the reader. ∎

It is an interesting fact that spinors in dimensions seven and eight are also related to certain exceptional geometries of subvarieties introduced in Harvey-Lawson [1], [2], [3]. The fundamental idea is this: Let φ be a differential p-form on a riemannian manifold X normalized so that

$$\varphi|_P \leqq \mathrm{vol}_P \tag{10.6}$$

for every oriented tangent p-plane P on X. This is equivalent to the condition that

$$\int_Z \varphi \leqq \mathrm{vol}(Z) \tag{10.7}$$

for every oriented p-dimensional C^1-submanifold Z of finite volume in X. If such a manifold Z has the property that

$$\int_Z \varphi = \mathrm{vol}(Z) \tag{10.8}$$

then Z is called a **φ-submanifold.** These creatures are interesting for the following reason. Let $Z \subset X$ be a compact oriented p-dimensional submanifold, possibly with boundary. We say that Z is **homologically volume-minimizing** if $\mathrm{vol}(Z) \leqq \mathrm{vol}(Z')$ for all such Z' with the property that $Z - Z'$ is a boundary (of some singular $(p+1)$-chain in X). When Z is not compact, it is called homologically volume-minimizing if this property holds for every compact subdomain-with-boundary in Z.

Proposition 10.10. *Suppose φ satisfies (10.6) and $d\varphi = 0$. Then every φ-submanifold is homologically volume-minimizing in X.*

Proof. Let Z be a compact φ-submanifold and Z' a competitor with $Z - Z' = \partial c$ for some singular chain c. Then by (10.7), (10.8) and Stokes' Theorem, we have

$$\mathrm{vol}(Z) = \int_Z \varphi = \int_{Z'} \varphi \leqq \mathrm{vol}(Z'). \quad \blacksquare$$

This proposition shows that every φ-submanifold is in particular a minimal submanifold, i.e., its mean curvature vector field is identically zero. It is therefore roughly as regular as X itself. For example, if X is real analytic, so is any φ-submanifold. The above definition extends immediately from submanifolds to integral currents (see Harvey-Lawson [3]). The "φ-subvarieties" can then have singularities, although they remain homologically volume-minimizing.

DEFINITION 10.11. If φ satisfies (10.6) and $d\varphi = 0$, then φ is called a **calibration.** The pair (X,φ) is a **calibrated manifold** and the family of φ-submanifolds is the associated **calibrated geometry.**

Calibrations are interesting for the special cases, not the general one. We know that few topological reductions of the structure group of a manifold are "integrable" in the sense that they are holonomy reductions. Similarly, few calibrations are "integrable" in the sense that the associated geometry of submanifolds is large. However, many interesting examples do exist.

EXAMPLE 10.12 (Kähler geometry; U_n-case). Let X be a Kähler manifold of complex dimension n. Fix p, $1 \leqq p \leqq n$, and consider the $2p$-form

$$\varphi_p = \frac{1}{p!}\,\omega^p$$

where ω is the Kähler form of X. Then φ_p is a calibration and the associated family of φ_p-submanifolds (φ_p-subvarieties) consists exactly of the complex analytic submanifolds (subvarieties) of dimension p in X. Proposition 10.8 says that every complex analytic subvariety is homologically volume-minimizing in X (a result due to H. Federer [1]).

EXAMPLE 10.13 (special lagrangian geometry; SU_n-case). Let X be a Ricci-flat Kähler manifold of complex dimension n, and assume that the canonical line bundle $\varkappa \equiv \Lambda_{\mathbb{C}}^{*} T^{*}X$ is trivial. Consider the n-form

$$\lambda = \mathrm{Re}(\Omega)$$

where Ω is a parallel holomorphic n-form (i.e., section of κ). Then λ is a calibration and the associated geometry is called **special lagrangian**. It is shown in Harvey-Lawson [3] that the special lagrangian subvarieties are as plentiful as the complex subvarieties.

There are three rich exceptional geometries which appear in dimensions seven and eight.

EXAMPLE 10.14 (associative and coassociative geometry; G_2-case). Consider euclidean space $\mathbb{R}^7 = \mathrm{Im}\,\mathbb{O}$, where $\mathbb{O}$ denotes the Cayley numbers, and define a parallel 3-form φ on $\mathbb{R}^7$ by setting

$$\varphi(x,y,z) = \langle x \cdot y, z\rangle$$

Then φ is a calibration with a rich geometry of submanifolds. It is called the **associative geometry** because the submanifolds are also defined by the vanishing of the **associator** $[x,y,z] = (xy)z - x(yz)$. The dual form

$$\psi = *\varphi$$

gives a geometry of 4-folds called the **coassociative geometry.**

These geometries exist (and are rich) on any riemannian 7-manifold with G_2-holonomy. Note that φ and ψ are clearly G_2-invariant since they are defined using Cayley multiplication. In fact φ and ψ *are the only nontrivial exterior p-forms,* $0 < p < 7$, *which are G_2-invariant.* Any riemannian 7-manifold with G_2-holonomy carries parallel forms φ and ψ which at each point are equivalent (over SO_7) to those given above. Furthermore, in any such manifold the geometry of associative 3-folds and coassociative 4-folds is vast. In particular, every analytic submanifold of dimension two is contained in an associative submanifold (whose germ along the surface is uniquely determined).

EXAMPLE 10.15 (Cayley geometry; Spin_7-case). Consider euclidean space $\mathbb{R}^8 = \mathbb{O}$ and define a parallel 4-form Φ on $\mathbb{R}^8$ by setting

$$\Phi(x,y,z,w) = \langle x \times y \times z, w\rangle$$

where the triple cross product is defined by $x \times y \times z = \frac{1}{2}\{x(\bar{y}z) - z(\bar{y}x)\}$. Then Φ is a calibration which, as shown in Harvey-Lawson [3], has a rich geometry of submanifolds, called the **Cayley geometry**. It includes as special cases all complex submanifolds and all special langrangian submanifolds for a 6-dimensional family of complex structures on $\mathbb{R}^8$. If we consider $\mathbb{R}^7 \subset \mathbb{R}^8$ as $\mathrm{Im}\,\mathbb{O} = 1^{\perp} \subset \mathbb{O}$, then $\Phi|_{\mathbb{R}^7} = \psi$ and $1 \llcorner \Phi \cong \varphi$.

The form Φ is Spin_7-invariant. *It is, in fact, the unique* Spin_7*-invariant* p*-form on* $\mathbb{R}^8$ *for* $0 < p < 8$. Any riemannian 8-manifold with Spin_7-holonomy carries a parallel 4-form of this type and has a large geometry of Cayley submanifolds.

The forms Φ and φ above were first discovered by Bonan [1].

There is a direct and intimate relationship between spinors and calibrations. It comes about by what can be roughly termed the process of "squaring." We shall examine this process to the point that meets our needs here. For further discussion the reader is referred to the book of Reese Harvey [1].

Let $n = 2m$ be an even integer and recall the fundamental algebra isomorphism

$$\mu : \mathbb{C}\ell_{2m} \longrightarrow \mathrm{Hom}_{\mathbb{C}}(\$_{\mathbb{C}}, \$_{\mathbb{C}}) \cong \$_{\mathbb{C}} \otimes \$_{\mathbb{C}}^* \tag{10.8}$$

We introduce on $\$_{\mathbb{C}}$ a Pin_{2m}-invariant hermitian metric $(\cdot,\cdot)$ by which we can identify $\$_{\mathbb{C}}^*$ with $\bar{\$}_{\mathbb{C}}$. ($\bar{\$}_{\mathbb{C}}$ is the same real vector space as $\$_{\mathbb{C}}$ but with the opposite complex structure, i.e., with scalar multiplication by t in $\$_{\mathbb{C}}$ replaced by $\bar{t}$.) The identification $\$_{\mathbb{C}} \otimes \bar{\$}_{\mathbb{C}} \cong \mathrm{Hom}_{\mathbb{C}}(\$_{\mathbb{C}}, \$_{\mathbb{C}})$ is then determined by associating to a pair of spinors $\sigma_1, \sigma_2 \in \$_{\mathbb{C}}$, the elementary endomorphism

$$L_{\sigma_1 \otimes \sigma_2}(\tau) \equiv (\tau, \sigma_2)\sigma_1 \qquad \text{for } \tau \in \$_{\mathbb{C}}.$$

Inverting (10.8) now gives the following:

Proposition 10.16. *There is an isomorphism*

$$\$_{\mathbb{C}} \otimes \bar{\$}_{\mathbb{C}} \xrightarrow{\approx} \mathbb{C}\ell_{2m}$$

which associates to a pair of spinors $\sigma_1, \sigma_2 \in \$_{\mathbb{C}}$, *the unique element,* $\varphi \in \mathbb{C}\ell_{2m}$ *such that* $\varphi \cdot \tau = (\tau, \sigma_2)\sigma_1$ *for all* $\tau \in \$_{\mathbb{C}}$. *Under this isomorphism we have for each* $\psi \in \mathbb{C}\ell_{2m}$, *an identification of endomorphisms*

$$\mu_\psi \otimes \mathrm{Id} \approx L_\psi \qquad \textit{and} \qquad \mathrm{Id} \otimes \mu_{\psi^*} \approx R_\psi$$

where L_ψ *and* R_ψ *denote respectively left and right multiplication by* ψ *on* $\mathbb{C}\ell_{2m}$, *and where* $\psi^* \equiv \overline{\alpha(\psi^t)}$. *Via I.1.3 the isomorphism above determines a canonical identification*

$$\$_{\mathbb{C}} \otimes \bar{\$}_{\mathbb{C}} \xrightarrow{\approx} \Lambda^*\mathbb{C}^{2m} \tag{10.9}$$

which is Pin_{2m}*-equivariant, where* $g \in \mathrm{Pin}_{2m}$ *acts on* $\$_{\mathbb{C}} \otimes \bar{\$}_{\mathbb{C}}$ *by* $\mu_g \otimes \mu_g$, *and on* $\Lambda^*\mathbb{C}^{2m}$ *by the complexification of the standard representation of* O_{2m} *on forms.*

Proof. All assertions but the last are easily checked. For the last, one need only recall that $\alpha(g^t) = g^{-1}$ for $g \in \mathrm{Pin}_{2m}$ and therefore $\mu_g \otimes \mu_g = \mu_g \otimes \mu_{(g^{-1})^*} \cong \mathrm{Ad}_g \cong$ the standard representation of g on $\Lambda^*\mathbb{C}^{2m}$.

There is a corresponding real version of this result. For simplicity we shall consider only the cases where $C\ell_n$ is a real matrix algebra. Suppose $n \equiv 6$ or $8 \pmod 8$ and let $\$$ denote an irreducible real $C\ell_n$-module. Then Clifford multiplication gives an isomorphism

$$\mu : C\ell_n \longrightarrow \mathrm{Hom}_{\mathbb{R}}(\$, \$).$$

When $n \equiv 7 \pmod 8$, there are two distinct isomorphisms

$$\mu^{\pm} : C\ell_n^{\pm} \longrightarrow \mathrm{Hom}_{\mathbb{R}}(\$^{\pm}, \$^{\pm})$$

where $\$^+$ (and $\$^-$) are the irreducible real $C\ell_n$-modules on which the volume form ω acts by Id (and -Id respectively), and where $C\ell_n^{\pm} = (1 \pm \omega) \cdot C\ell_n$. Inverses to these maps are given explicitly as follows:

Proposition 10.17. *Let $\$$ be an irreducible real $C\ell_n$-module where $n \equiv 6$ or 8 (mod 8), and assume $\$$ is provided with a* Pin_n*-invariant inner product $\langle \cdot, \cdot \rangle$. Then assigning to a pair $\sigma, \tau \in \$$ the endomorphism $L_{\sigma \otimes \tau}(\cdot) = \langle \cdot, \tau \rangle \sigma$ determines an isomorphism*

$$\$ \otimes \$ \xrightarrow{\approx} C\ell_n \tag{10.10}$$

under which $\mu_\varphi \otimes \mu_{\psi^*} \approx L_\varphi \otimes R_\psi$ where $\psi^* = \alpha(\psi^t)$. Via I.1.3 this gives a canonical Pin_n*-equivariant isomorphism*

$$\$ \otimes \$ \xrightarrow{\approx} \Lambda^* \mathbb{R}^n \tag{10.11}$$

where $g \in \mathrm{Pin}_n$ acts on $\$ \otimes \$$ by $\mu_g \otimes \mu_g$. The isomorphism (10.11) *decomposes into two isomorphisms:*

$$\odot^2 \$ \xrightarrow{\approx} \bigoplus_{p \equiv 3 \text{ or } 4 \pmod 4} \Lambda^p \mathbb{R}^n \tag{10.12}$$

$$\Lambda^2 \$ \xrightarrow{\approx} \bigoplus_{p \equiv 1 \text{ or } 2 \pmod 4} \Lambda^p \mathbb{R}^n \tag{10.13}$$

under the canonical decomposition $\$ \otimes \$ = (\odot^2 \$) \oplus (\Lambda^2 \$)$ into symmetric and skew-symmetric 2-forms respectively.

When $n \equiv 7$ (mod 8), *all analogous statements hold with $\$$ replaced by $\$^+$ (or $\$^-$) and with $C\ell_n$ replaced by $C\ell_n^+$ (or $C\ell_n^-$ respectively).*

Proof. The initial statements are similar to those in Proposition 10.16 and are proved analogously. For statements (10.12) and (10.13), we note first that under the isomorphism (10.10), symmetric tensors go to self-adjoint transformations and skew-symmetric tensors go to skew-adjoint transformations. On the other hand Clifford multiplication by a p-form φ has the property that $(\mu_\varphi)^* = (-1)^{\frac{1}{2}p(p+1)} \mu_\varphi$. This proves (10.12) and (10.13). Details of the case where $n \equiv 7 \pmod 8$ are left to the reader. ∎

For any spinor σ in $\$$ (or $\$^+$ if $n \equiv 7 \pmod 8$)), this proposition gives a decomposition of the "square":

$$\sigma \otimes \sigma = \varphi_0 + \varphi_3 + \varphi_4 + \varphi_7 + \varphi_8 + \ldots + \varphi_{4t-1} + \varphi_{4t} + \ldots \qquad (10.14)$$

with $\varphi_p \in \Lambda^p\mathbb{R}^n$ for each p. Under special assumptions about σ, certain of these components will vanish or be related to one another by the Hodge $*$-isomorphism. We shall first consider the case where $n = 8$.

Theorem 10.18. *Let $\$$ denote the irreducible real $C\ell_8$-module, and let $\$ = \$^+ \oplus \$^-$ be its standard decomposition into ± 1-eigenspaces for the volume form ω. Then for any unit-length spinor $\sigma \in \$^+$, the decomposition (10.14) becomes*

$$\sigma \otimes \sigma = 1 + \Phi + \omega$$

where $\Phi \in \Lambda^4\mathbb{R}^8$ is the Cayley 4-form defined in 10.13.

Proof. To begin consider $\sigma \otimes \sigma \in C\ell_8$. Since $\omega\sigma = \sigma$, we have by Proposition 10.17 that $\sigma \otimes \sigma = (\omega\sigma) \otimes \sigma = \omega(\sigma \otimes \sigma)$ and similarly that $\sigma \otimes \sigma = \sigma \otimes (\omega\sigma) = (\sigma \otimes \sigma) \cdot \alpha(\omega^t) = (\sigma \otimes \sigma)\omega$. Now left and right multiplication are related to the Hodge $*$-operator by (5.35) of Chapter II. This shows that if we write $\sigma \otimes \sigma = \sum \varphi_p$ as in (10.14), then

$$\sum (-1)^{\frac{p(p-1)}{2}} * \varphi_p = \sum \varphi_p = \sum (-1)^{\frac{p(p+1)}{2}} * \varphi_p.$$

Hence $\varphi_p = 0$ for $p \neq 0,4,8$, and furthermore $*\varphi_4 = \varphi_4$, $*\varphi_0 = \varphi_8$. The identity element $1 \in C\ell_8$ can be written as $1 = \sum \sigma_\alpha \otimes \sigma_\alpha$ where $\{\sigma_1, \ldots, \sigma_{16}\}$ is any orthonormal basis of $\$$. It follows by choosing $\sigma_1 = \sigma$, that $\langle 1, \sigma \otimes \sigma\rangle = 1$. Therefore $\sigma \otimes \sigma = 1 + \Phi + \omega$ for some $\Phi \in \Lambda^4\mathbb{R}^8$. (We are using here the inner product on $C\ell_8$ for which 1 and $e_{i_1} \cdots e_{i_p}$, $1 \leq i_1 < \ldots < i_p$, form an orthonormal basis. This differs from the standard one on $\mathrm{Hom}(\mathbb{R}^{16},\mathbb{R}^{16})$ by a factor of $\frac{1}{16}$.)

Observe now that since the isotropy subgroup of any nonzero element $\sigma \in \$^+$ is Spin_7^+, the image $\mathrm{Ad}(\mathrm{Spin}_7^+) \subset \mathrm{SO}_8$ must leave invariant each component of $\sigma \otimes \sigma$. In particular Φ is a Spin_7^+-invariant 4-form. This identifies it as a multiple of the Cayley 4-form. To determine which multiple, we proceed as follows. Fix orthonormal vectors $e_1, \ldots, e_4$. Then

$$\begin{aligned}\langle \sigma \otimes \sigma, e_1 \ldots e_4\rangle &= \langle \sigma \otimes \sigma, e_1 \ldots e_4 \cdot 1\rangle = \langle (e_4 \cdots e_1\sigma) \otimes \sigma, 1\rangle \\ &= \langle e_4 \cdots e_1\sigma, \sigma\rangle \leq \|\sigma\|^2 = 1. \end{aligned} \qquad (10.15)$$

This proves that the Λ^4-component of $\sigma \otimes \sigma$ is a calibration, i.e., it satisfies (10.6).

Note that we have equality in (10.15) if and only if $e_4 \cdots e_1\sigma = \sigma$. Set $w = e_1 \cdots e_4$. Since $w^2 = 1$ and $w\$^+ = \$^+$, we can decompose $\$^+ = \$^+_+ \oplus \$^+_-$ into the ± 1 eigenspaces of w. Multiplying by e_1e_5 maps $\$^+_+$ isomorphically onto $\$^+_-$, so $\dim \$^+_+ = 4$. The group Spin_7 is transitive on the unit sphere in $\$^+$ so we may assume $\sigma \in \$^+_+$. Hence equality does occur in (10.15) for some choices of $e_1, \ldots, e_4$. This proves that the Λ^4-component of $\sigma \otimes \sigma$ is exactly the Cayley 4-form. ∎

Entirely analogous arguments prove the following:

Theorem 10.19. *Let $\$$ be the irreducible real $C\ell_7$-module on which the volume form ω acts by $+1$. Then for any unit spinor $\sigma \in \$$, the decomposition (10.14) becomes*

$$\sigma \otimes \sigma = 1 + \varphi + \psi + \omega$$

where $\varphi \in \Lambda^3\mathbb{R}^7$ is the associative calibration and $\psi = *\varphi \in \Lambda^4\mathbb{R}^7$ is the coassociative calibration.

It is clear that squaring spinors is a natural way to produce interesting calibrations in all dimensions. (This interesting fact was first noticed and proved by Dadok and Harvey [1].) The method is efficient both locally and globally on a manifold. It is also efficient for analysing the condition $d\varphi = 0$ as we shall now see.

Let Φ be a Cayley 4-form and consider the orbit $\mathrm{GL}_8 \cdot \Phi \subset \Lambda^4\mathbb{R}^8$. This orbit is independent of the choice of Φ and of the underlying inner product. It consists precisely of all 4-forms whose isotropy group in GL_8 is isomorphic to Spin_7. We call the forms in $\mathrm{Ca} \equiv \mathrm{GL}_8 \cdot \Phi$, the 4-forms of **Cayley type.** Note that any such form Φ' determines uniquely an inner product on $\mathbb{R}^8$ which makes Φ' a Cayley calibration.

Theorem 10.20. *Let X be a smooth oriented 8-manifold, and suppose Φ is a smooth 4-form on X which is of Cayley type at each point. Then Φ is closed if and only if Φ is parallel in the riemannian metric it determines.*

Therefore, to find a closed 4-form of Cayley type is to find a metric with holonomy $\subset \mathrm{Spin}_7$.

Proof. Introduce the metric for which Φ is a Cayley 4-form. Let σ be a local spinor field so that $\sigma \otimes \sigma = 1 + \Phi + \omega$. Since $*\Phi = \Phi$ we see that $d\Phi = 0 \Leftrightarrow d^*\Phi = 0 \Leftrightarrow D\Phi = 0 \Leftrightarrow D(\sigma \otimes \sigma) = 0$ where $D \cong d + d^*$ is the Dirac operator on $C\ell(X)$ discussed in Chapter II, §5. Let $\not{D}$ denote the Atiyah-Singer operator on the real spinor bundle $\$$. Theorem 10.19 will be a direct consequence of the following lemma:

Lemma 10.21. *For any $\sigma \in \Gamma(\$)$ with $\|\sigma\| \equiv 1$, we have*

$$\|D(\sigma \otimes \sigma)\|^2 = \|\not{D}\sigma\|^2 + \|\nabla\sigma\|^2.$$

Proof. From 10.17 we see that

$$\begin{aligned} D(\sigma \otimes \sigma) &= \sum e_j \nabla_{e_j}(\sigma \otimes \sigma) = \left(\sum e_j \nabla_{e_j}\sigma\right) \otimes \sigma + \sum (e_j\sigma) \otimes (\nabla_{e_j}\sigma) \\ &= (\not{D}\sigma) \otimes \sigma + \sum (e_j\sigma) \otimes (\nabla_{e_j}\sigma). \end{aligned}$$

Since $\|\sigma\|^2 \equiv 1$, we see that $\langle \nabla\sigma, \sigma\rangle \equiv 0$. Furthermore $\langle e_i\sigma, e_j\sigma\rangle = -\langle e_je_i\sigma, \sigma\rangle = \delta_{ij}$. It follows that

$$\begin{aligned}\|D(\sigma \otimes \sigma)\|^2 &= \|\not{D}\sigma\|^2 + \sum \langle e_i\sigma, e_j\sigma\rangle\langle \nabla_{e_i}\sigma, \nabla_{e_j}\sigma\rangle \\ &= \|\not{D}\sigma\|^2 + \sum \|\nabla_{e_i}\sigma\|^2 = \|\not{D}\sigma\|^2 + \|\nabla\sigma\|^2. \quad \blacksquare\end{aligned}$$

In analogy with the above we define a 3-form $\varphi \in \Lambda^3\mathbb{R}^7$ to be of **associative type** if $G_\varphi \equiv \{g \in \mathrm{GL}_7 : g^*\varphi = \varphi\}$ is isomorphic to G_2. Bryant observed that the set of such 3-forms is **open** in $\Lambda^3\mathbb{R}^7$. Each such 3-form determines a unique inner product on $\mathbb{R}^7$ which makes it an associative 3-form. Arguing as in the proof of 10.20 gives the following:

Theorem 10.22. *Let X be a smooth oriented 7-manifold and suppose φ is a smooth 3-form on X which is of associative type at each point. Then for the riemannian metric determined by φ, the form φ is closed and coclosed if and only if it is parallel.*

Therefore, to find a metric with holonomy $\subset G_2$ it suffices to find a closed 3-form φ, pointwise of associative type, such that

$$d(*_\varphi\varphi) = 0$$

where $*_\varphi$ is the Hodge $*$-operator for the metric determined by φ.

Characterizations of this type were formulated and used in the fundamental work of Bryant [1],[2].

§11. Spinor Cohomology and Complex Manifolds with Vanishing First Chern Class

In this section we shall examine the geometry of Dirac bundles over complex manifolds. We shall recapture a number of classical results and prove some new ones. As an application we shall show the existence of compact manifolds with Sp_n-holonomy. The results here are due to the second author and can be found in Michelsohn [1].

Fix a compact Kähler manifold X of dimension n, and let S be a holomorphic bundle of left modules over $C\ell(X)$ equipped with a hermitian metric and its associated canonical hermitian connection ∇. (This is the riemannian connection determined uniquely by the condition that $\sigma \in \Gamma(S)$ is holomorphic if and only if $\nabla_{Jv}\sigma = i\nabla_v\sigma$ for all $v \in TX$; (see Wells [1].) We assume that this metric and connection make S a Dirac bundle over X.

We define operators $\mathscr{D}$ and $\bar{\mathscr{D}}$ on $\Gamma(S)$ by setting

$$\mathscr{D} = \sum_{j=1}^{n} \varepsilon_j\nabla_{\bar{\varepsilon}_j} \qquad \bar{\mathscr{D}} = \sum_{j=1}^{n} \bar{\varepsilon}_j\nabla_{\varepsilon_j} \tag{11.1}$$

where $\{\varepsilon_1, \ldots, \varepsilon_n\}$ is any local hermitian frame field defined as in §§8 and 9. The curvature R^S of the canonical hermitian connection is of type (1,1), i.e., $R^S_{\varepsilon_i,\varepsilon_j} = R^S_{\bar{\varepsilon}_i,\bar{\varepsilon}_j} = 0$ for all i,j. From this and the identities (8.3) it follows that

$$\mathscr{D}^2 = 0 \qquad \text{and} \qquad \bar{\mathscr{D}}^2 = 0 \tag{11.2}$$

These operators are easily shown to be formal adjoints of one another.

Proposition 11.1. *There is a parallel orthogonal direct sum decomposition of S into holomorphic sub-bundles*

$$S = S^0 \oplus S^1 \oplus \ldots \oplus S^n$$

with the property that

$$\mathscr{D}:\Gamma(S^i) \longrightarrow \Gamma(S^{i+1}) \qquad \text{and} \qquad \bar{\mathscr{D}}:\Gamma(S^{i+1}) \longrightarrow \Gamma(S^i)$$

for all i.

Proof. Let $e_1, Je_1, \ldots, e_n, Je_n$ be a local unitary frame field on X and define $\varepsilon_j = \frac{1}{2}(e_j - Je_j)$ as above. Consider the family of commuting elements $\omega_1, \ldots, \omega_n, \bar{\omega}_1, \ldots, \bar{\omega}_n$ defined as in (9.7) by setting $\omega_j = -\varepsilon_j\bar{\varepsilon}_j$ and $\bar{\omega}_j = -\bar{\varepsilon}_j\varepsilon_j$. To each (possibly empty) subset $I = \{i_1, \ldots, i_r\} \subset \{1, \ldots, n\}$ with complementary subset $\{j_1, \ldots, j_{n-r}\}$ we associate the element

$$\omega_I = \omega_{i_1} \cdots \omega_{i_r}\bar{\omega}_{j_1} \cdots \bar{\omega}_{j_{n-r}}$$

and we denote $|I| = r$. From the relations (9.9)–(9.11) we can write

$$1 = \prod_{j=1}^{n} (\omega_j + \bar{\omega}_j) = \sum_{r=0}^{n} \pi_r$$

where

$$\pi_r = \sum_{|I|=r} \omega_I \tag{11.3}$$

The operators π_r are independent of the choice of unitary frame field and are therefore globally defined on X. They have the following basic properties for all r, s and j:

$$\nabla\pi_r = 0 \tag{11.4}$$

$$\pi_r^2 = \pi_r \tag{11.5}$$

$$\pi_r\pi_s = 0 \qquad \text{if } r \neq s \tag{11.6}$$

$$\langle \pi_r\sigma, \tau \rangle = \langle \sigma, \pi_r\tau \rangle \qquad \text{for } \sigma, \tau \in S \tag{11.7}$$

$$\varepsilon_j\pi_{r-1} = \pi_r\varepsilon_j \qquad \text{and} \qquad \bar{\varepsilon}_j\pi_{r+1} = \pi_r\bar{\varepsilon}_j. \tag{11.8}$$

Properties (11.5)–(11.8) follow directly from (9.8)–(9.13), as does the fact that $\pi_r \in \mathbb{C}\ell^{0,2r-n}(X)$ for each r. Since $0 = \nabla(1) = \sum \nabla\pi_r$ and ∇ preserves sections of the subbundle $\mathbb{C}\ell^{p,q}(X)$, we then obtain property (11.4). We now define

$$S^r \equiv \pi_r \cdot S \qquad r = 0,1,\ldots,n$$

From the properties above we see that the π_r's form a parallel family of orthogonal projection operators. Therefore by (11.3) we have the orthogonal decomposition $S = \oplus_r S^r$.

We now show that each S^r is holomorphic. Let $\sigma \in \Gamma(S)$ be a (local) holomorphic section, and write $\sigma_r = \pi_r\sigma$. Then for any $v \in TX$ we have that $\nabla_{Jv}(\sigma_r) = \pi_r(\nabla_{Jv}\sigma) = \pi_r(i\sigma) = i\sigma_r$. Hence, each σ_r is holomorphic, and one sees immediately that the space of such local cross sections spans the fibre of S^r at each point. Therefore S^r is a holomorphic sub-bundle.

The fact that $\mathscr{D}(\Gamma(S^r)) \subset \Gamma(S^{r+1})$ and $\bar{\mathscr{D}}(\Gamma(S^r)) \subset \Gamma(S^{r-1})$ follows directly from (11.3) and (11.8). ■

We now have defined a complex

$$0 \longrightarrow \Gamma(S^0) \xrightarrow{\mathscr{D}} \Gamma(S^1) \xrightarrow{\mathscr{D}} \cdots \xrightarrow{\mathscr{D}} \Gamma(S^n) \longrightarrow 0 \tag{11.9}$$

with adjoint complex

$$0 \longleftarrow \Gamma(S^0) \xleftarrow{\bar{\mathscr{D}}} \Gamma(S^1) \xleftarrow{\bar{\mathscr{D}}} \cdots \xleftarrow{\bar{\mathscr{D}}} \Gamma(S^n) \longleftarrow 0. \tag{11.10}$$

This complex is elliptic, i.e., at any non-zero cotangent vector the symbol sequence is exact. To see this note that the principal symbol of $\mathscr{D}$ at $\xi \in TX$ is given by Clifford multiplication $\sigma_\xi(\mathscr{D}) = \xi_{\mathbf{C}}\cdot \equiv \frac{1}{2}(\xi - iJ\xi)\cdot$. Similarly we have $\sigma_\xi(\bar{\mathscr{D}}) = \bar{\xi}_{\mathbf{C}}\cdot$. The exactness of (11.9) follows from the identity $\xi_{\mathbf{C}}\bar{\xi}_{\mathbf{C}} + \bar{\xi}_{\mathbf{C}}\xi_{\mathbf{C}} = -\|\xi\|^2$.

DEFINITION 11.2. The **rth cohomology group of the Dirac bundle** S is the quotient

$$\mathscr{H}^r(X,S) = \frac{\ker(\mathscr{D}|_{\Gamma(S^r)})}{\mathscr{D}(\Gamma(S^{r-1}))}.$$

From the ellipticity of the complex (11.9) it follows that each of the vector spaces $\mathscr{H}^r(X,S)$ is finite dimensional.

Taking $S = \mathbb{C}\ell(X)$ and using both $\mathscr{D}$ and $\mathscr{D}\hat{}$ we essentially recover the Clifford cohomology groups of X (see §8).

This viewpoint gives a simple unified approach to the fundamental vanishing theorems of Kähler geometry. Let us define elliptic self-adjoint operators $\nabla^*\nabla$ and $\bar{\nabla}^*\bar{\nabla}$ on $\Gamma(S)$ by

$$\nabla^*\nabla = -\sum_{j=1}^n \nabla^2_{\varepsilon_j,\bar{\varepsilon}_j} \qquad \bar{\nabla}^*\bar{\nabla} = -\sum_{j=1}^n \nabla^2_{\bar{\varepsilon}_j,\varepsilon_j}$$

with $\varepsilon_1,\ldots,\varepsilon_n$ as above. They have the property that $\int \langle\nabla^*\nabla\sigma,\sigma\rangle = \int |\nabla\sigma|^2$ where ∇ is defined by $\nabla_v\sigma = \frac{1}{2}(\nabla_v - i\nabla_{Jv})\sigma$. Consequently we have that

$$\ker(\nabla^*\nabla) = \{\text{holo. sections of } S\}, \quad \ker(\bar{\nabla}^*\bar{\nabla}) = \{\text{antiholo. sections of } S\}.$$

We now define associated curvature operators

$$\mathfrak{R}_{\mathbb{C}} = \sum_{j,k=1}^{n} \varepsilon_j \cdot \bar{\varepsilon}_k \cdot R^S_{\bar{\varepsilon}_j,\varepsilon_k}, \qquad \bar{\mathfrak{R}}_{\mathbb{C}} = \sum_{j,k=1}^{n} \bar{\varepsilon}_j \cdot \varepsilon_k \cdot R^S_{\varepsilon_j,\bar{\varepsilon}_k}.$$

Applying the arguments of II.8 and using the fact that $R^s_{\varepsilon_j,\varepsilon_k} = R^s_{\bar{\varepsilon}_j,\bar{\varepsilon}_k} = 0$ for all i,j, gives the following (Michelsohn [1]).

Theorem 11.3.

$$\mathscr{D}\bar{\mathscr{D}} + \bar{\mathscr{D}}\mathscr{D} = \nabla^*\nabla + \mathfrak{R}_{\mathbb{C}} = \bar{\nabla}^*\bar{\nabla} + \bar{\mathfrak{R}}_{\mathbb{C}}.$$

Taking the average of these two formulas gives a formula of the type presented in §8 of Chapter II.

We now consider a holomorphic hermitian line bundle, Λ, over X. The curvature R^Λ of Λ is an imaginary valued, J-invariant 2-form (i.e., (1,1)-form) on X. In terms of a hermitian frame field $\varepsilon_1, \dots, \varepsilon_n$ as above, we have

$$\overline{R^\Lambda_{\bar{\varepsilon}_j,\varepsilon_k}} = -R^\Lambda_{\varepsilon_j,\bar{\varepsilon}_k} = R^\Lambda_{\bar{\varepsilon}_k,\varepsilon_j}$$

for all j,k; that is, the matrix $a_{jk} = R^\Lambda_{\bar{\varepsilon}_j,\varepsilon_k}$ is hermitian symmetric. If this matrix is positive definite at each point, then Λ is called a **positive** hermitian line bundle. (If a_{jk} is negative definite at each point, then Λ is called **negative**. In this case Λ^* is positive.) It is straightforward to see that any closed imaginary (1,1)-form representing the first Chern class of Λ is the curvature form of a hermitian metric on Λ. We now have a vanishing theorem of "Kodaira type."

Theorem 11.4. *Let X be a compact Kähler manifold and let S be a holomorphic Dirac bundle as above. Then for any positive line bundle Λ there exists an integer N such that for $m \geqq N$*

$$\mathscr{H}^r(X, S \otimes \Lambda^m) = 0$$

for all $r > 0$.

Proof. The curvature of the bundle $S \otimes \Lambda^m$ is given by

$$R^{S\otimes\Lambda^m}_{V,W}(\sigma \otimes \ell) = (R^S_{V,W}\sigma) \otimes \ell + m\sigma \otimes (R^\Lambda_{V,W}\ell).$$

Since Λ is positive we may choose local hermitian frames so that $R^\Lambda_{\bar{\varepsilon}_j,\varepsilon_k} = -\lambda_j\delta_{jk}$ where $0 < \lambda_1 \leqq \lambda_2 \leqq \cdots \leqq \lambda_n$. It follows that

$$\begin{aligned}\mathfrak{R}^{S\otimes\Lambda^m}_{\mathbb{C}}(\sigma \otimes \ell) &= \left(\sum_{j,k} \varepsilon_j\bar{\varepsilon}_k R^S_{\bar{\varepsilon}_j,\varepsilon_k}\sigma\right) \otimes \ell + m\sum_{j,k} (\varepsilon_j\bar{\varepsilon}_k\sigma) \otimes (R^\Lambda_{\bar{\varepsilon}_j,\varepsilon_k}\ell) \\ &= \mathfrak{R}^S_{\mathbb{C}}(\sigma) \otimes \ell - m\left(\sum_j \lambda_j\varepsilon_j\bar{\varepsilon}_j\sigma\right) \otimes \ell \\ &= [\mathfrak{R}^S_{\mathbb{C}}(\sigma) + m\tau\sigma] \otimes \ell\end{aligned}$$

where $\tau = -\sum \lambda_j \varepsilon_j \bar{\varepsilon}_j = \sum \lambda_j \omega_j$. Note that for any $\sigma \in S$ we have

$$(\tau\sigma, \sigma) = \sum \lambda_j(\omega_j\sigma, \sigma) = \sum \lambda_j(\omega_j^2\sigma, \sigma) = \sum \lambda_j \|\omega_j\sigma\|^2$$
$$\geqq \lambda_1 \sum \|\omega_j\sigma\|^2 = \lambda_1 \sum (\omega_j\sigma, \sigma) = \lambda_1(w\sigma, \sigma)$$

where $w = \sum \omega_j$. This element has the fundamental property that

$$w\pi_r = r\pi_r$$

which follows from the elementary fact that $\omega_i\omega_I = \omega_I$ if $i \in I$ and $\omega_i\omega_I = 0$ otherwise. Consequently, if $\sigma \in S^r$, i.e., if $\pi_r\sigma = \sigma$, then $w\sigma = r\sigma$. It follows that on $\Gamma(S^r \otimes \Lambda^m)$

$$\mathfrak{R}_{\mathbf{C}}^{S \otimes \Lambda^m} \geqq (\mathfrak{R}_{\mathbf{C}}^S + mr\lambda) \otimes 1$$

where $\lambda > 0$ is the minimum of λ_1 on X. For $r \geqq 1$ and m sufficiently large we have $\mathfrak{R}_{\mathbf{C}}^{S \otimes \Lambda^m} > 0$, and it follows from 11.3 that $\mathscr{D}\bar{\mathscr{D}} + \bar{\mathscr{D}}\mathscr{D} > 0$ on $\Gamma(S^r \otimes \Lambda^m)$. ∎

There is also a vanishing theorem of "Nakano type" for the Clifford bundle. We refer to Michelsohn [1] for the proof.

Theorem 11.5. *If Λ is a negative line bundle over a compact Kähler manifold X, then*

$$\mathscr{H}^{p,q}(X, \Lambda) = 0$$

for all $q < 0$.

Let us now suppose that X is a Kähler spin manifold, i.e., that $c_1(X) \equiv 0 \pmod 2$, and let $\$_{\mathbf{C}}$ be the bundle of complex spinors on X with its canonical riemannian connection. This bundle is holomorphic; in fact, it is equivalent as a complex vector bundle to $\Lambda_{\mathbf{C}}^* T^*X \otimes \delta^{1/2}$ where $\delta^{-1} = \Lambda_{\mathbf{C}}^n T^*X$ is the **canonical bundle** (see App. D.). Since X is Kähler, the canonical connection on $\$_{\mathbf{C}}$ is also the canonical hermitian one. We can therefore apply Theorem 11.3. To compute the terms $\mathfrak{R}_{\mathbf{C}}$ and $\bar{\mathfrak{R}}_{\mathbf{C}}$ which appear there, we need to know the curvature tensor of $\$_{\mathbf{C}}$. From (4.37) of Chapter II we see that this can be expressed in terms of the Riemann curvature tensor R of X by the formula

$$R_{V,W}^{\$} = \tfrac{1}{4} \sum_{i,j=1}^{2n} \langle R_{V,W}\eta_i, \eta_j \rangle \eta_i \cdot \eta_j \cdot$$

where $\eta_1, \ldots, \eta_{2n}$ is any real orthonormal basis of the tangent space. Choosing a unitary basis $e_1, Je_1, \ldots, e_n, Je_n$ and writing $\{\varepsilon_j\}$ as above, we can reexpress this as

$$R_{V,W}^{\$} = \sum_{j,k=1}^{n} \{\langle R_{V,W}\varepsilon_j, \bar{\varepsilon}_k \rangle \bar{\varepsilon}_j \cdot \varepsilon_k \cdot + \langle R_{V,W}\bar{\varepsilon}_j, \varepsilon_k \rangle \varepsilon_j \cdot \bar{\varepsilon}_k \cdot\}$$
$$= 2\sum_{j,k} \langle R_{V,W}\varepsilon_j, \bar{\varepsilon}_k \rangle \bar{\varepsilon}_j \cdot \varepsilon_k \cdot + \sum_j \langle R_{V,W}\varepsilon_j, \bar{\varepsilon}_j \rangle$$

where we use the fact that $\varepsilon_j\bar{\varepsilon}_k + \bar{\varepsilon}_k\varepsilon_j = -\delta_{jk}$. It follows that

$$\begin{aligned}\mathfrak{R}_{\mathbb{C}} &= \sum_{j,k=1}^{n} \varepsilon_j\bar{\varepsilon}_k R^{\$}_{\bar{\varepsilon}_j,\varepsilon_k} \\ &= 2\sum_{j,k,l,m} R_{\bar{j}kl\bar{m}}\, \varepsilon_j \cdot \bar{\varepsilon}_k \cdot \bar{\varepsilon}_l \cdot \varepsilon_m \cdot + \sum_{j,k,l} R_{\bar{j}kl\bar{l}}\varepsilon_j \cdot \bar{\varepsilon}_k \cdot\end{aligned} \tag{11.11}$$

where $R_{\bar{j}kl\bar{m}}$ denotes $\langle R_{\bar{\varepsilon}_j,\varepsilon_k}\varepsilon_l, \bar{\varepsilon}_m\rangle$. Now the Bianchi identity states that $R_{\bar{j}kl\bar{m}} = R_{\bar{j}lk\bar{m}} + R_{lk\bar{j}\bar{m}} = R_{\bar{j}lk\bar{m}}$. That is, $R_{\bar{j}kl\bar{m}}$ is symmetric in k and l. But $\bar{\varepsilon}_k\bar{\varepsilon}_l = -\bar{\varepsilon}_l\bar{\varepsilon}_k$, and so equation (11.11) becomes

$$\mathfrak{R}_{\mathbb{C}} = \sum_{j,k,l} \langle R_{\varepsilon_l,\bar{\varepsilon}_l}\bar{\varepsilon}_j, \varepsilon_k\rangle \varepsilon_j \cdot \bar{\varepsilon}_k \cdot.$$

This relates to the Ricci form of X, which for tangent vectors V and W is given by

$$\begin{aligned}\mathrm{Ric}(V,W) &= -\sum_j \{\langle R_{e_j,V}e_j, W\rangle + \langle R_{Je_j,V}Je_j, W\rangle\} \\ &= -2i\sum_j \langle R_{\varepsilon_j,\bar{\varepsilon}_j}JV, W\rangle.\end{aligned}$$

Thus, we are able to write $\mathfrak{R}_{\mathbb{C}}$ as

$$\mathfrak{R}_{\mathbb{C}} = -\frac{1}{2}\sum_{j,k} \mathrm{Ric}(\bar{\varepsilon}_j,\varepsilon_k)\varepsilon_j \cdot \bar{\varepsilon}_k \cdot.$$

Since Ric is hermitian symmetric we can choose our basis so that $\mathrm{Ric}(\bar{\varepsilon}_j,\varepsilon_k) = \frac{1}{2}\lambda_j\delta_{jk}$ where $\lambda_j = \mathrm{Ric}(e_j,e_j) = \mathrm{Ric}(Je_j,Je_j)$ for $j = 1,\ldots,n$ are the eigenvalues. As before, we write $-\varepsilon_j\bar{\varepsilon}_j = \omega_j$ and carry out a similar analysis for $\bar{\mathfrak{R}}_{\mathbb{C}}$. Thus we obtain the following:

Theorem 11.6. *Let X be a Kähler manifold equipped with a spin structure. Then on the bundle of spinors $\$_{\mathbb{C}}$*

$$\mathscr{D}\bar{\mathscr{D}} + \bar{\mathscr{D}}\mathscr{D} = \nabla^*\nabla + \frac{1}{4}\sum_j \lambda_j\omega_j = \bar{\nabla}^*\bar{\nabla} + \frac{1}{4}\sum \lambda_l\bar{\omega}_j,$$

where $\lambda_1,\ldots,\lambda_n$ are the eigenvalues of the Ricci tensor.

Taking the average of these two formulas gives the Lichnerowicz Theorem (II.8.8). To see this we observe that

$$\nabla^*\nabla + \bar{\nabla}^*\bar{\nabla} = \frac{1}{2}\nabla^*\nabla \qquad \text{and} \qquad \sum_j \lambda_j(\omega_j + \bar{\omega}_j) = \sum_j \lambda_j = \frac{1}{2}\kappa \tag{11.12}$$

where $\kappa = \mathrm{trace}_{\mathbb{R}}(\mathrm{Ric})$ is the scalar curvature of X. Theorem 11.6 easily gives the following:

Corollary 11.7. *Let Λ be a hermitian line bundle over a Kähler manifold X with curvature form R^Λ, and assume that X is spin. If* $\mathrm{Ric} > 2R^\Lambda$, *then*

$$\mathscr{H}^r(X, \$_{\mathbb{C}} \otimes \Lambda) = 0 \tag{11.13}$$

for all $r > 0$. Similarly, if $\mathrm{Ric} > -2R^\Lambda$, *then (11.13) holds for all $r < n$.*

Proof. Straightforward computation shows that the terms $\mathfrak{R}_{\mathbb{C}}$ and $\bar{\mathfrak{R}}_{\mathbb{C}}$ in Theorem 11.3 for the Dirac bundle $\$_{\mathbb{C}} \otimes \Lambda$ are exactly given by Clifford multiplication by the elements

$$\begin{aligned} \mathfrak{R}_{\mathbb{C}} &= \sum_{j,k} \left(-\frac{1}{2} \mathrm{Ric}_{j,k} + R^\Lambda_{j\bar{k}} \right) \varepsilon_j \bar{\varepsilon}_k \\ \bar{\mathfrak{R}}_{\mathbb{C}} &= \sum_{j,k} \left(-\frac{1}{2} \mathrm{Ric}_{\bar{j}k} - R^\Lambda_{\bar{j}k} \right) \bar{\varepsilon}_k \varepsilon_j \end{aligned} \tag{11.14}$$

Choosing a basis which diagonalizes the hermitian form $\frac{1}{2}\mathrm{Ric} - R$ with eigenvalues $\mu_1, \ldots, \mu_n$ we find that $\mathfrak{R}_{\mathbb{C}} = \sum \mu_j \omega_j$. This proves the first statement. The second is proved similarly. ■

Taking the average of the two formulas of Theorem 11.3 in this case and applying (11.12) gives the formula (cf. Theorem D.12)

$$4(\mathscr{D}\bar{\mathscr{D}} + \bar{\mathscr{D}}\mathscr{D}) = \nabla^*\nabla + \tfrac{1}{4}\kappa + \Omega^\Lambda \tag{11.15}$$

where Ω^Λ is the Clifford element corresponding to the curvature 2-form of Λ.

Theorem 11.8. *Let X be a compact Kähler manifold equipped with a spin structure, and let Λ be a hermitian line bundle over X. If the scalar curvature κ satisfies the inequality $\frac{1}{4}\kappa > |\lambda_1| + \cdots + |\lambda_n|$ at each point, where $\lambda_1, \ldots, \lambda_n$ are the eigenvalues of the curvature form of Λ, then the cohomology groups*

$$\mathscr{H}^r(X, \$_{\mathbb{C}} \otimes \Lambda) = 0$$

for all r. In particular, by the Atiyah-Singer Index Theorem,

$$\{\mathrm{ch}\,\Lambda \cdot \hat{\mathbf{A}}(X)\}[X] = 0.$$

It is interesting to specialize our discussion to the case where $\Lambda = \delta^{r/s}$ where $\delta = \Lambda^n_{\mathbb{C}} TX$ is the anticanonical line bundle of X (the line bundle generated by the global section $\varepsilon_1 \cdots \varepsilon_n$ in $\mathbb{C}\ell(X)$). An easy adaptation of the arguments given for D.2 shows the following (see Michelsohn [1]):

Lemma 11.9. *Suppose $c_1(X) = k\alpha$ where k is odd and $\alpha \in H^2(X;\mathbb{Z})$ is indivisible (i.e., not a non-trivial integral multiple of any other class in $H^2(X;\mathbb{Z})$). Then for any odd integer m, there is a bundle of the form $\$_{\mathbb{C}} \otimes \delta^{m/2k}$ globally* *defined on X where locally $\$_{\mathbb{C}}$ is the bundle of spinors with its canonical connection.*

For any rational number t, the curvature of δ^t with its natural metric is just tR^δ where for any unit vector e, $R^\delta_{e,Je} = \text{Ric}(e,e) = \text{Ric}(Je,Je)$. If we choose a unitary basis $e_1,Je_1,\ldots,e_n,Je_n$ so that $\text{Ric}(e_j,e_k) = \lambda_j\delta_{jk}$, then $R^{t\delta}_{\bar{\varepsilon}_j,\varepsilon_k} = t\text{Ric}(\bar{\varepsilon}_j,\varepsilon_k) = (t/2)\lambda_j\delta_{jk}$, and applying (11.14) gives the following:

Lemma 11.10. *On $\Gamma(\$_{\mathbb{C}} \otimes \delta^{t/2})$*

$$\mathscr{D}\bar{\mathscr{D}} + \bar{\mathscr{D}}\mathscr{D} = \nabla^*\nabla + \frac{1}{4}(1+t)\sum_{j=1}^{n}\lambda_j\omega_j$$
$$= \bar{\nabla}^*\bar{\nabla} + \frac{1}{4}(1-t)\sum_{j=1}^{n}\lambda_j\bar{\omega}_j$$

where $\lambda_1,\ldots,\lambda_n$ are the eigenvalues of the Ricci tensor on X.

REMARK 11.11. The operators which appear here can be diagonalized as follows. Suppose that $\sigma \in \$^r_{\mathbb{C}} \otimes \delta^{t/2}$, i.e., that $\pi_r\sigma = \sigma$. From the fact that $\pi_r = \sum_{|I|=r}\omega_I$ and that

$$\omega_i\omega_I = \begin{cases}\omega_I & \text{if } i \in I\\ 0 & \text{if } i \notin I\end{cases} \qquad \bar{\omega}_i\omega_I = \begin{cases}0 & \text{if } i \in I\\ \omega_I & \text{if } i \notin I,\end{cases}$$

we see that on $\$^r_{\mathbb{C}} \otimes \delta^{t/2}$ we have

$$\sum_j \lambda_j\omega_j = \sum_{|I|=r}\lambda_I\omega_I \qquad \text{and} \qquad \sum_j \lambda_j\bar{\omega}_j = \sum_{|I|=r}\lambda_{I'}\omega_I$$

where

$$\lambda_I = \lambda_{i_1} + \cdots + \lambda_{i_r} \qquad \text{and} \qquad \lambda_{I'} = \lambda_1 + \cdots + \lambda_n - \lambda_I \tag{11.16}$$

when I is the multiindex $\{i_1,\ldots,i_r\}$. It follows from (9.9)–(9.13) that multiplication by the elements ω_I constitutes a family of orthogonal projections onto mutually perpendicular subspaces of $\$_{\mathbb{C}} \otimes \delta^{t/2}$. These formulas give a delicate and precise analysis of what conditions are necessary for the positivity of $\sum\lambda_j\omega_j$ and $\sum\lambda_j\bar{\omega}_j$. Note in particular that the numbers $\{\lambda_I : |I| = r\}$ are precisely the eigenvalues of Ric acting on $\Lambda^r TX$ as a derivation.

We see from Lemma 11.9 that $\$^r_{\mathbb{C}} \otimes \delta^{1/2}$ always exists globally. It corresponds naturally to the bundle $\Lambda^{0,*}T^*X$ (see App. D), and there is an isomorphism $\mathscr{H}^r(X, \$_{\mathbb{C}} \otimes \delta^{1/2}) \cong H^r(X,\mathcal{O})$, where $\mathcal{O}$ is the structure sheaf of X.

Corollary 11.12. *On $\Gamma(\$^r_{\mathbb{C}} \otimes \delta^{1/2})$,*

$$\mathscr{D}\bar{\mathscr{D}} + \bar{\mathscr{D}}\mathscr{D} = \nabla^*\nabla + \frac{1}{2}\sum_{|I|=r}\lambda_I\omega_I = \bar{\nabla}^*\bar{\nabla}$$

In particular, if Ric > 0, *then* $\mathscr{H}^r(X, \$_{\mathbb{C}} \otimes \delta^{1/2}) = 0$ *for all* $r > 0$ *and*

$$\{\operatorname{ch} \delta^{1/2} \cdot \hat{\mathbf{A}}(X)\}[X] = \operatorname{Td}(X) = 1.$$

Arguing as above, we can conclude the following:

Theorem 11.13. *Let* Λ *be a hermitian line bundle over* X *and suppose that* $(1 + t)\text{Ric} + 2R^\Lambda > 0$. *If* $\$_{\mathbb{C}} \otimes \delta^{1/2} \otimes \Lambda$ *exists globally on* X, *then*

$$\mathscr{H}^r(X, \$_{\mathbb{C}} \otimes \delta^{1/2} \otimes \Lambda) = 0 \tag{11.17}$$

for all $r > 0$. *Similarly, if* $(1 - t)\text{Ric} - 2R^\Lambda > 0$, *then* (*11.17*) *holds for all* $r < n$.

We now average the formulas of Lemma 11.10 and obtain

Theorem 11.14. *Let* X *be a compact Kähler manifold with* $c_1(X) = k\alpha$, $k \in \mathbb{Z}^+$ *where* $\alpha \in H^2(X;\mathbb{Z})$ *is indivisible. Let* $\$_{\mathbb{C}} \otimes \delta^{p/2k}$ *be the twisted bundle of spinors where* $\delta^{1/2k}$ *is a local holomorphic 2kth root of the anti-canonical bundle and where* $p + k \equiv 0 \pmod 2$. *Then on* $\Gamma(\$^r_{\mathbb{C}} \otimes \delta^{p/2k})$

$$4(\mathscr{D}\bar{\mathscr{D}} + \bar{\mathscr{D}}\mathscr{D}) = \nabla^*\nabla + \frac{1}{2k} \sum_{|I| = r} [(k + p)\lambda_I + (k - p)\lambda_{I'}]\omega_I.$$

In particular, if Ric > 0, *then*

$$\mathscr{H}^r(X, \$_{\mathbb{C}} \otimes \delta^{p/2k}) = 0 \qquad \textit{for all } r$$

whenever $|p| < k$ *and* $p + k$ *is even.*

We apply the Atiyah-Singer Theorem to the complex $\$^*_{\mathbb{C}} \otimes \delta^{p/2k}$ to get

Theorem 11.15. *Let* X *be a compact complex manifold such that* $c_1(X) = k\alpha$, $k \in \mathbb{Z}^+$, *where* α *is indivisible. If* X *admits a Ricci-positive Kähler metric, then the Hilbert polynomial*

$$P_\alpha(t) = \{e^{\frac{1}{2}t\alpha} \cdot \hat{\mathbf{A}}(X)\}[X]$$

vanishes for all integers t *such that* $|t| < k$ *and* $t + k$ *is even.*

From Yau [2], [3] we know that X carries a Ricci-positive Kähler metric if and only if $c_1(\delta) = c_1(X)$ is represented by a closed positive (1,1)-form.

To illustrate Theorem 11.15 we consider complex projective n-space $\mathbb{P}^n(\mathbb{C})$ and let $\omega \in H^2(\mathbb{P}^n(\mathbb{C}); \mathbb{Z}) = \mathbb{Z}$ denote the standard generator. Then $c_1 = (n + 1)\omega$ and a direct calculation shows that

$$P_\omega(t) = \frac{1}{2^n n!} \prod_{j=1}^{n} (t - n + 2j - 1).$$

This is predicted by Theorem 11.15 since $\mathbb{P}^n(\mathbb{C})$ carries a Kähler metric of positive curvature.

From the work of Yau [2], [3] we know that the hypersurface $V^n(d) \subset \mathbb{P}^{n+1}(\mathbb{C})$ carries a Kähler metric of positive Ricci curvature for $d \leq n + 1$. Furthermore, for $n \geq 2$, $c_1(V^n(d)) = (n + 2 - d)\omega$ where ω is a generator of $H^2(V^n(d); \mathbb{Z})$. It follows that the polynomial $P_\omega(t)$ on $V^n(d)$ has zeros at $n - d - 2j$ for $j = 0, \ldots, n - d$.

Theorem 11.15 has the following consequence:

Corollary 11.16. *Let X be a compact complex manifold of dimension n. If X admits a Kähler metric of positive Ricci curvature, then $k \nmid c_1(X)$ for $k > n + 1$.*

Proof. The Hilbert polynomial of 11.15 is of degree $\leq n$ and can have therefore no more than n zeros. ∎

One case of Theorem 11.14 is of particular interest since it involves only the scalar curvature $\kappa = 2(\lambda_1 + \cdots + \lambda_n)$.

Corollary 11.17. *Let X, δ and k be as in Theorem* 11.14. *If the scalar curvature κ of X satisfies $\kappa > 0$, then*

$$\mathscr{H}^n(X, \$_{\mathbb{C}} \otimes \delta^{p/2k}) = 0 \qquad \text{for all } p > -k$$

$$\mathscr{H}^0(X, \$_{\mathbb{C}} \otimes \delta^{p/2k}) = 0 \qquad \text{for all } p < k$$

where $p + k$ is even. Similarly, if $\kappa < 0$, then

$$\mathscr{H}^n(X, \$_{\mathbb{C}} \otimes \delta^{p/2k}) = 0 \qquad \text{for all } p < -k$$

$$\mathscr{H}^0(X, \$_{\mathbb{C}} \otimes \delta^{p/2k}) = 0 \qquad \text{for all } p > k$$

where $p + k$ is even. In particular if $\kappa > 0$, then all the plurigenera of X are zero (cf. Yau [1]).

This last statement follows by setting $p = k - 2kg$ for $g \in \mathbb{Z}^+$. The **gth plurigenus** of X is exactly the dimension of $\mathscr{H}^0(X, \$_{\mathbb{C}} \otimes \delta^{\frac{1}{2}-g}) \cong H^0(X, \mathcal{O}(\delta^{-g}))$.

Note that the formula in Theorem 11.14 is particularly simple when $\mathrm{Ric} \equiv 0$. From Yau's proof of the profound conjectures of Calabi we know exactly when such metrics exist.

Theorem 11.18 (Yau [2]). *Let X be a compact complex manifold which admits at least one Kähler metric. Then X carries a Ricci-flat Kähler metric if and only if $c_1(X) = 0$.*

We can now apply the Splitting Theorem of Cheeger and Gromoll [2] to conclude that any compact Ricci flat Kähler manifold X has a finite covering $\tilde{X}$ which decomposes into a Kähler product

$$\tilde{X} = T \times X_0$$

where T is a flat complex torus and where X_0 is a compact Ricci-flat manifold which is *simply-connected.*

When Ric $\equiv 0$, Theorem 11.14 gives us the following simple identity between operators

$$\mathscr{D}\bar{\mathscr{D}} + \bar{\mathscr{D}}\mathscr{D} = \nabla^*\nabla \tag{11.18}$$

on $\$_{\mathbb{C}}$. In this case there is a canonical connection-preserving isometry $\$_{\mathbb{C}}^* = \Lambda^{0,*}(X) = \Lambda_{\mathbb{C}}^* TX$ of graded bundles, which carries $\mathscr{D}$ to $\bar{\partial}$ and therefore identifies $\mathscr{H}^r(X,\$_{\mathbb{C}})$ with $H^r(X,\mathcal{O})$ (cf. App. D). Equation (11.18) therefore implies the following:

Proposition 11.19. *On a compact Ricci-flat Kähler manifold, any harmonic spinor field is parallel. Equivalently, any harmonic (0,r)-form is parallel.*

Suppose now that X is a compact simply-connected Ricci-flat Kähler manifold of dimension n. Then the line bundles $\$_{\mathbb{C}}^0 \cong \Lambda^{0,0}$ and $\$_{\mathbb{C}}^n \cong \Lambda^{0,n}$ are trivial (they have parallel cross-sections). The index of the elliptic complex (11.9) in this case is just

$$\hat{A}(X) = \sum (-1)^r \dim \mathscr{H}^r(X,\$_{\mathbb{C}}) = \sum (-1)^r \dim H^r(X,\mathcal{O}) = \mathrm{Td}(X)$$

and $\dim \mathscr{H}^0 = \dim \mathscr{H}^1 = 1$. Moreover, by the generalized Hodge Decomposition Theorem (cf. III.5.5) we have $\mathscr{H}^r(X,\$_{\mathbb{C}}) \cong \ker(\mathscr{D}\bar{\mathscr{D}} + \bar{\mathscr{D}}\mathscr{D})|_{\Gamma(\$_{\mathbb{C}}^r)} = \ker(\nabla^*\nabla|_{\Gamma(\$_{\mathbb{C}}^r)})$. Consequently, if $\mathrm{Td}(X) \neq 1 + (-1)^n$, then there must exist non-trivial parallel sections of $\$_{\mathbb{C}}^r$ (i.e., parallel $(0,r)$-forms) for some $r \neq 0,n$. This implies that the local holonomy group G of X is properly contained in SU_n. If G is a product of two non-trivial groups, then X is a product manifold. Otherwise G belongs to the list of Berger discussed in §10, and we conclude that either X is locally symmetric or $G = \mathrm{Sp}_{n/2}$, for n even and >2. It is elementary that locally symmetric, Ricci-flat manifolds are flat (and therefore not simply-connected in the compact case). Combining the three paragraphs above proves the following general structure theorem.

Theorem 11.20 (Michelsohn [1]). *Let X be a compact simply-connected Kähler manifold with $c_1(X) = 0$. Then there is a finite covering manifold $\tilde{X}$ of X which is biholomorphically equivalent to a product of compact Kähler manifolds with vanishing first Chern class*

$$\tilde{X} = T \times X_1 \times \cdots \times X_k \times Y$$

where T is a complex torus and each of $X_1, \ldots, X_k, Y$ is simply-connected, where

$$\mathrm{Td}(X_j) = \begin{cases} 0 & \text{if } \dim X_j \text{ is odd} \\ 2 & \text{if } \dim X_j \text{ is even} \end{cases}$$

and where Y admits a metric with Sp_m-holonomy.

This decomposition theorem is a wonderfully effective device for detecting manifolds with Sp_m-holonomy.

Corollary 11.21. *Let X be a compact simply-connected Kähler manifold of dimension $2m$ with $c_1(X) = 0$. If* $\mathrm{Td}(X) \neq 0$ or 2^k *for some k, then X carries a metric with* Sp_m*-holonomy. Alternatively, if X is not a product and* $\mathrm{Td}(X) \neq 2$, *then the same conclusion holds.*

Theorem 11.20 with its corollary first appeared in Michelsohn [1] in prepublication form. Before printing, it was modified due to the publication of an announcement of the non-existence of compact manifolds with Sp_m-holonomy for $m \geqq 2$. This announcement was subsequently found to be in error. The first counterexample was produced by A. Fujiki using the results above. To construct this example we first take the symmetric square of a Kummer surface. This is the compact analytic space $X_0 = V^2(4) \times V^2(4)/\mathbb{Z}_2$ where $\mathbb{Z}_2$ is generated by the interchange of factors in the product. Resolving the singularities along the diagonal gives a compact complex 4-manifold X. Alternatively, we could construct X by first blowing up the diagonal (i.e., by replacing the diagonal by its projectivized normal bundle) and then dividing by the natural extension of the "flip" to this space. The manifold X is a certain component of the Hilbert scheme of 0-cycles on $V^2(4)$. It is straightforward to see that X is simply-connected and algebraic (hence Kähler) and that $c_1(X) = 0$. However, further computation shows that $\mathrm{Td}(X) = 3$. Consequently the Calabi-Yau metric on X, given by Theorem 11.18, has Sp_2-holonomy. (See Beauville [1] for further discussion and examples.)

§12. The Positive Mass Conjecture in General Relativity

The classical theory of gravity, as enunciated by Einstein, is a subject formulated in the language of differential geometry. In many ways it stands apart from other theories of modern physics. One of its curious features is that there seems to be no satisfactory way of defining an energy density for a gravitational field. Nevertheless, there is a concept of the total energy of a gravitating system, which is defined in terms of the asymptotic behavior of the field at large distances. It is in many ways essential to the theory that this energy be $\geqq 0$ (and $= 0$ only on flat Minkowski space).

The proof of this "positive mass conjecture" has a long history during which many special cases were established. The key case of space-times admitting a maximal space-like hypersurface was first proved by Schoen and Yau [3] using minimal surface techniques. Using Jang's equation they subsequently adapted their techniques to prove the general result [5].

Of interest here is the fact that an alternative proof of this theorem can be given using the Dirac operator on spinors. This proof was found by E. Witten [1]. Very roughly the idea is as follows: Consider a space-time X (i.e., a 4-dimensional Lorentzian manifold), with a properly embedded space-like hypersurface $M \subset X$. We assume that X satisfies Einstein's equations

$$\mathrm{Ric} - \tfrac{1}{2} g\kappa = T$$

where the energy-momentum tensor T is positive on time-like vectors. It is assumed that as one goes to infinity in M, the geometry of $M \subset X$ is, in a very explicit way, asymptotic to the standard geometry $\mathbb{R}^3 \subset \mathbb{R}^{3,1}$ in Minkowski space. In this context Witten considers the following problem. Let S denote the spinor bundle of X with its canonical connection ∇, and consider the restriction of (S,∇) to the hypersurface M. On this bundle one has a Dirac operator defined by

$$\not{\partial} = \sum_{j=1}^{3} e_j \cdot \nabla_{e_j}$$

where e_1,e_2,e_3 is any local orthonormal frame field on M. One now applies Bochner's method and writes

$$\not{\partial}^2 = \nabla^*\nabla + \mathfrak{I} \tag{12.1}$$

where, since ∇ is the metric connection of X (and not of M), the term $\mathfrak{I}$ involves the second fundamental form of the hypersurface M in X. Computation shows that the positivity of the tensor T on the normal vectors to M implies the positivity of $\mathfrak{I}$. This in turn shows that on suitable subspaces of $\Gamma(S)$ the operator $\not{\partial}$ is invertible. Since S is asymptotically flat, one can consider solutions of the equation $\not{\partial}\sigma = 0$ which are asymptotically constant. Witten shows that there are unique solutions for each constant value at infinity. Let σ be such a solution and consider a region $\Omega \subset M$ with smooth boundary $\partial\Omega$. From (12.1) we have

$$\begin{aligned} 0 &= \int_\Omega \{\langle \nabla^*\nabla\sigma, \sigma\rangle + \langle \mathfrak{I}(\sigma), \sigma\rangle\} \\ &= -E_\Omega(\sigma) + \int_\Omega \{\|\nabla\sigma\|^2 + \langle \mathfrak{I}(\sigma), \sigma\rangle\} \end{aligned}$$

where $E_\Omega(\sigma)$ is an integral over $\partial\Omega$ which satisfies

$$E_\Omega(\sigma) \geqq 0 \tag{12.2}$$

since $\mathfrak{I} \geqq 0$. Witten shows that the inequality (12.2) viewed asymptotically, establishes the positivity of the energy of the system.

For the complete story the reader is referred to the original paper of Witten [1] and also to the rigorous account of Parker and Taubes [1].

APPENDIX A

Principal G-Bundles

Let X be a paracompact Hausdorff space and G a topological group. A **principal G-bundle** over X is essentially a bundle of "affine G-spaces" over X. To be precise, it is a fibre bundle $\pi: P \to X$ together with a continuous, right action of G on P which preserves the fibres and acts simply and transitively on them. Thus, the fibres are exactly the orbits of G. Moreover, every point in X has a neighborhood U and a homeomorphism $h_U: \pi^{-1}(U) \xrightarrow{\approx} U \times G$ of the form $h_U(p) = (\pi(p), \gamma(p))$ and with the property that $h(pg) = (\pi(p), \gamma(p)g)$ for all $g \in G$. Thus, the bundle is locally of the form

$$\begin{array}{c} U \times G \\ \downarrow \\ U \end{array}$$

where G acts by multiplication on the right.

Two principal G-bundles $\pi: P \to X$ and $\pi': P' \to X$ are said to be **equivalent** if there is a homeomorphism $H: P \to P'$ so that

$$\begin{array}{ccc} P & \xrightarrow{H} & P' \\ & {\scriptstyle\pi}\searrow \quad \swarrow{\scriptstyle\pi} & \\ & X & \end{array}$$

commutes and so that $H(pg) = H(p)g$ for all $g \in G$. The equivalence classes of principal G-bundles over X will be denoted by $\mathrm{Prin}_G(X)$.

EXAMPLE. Let $\pi: P \to X$ be a 2-sheeted covering space of X, and let $G = \mathbb{Z}_2$. The group $\mathbb{Z}_2$ acts on P by interchanging the sheets. This is clearly a principal $\mathbb{Z}_2$-bundle. In fact it is not difficult to see that $\mathrm{Prin}_{\mathbb{Z}_2}(X) \cong \mathrm{Cov}_2(X)$ for any manifold X.

More generally, any normal covering of a manifold is a principal Γ-bundle where Γ is the group of deck transformations of the covering (with the discrete topology).

EXAMPLE. Let E be a real n-dimensional vector bundle over X, and let $P_{\mathrm{GL}}(E)$ be the **bundle of bases** in E, i.e., the bundle whose fibre at $x \in X$ is the set of all bases for the vector space E_x. This is a principal GL_n-bundle where GL_n is the group of invertible $n \times n$ real matrices. The action of GL_n on $P_{GL}(E)$ is defined as follows. Fix a matrix $g = ((a_{ij}))$ in GL_n. Then given a basis $p = (v_1, \ldots, v_n)$ of E_x at a point x, we set $pg \equiv (v'_1, \ldots, v'_n)$ where

$$v'_j = \sum_k v_k a_{kj}, \qquad j = 1, \ldots, n$$

This action is clearly continuous and is simple and transitive on the fibres.

If E is oriented, we may consider the **bundle $P_{GL^+}(E)$ of oriented bases** in E. The construction above makes this a principal GL_n^+-bundle where $GL_n^+ \equiv \{g \in GL_n : \det(g) > 0\}$.

If E is riemannian, we can consider the **bundle $P_O(E)$ of orthonormal bases.** This is a principal O_n-bundle where O_n is the orthogonal group. If X is oriented we get the **bundle $P_{SO}(E)$ of oriented orthonormal bases,** which is a principal SO_n-bundle, where $SO_n \equiv \{g \in O_n : \det(g) = 1\}$.

There are natural complex and quaternionic analogues of these constructions.

Recall that a general fibre bundle $B \xrightarrow{\pi} X$ with fibre F and structure group $G \subseteq \text{Homeo}(F)$ is given by the following data. There is an open cover $\mathscr{U} = \{U_\alpha\}_{\alpha \in A}$ of X, and over each U_α there is a **local trivialization** $\pi^{-1}(U_\alpha) \xrightarrow{h_\alpha} U_\alpha \times F$ so that $\text{pr} \circ h_\alpha = \pi$ (where pr is projection onto U_α). The change of trivialization over $U_\alpha \cap U_\beta$ is of the form $(U_\alpha \cap U_\beta) \times F \xrightarrow{h_\alpha \circ h_\beta^{-1}} (U_\alpha \cap U_\beta) \times F$ where $h_\alpha \circ h_\beta^{-1}(x, f) = (x, g_{\alpha\beta}(x)f)$ and $g_{\alpha\beta} : U_\alpha \cap U_\beta \to G$ are continuous functions called the **transition functions** of the bundle. They satisfy the **cocycle condition**

$$g_{\alpha\beta} g_{\beta\gamma} g_{\gamma\alpha} \equiv 1 \tag{A.1}$$

in $U_\alpha \cap U_\beta \cap U_\gamma$. The bundle can be reassembled from this data by pasting together the local products $\{U_\alpha \times F\}_{\alpha \in A}$ with these homeomorphisms.

Any fibre bundle B with structure group G as above has an **associated principal G-bundle** $P_G(B)$. It is obtained by simply replacing F by G in the local products and then pasting together the $\{U_\alpha \times G\}_{\alpha \in A}$ by the *same* transition functions, where $g_{\alpha\beta}(x)$ acts on G by multiplication on the left. Note that we are free to multiply on the right by elements of G. Since right and left multiplication commute, this multiplication by elements of G on the right makes sense when we assemble the bundle $P_G(B)$. Note, however, that while each fibre looks like G, there is no preferred element, i.e., no "identity."

The original fibre bundle B can be recaptured from $P_G(B)$ as an **associated** bundle, that is, $B \cong P_G(F) \times_\varphi F$ where $\varphi : G \to \text{Homeo}(F)$ (see II.2).

We now observe that *every principal G-bundle over X can be presented by transition functions $g_{\alpha\beta} : U_\alpha \cap U_\beta \to G$ multiplying G on the left as above.* Of course a principal G-bundle is a fibre bundle and therefore is always given by a family of transition functions $G_{\alpha\beta} : U_\alpha \cap U_\beta \to \text{Homeo}(G)$ over some open cover $\{U_\alpha\}$ of X. Since the bundle is principal, these functions must commute pointwise with right multiplication by G. Let $g_{\alpha\beta}(x) = G_{\alpha\beta}(x)(1)$. Then $G_{\alpha\beta}(x)(g) = G_{\alpha\beta}(x)(1)g = g_{\alpha\beta}(x)g$ and we are reduced to the simpler case, as claimed.

Thus, we have seen that every principal G-bundle on X is given by a pair $(\mathcal{U}, \{g_{\alpha\beta}\})$ where $\mathcal{U} = \{U_\alpha\}_{\alpha\in A}$ is an open cover of X and where $g_{\alpha\beta}: U_\alpha \cap U_\beta \to G$ are continuous functions satisfying the cocycle condition (A.1). Such a pair can be thought of as a Čech 1-cocycle with coefficients in G (or, more precisely, in the sheaf of germs of continuous maps to G). One can easily check that two such bundles, constructed from cocycles $\{g_{\alpha\beta}\}$ and $\{g'_{\alpha\beta}\}$ on $\mathcal{U}$ are equivalent if and only if there exist continuous maps $g_\alpha: U_\alpha \to G$ for each α such that

$$g'_{\alpha\beta} = g_\alpha^{-1} \cdot g_{\alpha\beta} \cdot g_\beta \tag{A.2}$$

in $U_\alpha \cap U_\beta$ for all α, β. (This can be considered a "Čech-coboundary" condition.)

Therefore, we define two 1-cocycles $\{g_{\alpha\beta}\}$ and $\{g'_{\alpha\beta}\}$ on $\mathcal{U}$ to be equivalent iff there is a "Čech 0-cochain" $\{g_\alpha\}_{\alpha\in A}$ such that (A.2) holds. The set of equivalence classes will be denoted by $H^1(\mathcal{U}; G)$. This set naturally represents the equivalence classes of principal G-bundles on X which can be trivialized over the open sets of $\mathcal{U}$.

Suppose now that $(\mathcal{V}, j)$ is a **refinement** of $\mathcal{U}$, i.e., $\mathcal{V}$ is an open cover of X and $j: \mathcal{V} \to \mathcal{U}$ is a map such that $V \subseteq j(V)$ for all $V \in \mathcal{V}$. Then by restriction we get a map $r_{\mathcal{V}\mathcal{U}}: H^1(\mathcal{U}; G) \to H^1(\mathcal{V}; G)$ which can be shown to be independent of the refinement function j. These maps satisfy the relation $r_{\mathcal{W}\mathcal{U}} = r_{\mathcal{W}\mathcal{V}} \circ r_{\mathcal{V}\mathcal{U}}$ for successive refinements $\mathcal{W} \to \mathcal{V} \to \mathcal{U}$. Thus we can take the direct limit

$$H^1(X; G) \equiv \varinjlim H^1(\mathcal{U}; G).$$

This limit naturally represents the equivalence classes of principal G-bundles on X. That is,

$$\mathrm{Prin}_G(X) \cong H^1(X; G).$$

If G is abelian, $H^1(X; G)$ is simply the first Čech cohomology group of X with coefficients in G.

EXAMPLE. Let $G = \mathbb{Z}_2$. Then $\mathbb{Z}_2$-bundles over X are precisely the twofold coverings of X, and the correspondence $\mathrm{Cov}_2(X) \xrightarrow{\approx} H^1(X; \mathbb{Z}_2)$ is just the isomorphism given in Lemma 1.1.

REMARK A.1. If X is a C^∞-manifold and G is a Lie group, we can require all the maps $g_{\alpha\beta}, g_\alpha$, etc. in the discussion above to be differentiable. The resulting set, denoted $H^1(X; G)_\infty$, represents classes of smooth principal G-bundles over X. The natural map $H^1(X; G)_\infty \to H^1(X; G)$ can be shown to be a bijection so we shall in general drop the reference to smoothness.

An analogous remark can be made concerning complex manifolds X and complex Lie groups G where the maps $g_{\alpha\beta}$ and g_α are required to be holomorphic. In this case, however, the corresponding map to $H^1(X;G)$ is far from a bijection in general. (An exception is the case where X is Stein.)

Note that if G is not abelian, $H^1(X;G)$ is not a group. It is merely a set with a distinguished element given by the trivial G-bundle. Nonetheless, if

$$1 \longrightarrow K \xrightarrow{\ i\ } G \xrightarrow{\ j\ } G' \longrightarrow 1 \tag{A.3}$$

is an exact sequence of topological groups, the standard arguments in Čech cohomology theory (cf. Hirzebruch [1, Ch.I, §2] show that for paracompact spaces X, there is an exact sequence:

$$\begin{aligned}\{*\} &\longrightarrow H^0(X;K) \xrightarrow{\ i_*\ } H^0(X;G) \xrightarrow{\ j_*\ } H^0(X;G') \\ &\longrightarrow H^1(X;K) \xrightarrow{\ i_*\ } H^1(X;G) \xrightarrow{\ j_*\ } H^1(X;G')\end{aligned}$$

of pointed sets. $H^0(X;G)$ is the set of 0-cocycles and is easily identified with the space of continuous maps from X to G. The maps i_* and j_* are given in the obvious way by coefficient homomorphisms. Note that for a principal G-bundle P, the corresponding principal G'-bundle $j_*(P)$ can be obtained by dividing P by the action of K on the right.

REMARK A.2. If the group K in (A.3) is abelian, then $H^2(X;K)$ is defined and the exact sequence (A.4) can be extended to

$$\ldots \longrightarrow H^1(X;K) \longrightarrow H^1(X;G) \longrightarrow H^1(X;G') \longrightarrow H^2(X;K) \tag{A.5}$$

EXAMPLE A.3. For the sequence

$$0 \longrightarrow \mathrm{SO}_n \longrightarrow \mathrm{O}_n \longrightarrow \mathbb{Z}_2 \longrightarrow 0,$$

the induced map

$$w_1 : H^1(X;\mathrm{O}_n) \longrightarrow H^1(X;\mathbb{Z}_2)$$

is just the first Stiefel-Whitney class. Clearly $w_1(P) = 0$ if and only if P comes from an SO_n-bundle, i.e., if and only if P is orientable.

EXAMPLE A.4. For the sequence

$$0 \longrightarrow \mathbb{Z}_2 \longrightarrow \mathrm{Spin}_n \xrightarrow{\ \xi_0\ } \mathrm{SO}_n \longrightarrow 0,$$

the induced cobundary map

$$w_2 : H^1(X;\mathrm{SO}_n) \longrightarrow H^2(X;\mathbb{Z}_2)$$

is the second Stiefel-Whitney class. Clearly $w_2(P) = 0$ if and only if P comes from a Spin_n-bundle, i.e., if and only if P carries a spin structure.

There is a caution, however. As shown in §1, distinct spin structures may give abstractly equivalent principal Spin_n-bundles.

Note that the definition of a Čech coboundary leads to a nice combinatorial definition of $w_2(P)$. Let $(\mathscr{U}, \{g_{\alpha\beta}\})$ be a cocycle representing P, where each set $U_\alpha \cap U_\beta$ is simply-connected. Lift each map $g_{\alpha\beta}$ to a map $\tilde{g}_{\alpha\beta}: U_\alpha \cap U_\beta \to \mathrm{Spin}_n$, and define

$$w_{\alpha\beta\gamma} = \tilde{g}_{\alpha\beta}\tilde{g}_{\beta\gamma}\tilde{g}_{\gamma\alpha} \tag{A.6}$$

in $U_\alpha \cap U_\beta \cap U_\gamma$. Since $\xi_0(w_{\alpha\beta\gamma}) \equiv 1$, we see that

$$w_{\alpha\beta\gamma}: U_\alpha \cap U_\beta \cap U_\gamma \to \mathbb{Z}_2.$$

This $\mathbb{Z}_2$-cocycle represents $w_2(P)$.

EXAMPLE A.5. Consider the sequence

$$0 \longrightarrow \mathbb{Z} \longrightarrow \mathbb{R} \longrightarrow S^1 \longrightarrow 0.$$

Since all groups here are abelian, the exact sequence in cohomology continues indefinitely. It is an elementary exercise to show that $H^i(X;\mathbb{R}) = 0$ for all $i > 0$ and for any manifold X. (Here $\mathbb{R}$ carries the standard, not the discrete, topology.) We thereby get an isomorphism

$$c_1 : H^1(X; S^1) \xrightarrow{\approx} H^2(X; \mathbb{Z}) \tag{A.7}$$

called the **first Chern class.** This shows that *the equivalence classes of principal S^1-bundles are in natural 1-to-1 correspondence with elements of $H^2(X;\mathbb{Z})$*

Note. For a full and clear discussion of the basic subject of fibre bundles the reader is referred to the classic book of Steenrod [1].

We conclude this section with some remarks about transferring bundles from one space to another. Let $P \xrightarrow{\pi} X$ be a principal G-bundle over a space X and let $f: Y \to X$ be a continuous map. Consider the set of points

$$f^*P \equiv \{(y, p) \in Y \times P : f(y) = \pi(p)\}, \tag{A.8}$$

in the product $Y \times P$, and note that projection of $Y \times P$ onto its factors induces maps on f^*P which make the following diagram commute:

$$\begin{array}{ccc} f^*P & \xrightarrow{\tilde{f}} & P \\ {\scriptstyle\tilde{\pi}}\downarrow & & \downarrow{\scriptstyle\pi} \\ Y & \xrightarrow{f} & X \end{array} \tag{A.9}$$

We now observe that $\tilde{\pi}: f^*P \to Y$ is a principal G-bundle over Y. It follows directly from definitions that G acts on f^*P and is simple and transitive on the fibres of $\tilde{\pi}$. The map $\tilde{f}$ commutes with this action. Any set of local trivializations of P over a family of open sets $\mathscr{U} = \{U_\alpha\}_{\alpha \in A}$ in X naturally determine local trivializations of f^*P over the family of open sets $f^*\mathscr{U} = \{f^{-1}U_\alpha\}_{\alpha \in A}$ in Y. The transition functions $\{g_{\alpha\beta}\}$ for this cover lift back to transition functions

$$f^*g_{\alpha\beta} \equiv g_{\alpha\beta} \circ f \tag{A.10}$$

for $f^*\mathscr{U}$.

DEFINITION A.6. The principal G-bundle $f^*P \to Y$ is called the **bundle induced from $P \to X$ by the mapping f.**

It is clear that if two bundles P and P' are equivalent on X, then f^*P and f^*P' are equivalent on Y. (Lift the equivalence.) Thus we have a map

$$f^* : \mathrm{Prin}_G(X) \longrightarrow \mathrm{Prin}_G(Y). \tag{A.11}$$

From equation (A.10) we see that the cocycle for f^*P is just pull-back via f of the cocycle for P. Thus, (A.11) corresponds to the usual induced mapping on Čech cohomology groups:

$$f^* : H^1(X; G) \longrightarrow H^1(Y; G).$$

This "pull-back" mapping on principal bundles has two important properties:

Proposition A.7. *For continuous maps* $X \xrightarrow{f} Y \xrightarrow{g} Z$, *the induced maps* $\mathrm{Prin}_G(Z) \xrightarrow{g^*} \mathrm{Prin}_F(Y) \xrightarrow{f^*} \mathrm{Prin}_G(X)$ *satisfy*

$$(g \circ f)^* = f^* \circ g^* \tag{A.12}$$

Proof. That $f^*(g^*P) = (g \circ f)^*P$ is obvious from the definition. ∎

Proposition A.8. *Let Y be a compact Hausdorff space. If two maps $f_0 : Y \to X$ and $f_1 : Y \to X$ are homotopic, then*

$$f_0^* = f_1^*.$$

We leave the proof as an exercise for the reader.

APPENDIX B

Classifying Spaces and Characteristic Classes

The point of this appendix is to briefly summarize the basic facts from the theory of classifying spaces and characteristic classes for principal G-bundles. Details can be found in Milnor-Stasheff [1] and Husemoller [1].

For simplicity we shall assume throughout this appendix that G is a Lie group. Furthermore all spaces will be assumed to have base points, and all maps will be base-point-preserving. All spaces will also be assumed to have the homotopy type of a countable CW-complex.

DEFINITION B.I. A **classifying space** for the group G is a connected topological space BG, together with a principal G-bundle $EG \to BG$, such that the following is true. For any compact Hausdorff space X, there is a one-to-one correspondence between the equivalence classes of principal G-bundles on X and the homotopy classes of maps from X to BG, given by associating to each map $f: X \to BG$, the induced bundle f^*EG over X (see the end of App. A).

Thus the induced bundle construction gives a natural bijection

$$\mathrm{Prin}_G(X) \cong [X, BG]. \tag{B.1}$$

Note that as a special case we have

$$\mathrm{Prin}_G(S^n) \cong \pi_n(BG). \tag{B.2}$$

Given two different classifying spaces for G, say B_0G and B_1G, there must exist mappings $f_0: B_0G \to B_1G$ and $f_1: B_1G \to B_0G$ such that $f_i^*E_{i+1}G \cong E_iG$ for $i \in \mathbb{Z}_2$. This implies that $(f_{i+1} \circ f_1)^*E_iG \cong E_iG$, and therefore $f_{i+1} \circ f_i$ is homotopic to the identity on B_iG for $i \in \mathbb{Z}_2$. Thus B_0G and B_1G are homotopy equivalent, i.e., BG is well defined up to homotopy type.

The bundle $EG \to BG$ is called the **universal principal G-bundle**. This bundle has the following homotopy characterization (see Steenrod [1] for example).

Theorem B.2. *Let $E \to B$ be a principal G-bundle with the property that the total space of E is contractible. Then (B, E) is a classifying space for G.*

The theory has definite interest due to the following:

Theorem B.3. *For any Lie group there exists a classifying space.*

The general construction of classifying spaces for principal bundles is due to Milnor [1,2] and proceeds as follows. Given a group G, we consider the n-fold join $G * \cdots * G$. (The **join** of two spaces X and Y is obtained from $X \times Y \times [0,1]$ by collapsing $\{x\} \times Y \times \{0\}$ to a point for each x and also collapsing $X \times \{y\} \times \{1\}$ to a point for each y.)

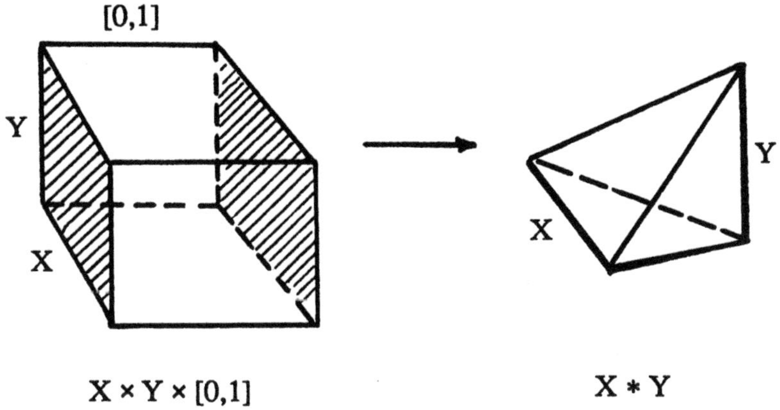

$$X \times Y \times [0,1] \longrightarrow X * Y.$$

The space $G * \cdots * G$ is $(n-1)$-connected. It also has a free G-action obtained by multiplying simultaneously on the right in each of the factors. We pass to the limit as $n \to \infty$, and define $EG \equiv G * G * G * \cdots$. This is infinitely-connected and has a free right G-action. We set $BG = EG/G$ with $\pi: EG \to BG$ the quotient map. From Theorem B.2, this bundle is classifying.

Of course this construction of Milnor can (and has) been carried out in enormous generality.

It should be noted that since EG is contractible, the long exact sequence of homotopy groups for a fibration implies that

$$\pi_{n-1}(G) \cong \pi_n(BG) \tag{B.3}$$

for all $n \geqq 1$. Thus, the homotopy groups of BG present nothing new. However, the homology of the space is quite interesting.

We begin the discussion with the following. Consider the singular cohomology $H^*(BG; \Lambda)$ with coefficients in a ring Λ.

Definition B.4. Each non-zero class in $H^*(BG; \Lambda)$ is a **universal characteristic class** for principal G-bundles.

Fix a class $c \in H^k(BG; \Lambda)$. Then for any principal G-bundle $P \to X$ there is a map $f_P : X \to BG$ so that $P \cong f_P^* EG$. We define the **c-characteristic**

class of P to be the class

$$c(P) = f_P^*(c) \in H^k(X; \Lambda). \tag{B.4}$$

Since f_P is well defined up to homotopy, the class $c(P)$ is uniquely defined.

Observe that any such characteristic class transforms naturally in the following sense. Given a principal G-bundle $P \to X$ and a continuous map $F: Y \to X$, the c-characteristic classes satisfy:

$$c(F^*P) = F^*c(P) \tag{B.5}$$

To see this we simply note that $f_{F^*P} = f_P \circ F$, and so $c(F^*P) = f_{F^*P}^*(c) = F^*f_P^*(c) = F^*c(P)$.

Given a continuous homomorphism $\varphi: H \to G$ of Lie groups, there is a corresponding continuous map

$$B\varphi: BH \longrightarrow BG \tag{B.6}$$

which classifies the principal G-bundle over BH associated to the representation φ. That is,

$$(B\varphi)^*(EG) = EH \times_\varphi G$$

where $EH \times_\varphi G$ denotes the quotient of $EH \times G$ by the action of H given by setting $\Phi_h(p, g) \equiv (ph^{-1}, hg)$ for $p \in EH$, $g \in G$ and $h \in H$. This association also transforms naturally. That is, for any composite homomorphism $H \xrightarrow{\varphi} K \xrightarrow{\psi} G$, we have

$$B(\psi \circ \varphi) = B(\psi) \circ B(\varphi).$$

If the homomorphism $\varphi: H \to G$ is a homotopy equivalence, then so is the mapping $B\varphi: BH \to BG$. This is easily seen by applying the 5-lemma to the induced map on the long exact homotopy sequences of the fibrations

$$H \longrightarrow EH \longrightarrow BH \qquad \text{and} \qquad G \longrightarrow EG \longrightarrow BG.$$

(Recall that BH and BG are assumed to be homotopy equivalent to countable CW-complexes.) This gives the following result:

Proposition B.5. *Let G be a connected Lie group and $K \subseteq G$ a maximal compact subgroup. Then $BG \cong BK$, that is, the classifying spaces of G and K are homotopy equivalent.*

Proof. It follows from the Iwasawa decomposition of G (cf. Helgason [1]) that the inclusion $K \hookrightarrow G$ is a homotopy equivalence. ∎

It follows, for example, that $\mathrm{BO}_n \cong \mathrm{BGL}_n(\mathbb{R})$, $\mathrm{BSO}_n \cong \mathrm{BGL}_n^+(\mathbb{R})$, $\mathrm{BU}_n \cong \mathrm{BGL}_n(\mathbb{C})$, $\mathrm{BSU}_n \cong \mathrm{BSL}_n(\mathbb{C})$, $\mathrm{BSp}_n \cong \mathrm{BSp}_n(\mathbb{C})$, etc.

Note that, in general, each mapping of the type $B\varphi: BH \to BG$ induces a transformation $(B\varphi)^*: H^*(BG; \Lambda) \to H^*(BH; \Lambda)$ of universal characteristic classes. Any non-zero element which goes to zero under this transformation is a universal obstruction to the reduction of the structure group from G to H. That is, if c is such a class and if $P \to X$ is a principal G-bundle with $c(P) \neq 0$, then P cannot be written as an associated bundle $P = P' \times_\varphi G$ (as above) for any principal H-bundle $P' \to X$.

The cases of basic interest here are those of the compact classical groups. For these cases there are useful alternative constructions of the classifying spaces.

For $K = \mathbb{R}, \mathbb{C}$ or $\mathbb{H}$, let $G^K(n, N)$ denote the Grassmann manifold of all n-dimensional K-linear subspaces of K^N. There are natural inclusions $G^K(n, N) \subset G^K(n, N + N')$ induced by the "first-coordinate" inclusions $K^N \subset K^{N+N'}$. Taking the direct limit, we get

$$G^K(n, \infty) \equiv \xrightarrow[N]{\lim} G^K(n, N),$$

the Grassmannian of K-planes in K^∞.

There is a **canonical K-vector bundle** $\mathbb{E}_n \to G^K(n, \infty)$ whose fibre at a plane $P \in G^K(n, \infty)$ consists of all the vectors in that plane. We shall show that

$$\begin{aligned} BO_n &\cong G^{\mathbb{R}}(n, \infty) \\ BU_n &\cong G^{\mathbb{C}}(n, \infty) \qquad \text{(B.7)} \\ BSp_n &\cong G^{\mathbb{H}}(n, \infty). \end{aligned}$$

The principal fibrations in each case are obtained by taking the appropriate bundle of bases (orthonormal, unitary, or symplectic) in the canonical bundle $\mathbb{E}_n$. Thereby, $\mathbb{E}_n$ becomes the **universal (real, complex or quaternionic) n-plane bundle** over the classifying space (BO_n, BU_n, or BSp_n respectively).

We begin our proof of these claims with the real case. For this we note that for any N, we have $G^{\mathbb{R}}(n, N) \cong O_N/(O_{N-n} \times O_n)$. The bundle of orthonormal frames for the canonical n-plane bundle $\mathbb{E}_n \to G^{\mathbb{R}}(n, N)$ is just the Stiefel manifold O_N/O_{N-n}. Hence, the bundle of orthonormal frames for $\mathbb{E}_n \to G^{\mathbb{R}}(n, \infty)$ is just the direct limit over N of the principal O_n-bundles

$$O_N/O_{N-n} \longrightarrow O_N/(O_{N-n} \times O_n)$$

It is an elementary exercise, using the fibrations $O_k \to O_{k+1} \to S^k$, to see that O_N/O_{N-n} is $(N - n - 2)$-connected. Hence the infinite Stiefel manifold $\varinjlim_N O_N/O_{N-n}$ is contractible, and the limiting fibration must represent the universal bundle $EO_n \to BO_n$ by Theorem B.2.

The arguments for BU_n and BSp_n are entirely analogous. Here one uses the facts that $G^{\mathbb{C}}(n, N) \cong U_N/(U_{N-n} \times U_n)$ and $G^{\mathbb{H}}(n, N) \cong Sp_N/(Sp_{N-n} \times Sp_n)$. We omit the details.

The reader has undoubtedly noticed that BO_n can be considered as the classifying space for real n-dimensional vector bundles. That is, for a compact Hausdorff space X, the equivalence classes of such bundles correspond one-to-one with elements of $[X, BO_n]$ by associating to $f: X \to BO_n$ the vector bundle $f^*\mathbb{E}_n$. This follows easily from the discussion above by passing to the bundles of orthonormal bases. The corresponding remarks hold for BU_n and BSP_n.

It should be noted that the fact that $G^K(n, \infty)$ is a classifying space for n-dimensional vector bundles (over K) can be proved by a direct geometric construction (see Milnor-Stasheff [1]).

We now examine the cohomology of the basic classifying spaces. These spaces are infinite-dimensional and have non-zero classes in arbitrarily high dimensions. Nevertheless, as **rings under the cup-product multiplication,** they are quite understandable. Basic references for all the following results are Milnor-Stasheff [1], Steenrod-Epstein [1], and Borel [2]. We begin with $\mathbb{Z}_2$-cohomology.

Theorem B.7. *The cohomology ring $H^*(BO_n; \mathbb{Z}_2)$ is a $\mathbb{Z}_2$-polynomial ring*

$$\mathbb{Z}_2[w_1, w_2, \dots, w_n]$$

on canonical generators $w_k \in H^k(BO_n; \mathbb{Z}_2)$ for $k = 1, \dots, n$.

Note. The classes $w_1, \dots, w_n$ are characterized inductively by the following fact. The kernel of the homomorphism $H^*(BO_n; \mathbb{Z}_2) \xrightarrow{i^*} H^*(BO_{n-1}; \mathbb{Z}_2)$ induced by the standard inclusion $O_{n-1} \xrightarrow{i} O_n$, is the principal ideal $\langle w_n \rangle$. Analogous statements characterize the canonical generators in the subsequent theorems.

The class w_k is called the universal **kth Stiefel-Whitney class.** Thus to any n-dimensional real vector bundle $E \to X$ classified by a map $f_E: X \to BO_n$, we have the associated kth Stiefel-Whitney class of E, $w_k(E) \equiv f_E^*(w_k)$. An important concept is that of the **total Stiefel-Whitney class** $w \equiv 1 + w_1 + \dots + w_n$. It satisfies the Whitney product formula for the sum of two bundles

$$w(E \oplus E') = w(E) \cup w(E') \tag{B.8}$$

which translates in longhand to the equations

$$w_k(E \oplus E') = \sum_{i=0}^{k} w_i(E) \cup w_{k-i}(E')$$

for $k = 1,2,\ldots$. This can be considered as a statement about the induced map on cohomology induced by $B(O_n \times O_{n'}) \to BO_{n+n'}$ under the usual "diagonal block" inclusion $O_n \times O_{n'} \subset O_{n+n'}$.

Given a smooth manifold X, we define the kth Stiefel-Whitney class of X to be $w_k(TX)$. It is an important fact that the *Stiefel-Whitney classes of a compact smooth manifold are invariants of the homotopy type of the manifold.* That is, given a homotopy equivalence $f: Y \to X$ between compact smooth manifolds, we have that $f^*w_k(X) = w_k(Y)$ (see Milnor-Stasheff [1], p. 131.) This will not be true of many other characteristic classes of manifolds.

Consider now the space BSO_n. This can be realized as above as the Grassmannian of oriented n-planes in $\mathbb{R}^\infty$. There are natural maps $BSO_n \to BO_n \to B\mathbb{Z}_2 = K(\mathbb{Z}_2, 1)$. This sequence is a fibration, and we have the following:

Theorem B.8. *The cohomology ring $H^*(BSO_n; \mathbb{Z}_2)$ is a $\mathbb{Z}_2$-polynomial ring*

$$\mathbb{Z}_2[w_2,\ldots,w_n]$$

where w_k denotes the lift of the universal kth *Stiefel-Whitney class by the map* $BSO_n \to BO_n$.

Finally we consider the space BSpin_n which sits in a fibration $\mathrm{BSpin}_n \to BSO_n \to K(\mathbb{Z}_2, 2)$.

REMARK B.9. The cohomology ring $H^*(\mathrm{BSpin}_n; \mathbb{Z}_2)$ is quite complex and since its precise structure is not used here, we refer the interested reader to the basic paper of D. Quillen [1]. (The "stable" groups $H^*(\mathrm{BSpin}; \mathbb{Z}_2)$ were first computed by E. Thomas [1].)

The $\mathbb{Z}_2$-cohomology of the spaces BU_n and BSp_n is merely the mod 2 reduction of the integral cohomology, and we have the following:

Theorem B.10. *The cohomology ring $H^*(BU_n; \mathbb{Z})$ is a $\mathbb{Z}$-polynomial ring*

$$\mathbb{Z}[c_1,c_2,\ldots,c_n]$$

on canonical generators $c_k \in H^{2k}(BU_n; \mathbb{Z})$ for $k = 1,\ldots,n$.

The class c_k is called the universal **kth Chern class.** Thus to any n-dimensional complex vector bundle $E \to X$ classified by a map $f_E: X \to BU_n$, we have the associated kth Chern class $c_k(E) = f_E^*(c_k)$. There is the concept of the **total Chern class** $c = 1 + c_1 + c_2 + \ldots + c_n$, and again there is a product formula

$$c(E \oplus E') = c(E) \cup c(E'). \tag{B.9}$$

Given any complex manifold X, the tangent bundle is a complex vector bundle. We call $c_k(X) \equiv c_k(TX)$ the **kth Chern class** of X.

There is now a ring isomorphism $H^*(\mathrm{BU}_n;\mathbb{Z}_2) \cong \mathbb{Z}_2[\tilde{c}_1,\ldots,\tilde{c}_n]$ where $\tilde{c}_k$ is of degree $2k$ for each k. The class $\tilde{c}_k$ is the mod 2 reduction of c_k. The natural inclusion $\mathrm{U}_n \subset \mathrm{O}_{2n}$ induces a map $\mathrm{BU}_n \to \mathrm{BO}_{2n}$. It can be shown that under this map $w_{2k} \mapsto \tilde{c}_k$ and, of course, $w_{2k+1} \to 0$. Hence, for any complex vector bundle E,

$$w_{2k}(E) \equiv c_k(E)(\text{mod } 2) \qquad \text{and} \qquad w_{2k+1}(E) = 0$$

for all k.

The picture is much the same for quaternion bundles.

Theorem B.11. *The cohomology ring $H^*(\mathrm{BSp}_n;\mathbb{Z})$ is a $\mathbb{Z}$-polynomial ring*

$$\mathbb{Z}[\sigma_1,\ldots,\sigma_n]$$

on canonical generators $\sigma_k \in H^{4k}(\mathrm{BSp}_n;\mathbb{Z})$ for $k = 1,\ldots,n$.

Under the natural map $\mathrm{BSp}_n \to \mathrm{BU}_{2n}$ we have that $c_{2k} \mapsto \sigma_k$ and $c_{2k+1} \mapsto 0$.

The integral cohomology of BO_n is a more complicated story due to the presence of 2-torsion. However, away from the prime 2, things become nice.

Theorem B.12. *Let Λ be any integral domain containing $\frac{1}{2}$* (e.g., $\mathbb{Z}[\frac{1}{2}]$ or $\mathbb{Q}$). *Then $H^*(\mathrm{BO}_n;\Lambda)$ is the polynomial algebra*

$$\Lambda[p_1,\ldots,p_{[n/2]}]$$

on canonical generators $p_k \in H^{4k}(\mathrm{BO}_n;\Lambda)$ for $k = 1,\ldots,[n/2]$.

Choose $\Lambda = \mathbb{Q}$. Then the class p_k is called the universal ***k*th rational Pontrjagin class.** Given a real vector bundle $E \to X$ classified by $f_E\colon X \to \mathrm{BO}_n$, we define the kth rational Pontrjagin class of E to be $p_k(E) = f_E^*(p_k)$. There is a **total rational Pontrjagin class** $p = 1 + p_1 + \ldots + p_{[n/2]}$ and a product formula

$$p(E \oplus E') = p(E) \cup p(E') \tag{B.10}$$

in $H^*(X;\mathbb{Q})$. There are rational Pontrjagin classes of a smooth manifold which, in the compact case, are homeomorphism invariants.

To understand the integral case we must examine the Bockstein homomorphism. The coefficient sequence $0 \to \mathbb{Z} \xrightarrow{2} \mathbb{Z} \to \mathbb{Z}_2 \to 0$ gives rise to a long exact sequence

$$\ldots \longrightarrow H^k(X;\mathbb{Z}) \xrightarrow{2} H^k(X;\mathbb{Z}) \longrightarrow H^k(X;\mathbb{Z}_2) \xrightarrow{\beta} H^{k+1}(X;\mathbb{Z}) \longrightarrow \cdots$$

where β is called the **Bockstein.** Note that the kernel of β is the set of classes in $H^*(X;\mathbb{Z}_2)$ which are mod 2 reductions of integral classes.

Theorem B.13. *The integral cohomology* $H^*(\mathrm{BO}_n;\mathbb{Z})$ *is an additive direct sum*

$$\mathbb{Z}[p_1,\ldots,p_{[n/2]}] \oplus \mathrm{Image}(\beta)$$

where $p_k \in H^{4k}(\mathrm{BO}_n;\mathbb{Z})$ *becomes the element* p_k *of Theorem* B.12 *under tensor product with* Λ, *and where* β *is the Bockstein homomorphism.*

The class p_k is called the universal **kth integral Pontrjagin class.** These classes have the following property. Consider the homomorphism $\mathrm{O}_n \to \mathrm{U}_n$ induced by complexification. The Chern classes lift back over the induced map $\mathrm{BO}_n \to \mathrm{BU}_n$. The classes c_{2k+1} go to zero, and

$$p_k = (-1)^k c_{2k}. \tag{B.11}$$

It is conventional to define the Pontrjagin classes of a real vector bundle E by this formula, i.e., by setting

$$p_k(E) = (-1)^k c_{2k}(E \otimes \mathbb{C}). \tag{B.12}$$

Observe now that the subset $\beta(H^*(\mathrm{BO}_n;\mathbb{Z}_2)) \subset H^*(\mathrm{BO}_n;\mathbb{Z})$ consists entirely of 2-torsion elements. Important among these are the classes

$$W_k \equiv \beta(w_{k-1}) \tag{B.13}$$

which measure whether the $(k-1)$st Stiefel-Whitney class is the mod 2 reduction of an integral class. Of course all the W_k's vanish when pulled back to BU_n or BSp_n under the maps $\mathrm{BU}_n \to \mathrm{BO}_{2n}$ and $\mathrm{BSp}_n \to \mathrm{BO}_{4n}$.

We now examine BSO_n.

Theorem B.14. *Let* Λ *be an integral domain containing* $\frac{1}{2}$. *Then for each n, there are ring homomorphisms*

$$H^*(\mathrm{BSO}_{2n+1};\Lambda) = \Lambda[p_1,\ldots,p_n]$$

$$H^*(\mathrm{BSO}_{2n};\Lambda) = \Lambda[p_1,\ldots,p_n,\chi]/\langle \chi^2 - p_n\rangle$$

where $p_k \in H^{4k}(\mathrm{BSO}_*;\Lambda)$ *and where* $\chi \in H^n(\mathrm{BSO}_{2n};\Lambda)$.

The elements p_k are the images of the Λ-Pontrjagin classes (of Theorem B.12) under the map $\mathrm{BSO}_n \to \mathrm{BO}_n$. The class χ is called the **universal Euler class.** It can be defined as usual for any oriented vector bundle $E \to X$ by setting $\chi(E) = f_E^*(\chi)$. (For odd-dimensional bundles it is defined to be zero.) There is a product formula

$$\chi(E \oplus E') = \chi(E) \cup \chi(E'). \tag{B.14}$$

There is a natural definition of χ as an integral cohomology class. (see Milnor-Stasheff [1]). Under the map $\mathrm{BU}_n \to \mathrm{BSO}_{2n}$, the class χ pulls back to c_n. Under mod 2 reduction, χ becomes w_n in $H^n(\mathrm{BSO}_{2n};\mathbb{Z}_2)$.

The analogue of Theorem B.13 holds for $H^*(\mathrm{BSO}_n;\mathbb{Z})$.

APPENDIX C

Orientation Classes and Thom Isomorphisms in K-Theory

The point of this appendix is to examine the notion of an orientation class for K-theory, KO-theory, or KR-theory on a vector bundle. Such classes exist only on bundles with an appropriate structure, and when they exist they determine a "Thom isomorphism" for the given theory. These can be compared with the standard Thom isomorphism for cohomology via the Chern character. For more elaboration of the matters discussed here the reader is referred to the book of Karoubi [2].

Let $\pi: E \to X$ denote a vector bundle over a locally compact space X. This bundle may be real or complex, or even Real (when X is provided with an involution). Let k^{-*} denote any of the theories K_{cpt}^{-*}, KO_{cpt}^{-*} or KR_{cpt}^{-*}. (This last theory is defined only for spaces with an involution. We shall let the reader make the obvious adaptations of the exposition to fit this case.)

Proposition C.1. *Via the projection $\pi: E \to X$, $k^{-*}(E)$ is canonically a $k^{-*}(X)$-module.*

Proof. Given any two locally compact spaces A and B, the outer tensor product induces a graded, bi-additive map

$$k^{-*}(A) \times k^{-*}(B) \longrightarrow k^{-*}(A \times B)$$

defined as follows. Fix elements $[V_0, V_1; \sigma] \in k(A \times \mathbb{R}^m) = k^{-m}(A)$ and $[W_0, W_1; \tau] \in k(B \times \mathbb{R}^n) = k^{-n}(B)$. Choose extensions of the isomorphisms $\sigma: V_0 \to V_1$ and $\tau: W_0 \to W_1$, which are defined outside compact subsets, to all of $A \times \mathbb{R}^m$ and $B \times \mathbb{R}^n$ respectively. Fix a metric in each of the bundles, and let $\sigma^*: V_1 \to V_0$ and $\tau^*: W_1 \to W_0$ be the adjoint morphisms to σ and τ. Let $V_i \boxtimes W_j$ denote the outer tensor product, i.e., $V_i \boxtimes W_j = (\text{pr}_1^* V_i) \otimes (\text{pr}_2^* W_j)$ where pr_1 and pr_2 are the projections of $A \times B \times \mathbb{R}^{n+m}$ onto $A \times \mathbb{R}^m$ and $B \times \mathbb{R}^n$ respectively. Then the product of $[V_0, V_1; \sigma]$ and $[W_0, W_1; \tau]$ is defined to be $[U_0, U_1; \rho]$ where

$$U_0 = (V_0 \boxtimes W_0) \oplus (V_1 \boxtimes W_1), \qquad U_1 = (V_1 \boxtimes W_0) \oplus (V_0 \boxtimes W_1)$$

and where

$$\rho = \begin{pmatrix} \sigma \boxtimes 1 & -1 \boxtimes \tau^* \\ 1 \boxtimes \tau & \sigma^* \boxtimes 1 \end{pmatrix}.$$

From the fact that

$$\rho\rho^* = \begin{pmatrix} \sigma\sigma^* \boxtimes 1 + 1 \boxtimes \tau^*\tau & 0 \\ 0 & \sigma^*\sigma \boxtimes 1 + 1 \boxtimes \tau\tau^* \end{pmatrix}$$

we see that ρ is an isomorphism at any point where either σ or τ is an isomorphism. Hence, ρ is an isomorphism outside a compact subset of $A \times B \times \mathbb{R}^{n+m}$. All choices involved in this definition are unique up to homotopy, so the product is well defined.

The desired module multiplication is now given by the composition

$$k^{-*}(X) \times k^{-*}(E) \longrightarrow k^{-*}(X \times E) \longrightarrow k^{-*}(E)$$

where the second homomorphism is induced by the proper map $(\pi \times \mathrm{Id})\colon E \to X \times E$. Verification that this multiplication is associative and distributive is left to the reader. ∎

From this point on we shall assume that X is compact.

DEFINITION C.2. A class $u \in k(E) = k^0(E)$ is said to be a **k-theory orientation** for the bundle E if $k^{-*}(E)$ is a free $k^{-*}(X)$-module with generator u.

EXAMPLE C.3. Let $E = X \times \mathbb{C}^m \xrightarrow{\pi} X$ be a trivialized hermitian vector bundle and define

$$u = [\Lambda_{\mathbb{C}}^{\text{even}}\underset{\sim}{\mathbb{C}}^m, \Lambda^{\text{odd}}\underset{\sim}{\mathbb{C}}^m; \sigma] \in K_{\text{cpt}}(E) \tag{C.1}$$

where $\underset{\sim}{\mathbb{C}}^m \cong \pi^*E$ denotes the trivial m-plane bundle on E and where

$$\sigma_{x,v}(\varphi) = v \wedge \varphi - v^* \llcorner \varphi$$

for $(x,v) \in X \times \mathbb{C}^m$ and $\varphi \in \Lambda_{\mathbb{C}}^{\text{even}}\underset{\sim}{\mathbb{C}}^m$. By using the identification $\mathbb{R}^{2m} \cong \mathbb{C}^m$ and taking the canonical orientation, this element can be rewritten as

$$u = [\$_{\mathbb{C}}^+, \$_{\mathbb{C}}^-; \mu] \tag{C.2}$$

where $\$_{\mathbb{C}} = \$_{\mathbb{C}}^+ \oplus S_{\mathbb{C}}^- \cong \pi^*\$_{\mathbb{C}}(E)$ is the irreducible complex graded $C\ell_{2m}$-module (extended trivially over E) and where

$$\mu_{x,v}(\varphi) = v \cdot \varphi$$

is given by Clifford multiplication. The fundamental assertion of the Bott Periodicity Theorem is that u gives a K-theory orientation on E (see 9.20, (9.8) and 9.28 of Chapter I).

EXAMPLE C.4. Let $E = X \times \mathbb{R}^{8m} \xrightarrow{\pi} X$ be a trivialized riemannian vector bundle, and define

$$u = [\$^+, \$^-; \mu] \in KO_{\text{cpt}}(E) \tag{C.3}$$

where $\$ = \$^+ \oplus \$^- = \pi^*\$(E)$ is the irreducible real graded $C\ell_{8m}$-module (extended trivially over E) and where

$$\mu_{x,v}(\varphi) = v \cdot \varphi$$

is given by Clifford multiplication. The Bott Periodicity Theorem (see 9.22, (9.8) and 9.28 of Chapter I) states that u is a KO-theory orientation for E.

EXAMPLE C.5. Let $E = X \times \mathbb{C}^m = X \times (\mathbb{R}^m \oplus i\mathbb{R}^m) \xrightarrow{\pi} X$ be a trivialized Real vector bundle over a compact space X with trivial involution. Set

$$u = [\mathbb{C}\ell_m^0, \mathbb{C}\ell_m^1; R + iL] \in KR_{\text{cpt}}(E) \tag{C.4}$$

where $\mathbb{C}\ell_m = \mathbb{C}\ell_m^0 \oplus \mathbb{C}\ell_m^1$ is extended trivially over E and where

$$(R + iL)_{x,u+iv}(\varphi) = \varphi \cdot u + iv \cdot \varphi$$

at any point $(x, u + iv) \in X \times (\mathbb{R}^m \oplus i\mathbb{R}^m)$. From the (1,1)-Periodicity Theorem of I.10 we see that u defines a KR-theory orientation on E.

We now return to the general case of a vector bundle over a compact space X.

DEFINITION C.6. A class $u \in k(E)$ is said to have the **Bott periodicity property** if u determines a k-theory orientation in any local trivialization of E over a closed subset $C \subset X$, i.e., if $k^{-*}(E|_C)$ is a free $k^{-*}(C)$-module generated by u, whenever $E|_C$ is trivial.

Theorem C.7 *Let $\pi: E \to X$ be a vector bundle over a compact space X. Then any class $u \in k(E)$ with the Bott periodicity property is a k-theory orientation for E.*

Proof. Let $\{C_j\}_{j=1}^N$ be a covering of X by closed subsets such that $E|_{C_j}$ is trivial for each j. We proceed by induction on N. For $N = 1$ the statement is obvious. Suppose we have proved the assertion for E restricted to $A = C_1 \cup \cdots \cup C_{N-1}$ (or to any closed subset of A). Set $B = C_N$. For each theory k^{-*} there exists a Mayer-Vietoris sequence with connecting homomorphisms of degree 1 (see Karoubi [2]). One easily checks that multiplication by the appropriate restrictions of u gives a morphism of exact sequences:

$$\begin{array}{ccccccccc}
\cdots \to k^{-i}(A \cup B) & \to & k^{-i}(A) \oplus k^{-i}(B) & \to & k^{-i}(A \cap B) & \to & k^{-i+1}(A \cup B) & \to \cdots \\
\downarrow & & \downarrow & & \downarrow & & \downarrow & \\
\cdots \to k^{-i}(E|_{A\cup B}) & \to & k^{-i}(E|_A) \oplus k^{-i}(E|_B) & \to & k^{-i}(E|_{A\cap B}) & \to & k^{-i+1}(E|_{A\cup B}) & \to \cdots
\end{array}$$

By inductioin the vertical arrows are isomorphisms from $k^{-*}(A)$, $k^{-*}(B)$ and $k^{-*}(A \cap B)$ (since $A \cap B$ is a closed subset of A). By the 5-lemma we conclude that the vertical arrows are isomorphisms also from $k^{-*}(A \cup B)$. The same argument clearly also applies to any closed subset of $A \cup B$. This completes the proof. ■

Theorem C.7 has the following immediate consequences:

Theorem C.8 (The Thom isomorphism in *K*-theory for U_m-bundles). *Let $\pi : E \to X$ be a complex hermitian vector bundle over a compact space X. Then the class*

$$\Lambda_{-1}(E) = [\pi^*\Lambda_{\mathbb{C}}^{\text{even}}E, \pi^*\Lambda_{\mathbb{C}}^{\text{odd}}E; \sigma] \in K_{\text{cpt}}(E)$$

where $\sigma_e(\varphi) = e \wedge \varphi - e^* \llcorner \varphi$, is a K-theory orientation for E. In particular, the map $i_! : K(X) \to K_{\text{cpt}}(E)$ given by

$$i_!(a) = (\pi^*a) \cdot \Lambda_{-1}(E)$$

is an isomorphism.

Proof. By Example C.3, $\Lambda_{-1}(E)$ has the Bott periodicity property. ■

Theorem C.9 (The Thom isomorphism in *KO*-theory for Spin_{8m}-bundles). *Let $\pi : E \to X$ be a real 8m-dimensional bundle with a spin structure, over a compact space X. Consider the class*

$$S(E) = [\pi^*\$^+, \pi^*\$^-; u] \in KO_{\text{cpt}}(E)$$

where $\$ = \$^+ \oplus \$^-$ is the irreducible graded real spinor bundle of E and where $\mu_e(\varphi) = e \cdot \varphi$ is given by Clifford multiplication. Then S(E) is a KO-theory orientation for E. In particular, the map $i_! : KO(X) \to KO_{\text{cpt}}(E)$ given by

$$i_!(a) = (\pi^*a) \cdot S(E)$$

is an isomorphism.

Proof. By Example C.4, $S(E)$ has the Bott periodicity property. ■

Theorem C.10 (The Thom isomorphism in *KR*-theory for Real bundles). *Let $\pi : E_0 \to X$ be a real vector bundle over a compact space, and let $E = E_0 \otimes \mathbb{C} = E_0 \oplus iE_0$ be the associated Real bundle (with trivial involution on X). Then the class*

$$U(E) = [\pi^*C\ell^0(E_0), \pi^*C\ell^1(E_0); R + iL] \in KR_{\text{cpt}}(E)$$

is a KR-theory orientation for E (where $(R + iL)(\varphi) = \varphi e - ie'\varphi$ at $e + ie' \in E_0 \oplus iE_0$ above a point in X). In particular, the map $i_! : KR(X) \to KR_{\text{cpt}}(E)$ given by

$$i_!(a) = (\pi^*a) \cdot U(E)$$

is an isomorphism.

Proof. By Example C.5, U(E) has the Bott periodicity property. ■

REMARK C.11. For a general Real bundle $\pi: E \to X$ over a compact space X with involution, the class $\Lambda_{-1}(E)$ given in C.8 defines an element in $KR_{\text{cpt}}(E)$ which is always a KR-theory orientation for E (see Atiyah [2]).

Theorem C.12 (The Thom isomorphisms in K-theory for SO_{2m}-, $Spin_{2m}$- and $Spin^c_{2m}$-bundles). *Let $\pi: E \to X$ be an oriented real vector bundle of dimension $2m$ over a compact space X. Then the class*

$$\delta(E) = [\pi^*\mathbb{C}\ell^+(E), \pi^*\mathbb{C}\ell^-(E); \mu] \in K_{\text{cpt}}(E) \otimes \mathbb{Q}$$

defined in (III.12.12) is a ($K \otimes \mathbb{Q}$)-theory orientation for E. If E has a spin structure, then the class

$$\mathbf{s}(E) = [\pi^*\$^+_{\mathbb{C}}(E), \pi^*\$^-_{\mathbb{C}}(E); \mu] \in K_{\text{cpt}}(E)$$

defined by (III.12.13) is a K-theory orientation for E. This remains true for any Spin^c-structure on E where $\$_{\mathbb{C}}(E) = \$^+_{\mathbb{C}}(E) \oplus \$^-_{\mathbb{C}}(E)$ is the associated irreducible spinor bundle (see App. D). The associated maps

$$i_!: K(X) \otimes \mathbb{Q} \longrightarrow K_{\text{cpt}}(E) \otimes \mathbb{Q} \qquad \textit{given by} \qquad i_!(a) = (\pi^*a) \cdot \delta(E)$$

and

$$i_!: K(X) \longrightarrow K_{\text{cpt}}(E) \qquad \textit{given by} \qquad i_!(a) = (\pi^*a) \cdot \mathbf{s}(E)$$

(in the Spin^c*-case) are isomorphisms.*

Proof. Consider first the spin (or Spin^c) case. In any local trivialization of E, say $E|_C \xrightarrow{\approx} C \times \mathbb{R}^{2m}$, the class $\mathbf{s}(E)$ becomes the element

$$\mathbf{s}(E|_C) = [\$^+_{\mathbb{C}}, \$^-_{\mathbb{C}}; \mu] \in K_{\text{cpt}}(C \times \mathbb{R}^{2m})$$

which is the pull-back to the product of the canonical generator of $K_{\text{cpt}}(\mathbb{R}^{2m})$ given by the Atiyah-Bott-Shapiro Isomorphism. Hence, $\mathbf{s}(E)$ has the Bott periodicity property for K-theory.

Suppose now that E is not necessarily spin. From the isomorphism: $\mathbb{C}\ell_{2m} = \text{Hom}_{\mathbb{C}}(\$_{\mathbb{C}}, \$_{\mathbb{C}}) = \$_{\mathbb{C}} \otimes \$^*_{\mathbb{C}}$, we see that in any local trivialization $E|_C \xrightarrow{\approx} C \times \mathbb{R}^{2m}$, the class $\delta(E)$ becomes

$$\begin{aligned} \delta(E|_C) &= [\$^+_{\mathbb{C}} \otimes \$^*_{\mathbb{C}}, \$^-_{\mathbb{C}} \otimes \$^*_{\mathbb{C}}; \mu] \\ &= [\$^+_{\mathbb{C}}, \$^-_{\mathbb{C}}; \mu] \cdot [\$^*_{\mathbb{C}}] \\ &= 2^m[\$^+_{\mathbb{C}}, \$^-_{\mathbb{C}}; \mu] \\ &= 2^m\mathbf{s}(E). \end{aligned}$$

Hence, after tensoring with $\mathbb{Q}$, $\delta(E)$ has the Bott periodicity property, and the proof is complete. ■

The Thom isomorphisms given above can be compared to the standard Thom isomorphism in cohomology via the Chern character. Formulas for this are given in Chapter III, §12 (see (III.12.14) and (III.12.15)).

REMARK C.13. Each of the Thom isomorphisms given above can be extended to bundles defined over a *locally* compact space X. The orientation classes given explicitly in the Theorems C.8–C.12 do not define classes in $k(E)$ when X is not compact. Nevertheless, one can easily show that they do pair with elements in $k(X)$ to give elements in $k(E)$. Basically, these orientation classes have compact support in "vertical slices" and elements of $k(X)$ lift to classes with compact support in the "horizontal directions," and so the product has compact support on E.

The reader has probably noted that if X is a manifold and $E = T^*X$ is its cotangent bundle, then the orientation classes given in the theorems above are the principal symbols of certain fundamental differential operators. In particular, Theorem C.8 shows that on any $8m$-dimensional spin manifold, the principal symbol of the Atiyah-Singer operator is an orientation class for the KO-theory of the cotangent bundle. Similarly, if X is any even-dimensional spin (or Spin^c) manifold, then by Theorem C.12 the principal symbol of the complex Atiyah-Singer operator gives an orientation class for the K-theory of the cotangent bundle. Viewed in this way, it is clear that the Atiyah-Singer operator is a fundamental one. Twisting this operator with a general coefficient bundle generates the KO-theory (or K-theory) of T^*X freely as a module over $KO(X)$ (or $K(X)$ respectively). Otherwise stated, at the level of principal symbols, every elliptic operator on X is an Atiyah-Singer operator with coefficients in some bundle.

The proof of Theorem C.7 carries over directly to give the following useful case of the Leray-Hirsch Theorem. Suppose $E \to X$ is a complex vector bundle over a compact space X, and let $p: \mathbb{P}(E) \to X$ be the associated projective bundle, i.e., the bundle whose fibre at each point x is the projective space of all complex lines through the origin in E_x. Note that $H^*(\mathbb{P}(E); \mathbb{Z})$ becomes an $H^*(X; \mathbb{Z})$-module under the homomorphism p^*. Consider now the "tautological" complex line bundle $\ell \to \mathbb{P}(E)$ whose fibre at $\ell_0 \in \mathbb{P}(E_x) \subset \mathbb{P}(E)$ consists of all vectors $v \in \ell_0 \subset E_x$. Set $u = c_1(\ell) \in H^2(\mathbb{P}(E); \mathbb{Z})$.

Theorem C.14. *Suppose $E \to X$ is a complex vector bundle of rank k. Then $H^*(\mathbb{P}(E); \mathbb{Z})$ is a free $H^*(X; \mathbb{Z})$-module with basis $1, u, u^2, \ldots, u^{k-1}$.*

Proof. When E is the trivial bundle, this follows directly from the Kunneth formula. In general one chooses a covering $\{C_j\}_{j=1}^N$ of X and proceeds by induction using the Mayer-Vietoris sequence exactly as in the proof of Theorem C.7. ∎

APPENDIX D

$Spin^c$-Manifolds

The notion of a Spin^c-manifold is an important one, but it was not treated in the main body of the text to avoid too much congestion in the exposition. It is essentially the "complex analogue" of the notion of a spin manifold, and much of the discussion of this case can be carried out in parallel with the spin case.

We begin with the complex version of a question mentioned in the introduction. Let X be an oriented riemannian manifold, and ask whether there exists a bundle of irreducible *complex* modules for the bundle $C\ell(X)$. If X is spin, the answer is certainly yes. (One merely takes the associated bundle $\$_{\mathbb{C}} \equiv P_{\mathrm{Spin}}(X) \times_{\Delta_{\mathbb{C}}} V$ where V is an irreducible complex $C\ell_n$-module.) However, the existence of a spin-structure is not necessary for the construction of such bundles. To see this we examine the complex representations of Spin_n more closely.

Suppose $\Delta_{\mathbb{C}}: \mathrm{Spin}_n \to \mathrm{U}_N$ is a complex spinor representation, i.e., one coming from an irreducible $C\ell_n$-module. Let $z: \mathrm{U}_1 \hookrightarrow \mathrm{U}_N$ denote the center, i.e., the scalar multiples of the identity. Then we get a homomorphism $\Delta_{\mathbb{C}} \times z: \mathrm{Spin}_n \times \mathrm{U}_1 \to \mathrm{U}_N$ which clearly has the element $(-1,-1)$ in its kernel. Dividing by this element gives the group

$$\mathrm{Spin}_n^c \equiv \mathrm{Spin}_n \times_{\mathbb{Z}_2} \mathrm{U}_1. \tag{D.1}$$

Note that there is a short exact sequence

$$0 \longrightarrow \mathbb{Z}_2 \longrightarrow \mathrm{Spin}_n^c \xrightarrow{\ \xi\ } \mathrm{SO}_n \times \mathrm{U}_1 \longrightarrow 1 \tag{D.2}$$

where the subgroup $\mathbb{Z}_2 \subset \mathrm{Spin}_n^c$ is generated by the element $[(-1,1)] = [(1,-1)]$. Now we have

$$\mathrm{Spin}_n^c \subset C\ell_n \otimes \mathbb{C}$$

as a multiplicative subgroup of the group of units. (In fact it is obtained by "tensoring" Spin_n with the unit complex numbers.) To construct the bundle of irreducible complex modules over X we proceed as in the spin case. With the sequence (D.2) in mind, we ask: Does there exist a principal Spin_n^c-bundle over X which admits a Spin_n^c-equivariant bundle mapping:

$$P_{\mathrm{Spin}^c} \xrightarrow{\ \xi\ } P_{\mathrm{SO}}(X) \times P_{\mathrm{U}_1} \tag{D.3}$$

for some principal U_1-bundle P_{U_1} over X? By "equivariant" we mean that

$\xi(pg) = \xi(p)\xi(g)$ for all $p \in P_{\text{Spin}^c}$ and all $g \in \text{Spin}_n^c$. We analyse this in the spirit of Appendix A. Consider the exact sequence

$$H^1(X; \text{Spin}_n^c) \xrightarrow{\xi} H^1(X; \text{SO}_n) \oplus H^1(X; \text{U}_1) \xrightarrow{w_2 + \tilde{c}_1} H^2(X; \mathbb{Z}_2) \quad \text{(D.4)}$$

determined by the coefficient sequence (D.2). The coboundary map associates to a pair $(P_{\text{SO}}, P_{\text{U}_1})$ the element $w_2(P_{\text{SO}}) + \tilde{c}_1(P_{\text{U}_1})$ where $\tilde{c}_1$ is the mod 2 reduction of the first Chern class of P_{U_1} (see Example A.5). Under the isomorphism (A.7) this coboundary map can be rewritten as

$$H^1(X; \text{SO}_n) \oplus H^2(X; \mathbb{Z}) \xrightarrow{w_2 + \rho} H^2(X; \mathbb{Z}_2) \quad \text{(D.5)}$$

where $\rho: H^2(X; \mathbb{Z}) \to H^2(X; \mathbb{Z}_2)$ is mod 2 reduction. Consequently, given the frame bundle $P_{\text{SO}}(X)$, we can find the bundle (D.3) provided that $w_2(P_{\text{SO}}(X)) = \rho(u)$ for some $u \in H^2(X; \mathbb{Z})$, i.e., provided that $w_2(X)$ is the mod 2 reduction of an integral class.

Of course this argument carries over to any principal SO_n-bundle.

DEFINITION D.1. Let P_{SO_n} be a principal SO_n-bundle over X. A **Spin^c-structure** on P_{SO_n} consists of a principal U_1-bundle P_{U_1} and also a principal Spin_n^c-bundle $P_{\text{Spin}_n^c}$ with a Spin_n^c-equivariant bundle map

$$P_{\text{Spin}_n^c} \longrightarrow P_{\text{SO}_n} \times P_{\text{U}_1}.$$

The class $c \in H^2(X; \mathbb{Z})$ corresponding to P_{U_1} under the isomorphism $H^2(X; \mathbb{Z}) \cong \text{Prin}_{\text{U}_1}(X)$ (cf. A.5), is called the **canonical class** of the Spin^c-structure.

The argument given above proves the following:

Theorem D.2. *A principal* SO_n*-bundle* P *carries a* Spin_n^c*-structure if and only if* $w_2(P)$ *is the* mod 2 *reduction of an integral class, that is, if and only if*

$$W_3(P) = 0.$$

(See B.13 for a discussion of the classes W_k.)

It should be noted that Theorem D.2 could be established in the spirit of Theorem 1.4ff by considering the appropriate 2-fold coverings of the spaces $P_{\text{SO}_n} \times P_{\text{U}_1}$.

DEFINITION D.3. An oriented riemannian manifold with a Spin^c-structure on its tangent (frame) bundle is called a **Spin^c-manifold.**

The theorem above has the following immediate corollary:

COROLLARY D.4. *An orientable manifold* X *carries a Spin^c-structure* (i.e., *X can be made into a Spin^c-manifold), if and only if the second Stiefel-Whitney class* $w_2(X)$ *is the* mod 2 *reduction of an integral class, i.e., if and only if* $W_3(X) = 0$.

If a manifold carries one Spin^c-structure, it often carries many. They are parameterized by the elements in $2H^2(X;\mathbb{Z}) \oplus H^1(X;\mathbb{Z}_2)$.

There are two important cases where a Spin^c-structure exists and is canonically determined:

EXAMPLE D.5. *Any bundle with a spin structure carries a canonically determined Spin^c-structure.* The Spin^c-bundle is obtained as $P_{\text{Spin}^c} \equiv P_{\text{Spin}} \times_{\mathbb{Z}_2} U_1$ where $\mathbb{Z}_2$ acts diagonally by $(-1,-1)$ and where U_1 of course denotes the trivial circle bundle. We see then that *any spin manifold is canonically a Spin^c-manifold.*

EXAMPLE D.6. *Any complex vector bundle E carries a canonically determined Spin^c-structure.* To see this, note first that $w_2(E) \equiv c_1(E) \pmod 2$, and so Theorem D.1 implies that E carries Spin^c-structures. To see that there is a canonical one we proceed as follows. Introduce a hermitian metric in E and let $P_{\mathrm{U}_n}(E)$ be the principal U_n-bundle of unitary frames. There is a canonical homomorphism

$$j: \mathrm{U}_n \hookrightarrow \mathrm{Spin}^c_{2n} \tag{D.6}$$

which we shall construct explicitly below. It is a lifting of the homomorphism $\mathrm{U}_n \to \mathrm{SO}_{2n} \times \mathrm{U}_1$ given by $g \mapsto (i(g), \det(g))$ where $i: \mathrm{U}_n \hookrightarrow \mathrm{SO}_{2n}$ is the standard inclusion. Thus, we have a commutative diagram

$$\begin{array}{ccc} & \overset{j}{\nearrow} & \mathrm{Spin}^c_{2n} \\ \mathrm{U}_n & & \big\downarrow \xi \\ & \underset{i \times \det}{\searrow} & \mathrm{SO}_{2n} \times \mathrm{U}_1 \end{array}$$

The canonical principal Spin^c bundle for E is now constructed as the associated bundle.

$$P_{\text{Spin}^c}(E) \equiv P_{\mathrm{U}_n}(E) \times_j \mathrm{Spin}^c_{2n} \tag{D.8}$$

Note, incidentally, that the associated U_1-bundle in this case is

$$P_{\mathrm{U}_1}(E) \equiv P_{\mathrm{U}_n}(E) \times_{\det} U_1. \tag{D.9}$$

This is the principal bundle of the complex line bundle $\Lambda^n E$ whose first Chern class is $c_1(E)$, i.e., $c_1(\Lambda^n E) = c_1(E)$.

One consequence of this discussion is that *every complex manifold, in fact, every almost complex manifold, is canonically a Spin^c-manifold.*

Before moving on we shall provide the details of the homomorphism (D.6). It can be shown to exist by proving the existence of the lifting (D.7) using covering space theory. However, it can be given explicitly as follows. Let $g \in \mathrm{U}_n$ and choose a unitary basis $\{e_1, \ldots, e_n\}$ of $\mathbb{C}^n$ in which g has the form: $g \cong \mathrm{diag}\{e^{i\theta_1}, \ldots, e^{i\theta_n}\}$. Let $\{e_1, Je_1, \ldots, e_n, Je_n\}$ be the canonically associated orthonormal basis of $\mathbb{R}^{2n} = \mathbb{C}^n$. Then

$$j(g) \equiv \prod_{k=1}^{n} \left(\cos \frac{\theta_k}{2} + \sin \frac{\theta_k}{2} e_k J e_k \right) \times e^{\frac{i}{2}\Sigma\theta_k} \tag{D.10}$$

in $\text{Spin}^c_{2n} = \text{Spin}_{2n} \times_{\mathbb{Z}_2} S^1$.

We have seen that spin manifolds and complex manifolds are all Spin^c-manifolds. In fact, it requires some searching about to find an orientable manifold which is not Spin^c. Note, for example, that $\mathbb{P}^{4n+1}(\mathbb{R})$, although neither spin nor complex, *is* Spin^c.

It was observed by Landweber and Stong that perhaps the simplest manifold which is not Spin^c is the orientable 5-manifold SU_3/SO_3 whose only non-zero mod 2 cohomology classes are 1, w_2, w_3 and $w_2 \cdot w_3$. Since $\text{Sq}^1(w_2) = w_3 \neq 0$, one concludes that $W_3 \neq 0$.

EXAMPLE D.7. (Universal non-Spin^c-manifolds) We shall construct here a family of open simply-connected n-manifolds $\mathfrak{U}^n(p)$ which are not Spin^c for each $n \geqq 9$. These examples are universal in the sense that every simply connected non-Spin^c-manifold of dimension $n \geqq 9$, carries some $\mathfrak{U}^n(p)$ as an open submanifold.

We fix a positive integer p and consider the cell complex

$$\Sigma \equiv S^2 \cup_\varphi D^3$$

where the map $\partial D^3 = S^2 \xrightarrow{\varphi} S^2$ is of degree 2^p. We can embed this complex into $\mathbb{R}^6$. In fact there is a natural *PL*-embedding given as follows. Consider $S^2 \cong \partial\Delta^3$, where Δ^3 denotes the standard 3-simplex, and embed the mapping cone $C_\varphi = S^2 \cup_\varphi S^2 \to S^2 * S^2 \approx S^5$ in the obvious way. Here "$*$" denotes the join. This clearly extends to an embedding $S^2 \cup_\varphi D^3 \hookrightarrow S^2 * D^3 \approx D^6$.

Let U_Σ be a regular neighborhood of Σ in $\mathbb{R}^6$. This is an open parallelizable manifold. We now claim that there is a 3-dimensional real vector bundle $E \to U_\Sigma$ which restricts to be the non-trivial bundle on $S^2 \subset \Sigma$. Since U_Σ retracts onto Σ, we need only find a map $F: \Sigma \to BSO_3$, which when restricted to S^2 represents the non-zero element in $\pi_2(BSO_3)$. However, since $\pi_2(BSO_3) \cong \pi_1(SO_3) \cong \mathbb{Z}_2$, any map $f: S^2 \to BSO_3$ extends to Σ. Therefore this bundle exists.

We define the manifold $\mathfrak{U}^n(p)$ as follows. Consider Σ embedded in the total space of E as a subset of the zero section; i.e., $\Sigma \subset U_\Sigma \subset E$. For each $n \geqq 9$, let $\mathfrak{U}^n(p)$ be a regular neighborhood of $\Sigma \times \{0\} \subset E \times \mathbb{R}^{n-9}$.

Theorem D.8. *The manifolds $\mathfrak{U}^n(p)$, $p \geqq 1$, are not Spin^c. Furthermore, any simply-connected manifold X^n of dimension $n \geqq 9$ is not Spin^c if and only if for some $p \geqq 1$, there exists an embedding*

$$\mathfrak{U}^n(p) \hookrightarrow X^n$$

as an open submanifold.

Proof. Fix $p \geqq 1$ and $n \geqq 9$ and set $\mathfrak{U} = \mathfrak{U}^n(p)$. Let $S^2 \subset \mathfrak{U}$ be the non-trivial 2-sphere. Then $T\mathfrak{U}|_{S^2} \cong E|_{S^2} \oplus$ (trivial), and so $w_2(\mathfrak{U})[S^2] = w_2(E)[S^2] \neq 0$ by construction. Since $H^2(\mathfrak{U}; \mathbb{Z}) = 0$ and $\pi_1(\mathfrak{U}) = 0$, we see that $w_2(\mathfrak{U})$ is not the mod 2 reduction of an integral class. (Note that $\mathfrak{U}$ is homotopy equivalent to Σ.) Hence, $\mathfrak{U}$ is not Spin^c.

Of course any open submanifold of a Spin^c-manifold is Spin^c. Hence, we need only show that a simply-connected n-manifold X^n which is not Spin^c must contain some $\mathfrak{U}^n(p)$ as an open submanifold. Since $\pi_1 X = 0$, we have an isomorphism $\pi_2 X \xrightarrow{\approx} H_2(X; \mathbb{Z})$. Since X is not Spin^c, the homomorphism $w_2 \in H^2(X; \mathbb{Z}_2) \cong \mathrm{Hom}(\pi_2 X, \mathbb{Z}_2)$ is not the mod 2 reduction of an element in $H^2(X; \mathbb{Z}) \cong \mathrm{Hom}(\pi_2 X, \mathbb{Z})$. Hence, w_2 is non-zero on some torsion class α of order 2^p in $\pi_2 X$. Now α is represented by a map $f: S^2 \to X$ which extends to a map $F: \mathfrak{U}^n(p) \to X$ (since $2^p\alpha = 0$ in $\pi_2 X$). The induced bundle F^*TX is clearly equivalent to $E \oplus$ (trivial) $\cong T\mathfrak{U}^n(p)$. Hence, by Smale-Hirsch immersion theory, F is homotopic to an immersion $\mathfrak{U}^n(p) \hookrightarrow X^n$. The immersion can be made an embedding by putting Σ into general position and shrinking $\mathfrak{U}^n(p)$ to a small neighborhood of Σ. This completes the proof. ■

We did not need to invoke Smale-Hirsch Theory here. The embedding $\mathfrak{U}^n(p) \hookrightarrow X^n$ which is homotopic to F can be built explicitly. A tubular neighborhood of $S^2 \subset \mathfrak{U}^n(p)$ is diffeomorphic to a tubular neighborhood of $S^2 \subset X^n$, since the normal bundles are equivalent. Completing the embedding to a neighborhood of $D^3 \subset \mathfrak{U}^n(p)$ is straightforward.

For compact examples we add a boundary to $\mathfrak{U}^n(p)$ and consider the *double*

$$X^n(p) \equiv \bar{\mathfrak{U}}^n(p) \cup_\partial \bar{\mathfrak{U}}^n(p)$$

where $\bar{\mathfrak{U}}^n(p) \equiv \mathfrak{U}^n(p) \cup \partial\mathfrak{U}^n(p)$.

Let us return our attention to manifolds which are Spin^c.

There is an alternative approach to this concept which comes from considering vector bundles.

DEFINITION D.9. Let X be a Spin^c-manifold of dimension n. By a **complex spinor bundle** for X we mean a vector bundle S associated to a representation of Spin^c by Clifford multiplication, i.e.,

$$S \equiv P_{\mathrm{Spin}^c}(X) \times_\Delta V$$

where V is a complex $\mathbb{C}\ell_n$-module and $\Delta: \mathrm{Spin}^c_n \to \mathrm{GL}(V)$ is given by restriction of the $\mathbb{C}\ell_n$-representation to $\mathrm{Spin}^c_n \subset \mathbb{C}\ell_n \otimes \mathbb{C}$. If the representation of $\mathbb{C}\ell_n$ is irreducible, we say that S is **fundamental.**

These spinor bundles are bundles of complex modules over $\mathbb{C}\ell(X)$.

When n is even there exists only one fundamental spinor bundle for X, denoted $S(X)$. It splits into a direct sum

$$S(X) = S^+(X) \oplus S^-(X) \qquad \text{(D.11)}$$

where $S^{\pm}(X) = (1 \pm \omega_{\mathbf{C}})S(X)$ and where $\omega_{\mathbf{C}} = i^{n/2}e_1 \cdots e_n$ is, as usual, the volume form. Multiplication by a non-zero tangent vector maps $S^{\pm}(X)$ to $S^{\mp}(X)$ and is invertible.

When n is odd, there are two irreducible complex representations of $\mathbf{C}\ell_n$. However they are equivalent when restricted to Spin_n^c. Hence, *there is only one fundamental spinor bundle $S(X)$ for any Spin^c-manifold X.*

For a spin manifold X, the bundle $S(X)$ is just the usual complex spinor bundle.

For a complex manifold X with its canonical Spin^c-structure, we have that

$$S(X) = \Lambda_{\mathbf{C}}^* TX \cong \Lambda^{0,*}$$

that is, $S(X)$ is simply the direct sum of the complex exterior powers of the tangent bundle, considered as a complex bundle (which, in turn, is isomorphic to the direct sum of the bundles of $(0,q)$-forms, $0 \leq q \leq n$). This follows directly from the interpretation of the complex spinor representation given in I.5 (see I.5.25 and the attending discussion). In particular, from this we see that the Clifford multiplication $\mathbf{C}\ell(X) \otimes_{\mathbf{R}} S(X) \to S(X)$ is generated as follows. For each tangent vector v, consider the linear map $\Lambda_{\mathbf{C}}^* T_x X \overset{\mu_v}{\to} \Lambda_{\mathbf{C}}^* T_x X$ given by

$$\mu_v(\varphi) = v \wedge \varphi - v^* \llcorner \varphi$$

where v^* is the 1-form corresponding to v under the hermitian metric on X. Repeating this operation gives

$$\mu_v(\mu_v \varphi) = -\|v\|^2 \varphi,$$

and so by the universal property of Clifford algebras, the map μ extends to a representation of $\mathbf{C}\ell(X)$. Since each μ_v is complex linear the representation is complex.

It is instructive to examine these fundamental spinor bundles as one varies the Spin^c-structure on a given manifold. To do this it is good to have clearly in mind the elementary isomorphisms:

$$\mathrm{Vect}_1(X) \cong \mathrm{Prin}_{\mathrm{U}_1}(X) \cong H^2(X; \mathbf{Z}) \tag{D.12}$$

where $\mathrm{Vect}_1(X)$ denotes the set of equivalence classes of complex line bundles on X and where the map to H^2 is given by the first Chern class (see Example A.5).

EXAMPLE D.10. Let X be a spin manifold with its canonical Spin^c-structure. We can change this structure by changing the U_1-bundle as follows. Pick an element $\alpha \in H^2(X; \mathbf{Z})$, and let $P_{\mathrm{U}_1}(\alpha) \in \mathrm{Prin}_{S^1}(X)$ and $\lambda_\alpha \in \mathrm{Vect}_1(X)$ be the corresponding elements under (D.12). Then we define the αth Spin^c-structure by

$$P_{\mathrm{Spin}^c}(\alpha) \equiv P_{\mathrm{Spin}}(X) \times_{\mathbf{Z}_2} P_{\mathrm{U}_1}(\alpha) \tag{D.13}$$

Note that this has a Spin_n^c-equivariant map

$$P_{\mathrm{Spin}^c}(\alpha) \longrightarrow P_{\mathrm{SO}}(X) \times P_{\mathrm{U}_1}(2\alpha) \tag{D.14}$$

where $P_{\mathrm{U}_1}(2\alpha) = P_{\mathrm{U}_1}(\alpha)/\mathbb{Z}_2$ is the "square" of $P_{\mathrm{U}_1}(\alpha)$. (In particular $\lambda_{2\alpha} = \lambda_\alpha^2$.)

We now let $S_\alpha(X)$ be the fundamental spinor bundle for the αth Spin^c-structure on X. Then it is easy to see that

$$S_\alpha(X) = S(X) \otimes \lambda_\alpha. \tag{D.15}$$

Thus, changing the Spin^c-structure on X by an element $\alpha \in H^2(X; \mathbb{Z})$ amounts to twisting the fundamental spinor bundle by the complex line bundle λ_α.

This is a general fact. The group $H^2(X; \mathbb{Z})$ acts on the set of Spin^c-structures on a Spin^c-manifold X, by tensoring the fundamental spinor bundle with complex line bundles (modulo this action, the Spin^c-structures correspond one-to-one with elements of $H^1(X; \mathbb{Z}_2)$.) Notice that tensoring with line bundles preserves the fibre-dimension and thus takes one bundle of irreducible modules over $C\ell(X)$ to another.

A good example of this phenomenon is the following. Suppose X is a complex manifold which is also spin. This means that $w_2(X) \equiv c_1(X) \equiv 0$ (mod 2). Now the class $-c_1(X) \in H^2(X; \mathbb{Z})$ corresponds to the complex line bundle $\kappa \equiv \Lambda_{\mathbb{C}}^n T^*X$, called the **canonical** bundle. Since $-c_1(X) = c_1(\kappa)$ is even, there exist square roots $\kappa^{1/2}$ for κ. The different square roots are parameterized by elements of $H^1(X; \mathbb{Z}_2)$, corresponding to choices of spin structure. Now the canonical spinor bundle $S(X) = \Lambda_{\mathbb{C}}^* TX$ for X as a Spin^c-manifold, and the canonical spinor bundle $\$_{\mathbb{C}}(X)$ for X as a spin manifold are related by the equation

$$\$_{\mathbb{C}}(X) = S(X) \otimes \kappa^{1/2}. \tag{D.16}$$

Much of this business of Spin^c-manifolds becomes transparent when viewed in terms of the fundamental spinor bundles. For example, let us return to the original problem of trying to construct a fundamental spinor bundle (i.e., a bundle of irreducible complex modules for $C\ell(X)$) over a manifold X. Locally of course we can always do it. Let $\{U_\alpha\}_{\alpha \in A}$ be a covering of X so that $U_{\alpha_1} \cap \cdots \cap U_{\alpha_k}$ is contractible for all $\alpha_1, \ldots, \alpha_n$. On each U_α, bundles can be trivialized and we can find a fundamental complex spinor bundle of the form $U_\alpha \times V$ where V is an irreducible $C\ell_n \otimes \mathbb{C}$-module. Now in passing from U_α to U_β we look for transition functions $\tilde{g}_{\alpha\beta}: U_\alpha \cap U_\beta \to \mathrm{Spin}_n$ so that $\xi \circ \tilde{g}_{\alpha\beta} = g_{\alpha\beta}: U_\alpha \cap U_\beta \to \mathrm{SO}_n$ are the corresponding transition functions for $P_{\mathrm{SO}}(X)$. Now the existence of a compatible set of choices is equivalent to the vanishing of the Čech cocycle

$$w_{\alpha\beta\gamma} \equiv \tilde{g}_{\alpha\beta}\tilde{g}_{\beta\gamma}\tilde{g}_{\gamma\alpha}: U_\alpha \cap U_\beta \cap U_\gamma \longrightarrow \mathbb{Z}_2 = \ker(\xi) \tag{D.17}$$

in $H^2(X; \mathbb{Z}_2)$.

Suppose now that the class $[w] \in H^2(X; \mathbb{Z}_2)$ is the mod 2 reduction of an integral class $\mathscr{W} \in H^2(X; \mathbb{Z})$, and let λ be the complex line bundle corresponding to $\mathscr{W}$. Consider the problem of finding a square root of λ, i.e., a line bundle $\lambda^{1/2}$ with $(\lambda^{1/2})^2 = \lambda$. Let $\gamma_{\alpha\beta}: U_\alpha \cap U_\beta \to S^1$ be the transition functions for λ. Since $U_\alpha \cap U_\beta$ is contractible we can extract a square root $\tilde{\gamma}_{\alpha\beta} \equiv \gamma_{\alpha\beta}^{1/2}: U_\alpha \cap U_\beta \to S^1$. However, the compatibility is just the Čech cocycle

$$w'_{\alpha\beta\gamma} \equiv \tilde{\gamma}_{\alpha\beta}\tilde{\gamma}_{\beta\gamma}\tilde{\gamma}_{\gamma\alpha}: U_\alpha \cap U_\beta \cap U_\gamma \longrightarrow \mathbb{Z}_2 = \ker(\sigma) \tag{D.18}$$

where

$$0 \longrightarrow \mathbb{Z}_2 \longrightarrow S^1 \xrightarrow{\ \sigma\ } S^1 \longrightarrow 0$$

and $\sigma(z) \equiv z^2$. The class $[w'] \in H^2(X; \mathbb{Z}_2)$ is just the coboundary of $\lambda \in H^1(X; S^1)$ under the associated long exact sequence in cohomology. In fact, consider the following commutative diagram:

$$\begin{array}{ccccc}
H^1(X; S^1) & \xrightarrow{\ \sigma\ } & H^1(X; S^1) & \xrightarrow{\ w'\ } & H^2(X; \mathbb{Z}_2) \\
\downarrow \wr & & \downarrow \wr & & \| \\
H^2(X; \mathbb{Z}) & \xrightarrow{\ 2\ } & H^2(X; \mathbb{Z}) & \xrightarrow{\ \rho\ } & H^2(X; \mathbb{Z}_2)
\end{array}$$

It is clear that our obstructions agree, i.e., $[w'] = \rho(c_1(\lambda)) = \rho(\mathscr{W}) = [w]$. And so $[w] + [w'] = 0$ in the wonderful world of $\mathbb{Z}_2$.

This means essentially that *while we cannot construct the spinor bundle and we cannot construct* $\lambda^{1/2}$, *we can construct their product.* Adjusting by coboundaries we can choose $\tilde{g}_{\alpha\beta}$ and $\tilde{\gamma}_{\alpha\beta}$ so that $w_{\alpha\beta\gamma} \equiv w'_{\alpha\beta\gamma}$. Thus the transition functions

$$G_{\alpha\beta}: U_\alpha \cap U_\beta \longrightarrow \mathrm{Spin}_n \times_{\mathbb{Z}_2} S^1$$

defined by $G_{\alpha\beta} = \tilde{g}_{\alpha\beta} \times \tilde{\gamma}_{\alpha\beta}$ have $w_{\alpha\beta\gamma} \equiv G_{\alpha\beta}G_{\beta\gamma}G_{\gamma\alpha} \equiv 0$, and thus determine a global bundle we may think of as

$$S(X) = S_0(X) \otimes \lambda^{1/2} \tag{D.19}$$

where $S_0(X)$ is the fundamental spinor bundle for the possibly non-existent spin structure on X, and where $\lambda^{1/2}$ is the possibly non-existent square root of λ with $c_1(\lambda) \equiv w_2(X)$ (mod 2).

Since $S(X)$ is a bundle of modules over $C\ell(X)$ we need only introduce an adapted metric and connection to make it a Dirac bundle. We do this as follows. Fix any line bundle λ as above and choose a unitary connection on λ. This induces a connection on the bundle $\lambda^{1/2}$ when it exists. Locally both $S_0(X)$ and $\lambda^{1/2}$ exist and $S_0(X)$ carries a canonical (riemannian) connection. We give the bundle $S(X) = S_0(X) \otimes \lambda^{1/2}$ the tensor product connection. It is not difficult to see that this connection is well defined globally.

It is easy to check, using the arguments of §4, that this connection makes $S(X)$ into a Dirac bundle.

Notice that the line bundle λ corresponds to the principal U_1-bundle P_{U_1} associated to the Spin^c-structure, i.e., $P_{U_1}(\lambda) = P_{Spin^c}(X) \times_p U_1$ where $p: Spin_n^c \to U_1$ is induced from the projection $Spin_n \times U_1 \to U_1$. Let us summarize:

Proposition D.11. *Let X be a Spin^c-manifold with associated complex line bundle λ. Then to each U_1-connection ω on λ there is a canonically associated connection on P_{Spin^c}. It is just the lift of the product of the canonical riemannian connection with ω via the covering map $P_{Spin^c} \to P_{SO} \times P_{U_1}(\lambda)$. This connection makes any complex spinor bundle for X into a Dirac bundle.*

Associated to any unitary connection on λ there is the **curvature 2-form** Ω. This is a closed 2-form, and the class $[(2\pi)^{-1}\,\Omega] \in H^2(X;\mathbb{R})$ represents $c_1(\lambda)$ or more precisely the image of $c_1(\lambda)$ under the map $H^2(X;\mathbb{Z}) \to H^2(X;\mathbb{R})$. We call this the **real Chern class of λ.**

The curvature form is defined as follows. For a local trivialization of $P_{U_1}(\lambda)$, the connection 1-form becomes simply a real-valued 1-form ω. Then

$$\Omega \equiv d\omega.$$

If we change trivializations by a transition function $g: V \cap V' \to U_1$, then the new connection 1-form is $\omega' = \omega + g^{-1}\,dg$. Locally we have $g = e^{i\theta}$ for some function θ. The equation above becomes $\omega' = \omega + i\,d\theta$. It follows that $d\omega' = d\omega$, that is, the curvature 2-form Ω is well defined globally.

Suppose now that ω_1 and ω_2 are two different connections on λ with associated curvature forms Ω_1 and Ω_2. Then it is easy to see that $\omega_1 - \omega_2 = \alpha$ for a globally defined 1-form α. Hence, $\Omega_1 - \Omega_2 = d\alpha$, and we see that the de Rham class $[\Omega]$ is independent of choice of connection.

We are now prepared to derive the Bochner-type identity for these spinor bundles.

Theorem D.12. *Let X be a Spin^c-manifold with associated line bundle λ, and fix a connection on λ with curvature 2-form Ω. Let S be a bundle of complex spinors on X with the canonical connection, and let D be the Dirac operator on S. Then*

$$\boxed{D^2 = \nabla^*\nabla + \tfrac{1}{4}\kappa + \tfrac{1}{2}\Omega}$$

where κ denotes the scalar curvature of X and where Ω denotes Clifford multiplication by the 2-form Ω.

Proof. The computation is local, so we can assume $S(X) = S_0(X) \otimes \lambda^{1/2}$ where $S_0(X)$ is a spinor bundle for the local spin structure. We may now apply Theorem II.8.17, and it is sufficient to compute the term $\mathfrak{R}^{\lambda^{1/2}}$. Recall that

$$\begin{aligned}\mathfrak{R}^{\lambda^{1/2}}(\sigma \otimes z) &= \sum_{j<k} (e_j e_k \sigma) \otimes R^{\lambda^{1/2}}_{e_j,e_k}(z) \\ &= \sum_{j<k} (e_j e_k \sigma) \otimes (\tfrac{1}{2}\Omega(e_j,e_k) iz) \\ &= \left(\frac{i}{2} \sum_{j<k} \Omega(e_j,e_k) e_j e_k \sigma\right) \otimes z \\ &= \left[\frac{i}{2}\left(\sum_{j<k} \Omega(e_j,e_k) e_j \wedge e_k\right)\sigma\right] \otimes z \\ &= \tfrac{i}{2}(\Omega \cdot \sigma) \otimes z \quad \blacksquare\end{aligned}$$

We used here the fact that the curvature of the induced connection on $\lambda^{1/2}$ is $\frac{1}{2}\Omega$. We also use the fact that $R^{\lambda}_{v,w}(Z) = \Omega(v,w)iz$. Note that Clifford multiplication by $i\Omega$ is a **hermitian symmetric** operation.

Given this theorem, the following lemma is relevant:

Lemma D. 13. *Let λ be a complex line bundle on a manifold X. Then any closed 2-form Ω which represents the real Chern class of λ can be realized as the curvature form of a* U_1*-connection on λ.*

NOTE D.14. This means, in particular, that if $c_1(\lambda)$ is a torsion class in $H^2(X; \mathbb{Z})$, then λ admits a *flat* connection.

Proof. Choose any U_1-connection ω_0 on η and let Ω_0 be its curvature form. Since Ω and Ω_0 represent the same de Rham class, there is a 1-form α so that $d\alpha = \Omega - \Omega_0$. Let $\omega = \omega_0 + \alpha$. Then the curvature of ω is $\Omega_0 + d\alpha = \Omega$. ■

Suppose now that X has even dimension and consider the Dirac operator

$$D^+ : \Gamma S^+(X) \longrightarrow \Gamma S^-(X)$$

for the splitting (D.11). Using (D.19) and the Atiyah-Singer Formula, we obtain the following:

Theorem D.15. *Let X be a compact Spin^c-manifold of even dimension, and let D denote the Dirac operator of the fundamental complex spinor bundle.*

Then

$$\operatorname{ind}(D^+) = \left\{e^{\frac{1}{2}c} \cdot \hat{\mathbf{A}}(X)\right\}[X] \tag{D.20}$$

where c is the canonical class of the Spin^c-structure, i.e., $c = c_1(\lambda)$ for the associated complex line bundle λ.

Combining this with D.13 and D.14 gives the following:

Corollary D.16. *Let X be a compact Spin^c-manifold such that $w_2(X)$ is the* mod 2 *reduction of a torsion class. If X carries a metric of positive scalar curvature, then $\hat{A}(X) = 0$.*

Corollary D.17. *Let X be a compact Spin^c-manifold with canonical class $c \in H^2(X;\mathbb{Z})$. Let Ω be any 2-form representing the de Rham class of c in $H^2(X;\mathbb{Z}) \otimes \mathbb{R}$. Then there is a riemannian metric on X whose scalar curvature satisfies*

$$\kappa > 2\|\Omega\|$$

only if $\{e^{\frac{1}{2}c} \cdot \hat{\mathbf{A}}(X)\}[X] = 0$.

Here the norm $\|\cdot\|$ is taken in the same metric. It is given by $\|\Omega\| = \sum |\lambda_j|$ where $\Omega = \sum \lambda_j e_{2j-1} \wedge e_{2j}$ for the diagonalizing orthonormal frame $e_1, e_2, \ldots$ at the point.

It is a nice exercise to apply this theorem to $\mathbb{P}^n(\mathbb{C})$ with the standard Fubini-Study metric, and then to verify directly the vanishing of the "Hilbert polynomial" at certain integers.

It is interesting to examine the Dirac operator on a complex hermitian manifold X, considered as a Spin^c-manifold. Recall that $S(X) = \Lambda^{0,*}$. The splitting $S(X) = S^+(X) \oplus S^-(X)$ corresponds to the even-odd decomposition $\Lambda^{0,*} = \Lambda^{0,\text{even}} \oplus \Lambda^{0,\text{odd}}$. The Dirac operator can be identified with

$$\bar{\partial} + \bar{\partial}^* : \Lambda^{0,\text{even}} \longrightarrow \Lambda^{0,\text{odd}}$$

where $\bar{\partial}^*$ is the hermitian adjoint of the operator $\bar{\partial}$. The index of this operator is the Todd genus of X.

If $c_1(X) \equiv 0 \pmod 2$, we can choose a spin structure on X, i.e., we choose a square root $\kappa^{1/2}$ of the line bundle $\kappa = \Lambda^{n,0}$. Then the associated Dirac operator can be identified with

$$\bar{\partial} + \bar{\partial}^* : \Lambda^{0,\text{even}} \otimes \kappa^{1/2} \longrightarrow \Lambda^{0,\text{odd}} \otimes \kappa^{1/2}.$$

Here $\bar{\partial}$ is the standard operator on $(0,*)$-forms with values in the holomorphic line bundle $\kappa^{1/2}$ (cf. (D.16)).

Observe that Theorem D.15 has the following immediate corollary which was first proved by Atiyah and Hirzebruch [1].

Corollary D.18. *Let X be a compact Spin^c-manifold. Then for any class $c \in H^2(X,\mathbb{Z})$ such that $c \equiv w_2(X)$* (mod 2), *the* (*rational*) *number*

$$\{e^{\frac{1}{2}c}\hat{\mathbf{A}}(X)\}[X]$$

is an integer.

As a final remark we point out that Spin^c-bundles are bundles which have natural K-theory orientation classes, just as spin bundles have natural KO-theory orientation classes. This makes them sometimes useful in general theories.

Bibliography

Adams, J. F.

[1] *Lectures on Lie Groups*, Benjamin Press, N.Y., 1969.

[2] Vector fields on spheres, *Ann. of Math. 75* (1962), 603–632.

[3] On the groups J(X): IV, *Topology 5* (1966), 21–71.

[4] Spin(8), triality, F_4, and all that, in *Superspace and Supergravity*, ed. S. W. Hawking and M. Rocek, Cambridge University Press, Cambridge, 1981.

Alvarez-Gaume, L.

[1] Supersymmetry and the Atiyah-Singer index theorem, *Commun. Math. Phys. 90* (1983), 161–173.

Anderson, D. W., Brown, Jr. E. H., and Peterson, F. P.

[1] The structure of the spin cobordism ring, *Ann. of Math. 86* (1967), 271–298.

Artin, E.

[1] *Geometric Algebra*, Interscience, New York, 1957.

Atiyah, M. F.

[1] La formule de l'indice pour les varietes a bord, in *Seminaire Cartan-Schwartz 16 (1963–64)*, Ecole Normale Superieure, (1965), 25.01–25.09.

[2] K-theory and Reality, *Quart. J. Math. Oxford (2) 17* (1966), 367–386.

[3] Algebraic topology and elliptic operators, *Comm. on Pure and App. Math. 20* (1967), 237–249.

[4] *K-Theory*, Benjamin Press, New York, 1967.

[5] Bott periodicity and the index of elliptic operators, *Quart. J. Math. Oxford (2) 19* (1968), 113–140.

[6] Algebraic topology and operators in Hilbert space, in *Lectures in Modern Analysis and Applications I*, Lect. Notes in Math. 103, Springer-Verlag, Berlin, 1969.

[7] Global theory of elliptic operators, *Proceedings of the International Conference on Functional Analysis and Related Topics (Tokyo 1969)*, University of Tokyo Press, 1970, 21–30.

[8] Topology of elliptic operators, *A.M.S. Proc. Symp. Pure Math. 16* (1970), 101–119.

[9] Vector fields on manifolds, *Arbeitsgemeinschaft fur Forschung des Landes Nordrhein-Westfalen 200* (1970), 7–23.

[10] *Elliptic Operators and Compact Groups*, Lect. Notes in Math. 401, Springer-Verlag, Berlin 1974.

[11] Elliptic operators, discrete groups and Von Neumann algebras, in *Colloque Analyse et Topologie en l'honneur de H. Cartan*, Astérisque 32–33, Soc. Math. de France, 1976, 43–72.

[12] Eigenvalues of the Dirac operator, *Arbeitstagung, Bonn, 1984*, Lect. Notes in Math. 1111, Springer-Verlag, Berlin, 1985, 251–260.

[13] Circular symmetry and stationary phase approximation, *Colloque en l'honneur de Laurent Schwartz, 2*, Astérisque 132, Soc. Math. de France, 1985, 43–60.

Atiyah, M. F., and Bott, R.

[1] On the periodicity theorem for complex vector bundles, *Acta Math. 112* (1964), 229–247.

[2] The index theorem for manifolds with boundary, *Bombay Colloquium on Differential Analysis*, Oxford, 1964, 175–186.

[3] A Lefschetz fixed point formula for elliptic complexes I, *Ann. of Math. 86* (1967), 374–407.

[4] A Lefschetz fixed point formula for elliptic complexes II, *Ann. of Math. 87* (1968), 415–491.

Atiyah, M. F., Bott, R., and Patodi, V. K.

[1] On the heat equation and the index theorem, *Invent. Math. 19* (1973), 279–330. (See also the *Errata*, same journal *28* (1975), 277–280.)

Atiyah, M. F., Bott, R., and Shapiro, A.

[1] Clifford modules, *Topology 3, Suppl. 1*, (1964), 3–38.

Atiyah, M. F., Donnelly, H., and Singer, I. M.

[1] Geometry and Analysis of Shimizu L-Functions, *Proc. Nat. Acad. Sci. USA 79* (1982), p. 5751.

[2] Eta Invariants. Signature of Cusps and Values of L-Functions, *Ann. of Math. 118* (1983), 131–171.

Atiyah, M. F., and Hirzebruch, F.

[1] Riemann-Roch theorems for differentiable manifolds, *Bull. A.M.S. 65* (1959), 276–281.

[2] Analytic cycles on complex manifolds, *Topology 1* (1962), 25–47.

[3] Spin manifolds and group actions, in *Essays in Topology and Related Topics; memoires dediès à Georges de Rham*, A. Haefliger and B. Narasimhan (eds.), Springer-Verlag, Berlin, 1970, 18–28.

Atiyah, M. F., Hitchin, N., and Singer, I. M.

[1] Self-duality in four-dimensional Riemannian geometry, *Proc. Roy. Soc. London Ser. A. 362* (1978), 425–461.

Atiyah, M. F., Patodi, V. K. and Singer, I. M.

[1] Spectral asymmetry and Riemannian geometry I, *Math. Proc. Camb. Phil. Soc. 77* (1975), 43–69; II, *78* (1975), 405–432; III, *79* (1976), 71–99.

Atiyah, M. F., and Segal, G. B.

[1] The index of elliptic operators II, *Ann. of Math. 87* (1968), 531–545.

Atiyah, M. F. and Singer, I. M.

[0] The index of elliptic operators on compact manifolds, *Bull. A.M.S. 69* (1963), 422–433.

[1] The index of elliptic operators I. *Ann. of Math. 87* (1968), 484–530.
[2] The index of elliptic operators III, *Ann. of Math. 87* (1968), 546–604.
[3] The index of elliptic operators IV, *Ann. of Math. 93* (1971), 119–138.
[4] The index of elliptic operators V, *Ann. of Math. 93* (1971), 139–149.
[5] Index theory for skew-adjoint Fredholm operators, *Publ. Math. I.H.E.S. 37* (1969), 305–326.
[6] Dirac operators coupled to vector potentials, *Proc. Nat. Acad. Sci. USA 81* (1984), 2597–2599.

Barratt, M. G.
[1] Track groups I, *Proc. London Math. Soc. (3) 5* (1955), 71–106.

Baum, H.
[1] *Spin-Strukturen und Dirac-Operatoren über Pseudoriemannschen Mannigfaltigkeiten*, Teubner, Leipzig, 1981.

Baum, P., and Douglas, R.
[1] Index theory, bordism, and *K*-homology, *Contemp. Math. 10* (1982), 1–31.
[2] *K* homology and index theory, *Proc. Symp. Pure Math. 38, Part 1*, 1982, 117–173.

Baum, P., Douglas, R., and Taylor, M.
[1] Cycles and relative cycles defined by first order elliptic operators (manuscript).

Beauville, A.
[1] Varietes kahleriennes dont la premiere classe de Chern est null, *J. Diff. Geom. 18* (1983), 755–782.

Bérard-Bergery, L.
[1] La courbure scalaire des variétés riemanniennes, Séminaire Bourbaki, 32e année (1979–80), no. 556.
[2] Scalar curvature and isometry group, in *Spectra of Riemannian Manifolds*, Kaigai Publications, Tokyo, 1983, 9–28.

Berger, M.
[1] Sur les groupes d'holonomie homogène des variétés a connexion affine et des variétés Riemanniennes, *Bull. Soc. Math. France 83* (1955), 279–330.

Berline, N., and Vergne, M.
[1] A computation of the equivariant index of the Dirac operator, *Bull. Soc. Math. France 113* (1985), 305–345.
[2] A proof of Bismut local index theorem for a family of Dirac operators, *Topology 26* (1987), 435–463.

Binz, R., and Pferschy, R.
[1] The Dirac operator and the change of the metric, *C.R. Math. Rep. Acad. Sci. Canada* V (1983), 269–274.

Bismut, J.-M.
[1] The Atiyah-Singer theorems: A probabilistic approach I, II, *J. Funct. Anal. 57* (1984), 56–99, 329–348.
[2] The Atiyah-Singer theorem for families of Dirac operators: Two heat equation proofs, *Invent. Math. 83* (1986), 91–151.

[3] Localization formulas, superconnections and the Index Theorem for families, *Comm. Math. Phys. 103* (1986), 127–166.
[4] The infinitesimal Lefschetz formulas: a heat equation proof, *J. Funct. Anal. 62* (1985), 435–457.

Bismut, J.-M., and Cheeger, J.
[1] Eta invariants and their adiabatic limits, *J.A.M.S. 2* (1989), 33–70.
[2] Families index for manifolds with boundary, superconnections and cones, *J. Funct. Anal.* (to appear).

Bismut, J.-M., and Freed, D.
[1] The analysis of elliptic families II, Dirac operators, eta invariants and the holonomy theorem, *Comm. Math. Phys. 107* (1986), 103–163.

Bochner, S.
[1] Curvature and Betti numbers I, II, *Ann. of Math. 49* (1948), 379–390, and *50* (1949), 79–93.

Bochner, S., and Yano, K.
[1] *Curvature and Betti Numbers*, Ann. of Math. Studies 32, Princeton University Press, Princeton, 1953.

Bokobza-Haggiag, J.
[1] Opérateurs pseudo-différentiels sur une variété différentiable, *Ann. Inst. Four. (Grenoble) 19* (1969), 125–177.

Bonan, E.,
[1] Sur des variétés riemanniennes à group d'holonomie G_2 ou $Spin_7$, *Compt. Rend. Acad. Sci., Paris, 262* (1966), 127–129.

Borel, A.
[1] Le plan projectif des octaves et les sphères comme espaces homogènes, *Compt. Rend., 230* (1950), 1378–1380.
[2] *Topics in the Homology Theory of Fibre Bundles.* Lecture notes, Chicago (1954). (Reprinted as Lect. Notes in Math. 36, Springer-Verlag, Berlin, 1967.)
[3] Sur la cohomologie des espaces fibrés principaux et des espaces homogènes de groupes de Lie compacts, *Ann. of Math. 57* (1953), 165–197.

Borel, A., and Hirzebruch, F.
[1] Characteristic classes and homogeneous spaces I, II, III. *Amer. J. Math. 80* (1958), 458–538; *81* (1959), 315–382; *82* (1960), 491–504.

Bott. R.
[1] Stable homotopy of the classical groups, *Ann. of Math. 70* (1959), 313–337.
[2] Some remarks on the periodicity theorems, *Colloque de Topologie*, Lille, 1959.
[3] The index theorem for homogeneous differential operators, in S. S. Cairns (ed.), *Differential and Combinatorial Topology, a Symposium in Honor of Marston Morse*, Princeton University Press, Princeton, 1965, 167–186.
[4] *Lectures on K(X)*, Benjamin Press, New York, 1969.

Bourguignon, J.-P.
[1] L'operateur de Dirac et la géometrie riemannienne, *Rend. Sem. Mat. Univ. Politec. Torino, 44* (1986), 317–359.

[2] Groupes d'holonomie des variétés Riemanniennes, *Astérisque 126* (1985), 169–180.

Bourguignon, J.-P., and Gauduchon, P.

[1] Operateur de Dirac et variations de metriques (to appear).

Brauer, R., and Weyl, H.

[1] Spinors in n dimensions, *Amer. J. Math. 57* (1935), 425–449.

Bredon, G. E.

[1] *Introduction to Compact Transformation Groups*, Academic Press, New York, 1972.

Browder, W., and Hsiang, W. C.

[1] *G*-actions and the fundamental group, *Invent. Math. 65* (1982), 411–424.

Bryant, R.

[1] Metrics with holonomy G_2 or Spin (7), *Arbeitstagung, Bonn, 1984*, Lect. Notes in Math. 1111, Springer-Verlag, Berlin, 1985, 269–277.

[2] Metrics with exceptional holonomy, *Ann of Math. 126* (1987), 525–576.

Calabi, E.

[1] Closed, locally euclidean, 4-dimensional manifolds, *Bull. A.M.S. 63* (1957), 135.

[2] Minimal immersions of surfaces in euclidean spheres, *J. Diff. Geom. 1* (1967), 111–127.

[3] Quelques applications de l'analyse complexe aux surfaces d'aire minima (together with *Topics in complex manifolds* by H. Rossi), Les Presses de l'Université de Montreal, 1968.

Carr, R.

[1] Construction of metrics of positive scalar curvature, *Trans. A.M.S. 307* (1988), 63–74.

Cartan, É.

[1] *La Théorie des Spineurs*, Hermann, Paris 1937. Second edition, *The Theory of Spinors*, MIT Press, Cambridge, Mass., 1966.

Cheeger, J.

[1] On the Hodge Theory of Riemannian Pseudo-Manifolds, *Proc. Symp. in Pure Math. 26* (1980), 91–146.

[2] Analytic Torsion and Reidemeister Torsion, *Proc. Nat. Acad. Sci. USA 74* (1977), 2651–2654.

[3] Spectral Geometry of Singular Riemannian Spaces, *J. Diff. Geom. 18* (1983), 575–651.

[4] On the Spectral Geometry of Spaces with Cone-like Singularities, *Proc. Nat. Acad. Sci. USA 76, 5* (1979), 2103–2106.

Cheeger, J., and Ebin, D. G.

[1] *Comparison Theorems in Riemannian Geometry*, American Elsevier, New York, 1982.

Cheeger, J., and Gromoll, D.

[1] On the structure of complete open manifolds of non-negative curvature, *Ann. of Math. 96* (1972), 413–443.

[2] The splitting theorem for manifolds of non-negative Ricci curvature, *J. Diff. Geom.* *6* (1971), 119–128.

Cheeger, J., and Simons, J.

[1] Differential characters and geometric invariants, in *Geometry and Topology*, ed. J. Alexander and J. Harer, Lect. Notes in Math. 1167, Springer-Verlag, New York, 1985, 50–80.

Chern, S. S.

[1] Characteristic classes of Hermitian manifolds, *Ann. of Math.* *47* (1946), 85–121.

[2] On the characteristic classes of Riemannian manifolds, *Proc. Natl. Acad. Sci. U.S.A.* *33* (1947), 78–82.

[3] On a generalization of Kähler geometry, *Algebraic Geometry and Topology: A Symposium in Honor of S. Lefshetz*, Princeton University Press, Princeton 1957, 103–121.

Chern, S. S., and Simons, J.

[1] Characteristic forms and geometric invariants, *Ann. of Math.* *99* (1974), 48–69.

Chevalley, C.

[1] *The Algebraic Theory of Spinors*, Columbia University Press, New York, 1954.

Chou, A. W.

[1] The Dirac operator on spaces with conical singularities and positive scalar curvature, *Trans. A.M.S.* *28* (1985), 1–40.

Chudnovsky, D. V., and Chudnovsky, G. V.

[1] Elliptic modular functions and elliptic genera, *Topology* *27* (1988), 163–170.

Dadok, J., and Harvey, R.

[1] *Calibrations and spinors* (to appear).

Dirac, P.A.M.

[1] The quantum theory of the electron, *Proc. Roy. Soc. of Lond. A.* *117* (1928), 610–624.

[2] *The Principles of Quantum Mechanics*, Oxford University Press, Oxford, 1930.

Donnelly, H.

[1] Eta-invariant of a fibered manifold, *Topology* *15* (1976), 247–252.

[2] Spectrum and the fixed point sets of isometries, I, *Math. Ann.* *244* (1976), 161–170.

[3] Essential spectrum and heat kernel, *Journal of Func. Anal.* *75* (1987), 362–381.

Donnelly, H., and Patodi, V. K.

[1] Spectrum and the fixed point set of isometries, II, *Topology* *16* (1977), 1–11.

Duistermaat, J. J., and Heckman, G. J.

[1] On the variation in the cohomology of the symplectic form of the reduced phase space, *Invent. Math.* *69* (1982), 259–269.

Dynin, A. S.

[1] On the index of families of pseudodifferential operators on manifolds with a boundary, *Soviet Math. Dokl.* *10* (1969), No. 3, 614–617.

Edmonds, A. L.

[1] Orientability of fixed point sets, *Proc. A.M.S.* *82* (1981), 120–124.

Federer, H.
[1] Some theorems on integral currents, *Trans. A.M.S. 177* (1965), 43–67.

Freedman, M.
[1] The topology of 4-manifolds, *J. Diff. Geom. 17* (1982).

Friedrich, T.
[1] Zur Existenz Paralleler Spinor-felder Über Riemannschen Mannigfaltigkeiten, *Coll. Math. XLIV* (1981), 277–290.

Gallot, S., amd Meyer, D.
[1] Operateur de courbure et Laplacien des formes differentialles d'une variété riemannienne, *J. Math. Pures et Appl. 54*, 975, 285–304.

Gel'fand, I. M.
[1] On elliptic equations, *Russian Math. Surveys 15* (1960), no. 3, 113–123.

Getzler, E.
[1] Pseudodifferential operators on supermanifolds and the Atiyah-Singer index theorem, *Comm. Math. Phys. 92* (1983), 163–178.
[2] A short proof of the local Atiyah-Singer index theorem, *Topology 25* (1986), 111–117.

Gilkey, P. B.
[1] Curvature and the eigenvalues of the Laplacian for elliptic complexes, *Adv. in Math. 10* (1973), 344–282.
[2] *Invariance Theory, the Heat Equation and the Atiyah-Singer Index Theorem*, Mathematical Lecture Series 4, Publish or Perish Press, Berkeley, Ca., 1984.
[3] The boundary integrand in the formula for the signature and Euler characteristic of a riemannian manifold with boundary, *Adv. in Math. 15* (1975), 334–360.
[4] Lefschetz Fixed Point Formulas and the Heat Equation, in *Partial Differential Equations and Geometry. Proceedings of the Park City Conference*, Marcel-Dekker, 1979, 91–147.
[5] Curvature and the heat equation for the de Rham complex, in *Geometry and Analysis*, Indian Academy of Sciences, 1980, 47–80.

Goldberg, R.
[1] *Curvature and homology*, Academic Press, New York and London, 1962.

Gromoll, D., and Meyer, W.
[1] The structure of complete manifolds of non-negative curvature, *Ann. of Math. 90* (1969), 75–90.
[2] An exotic sphere of non-negative curvature, *Ann. of Math. 96* (1972), 413–443.

Gromov, M.
[1] Curvature, diameter and Betti numbers, *Comm. Math. Helv. 56* (1981), 179–195.
[2] Volume and bounded cohomology, *Publ. Math. I.H.E.S. 56* (1983), 213–307.
[3] Filling Riemannian manifolds, *J. Diff. Geom. 18* (1983), 1–147.
[4] Large Riemannian manifolds, in *Curvature and Topology of Riemannian Manifolds*, Lect. Notes in Math. 1201, Springer-Verlag, New York, 1985, 108–121.
[5] *Partial Differential Relations*, Springer-Verlag, New York, 1986.

Gromov, M., and Lawson, H. B., Jr.

[1] Spin and scalar curvature in the presence of a fundamental group I, *Ann. of Math. 111* (1980), 209–230.

[2] The classification of simply-connected manifolds of positive scalar curvature, *Ann. of Math. 111* (1980), 423–434.

[3] Positive scalar curvature and the Dirac operator on complete Riemannian manifolds, *Publ. Math. I.H.E.S. 58* (1983). 295–408.

Habegger, N.

[1] Une variété de dimension 4 avec forme d'intersection paire et signature 8, *Comm. Math. Helv. 57* (1982), 22–24.

Harvey, R.

[1] *Spinors and Calibrations* (forthcoming).

Harvey, R., and Lawson, H. B., Jr.

[1] Geometries associated to the group SU_n and varieties of minimal submanifolds arising from the Cayley arithmetic, in *Minimal submanifolds and geodesics*, Kaigai Publications, Tokyo, 1978, 43–59.

[2] A constellation of minimal varieties defined over the group G_2. *Proceedings of Conference on Geometry and Partial Differential Equations*, Lecture Notes in Pure and Applied Mathematics, 48, Marcel-Dekker, 1978, 43–59.

[3] Calibrated geometries, *Acta Math. 148* (1982), 47–157.

Hattori, A., and Taniguchi, H.

[1] Smooth S^1-actions and bordism, *J. Math. Soc. Japan 24* (1972), 701–731.

Helgason, S.

[1] *Differential Geometry and Symmetric Spaces*, Academic Press, New York, 1962.

Hempel, J.

[1] *3-Manifolds*, Ann. of Math. Studies 86, Princeton University Press, Princeton, 1976.

Hernandez, H.

[1] A class of compact manifolds with positive Ricci curvature, *Differential Geometry. A.M.S. Proc. Symp. Pure Math. 27* (1975) 73–87.

Hijazi, O.

[1] A conformal lower bound for the smallest eigenvalue of the Dirac operator and Killing spinors, *Comm. Math. Phys. 104* (1986), 151–162.

[2] Caractérisation de la sphère par les premières valeurs propres de l'opérateur de Dirac en dimensions 3, 4, 7 et 8, preprint, Max Planck Institut für Mathematik, Bonn. 1986.

Hirzebruch, F.

[1] *Topological Methods in Algebraic Geometry*, 3rd ed. Springer-Verlag, Berlin-Heidelberg-New York, 1966.

Hirzebruch, F., and Mayer, K. H.

[1] *O(n) Mannigfaltigkeiten, exotische Spharen, und Singularitaten*, Lect. Notes in Math. 57, Springer-Verlag, New York, 1968.

Hirzebruch, F., and Zagier, D. B.

[1] *The Atiyah-Singer Theorem and Elementary Number Theory*, Mathematics Lectures Series 3, Publish or Perish Press, Berkeley, Ca., 1974.

Hitchin, N.
[1] Harmonic spinors, *Adv. in Math. 14* (1974), 1–55.

Hörmander, L.
[1] *Linear Partial Differential Operators*, Springer-Verlag, New York, 1963.
[2] Pseudo-Differential Operators, *Comm. in Pure Appl. Math. 18* (1965), 501–517.
[3] Pseudo-differential operators and hypoelliptic equations, *A.M.S. Proc. Symp. Pure Math. 10* (1967), 138–183.

Hsiang, W.-C. and Hsiang, W.-Y.
[1] On the degree of symmetry of homotopy spheres, *Ann. of Math. 89* (1969), 52–67.

Husemoller, D.
[1] *Fibre bundles*. McGraw-Hill, New York, 1966. 2nd ed., GTM, vol. 20, Springer-Verlag, Berlin-Heidelberg-New York, 1975.

Jänich, K., and Ossa, E.
[1] On the signature of an involution, *Topology 8* (1969), 27–30.

Kähler, E.
[1] Der innere Differentialkalkül, *Rend. di Mat. 21* (1962), 425–523.

Karoubi, M.
[1] Algèbres de Clifford et K-théorie, *Ann. Sci. Ec. Norm. Sup. 4e ser. 1* (1968), 161–270.
[2] *K-Theory, An Introduction*, Springer-Verlag, Berlin, 1978.

Kasparov, G.
[1] Operator K-theory and its application: elliptic operators, group representations, higher signature, C^*-extensions. Preprint, Chernogolovka 1983.

Kawakubo, K., and Raymond, F.
[1] The index of manifolds with toral actions and geometric interpretations of the $\sigma(\infty,(S^1,M^n))$ invariant of Atiyah and Singer, *Invent. Math. 15* (1972), 53–66.

Kazdan, J.
[1] Deforming to positive scalar curvature on complete manifolds, *Math. Ann. 261* (1982), 227–234.

Kazdan, J., and Warner, F.
[1] Prescribing curvatures, *Proc. of Symp. in Pure Math. 27* (1975), 309–319.

Kervaire, M., and Milnor, J.
[1] Groups of homotopy spheres, *Ann. of Math.* 77 (1963), 504–537.

Kirchberg, K. D.
[1] An estimation for the first eigenvalue of the Dirac operator on closed Kähler manifolds of positive scalar curvature, Preprint, Humboldt Univ. zu Berlin, 1985.

Kobayashi, S.
[1] Fixed points of isometries, *Nagoya Math. J. 13* (1958), 63–68.

Kobayashi, S., and Nomizu,
[1] *Foundations of Differental Geometry I, II*, Interscience, John Wiley, New York, 1963, 1969.

Kohn, J. J., and Nirenberg, L.
[1] An Algebra of pseudo-differential operators, *Comm. Pure Appl. Math. 18* (1965), 269–305.

Koppelman, W.
[1] On the index of elliptic operators on closed surfaces, *Amer. J. Math. 85* (1963), 597–632.

Kosmann, Y.
[1] Derivées de Lie des spineurs. *Ann. di Mat. ed Appl. 91* (1972), 317–395.

Ku, H. T.
[1] A generalization of the Conner inequalities, of *Proceedings of the Conference on Transformation Groups, New Orleans*, Springer-Verlag, New York, 1967, 401–413.

Kuiper, N.
[1] Contractibility of the Unitary Group in Hilbert Space, *Topology 3* (1963), 19–30.

Landweber, P., and Stong, R.
[1] Circle actions on spin manifolds and characteristic numbers, *Topology 27* (1988), 145–162.
[2] A bilinear form for spin manifolds, *Trans. A.M.S. 300* (1987), 625–640.

Lang, S.
[1] *Algebra*, Addison Wesley, Reading, Mass., 1965.

Lawson, H. B., Jr.
[1] *Lectures on Minimal Submanifolds, Vol. 1*, Publish or Perish Press, Berkeley, Ca., 1980.
[2] *The Theory of Gauge Fields in Four Dimensions*, American Mathematical Society, Providence, 1985.
[3] Complete manifolds of positive scalar curvature, in *Proceedings of The 1981 Symposium on Differential Geometry and Differential Equations* (Shanghai-Hefei), ed. Gu Chaohao, Science Press, Beijing, 1984, pp. 147–182.

Lawson, H. B., Jr., and Michelsohn, M. L.
[1] Clifford bundles, spinors and Dirac operators, *J. Diff. Geom. 15* (1980), 237–267.

Lawson, H. B., Jr., and Yau, S. T.
[1] Scalar curvature, nonabelian group actions and the degree of symmetry of exotic spheres, *Comm. Math. Helv. 49* (1974), 232–244.

Lichnerowicz, A.
[1] Laplacien sur une variété riemannienne et spineure, *Atti Accad. Naz. dei Lincei, Rendiconti 33* (1962), 187–191.
[2] Spineurs harmoniques, *C.R. Acad. Sci. Paris, Ser A–B 257* (1963), 7–9.
[3] Champs spinoriels et propagateurs en relativité générale, *Bull. Math. Soc. France 92* (1964), 11–100.

Lusztig, G.
[1] Novikov's higher signature and families of elliptic operators, *J. Diff. Geom. 7* (1971), 229–256.

Mathai, V., and Quillen, D.
[1] Superconnections, Thom classes, and equivariant differential forms. *Topology 25* (1986), 85–110.

Mayer, K. H.
[1] Elliptische Differdntialoperatoren und ganzzahligkeitsatze für characterische Zahlen, *Topology 4* (1985), 295–313.

Michelsohn, M. L.
[1] Clifford annd spinor cohomology of Kähler manifolds, *Amer. J. Math. 102* (1980), 1083–1146.
[2] On the existence of special metrics in complex geometry, *Acta Math. 149* (1982). 261–295.
[3] Spinors, twistors and reduced holonomy, in *Geometry of moduli and 4-dimensional manifolds, R.I.M.S. Publications 616 Kyoto* (1987), 1–3.

Milnor, J.
[1] Construction of universal bundles: I, *Ann. of Math.* (*2*) *63* (1956), 272–284.
[2] Construction of universal bundles: II, *Ann. of Math.* (*2*) *63* (1956), 430–436.
[3] On manifolds homeomorphic to the 7-sphere, *Ann. of Math.* (*2*) *64* (1956), 399–405.
[4] A unique factorization theorem for 3-manifolds, *Amer. J. Math. 84* (1962), 1–7.
[5] Spin structures on manifolds, *L' Enseignement Math 9* (1963) 198–203.
[6] *Morse Theory*, Ann. of Math. Studies 51, Princeton University Press, Princeton, 1963.
[7] Remarks concerning spin manifolds, in S. S. Cairns (ed.), *Differential and Combinatorial Topology, a symposium in honor of Marston Morse*, Princeton University Press, Princeton, 1965, 55–62.
[8] *Lectures on the h-corbordism theorem*, Princeton University Press., Princeton, 1965.

Milnor, J., and Husemoller, D.
[1] *Symmetric Bilinear Forms*, Springer-Verlag, New York, 1973.

Milnor, J., and Stasheff, J. D.
[1] *Characteristic Classes*, Ann. of Math. Studies 76, Princeton University Press, Princeton, 1974.

Mishchenko, A. S., and Fomenko, A. T.
[1] The index of elliptic operators over C*-algebras, *Izv. Akad. Nauk. USSR. Ser. Mat. 43* (1979), 831–859, *Math. USSR-Izv.* 15 (1980), 87–112.

Miyazaki, T.
[1] On the existence of positive scalar curvature metrics on non-simply-connected manifolds, *J. Fac. Sci. Univ. Tokyo, Sect. IA 30* (1984), 549–561.
[2] Simply connected spin manifolds and positive scalar curvature, *Proc. A.M.S. 93* (1985), 730–734.

Myers, S. B.
[1] Riemannian manifolds with positive mean curvature, *Duke Math. J. 8* (1941), 401–404.

Newlander, A., and Nirenberg, L.
[1] Complex analytic coordinates in almost complex manifolds, *Ann. of Math. 65* (1957), 391–404.

Nirenberg, L.
[1] Pseudo-differential operators, *A.M.S. Proc. Symp. Pure Math. 16* (1970), 149–167.
[2] *Lectures on Linear Partial Differential Equations*, Regional Conference Series in Math. 17, American Mathematical Society, Providence, 1972.

Ochinine, S.
[1] Sur les genres multiplicatifs definis par des intégrales elliptiques, *Topology 26* (1987), 143–151.

O'Neill, B.
[1] The fundamental equations of a submersion, *Michigan Math. J. 13* (1966), 459–469.

Palais, R. S.
[1] *Seminar on the Atiyah-Singer Index Theorem*, Ann. of Math. Studies 57, Princeton University Press, Princeton, 1965.

Parker, T., and Taubes, C.
[1] On Witten's proof of the positive energy theorem, *Commun. Math. Phys. 84* (1982), 223–238.

Patodi, V. K.
[1] Curvature and the eigenforms of the Laplace operator, *J. Diff. Geom. 5* (1971), 233–249.
[2] An analytic proof of the Riemann-Roch-Hirzebruch theorem for Kaehler manifolds, *J. Diff. Geom. 5* (1971), 251–283.
[3] Holomorphic Lefschetz fixed point formula, *Bull. A.M.S. 79* (1973), 825–828.
[4] Curvature and the Fundamental Solution of the Heat Operator, *J. Indian Math. Soc. 34* (1970), 269–285.
[5] Holomorphic Lefschetz fixed point formula, *Bull. A.M.S. 79* (1973), 825–828.

Penrose, R., and Rindler, W.
[1] *Spinors and space-time, I and II*, Cambridge University Press, 1984.

Pferschy, R.
[1] *Die Abhängigkeit des Dirac-Operators von der Riemannschen Metrik*, Diss. Techn. Univ. Graz, 1983.

Quillen, D.
[1] The mod 2 cohomology rings of extra-special 2-groups and the spinor groups, *Math. Ann. 194* (1971), 197–212.
[2] Superconnections and the Chern character, *Topology 24* (1985), 89–95.

Raghunathan, M. S.
[1] *Discrete Subgroups of Lie Groups*, Springer-Verlag, New York, 1972.

Rempel, S., and Schulze, B.-F.
[1] *Index Theory of Elliptic Boundary Problems*, Akademie-Verlag, Berlin, 1982.

Rosenberg, J.

[1] C*-algebras, positive scalar curvature, and the Novikov conjecture. *Publ. Math. I.H.E.S.* *58* (1983) 197–212.

[2] C*-algebras, positive scalar curvature, and the Novikov conjecture, II, in *Proc. U.S.-Japan Seminar on Geometric Methods in Operator Algebras, Kyoto 1983*, H. H. Araki and E. G. Efros (eds.).

[3] C*-algbera, positive scalar curvature, and the Novikov conjecture, III, *Topology 25* (1986), 319–336.

Schoen, R., and Yau, S.-T.

[1] Existence of incompressible minimal surfaces and the topology of three dimensional manifolds of non-negative scalar curvature, *Ann. of Math. 110* (1979), 127–142.

[2] On the structure of manifolds with positive scalar curvature, *Manuscripta Math. 28* (1979), 159–183.

[3] On the proof of the positive mass conjecture in general relativity, *Comm. Math. Phy. 65* (1979), 45–76.

[4] The energy and the linear momentum of space-times in general relativity, *Comm. Math. Phy. 79* (1981), 47–51.

[5] Proof of the positive mass theorem, II, *Comm. Math. Phy. 79* (1981), 231–260.

[6] Complete three dimensional manifolds with positive Ricci curvature and scalar curvature, in *Seminar on Differential Geometry*, Ann. of Math. Studies 102, Princeton University Press, Princeton, 1982, 209–227.

Schultz, R.

[1] Circle actions on homotopy spheres bounding plumbing manifolds, *Proc. A.M.S. 36* (1972), 297–300.

Seeley, R. T.

[1] Complex powers of an elliptic operator, *A.M.S. Proc. Symp. Pure Math. 10* (1967), 288–307.

[2] Topics in Pseudo-Differential Operators, in *Pseudo-Differential Operators*, C.I.M.E., (1968), 168–305.

[3] A proof of the Atiyah-Bott-Lefschetz fixed point formula, *An. Acad. Brasil Cienc. 41* (1969), 493–501.

Segal, G. B.

[1] Equivariant *K*-theory *Publ. Math. I.H.E.S. 34* (1968), 129–151.

Sha, J.-P., and Yang, D.-G.

[1] Examples of manifolds of positive Ricci curvature, *J. Diff. Geom. 29* (1989), 95–104.

[2] Positive Ricci curvature on the connected sums of $S^n \times S^m$. (to appear).

Shanahan, P.

[1] *The Atiyah-Singer Index Theorem, An Introduction*, Lect. Notes in Math. 638, Springer-Verlag, New York, 1978.

Simons, J.

[1] Characteristic forms and transgression II-Characters associated to a connection (preprint).

Singer, I. M.
[1] Elliptic operators on manifolds, in *Pseudo-Differential Operators* (C.I.M.E., Stresa, 1968), Edizione Cremonesa, 1969, 333–375.
[2] Future extensions of index theory and elliptic operators, in *Prospects in Mathematics*, Ann. of Math. Studies 70, Princeton University Press, Princeton, 1971, 171–185.
[3] Recent applications of index theory for elliptic operators, in *Partial Differential Equations, Proc. Sympos. Pure Math. 23* (Berkeley 1971), 11–31.
[4] Families of Dirac operators with applications to physics, in *Élie Cartan et les Mathèmatiques d'Aujourd'hui*, Astérisque, Soc. Math. de France, Paris, 1985, 323–340.

Smith, D. E.
[1] The Atiyah-Singer invariant, torsion invariants, and group actions on spheres, *Trans. A.M.S. 177* (1983), 469–488.

Steenrod, N.
[1] *The Topology of Fibre Bundles*, Princeton University Press, Princeton, 1951.

Steenrod, N., and Epstein, D.
[1] *Cohomology Operations*, Ann. of Math. Studies 50, Princeton University Press, Princeton, 1962.

Stong, R. E.
[1] *Notes on Cobordism Theory*, Princeton University Press, Princeton, 1968.

Taylor, M.
[1] *Pseudodifferential Operators*, Lect. Notes in Math. 416, Springer-Verlag, New York 1974.
[2] *Pseudodifferential Operators*, Princeton University Press, Princeton, 1981.

Thom, R.
[1] Quelques propriétés globales des variétés différentiables, *Comm. Math. Helv. 28* (1954), 17–86.

Thomas, E.
[1] On the cohomology groups of the classifying space for the stable spinor group, *Bol. Soc. Mex* (1962), 57–69.

Toledo, D.
[1] On the Atiyah-Bott formula for isolated fixed points, *J. Diff. Geom. 8* (1973), 401–436.

Tolendo, D., and Tong, Y. L.
[1] The Holomorphic Lefschetz formula, *Bull. A.M.S. 81* (1975), 1133–1135.

Vafa, C., and Witten, E.
[1] Eigenvalue inequalities for fermions in gauge theories, Princeton University preprint, April 1984. *Commun. in Math. Physics, 95*, No. 3 (1984), 257–276.

Ward, R., and Wells, R. O., Jr.
[1] *Twistor Geometry*, Cambridge University Press, Cambridge, 1985.

Warner, F.
[1] *Foundations of Differentiable Manifolds and Lie Groups*, Scott Foresman, New York, 1971.

Weil, A.

[1] Correspondence, by R. Lipschitz, *Ann. of Math. 69* (1959), 247–251.

Wells, R.

[1] *Differential Analysis on Complex Manifolds*, Prentice Hall, Englewood Cliffs, N.J., 1973, 2nd ed. Springer-Verlag.

Widom, H.

[1] A complete symbolic calculus for pseudodifferential operators, *Bull. Sc. Math. 104* (1980), 19–63.

Witten, E.

[1] A new proof of the positive energy theorem, *Comm. Math. Phy. 80* (1981), 381–402.

[2] The index of the Dirac operator in loop space (forthcoming).

Wolf, J.

[1] Essential self-adjointness for the Dirac operator and its square, *Indiana Univ. Math. J. 22* (1972/73), 611–640.

Wu, H.

[1] *The Bochner Technique in Differential Geometry*, Mathematical Reports, Harwood Academic Publishers, *3* (1988), 298–542.

Yau, S.-T.

[1] On the curvature of compact hermitian manifolds, *Inventiones Math. 25* (1974), 213–240.

[2] Calabi's conjecture and some new results in algebraic geometry, *Proc. Natl. Acad. Sci. USA 74* (1977), 1798–1799.

[3] On the Ricci curvature of a compact Kähler manifold and the complex Monge-Ampère equation, I, *Comm. Pure and Appl. Math. 31* (1978), 339–411.

Index

Notation Index

ERRATA

1. In Proposition 1.3, assume *k* has characteristic zero.

2. On page 11, line 24, assume that $\mathscr{A}$ and $\mathscr{B}$ are filtered.

3. On page 43, line 10, the Lie group is also connected.

4. On page 192, line 22, change "Sobolev Embedding" to "Rellich."

5. On page 257, lines 2 and 3 of Remark 13.11, add the following omitted line: "Any elliptic operator *P* can be converted to a pseudodifferential operator of degree zero."

6. On page 278, line 33, ". . . compact *connected* spin manifolds."

7. In Definition 3.2 of Chapter IV, the action should be effective on each connected component of *X*.

GPSR Authorized Representative: Easy Access System Europe - Mustamäe tee
50, 10621 Tallinn, Estonia, gpsr.requests@easproject.com

www.ingramcontent.com/pod-product-compliance
Ingram Content Group UK Ltd.
Pitfield, Milton Keynes, MK11 3LW, UK
UKHW021414150226
468034UK00008B/70
* 9 7 8 0 6 9 1 0 8 5 4 2 5 *